VOLUME XLIX CONSISTS OF THREE PARTS:

I. *Natural Science*

II. *Social Science*

III. *Literature and Art*

Papers of the

MICHIGAN ACADEMY

OF SCIENCE, ARTS, AND LETTERS

VOLUME XLIX

(1963 Meeting)

Editor

RALPH A. LOOMIS

PART I. NATURAL SCIENCE

Published for the Michigan Academy of Science, Arts, and Letters by

THE UNIVERSITY OF MICHIGAN PRESS

ANN ARBOR

1964

Printed and bound by CPI Group (UK) Ltd, Croydon, CR0 4YY

Paperback ISBN: 978-0-472-75153-2

CONTENTS

Part I. Natural Science

BOTANY

FORESTRY

GEOLOGY

ZOOLOGY

ENTOMOLOGY

FISHERIES AND WILDLIFE MANAGEMENT

PART I

NATURAL SCIENCE

narrow spacing relative to an earlier cover was offset by the development of a serious chlorosis. Diseased trees appeared in patches or islands throughout the planting and were conspicuous because of their extreme variations in heights from those of the healthy spruces (Baxter 1952). In some areas development of the fungus *Peniophora byssoidea* (Pers.) von Hohn, and Litsch.[1] was an associated element involved (Table I).

This organism formed a dense felty mat over the ground that consisted of fallen spruce needles bound together with fungus mycelium. The mat was of such a nature that it could be lifted off the soil surface in chunks or sheets. Walking over the covering when dry produced a sound similar to that made by trampling over building paper. It is believed that this mat, which has also been observed in European spruce plantings, affected soil aeration and that it prevented light rains from reaching tree roots.

The indirect effects of tree-spacing upon the development of the mat were most evident in the part of the plantation where the narrow spacing was used. The *Peniophora* mat did not form so conspicuously where the trees were spaced five by five feet apart, and it did not form at all in the nearby mixed plantation of Norway spruce and black locust.

After the factors that influenced unsatisfactory growth of spruce in the pure stand at the Saginaw Forest were diagnosed, the entire stand was thinned in 1933. This reduced the competition in the stagnating plantation as it aged. Opening the stand eliminated the fungus mat within three or four years, and chlorosis was mostly corrected within six years. 30 years later, however, long-time carry-over effects of the chlorosis of early life could be spotted by the generally small diameters of the once affected spruces. Although emphasis is placed upon the fact that the trees in this forest are healthy today (Table I and Fig. 1), it is believed that earlier diagnoses and stand thinning would have shortened the complete recovery period by several years.

[1] Specimens of this mat-causing fungus from the Saginaw Forest were sent to Torsten Lagerberg, professor of forest pathology in Stockholm, Sweden, to compare with European specimens. Professor Lagerberg confirmed the identity of the fungus *Peniophora byssoidea*, an organism likewise often associated with stagnated spruce stands in Europe.

TABLE I

EVENTS IN THE BIOLOGICAL CASE HISTORY OF A PURE SPRUCE PLANTING ESTABLISHED IN 1904

Date of Appearance; Number of Years after Planting	Disease, Fungus or Injury	Disease Preventive	Comments
1912 (8 yrs.)	Chlorosis	First general thinning of plantation in 1933	Chlorosis gradually disappeared within about 6 years after thinning
1918 (14 yrs.)	*Amanita muscaria*		Abundant fruiting each fall 1918 to mid-1930's. Disappeared after stand thinning in 1933
1929 (25 yrs.)	*Stereum sanguinolentum*		Common on slashings and stumps. Not known to cause heartrot at Saginaw. Mostly disappeared after stand thinning in 1933
c. 1929 (c. 25 yrs.)	*Peniophora byssoidea*	First general thinning of plantation in 1933	Fungus mat over soil disappeared within few years after thinning
1928 to present	(No injury, but see Table II for 1928-1935)		No trunk splitting as in mixed stand with locust
c. 1929	Drought		Many trees, and especially those in section of plantation with 3 × 3 spacing, died
1948 (44 yrs.)	Valsa canker	Infected trees felled when atmosphere was relatively dry to insure unfavorable conditions for spore germination	Disease controlled
c. 1948 (c. 44 yrs.)	(No fungus, but see Table II for 1960)		No trunk splitting. Therefore no infection courts for *Polyporus abietinus*. *Fomes pini* not present

Fungi of the Forest Floors

The establishment of a pure forest of spruce and a mixed stand of spruce and locust resulted in the development of entirely different

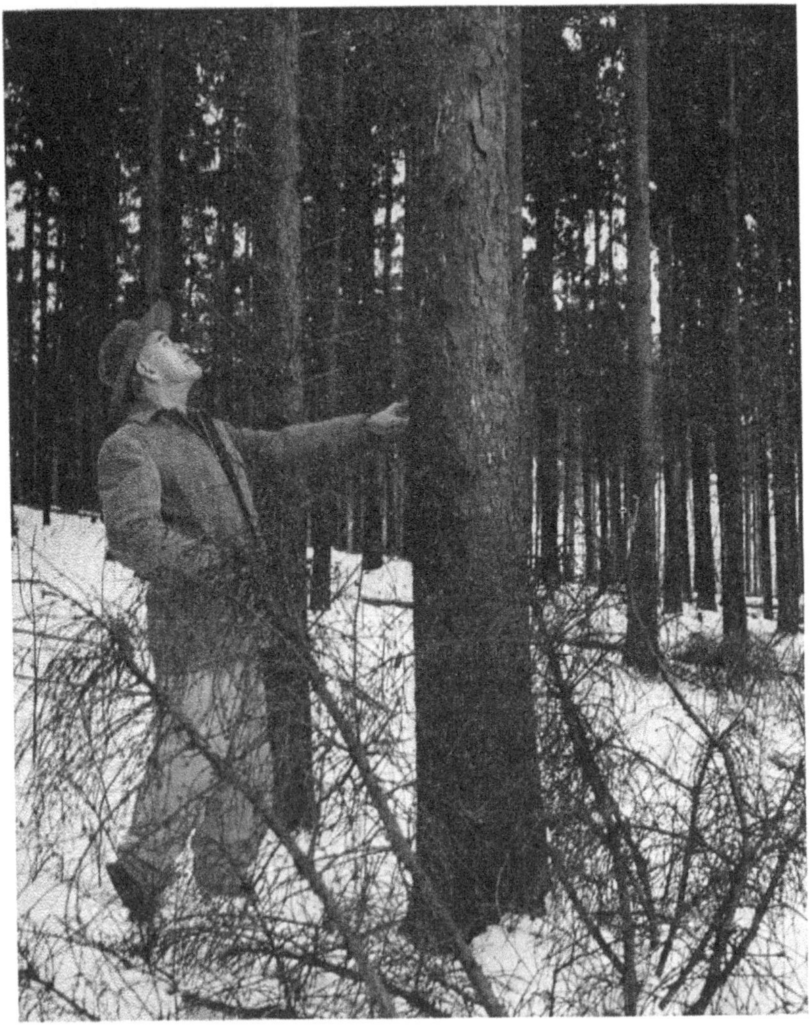

Fig. 1. Pure stand of Norway spruce established at the Saginaw Forest in 1904. Photo 1963.

fungus floras. Furthermore, the presence of chlorosis in part of the pure planting actually delayed the formation of a forest floor (the accumulation of needles as they fall from the trees) (Hawley and Lutz 1943; Hosley 1936) for more than 30 years. The disappearance of the original old-field vegetation elsewhere in the pure stand and the appearance of fleshy fungi took place within 14 years after tree planting.

The mushroom flora in the pure plantation soon became rich in various species of the genera *Cortinarius*, *Mycena*, and *Inocybe*. *Tricholoma myomyces* (Pers.) Lange and species of the brown-spored fungi were characteristic of the forest floor.

My first record of *Amanita muscaria* Fr. (the orange form) was made in 1918 in the pure plantation of spruce. Its abundance increased until the time the 1933 thinnings were made in the plantation. By the middle of the 30's, *Amanita* ceased to fruit. Consequently, it has not been seen here for almost 30 years—an observation corroborated by Alexander Smith, who has made numerous studies of the fleshy fungi of the forest. The sensitiveness of *Peniophora byssoidea* to the thinning has been discussed.

The mushroom flora of the mixed, Norway spruce-black locust plantation, on the other hand, has been scant during the entire life of this forest. The scanty forest floor has been singularly free from species of *Cortinarius* and most other fungi so abundant in the pure stand. The ubiquitous *Collybia dryophila* Fr. has been present and, according to Dr. Smith, *Laccaria ochropurpurea* (Berk.) Pk. has been one of the few conspicuous mushrooms of this mixed hardwood-Norway spruce plantation. *Amanita muscaria* has never been seen in this stand.

Polyporus abietinus
Destroys the Spruce; Mixed Planting

Advantages are to be gained by increasing the amount of plant-available nitrogen in the soil for a number of crops, but responses of Norway spruce planted under locusts at the Saginaw Forest led indirectly to their destruction (Table II).

At first, the resulting accelerated growth of the spruces seemed to demonstrate the advisability of mixing locusts with Norway spruce for quick rehabilitation of wastelands. Not until 1929 did it become apparent that this accelerated growth was undesirable (Fig. 2).

Trunk splitting became so severe in this mixed planting that

TABLE II

EVENTS IN THE BIOLOGICAL CASE HISTORY OF A 1915 SPRUCE PLANTING MIXED WITH LOCUST, ESTABLISHED IN 1906

Date of Appearance Number of Years after Planting	Disease, Fungus, or Injury	Disease Preventive	Comments
1915-63 (entire life)	No chlorosis as in pure planting	Spruce planted with a legume—on an infertile site	No trace of chlorosis during entire life
Especially serious relatively soon after establishment, to c. 1929	Borer damage in locust		Since 1929, decrease in borer damage, although some present
Infection at old borer wounds increased after 1929	*Fomes rimosus* in locust		Population of fruiting bodies increases. Previously to the 1930's, fungus rare in Michigan
	No *Amanita muscaria*, No *Stereum sanguinolentum* No *Peniophora byssoidea*		Unlike the flora of the pure stand, these fungi have never appeared
c. 1928-35	Trunk splitting	No trunk splitting in pure stand	91 spruces with trunk cracks the first season; 338 spruces with trunk cracks by 1935
1940-63 (c. 25 yrs.)	Trunk rot caused by *Polystictus abietinus*	No locusts; no trunk cracks, no rot	
1960 first fruiting body observed	*Fomes pini*		Small branch stubs and not trunk cracks provided the infection courts

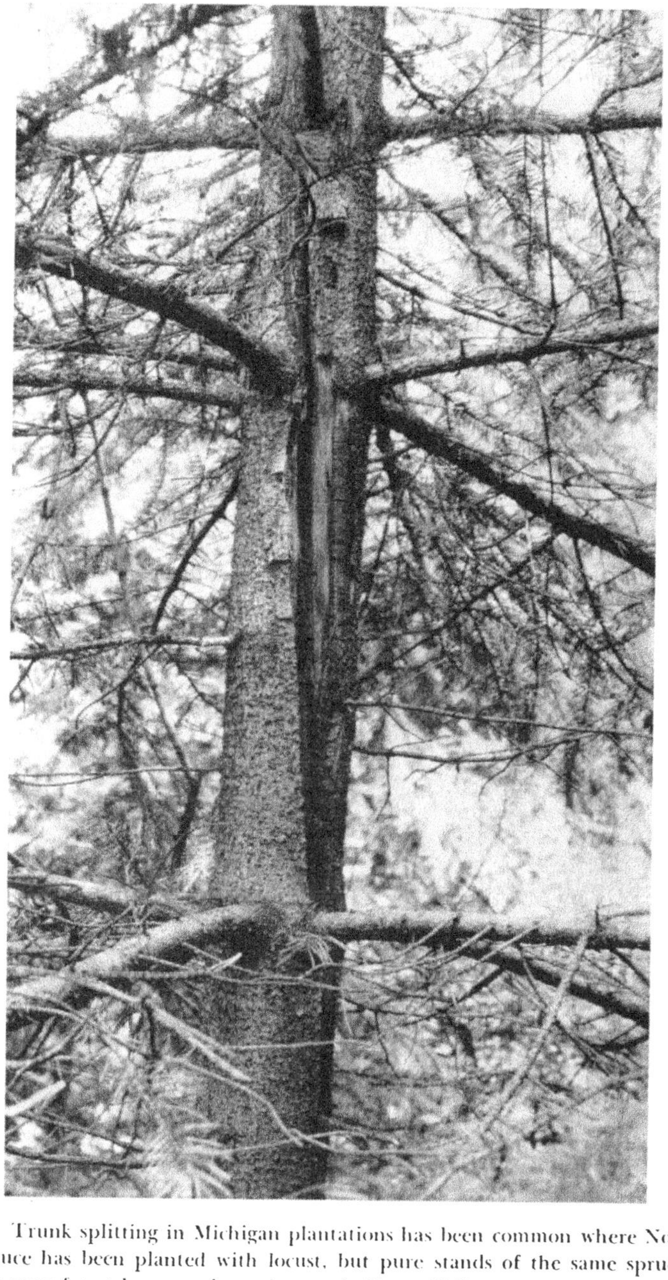

FIG. 2. Trunk splitting in Michigan plantations has been common where Norway spruce has been planted with locust, but pure stands of the same spruce in the same forest have not been damaged. Photo 1935.

5% of the boles were made worthless during this one winter. The number of damaged trunks increased during the winters that followed, so that by 1963 relatively few spruces in the stand had escaped injury.

The trunk cracks served as infection courts for the wood-destroying fungus *Polyporus abietinus* (Dicks.) Sacc. & Cub. This omnivorous species, ordinarily regarded as a slash-rotting organism or one of importance on stored coniferous products, has destroyed all but a few of the Norway spruces in the plantation. The trunks have become so thoroughly decayed that windfall has been the indirect result of fungus attack (Fig. 3). There has been no trunk rot or windfall in the nearby pure spruce planting (Fig. 1).

Red Ring Rot in Mixed Planting

Foresters know that losses from red ring rot caused by *Fomes pini* (Thore) Lloyd are greatest in overmature timber and that the volume of rot exceeds that caused by any other fungus. But at the Saginaw Forest, it has appeared in 48-year-old spruce mixed with locust. In spite of the prevalence of trunk cracks in the spruces described in the previous section, however, none served as infection courts for *F. pini*. Pencil-size branch stubs provided the points of infection. Conks issued from branch stubs only. No *F. pini* has been observed in the pure planting of spruce at the Saginaw Forest.

Fomes annosus at the Saginaw Forest

That first generation spruce planted on old fields is especially subject to fungus attack—chiefly *Fomes annosus*—is a widely expressed opinion in Europe (Hosley 1936). In America, too, *F. annosus* is currently attracting attention among plantation owners because of their awareness of the damage it causes in so many European areas where intensive forestry is practiced.

Fomes annosus was discovered on a Norway spruce stump in a row-planting at the Saginaw Forest in 1961. Other conifers exotic to Michigan that have been attacked here are Austrian pines, Scots pines, and Ponderosa pines. Two Austrian pines were killed by the fungus in 1962, and two small wild black cherry trees in the Austrian pine plantation were killed in 1961. Wild black cherry is likewise a previously unreported host for the fungus in Michigan.

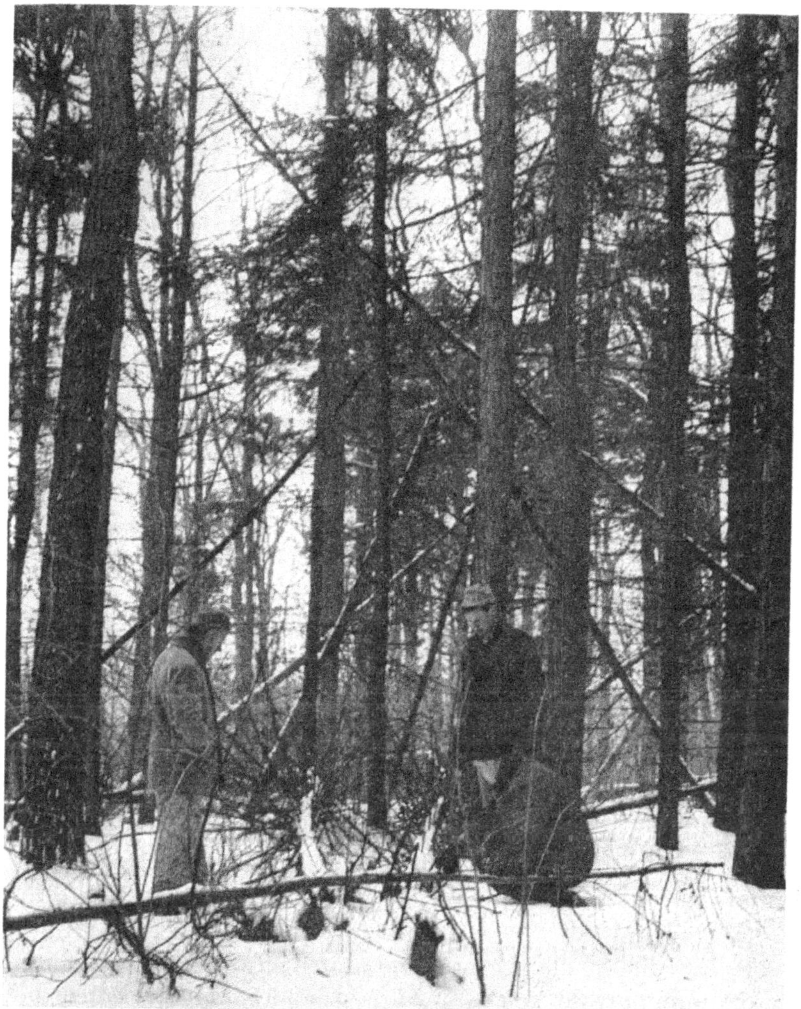

FIG. 3. Trunk splitting in Norway spruce provided infection courts for *Polyporus abietinus*. The trunk rot that followed is indirectly responsible for the windfall and destruction of the spruce in the forest. Photo 1963.

Summary

1. Some of the longest, if not the longest, continuous records of disease patterns in the nation to accompany silvicultural treatments and tree responses to site are reported in this paper.

2. Fungus and disease patterns in a pure planting of Norway spruce are compared and contrasted with those encountered in a nearby Norway spruce stand mixed with locust. The 1904 planting of Norway spruce is one of the oldest in Michigan.

3. "Spruce sickness," an ailment encountered at the Saginaw Forest in the pure planting, is believed to be generally similar to that encountered in Saxony and due indirectly to a deteriorated soil environment.

4. Insufficient amount of thinning at a relatively early age is believed to be one factor that led to chlorosis in the pure forest. Chlorosis was corrected by thinning, but several years had to pass to accomplish the desired results.

5. Carry-over effects of the disease are still present in healthy trees after approximately 30 years. It is believed that an earlier thinning would have reduced the length of the recovery period.

6. The sensitiveness of the fungi on the forest floor to man's silvicultural operations is striking. *Amanita muscaria*, for example, was a conspicuous element in the flora each fall from 1918 to the middle of the 1930's. Fruiting stopped soon after a thinning was made in 1933. It has not been observed in the plantation during the 25 or more years that have followed.

7. Trunk splitting in the spruces occurred only in the mixed planting with locust. Eventually, the majority of the trees were made worthless. The cracks provided infection courts for *Polyporous abietinus*. Decay followed infection, and this indirectly resulted in windfall.

8. In spite of the presence and abundance of infection courts in the form of trunk cracks, small dead branch stubs presented the only means of infection for *Fomes pini*. *F. pini* has not been observed in the pure spruce planting.

9. Natural pruning did not take place in the plantations, even where spacings were three by three feet. Small live branches must be pruned from the crop trees if *Fomes pini* is to be avoided, even in relatively young plantations.

10. Norway spruce is one of the several host trees exotic to Michigan for *Fomes annosus* at the Saginaw Forest.

LITERATURE CITED

BAXTER, DOW V. 1952. Pathology in Forest Practice. 2nd ed. New York: John Wiley & Sons, Inc. 601 pp.

HAWLEY, RALPH C., AND HAROLD J. LUTZ. 1943. Establishment, Development, and Management of Conifer Plantations in the Eli Whitney Forest, New Haven, Conn. Yale Univ., School of For., Bull. 53: 1-81.

HOSLEY, N. W. 1936. Norway Spruce in the North Eastern United States. Harvard For., Bull. 19: 1-80.

Papers of the Michigan Academy of Science, Arts, and Letters
Vol. XLIX, 1964 (1963 Meeting)

RESUPINATE HYDNACEOUS FUNGI OF NORTH AMERICA. III. ADDITIONAL TYPE STUDIES[1]

ROBERT L. GILBERTSON
State University of New York College of Forestry

This paper continues type studies of the approximately 115 species of resupinate hydnaceous fungi originally described from North America. Previous papers in this series (Gilbertson 1962, 1963) concern species originally described by Peck, and by Bresadola, Overholts, and Lloyd. The 15 species discussed in this paper were proposed by several workers, most of whom are author or co-author of descriptions of only one or two new species. The only exception is J. B. Ellis, who is co-author for five species.

H. J. Banker, who contributed an excellent series of type studies in the Hydnaceae, restricted his researches almost entirely to genera characterized by pileate basidiocarps. He did propose two new species of resupinate hydnaceous fungi, *Hericeum fimbriatum* from Pennsylvania and *Hydnum macrescens* from New York. E. A. Burt, the able student of the Thelephoraceae, proposed two new species of *Odontia*, both found on sugarcane. These are *O. sacchari* from Cuba and *O. saccharicola* from Puerto Rico.

J. B. Ellis corresponded with the English mycologist M. C. Cooke and sent many specimens from North America to Cooke for identification. Among the species described as new by Cooke and Ellis are four resupinate hydnaceous fungi. These are *Radulum concentricum* from Oregon and *Odontia fusca, Hydnum pallidum,* and *Grandinia tabacina* from the vicinity of Ellis' home in Newfield, New Jersey. Ellis and B. M. Everhart proposed the species *G. fuscolutea,* also from New Jersey.

During his researches on the Hydnaceae of Iowa, L. W. Miller described *Odontia crustula* and *O. laxa,* both from Iowa. A. P. Morgan proposed two species, *Hydnum casearium* and *H. alboviride,* from Ohio. As a result of his investigation of Texas root rot of cotton, C. L. Shear described *H. omnivorum* as the cause of the disease. John Torrey, one of America's earliest naturalists, gave a

[1] Contribution from the Department of Forest Botany and Pathology, State University College of Forestry at Syracuse University, Syracuse, New York.

description of *H. chrysorhizum* in the third edition of Eaton's *Manual of Botany*. *H. chrysocomum*, the last species to be considered here, was proposed by L. W. Underwood from New York.

Type material of all of the species listed above was available for study, with the exception of *Hydnum alboviride* and *Grandinia fuscolutea*. Capitalized color names are taken from Ridgway (1912).

HERICEUM FIMBRIATUM Banker, Torrey Bot. Club Mem. 12:122. 1906.

TYPE: Chamokin, Pa.; on dead stump; Banker No. 860; Oct. 23, 1903 (NY).

Basidiocarp annual, adherent, up to 8 mm thick; margin Light Buff, fibrillose, showing indication of tooth formation to the edge; toothed surface Hazel, darker than margin, the teeth brittle, up to 8 mm long, cylindric, tapering to the tip, appearing horny or cartilaginous, the apices entire to fimbriate; subiculum pale buff, very thin; teeth in clusters from nodules of subiculum tissue.

Subiculum hyphae closely interwoven, not easily separated, thick-walled, with clamp connections, 2.5-3.5 μ in diam.; cystidia none; some undifferentiated hyphae projecting from the hymenium, these 2.5-3 μ in diam.; basidia narrowly clavate, 4-sterigmate, 20×4 μ; sterigmata up to 4 μ long; spores hyaline, smooth, cylindric, straight to slightly curved, $4\text{-}4.5 \times 2\text{-}2.5$ μ.

Macroscopically, this fungus is similar to *Steccherinum setulosum* (Berk. & Curt.) Miller. The heavily incrusted cystidia typical of *S. setulosum*, however, are not present in *H. fimbriatum*. It also resembles *Mycoacia setosa* (Pers.) Donk but has distinctly different spores.

HYDNUM MACRESCENS Banker, *in* Peck, New York State Mus. Bull. 75, p. 15. 1904.

TYPE: Mt. Macgregor, Saratoga County, New York; on *Stereum frustulosum*; C. H. Peck; July (NYS).

Basidiocarp annual, adherent, the margin farinaceous to arachnoid, up to 0.5 mm wide; toothed surface Salmon Color on drying, the teeth wart-like, soft, crumbly, the apices tufted; context pale buff, soft, very thin.

Subiculum hyphae loosely interwoven, thin-walled, with clamp connections, often branched, 2-3 μ in diam.; cystidia of two types, some hyphoid, tapering to the tip, 2-3.5 μ in diam. and projecting to 15 μ, others bowling pin shaped or with a globose apical swelling, 3.5-4.5 μ in diam. and barely projecting; basidia clavate, 4-sterigmate, 4-5 μ in diam., the sterigmata rounded and knoblike on immature basidia; spores hyaline, smooth, ellipsoid, 4-6 × 3-4 μ.

This species is very similar to *Odontia burtii* (Peck) Gilbertson but differs in spore shape. *O. burtii* has cylindric-ellipsoid spores that are 6-7 × 3-3.5 μ. Banker believed that *Hydnum macrescens* was probably found only on *Stereum frustulosum*.

ODONTIA SACCHARI Burt, Mo. Bot. Gard. Ann. 4:233. 1917.

TYPE: Santiago de las Vegas, Cuba; on *Saccharum officinarum*; W. T. Horne; Aug. 24, 1905 (F).

Basidiocarp annual, adherent when dry, Light Buff; margin thinning out, fertile; teeth represented by minute, barely discernible granules up to 150 μ long; subiculum very thin, soft.

Subiculum hyphae closely interwoven, thin-walled, frequently branched, 2.5-3 μ in diam., with clamp connections and also some simple septa; cystidia scattered but conspicuous, smooth or lightly incrusted, cylindric, clavate, or narrowed at the tip, occasionally with minute clamp connections, 6-15 μ in diam. and projecting 25-45 μ; basidia abundant but rarely sterigmate, a few seen distinctly 2-sterigmate, 5-5.5 μ in diam., no spores seen attached; free spores present in abundance, these hyaline, smooth, cylindric, nonamyloid, 11-13 × 4-5 μ.

Additional specimens studied: PUERTO RICO—Rio Piedras; J. R. Johnston; on *Saccharum officinarum*; July 25, 1912; comm. by J. A. Stevenson No. 4509 (BPI): Rio Piedras; on *Saccharum officinarum*; J. A. Stevenson Nos. 6382, 2908, and 5628 (BPI).

The large cylindric spores are considered to be conidia by Burt, and he describes them as developing on short branches of the subiculum hyphae. Burt gives the basidiospores as "hyaline, even, subglobose, $3\frac{3}{4} \times 3\text{-}3\frac{3}{4}$ μ." I have not found spores of this type attached to basidia.

Burt makes no reference to *Hydnum sacchari* Spreng. ex Fries (Syst. Myc. 1:416. 1821). This species was originally described from sugarcane in Guadeloupe in 1820 (K. Svenska Vet. Akad. Handl., p. 51, 1820). I have not seen authentic specimens of *H. sacchari* Spreng. ex Fries. One specimen in the National Fungus Collections has been so identified by Anna E. Jenkins. It was collected on sugarcane at Rio Piedras, Puerto Rico, by J. R. Johnston in 1914. It is in excellent condition with well-developed cylindric teeth, loosely arranged, thin-walled hyphae with clamp connections, large cylindric cystidia that are 5-6 μ wide and project up to 55 μ, and short cylindric spores that are 4-4.5 × 2-2.5 μ. It is not the same as *O. sacchari* Burt.

ODONTIA SACCHARICOLA Burt, Mo. Bot. Gard. Ann. 4:235. 1917.

TYPE: Rio Piedras, Puerto Rico; on *Saccharum officinarum*; J. A. Stevenson No. 3176; Oct. 5, 1915 (BPI).

Basidiocarp annual, up to 150 μ thick, white to Pale Ochraceous-Buff, apparently separable before drying; margin thinning out, farinaceous; teeth granular or wartlike, soft, very short; subiculum soft, up to 150 μ thick.

Subiculum hyphae not readily separable, very difficult to distinguish, some very slender, 1-1.5 μ in diam., others apparently thin-walled, collapsed; globose masses of crystals present throughout tissues; some cylindric or hyphoid structures projecting from apices of teeth, these septate, 2-3 μ in diam. and projecting to 25 μ; basidia not seen; hymenial surface covered with darkly staining, subglobose to ovoid structures that become detached and resemble chlamydospores, these 7-8 × 5-6 μ; subglobose spores 2.5-3 μ in diam. are quite abundant; also short cylindric spores 4.5-5 × 2-2.5 μ are present, like those described by Burt as basidiospores.

Additional specimens studied: PUERTO RICO—Rio Piedras; J. A. Stevenson Nos. 2657, 2657a, and 3677; all on *Saccharum officinarum*; J. A. Stevenson No. 6213; on *Paspalum* (BPI).

These additional specimens all seem to have the same characters as the type, and none have hymenia in sporulating condition. Some 4-sterigmate basidia were seen, these 5-6 μ in diam.

RADULUM CONCENTRICUM Cooke and Ellis, Grevillea 14:13. 1885.

Irpex owensii Lloyd, Myc. Writings 5:616. 1916.

Radulum owensii (Lloyd) Lloyd, Myc. Writings 5: Radulum, p. 10. 1917.

ISOTYPE: Coos River, Oregon; on dead myrtle; W. S. Carpenter (NY). Material from the type collection was distributed as No. 1710, North American Fungi, by Ellis and Everhart.

Basidiocarp annual, adherent, up to 5 mm thick, Fawn Color to Verona Brown, the teeth flattened to cylindric, in some areas concentrically arranged near the margin, up to 3 mm long; inner tissue of teeth dark brown and continuous with dark brown tissue of inner subiculum, outer layer of teeth and subiculum pinkish in section; entire subiculum up to 2 mm thick.

Subiculum hyphae of outer pale layer hyaline, thin-walled, with clamp connections, 3-3.5 μ in diam., those of inner brown layer similar but pale to dark brownish, also with abundant clamp connections; short hyphoid cystidioles present, barely projecting, 2-2.5 μ in diam.; basidia clavate, 4-sterigmate, 5-6 μ in diam.; spores hyaline, smooth, cylindric-ellipsoid, 6-8.5 × 3-4 μ.

The isotype specimen at the New York Botanical Garden is annotated by M. C. Cooke as follows: ". . . *Radulum concentricum* C & E from the tendency of the teeth to form concentric zones especially near the margin, an unusual feature in *Radulum*." On the basis of other specimens examined, I do not consider this concentric arrangement to be typical of the species.

This fungus is better known as *Radulum owensii* (Lloyd) Lloyd (*Irpex owensii* Lloyd), which is here placed in synonymy with *Radulum concentricum* Cooke and Ellis.

A list of 17 specimens from British Columbia, California, Montana, Oregon, and Washington is given by Gilbertson (1963) under *Irpex owensii*.

Odontia fusca Cooke and Ellis, Grevillea 9:103. 1881.

Isotype: Newfield, New Jersey; presumably collected by Ellis; Oct. 1880; Cooke No. 3429 (NY).

Basidiocarp annual, Snuff Brown to Bister; hymenophore consisting of intertwining ropelike strands of hyphae which give rise to granular or papillate teeth; hyphae thin- to thick-walled, with frequent clamp connections, 2-3 μ in diam.; basidia 5-6 μ in diam.; spores hyaline, ovoid, echinulate, 4.5-5 × 2.5-3 μ.

This species was placed in synonymy with *Hypochnus fumosus* Fries by Burt (1916). Rogers and Jackson (1943) list *H. fumosus* sensu Burt and *O. fusca* as synonyms of *Corticium sulphureum* (Pers. ex Fries) Fries. Donk (1957) proposed the new combination *Cristella sulphurea* for this species.

Hydnum pallidum Cooke and Ellis, Grevillea 9:103. 1881.

Isotype: Newfield, New Jersey; on old oak stump; presumably collected by Ellis; Nov. 18 (NY). The number 1840 is on the accompanying label, and also Cooke No. 3118, the number included in the original description.

Basidiocarp annual, adherent, Light Buff; the margin farinaceous; teeth simple or laterally fused at the base, brittle, cylindric, tapering to the tip, the apices entire, pointed; subiculum pale buff, very thin.

Subiculum hyphae thin-walled, 2-3 μ in diam., with clamp connections; hyphae tufted at apices of teeth; basidia clavate, 4-sterigmate, 4-5 μ in diam.; spores subglobose, hyaline, smooth, 4-5 $\times$ 3-4 μ.

This species is placed in synonymy with the common *Odontia stipata* (Fries) Quél. (basionym = *Hydnum stipatum* Fries, Syst. Myc. 1: 425. 1821).

GRANDINIA TABACINA Cooke and Ellis, Grevillea 9: 103. 1881.

ISOTYPE: Newfield, New Jersey; on rotten red cedar; presumably collected by Ellis; Aug. 28, 1878 (NY). The number 3086, given in the original description, is on the sheet to which the specimens are attached, as is 3149.

Two species are represented on the sheet which bears the number given in the original description, 3086. The number 3149 has also been written on the sheet, but it is not clear to which of the two species the numbers apply. At the top of the sheet are glued 8 pieces of *Corticium sulphureum* (Pers. ex Fries) Fries. A pencil line has been drawn separating these from 5 pieces of a *Tomentella* species. The number 182 has been written in with the *C. sulphureum* and 183 with the *Tomentella.*

Another packet in the same folder is numbered 3149 and contains only the *Tomentella.* This fact, plus characters given in the original description of *G. tabacina,* "Ceracea" and "ambitu subradiante," indicate that no. 3086, the basis of Cooke's description of *G. tabacina,* is *Corticium sulphureum* and not the *Tomentella,* as Burt (1916) has stated. Burt places *G. tabacina* in synonymy with *Hypochnus granulosus* (Peck) Burt.

ODONTIA CRUSTULA Miller, Mycologia 26: 29. 1934.

TYPE: Little Sioux, Milford, Iowa; L. W. Miller No. 15; on linden; June 16, 1931 (IA).

Basidiocarp annual, crustlike, adherent, Light Buff; teeth minute, cylindric, branched and appearing fimbriate at apices, up to 200 μ long; subiculum very thin, appearing farinaceous or arachnoid.

Subiculum hyphae closely interwoven, not easily separating, with abundant crystalline material, frequently branched, thin- to thick-walled, 2-3 μ in diam., with clamp connections; incrusted hyphae projecting at apices of teeth, usually traceable to core of teeth where some incrusted hyphae are present; basidia clavate, 4-sterigmate, 4-5 μ in diam., with a basal clamp connection; spores hyaline, smooth, cylindric-ellipsoid, 5-6 × 2.5-3 μ.

Additional specimens studied: IOWA—Little Sioux; L. W. Miller No. 21; on linden; June 16, 1931 (IA): West Okoboji; G. W. Martin; July 15, 1931 (IA): Milford; E. T. Rose; July 4. 1934 (IA).

This species is similar to *Odontia pruni* Lasch but differs in its smaller spores and clamp connections.

ODONTIA LAXA Miller, Mycologia 26: 19. 1934.

TYPE: Wellman, Iowa; on old bark of deciduous wood; L. W. Miller No. 338; Oct. 8, 1932 (IA).

Basidiocarps annual, effused in small patches up to 2 cm wide, readily separable, white, very soft and thin, appearing farinaceous or arachnoid; teeth soft, up to 250 μ long, cylindric, fimbriate along their entire length and at apices; subiculum white, soft, very thin.

Subiculum hyphae loosely interwoven, often branched, thin-walled, with abundant simple septa, 3-6 μ in diam., partly to completely incrusted with large, coarse crystals; hyphae in teeth similar; cystidia abundant, heavily incrusted, 8-10 μ in diam. and projecting up to 40 μ; basidia difficult to discern, apparently 3.5-4 × 8 μ; spores hyaline, smooth, ovoid to ellipsoid, 2.5-3 × 2-2.5 μ.

Microscopically *Odontia laxa* is similar to *O. subcrinale* (Peck) Gilbertson (Gilbertson 1962). However, *O. subcrinale* has a membranous, separable basidiocarp that is pinkish and rhizomorphic.

HYDNUM CASEARIUM Morgan, Jour. Cinc. Soc. Nat. Hist. 18: 11. 1895.

Radulum casearium (Morg.) Lloyd, Myc. Writings 5: Radulum, p. 8. 1917.

ISOTYPE: Cincinnati, Ohio; on hickory; A. P. Morgan; no date (CGL No. 16737).

Basidiocarp annual, adherent, no marginal tissue present, toothed surface cinereous to buff on drying, the teeth cartilaginous in appearance, brittle, smooth, up to 8 mm long and 0.5 mm in diam., cylindric, tapering to the tip, densely crowded and often laterally

septa, lightly to heavily incrusted, 4-8 μ in diam.; hymenial layer deteriorated, cystidia and basidia absent; probable basidiospores hyaline, smooth, short-cylindrical, 3.5-4.5 × 2-2.5 μ.

Conidiophores and subglobose conidia that are 4-5 × 3-4 μ are abundant in all mounts made from the lectotype of *Hydnum chrysorhizum*. This is a widely distributed and common species in the southern and eastern U. S. and is also reported from the USSR by Nikolajeva (1961) as *Sarcodontia fragilissima* (Berk. & Curt.) Nikol.

Hydnum chrysocomum Underwood, Torrey Bot. Club Bull. 24: 82. 1897.

Type: New Dorp, Staten Island, New York; L. M. Underwood; Oct. 17, 1896 (NY).

Basidiocarp annual, separable; margin rhizomorphic, cream color; toothed surface now pinkish, the teeth brittle, cylindric, crowded, apices smooth; subiculum soft.

Subiculum hyphae closely interwoven, thin- to thick-walled, with abundant simple septa, some heavily incrusted, 3-10 μ in diam.; basidia not seen; probable basidiospores hyaline, smooth, short-cylindric, 4-4.5 × 2-2.5 μ.

This type specimen is heavily overgrown by a hyphomycete with conidiophores and large numbers of ovoid to ellipsoid, rough conidia. Although the specimen is in poor condition and hymenial characters are not discernible, it is quite probable that it is the same fungus as *Oxydontia chrysorhizon* (Torrey) Rogers and Martin as a number of authors have stated.

Acknowledgments

The type specimens on which this paper is based were made available through the kindness of the curators of the herbaria of the following institutions: New York Botanical Garden (NY); New York State Museum, Albany (NYS); National Fungus Collections (BPI) and C. G. Lloyd Mycological Collection (CGL), Beltsville, Maryland; Farlow Library and Herbarium of Cryptogamic Botany, Cambridge, Mass. (F); and State University of Iowa, Iowa City (IA). This research was supported in part by Grant No. G24382 from the National Science Foundation.

LITERATURE CITED

BASHAM, J. T. 1958. Decay of Trembling Aspen. Canad. Jour. Bot. 36: 491-505.

BURT, A. E. 1916. The Thelephoraceae of North America. VI. Hypochnus. Mo. Bot. Gard. Ann. 3: 203-241.

DONK, M. A. 1957. Notes on Resupinate Hymenomycetes. IV. Fungus 27: 1-29.

GILBERTSON, R. L. 1962. Resupinate Hydnaceous Fungi of North America. I. Type Studies of Species Described by Peck. Mycol. 54: 658-677.

——. 1963. ——. II. Type Studies of Species Described by Bresadola, Overholts, and Lloyd. Mich. Acad. Sci. 48: 137-149.

MILLER, L. W. 1933. The Hydnaceae of Iowa. I. The Genera *Grandinia* and *Oxydontia*. Mycol. 25: 356-368.

NIKOLAJEVA, T. L. 1961. Flora plantarum cryptogamarum URSS. VI. Fungi (2). Familia Hydnaceae. Leningrad. 339 pp.

RIDGWAY, ROBERT. 1912. Color Standards and Color Nomenclature. Washington, D.C. Published by the author.

ROGERS, D. P., AND H. S. JACKSON. 1943. Notes on the Synonymy of Some North American Thelephoraceae and Other Resupinates. Farlowia 1: 263-336.

PAPERS OF THE MICHIGAN ACADEMY OF SCIENCE, ARTS, AND LETTERS
VOL. XLIX, 1964 (1963 MEETING)

THE GENERA *GRAMMOTHELE* AND *POROGRAMME*[1]

JOSIAH L. LOWE
State University of New York College of Forestry

MANY resupinate Hymenomycetes have been described from the tropics and the types are deposited in scattered herbaria, chiefly in the North Temperate Zone. As few of these have been restudied and modern descriptions published, accurate identifications can be made only by persons having access to the type specimens. Comprehensive studies of all described species in their respective genera and of species in other genera of similar aspect, will make possible identification by mycologists generally, and will greatly stimulate interest in the tropical fungus flora. The present genera have been much neglected because a modern treatment has not been available.

The genus *Grammothele* was proposed by Berkeley and Curtis (1868) to accommodate four new species from Cuba. The authors did not select the type for their new genus; subsequent students, however, have uniformly designated *G. lineata* to stand as the type (Donk 1960). Two species and one variety were described later by Bresadola. *Porogramme* was first proposed by Patouillard (1899) as a subgenus to group together three new species differing from *Poria* by the small size of their pores, by their shallowness, and by partitions which divide the trama into minute fragments. The subgenus was elevated to generic rank the next year (Patouillard 1900) and five additional species were listed, mostly by transfer of species formerly placed in *Poria*. The genus *Grammothele* was accorded an intermediate position between *Hymenogramme* and *Porogramme* in this paper. In 1913 Patouillard described two more species of *Porogramme*.

Both *Grammothele* and *Porogramme* have sterile tissue which extends below the hymenium (Pl. I, Fig. G) and forms a more or less poroid lower surface. *Grammothele* has hyphal fascicles (Pl. I, Fig. A), structures which rise in both the context and the trama and are visible externally under slight magnification. The poroid

[1] Contribution from the Department of Forest Botany and Pathology, State University College of Forestry at Syracuse University, Syracuse, New York.

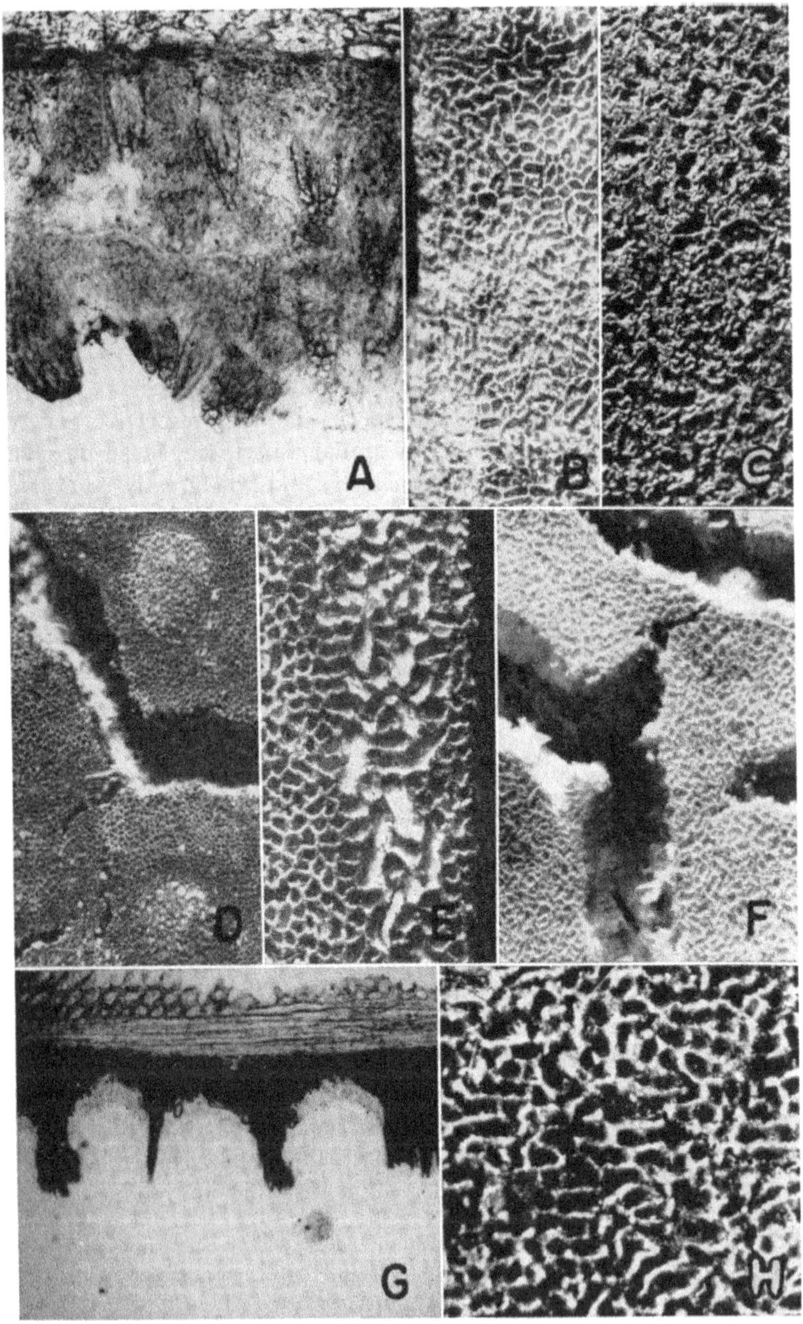

PLATE I

EXPLANATION OF PLATE I

Structural features of *Grammothele* and *Porogramme*. FIG. A. Vertical section through the type of *G. lineata* showing hyphal fascicles arising throughout the tissue. × 110. FIG. B. Pore surface of type material of *P. graphica*, × 15. FIG. C. Contrasting pore surface of type of *P. lateritia*, × 15. FIGS. D AND E. Pore surfaces of two specimens of *P. fuligo* showing variable pore size, × 15. Fig. D from No. 574, Farlow Herbarium; Fig. E from Farquharson collection at Stockholm. FIG. F. Pore surface of type of *P. richeriae*, × 15. FIG. G. Vertical section of *P. fuligo* showing hymenium at the base of the tubes, × 110. From Sydow, F. Exot. Exsic. No. 204 at Stockholm. FIG. H. Pore surface of *Kneiffia grisea*, × 15, from Theissen Exsic. No. 5, at Stockholm.

fruiting surface is a strong argument for placing these genera in the Polyporaceae, and Donk (1960) has assigned them to this position. In the polypores, however, a hymenium normally lines the downwardly projecting tissue of the tubes. As *Grammothele* and *Porogramme* do not conform to this typical structure, the generic position of these genera is somewhat uncertain. *Grammothele* was originally assigned to the Hydnaceae, and this disposition has been followed by Killermann (1928). Patouillard (1899), at first assigned *Porogramme* to the Polyporaceae; he later (1900) placed it in a new group, the Odontiés. Both genera seem better placed in the Polyporaceae than in the Hydnaceae, but a satisfactory position must await a comprehensive study of many somewhat similar tropical species now placed in still other genera.

In this paper two species of *Grammothele* and eight of *Porogramme* are considered valid. The following are reduced to synonymy for the first time: *G. hydnopora* var. *bahianensis* = *P. delicatula; G. hydnopora* and *G. cineracea* = *G. lineata; Polyporus ravenalae* = *Porogramme fuligo*. The following are recombinations in the genus *Porogramme* for the first time: *P. calcicolor* (Sacc. & Syd.), *P. delicatula* (P. Henn.), and *P. delicata* (Bres.).

Grateful acknowledgment is made to the directors or curators at the following herbaria for the opportunity to study specimens in their institutions or through loan: Farlow Library and Herbarium of Cryptogamic Botany at Harvard University, Cambridge, Mass.; Kew Herbarium at the Royal Botanic Gardens, Kew, England; National Fungus Collections, Beltsville, Maryland; Naturhistoriska Riksmuseet, Stockholm, Sweden; and the New York Botanical Garden, New York City. Acknowledgment is also made to the National Science Foundation for support during various segments of this study.

GRAMMOTHELE Berk. & Curt.

Jour. Linn. Soc. 10: 327. 1868.

Sporophore resupinate, with a poroid hymenial surface; tubes well or imperfectly formed, with the hymenium restricted to the base of the tubes; hyphal fascicles present; basidia nonseptate; spores hyaline.

Key to the Species

1a. Tubes well formed, 1-3 per mm; spores 4-6 × 1.5-2.5 μ
1. *G. lineata*

1b. Tubes imperfectly formed, about 3 per mm; spores 4-4.5 × 1.5-2 μ
2. *G. mappa*

1. Grammothele lineata Berk. & Curt.

Jour. Linn. Soc. 10: 327. 1868. *Grammothele grisea* Berk. & Curt., Jour. Linn. Soc. 10: 327. 1868. *Grammothele polygramma* Berk. & Curt., Jour. Linn. Soc. 10: 327. 1868. *Polyporus hydnoporus* Berk., Videnskab. Medd. For. Kjebenh. for 1879-80, p. 751 (31 in reprint). *Porogramme duporti* Pat., Soc. Mycol. France Bul. 29: 208. 1913. *Grammothele cineracea* Bres., Hedwigia 56: 299. 1915.

Annual or occasionally reviving once or twice, becoming widely effused, up to 2 mm thick, without distinctive taste; margin when present white, arachnoid or mycelioid, up to 2 mm wide; pore surface probably white to grayish and with a purplish tinge when fresh, when dry varying from cream and pale to dark sordid gray in weathered specimens, the tubes arising from hyphal pegs arranged in lines about which tissue grows, and by lateral expansion these lines intersect each other to form a poroid surface, the tubes usually rigid and rather brittle, sometimes rather soft and fragile, on horizontal surfaces up to 2 mm long, on vertical surfaces (as in types of *G. lineata* and *G. grisea*) much elongated, to 10 mm or more, the mouths subangular, 1-2 (-3) per mm or elongated and linear, the edges rather thin to rather wide, studded with whitish or brownish blunt pegs which are also along the sides of the tubes; context apparently lacking or up to 0.2 mm thick, white to pale brown, corky-fibrous, the tramal tissue often apparently rising directly from the substratum, with a variable amount of dark tissue, this sometimes appearing continuous with context tissue, with dark streaks (the hyphal fascicles); tube cavities filled with cream-colored tissue as the dissepiments elongate.

White tissue not changing color in KOH, the dark tissue becoming more intensely colored, the context hyphae rarely branched, thick-walled to apparently solid, not incrusted, nonseptate, 1.5-2 μ in diam., continued relatively unchanged into the tramal tissue;

tube cavities filled with tissue that often appears alveolar, the hyphae much branched, solid, mostly 0.5-1 μ in diam.; hymenium 12-19 μ thick; hyphal fascicles (Pl. I, Fig. A) in KOH varying from hyaline to yellowish or brownish, with or without coarse incrustations, blunt, 15-50 μ in diam.; cystidia none; basidia slender-clavate, 12-15 $\times$ 4-5 μ; spores hyaline, smooth, I –, oblong, apiculate, 4-6 $\times$ 1.5-2.5 μ.

On angiosperms, associated with a white rot, in Cuba, Guadeloupe, Jamaica, Brazil, Tonkin, Malaya, and the Philippines.

G. polygramma appears to be based on an immature specimen with poorly formed pores. *G. mappa* is perhaps an immature growth form of *G. lineata*. The type of *Polyporus hydnoporus* is sterile; the spore measurements given by Killermann (1928, p. 169) as 12-14 $\times$ 4-5 μ were probably based on a specimen now in the Bresadola Herbarium at Stockholm, collected by Lloyd in Java and misnamed *Grammothele hydnopora* by Bresadola. The plant is *Trametes subserpens* Murr.

G. lineata has been selected by several students to stand for the type of *Grammothele* (Donk 1960, p. 221).

Berkeley and Curtis (1868) described *Kneiffia grisea* from Cuba on the same page as they described *G. grisea*. Patouillard (1900) considered that these two species were synonymous, and his decision has been accepted by other students. Bresadola, in herbarium notes, believed that *K. grisea* was a juvenile state of *G. grisea*. These two plants are very similar to each other; they differ markedly, however, in their pore size and shape. The type of *K. grisea* has poorly outlined pores that average about 6 per mm. In a Brazilian specimen, collected by Rick and issued by Theissen in his Dec. Fung. Brasil., No. 5, the pores are well formed, somewhat daedaloid, and measure about 4 per mm (Pl. I, Fig. H). The type of *K. grisea* is sterile; spores on the Rick collection and in several others are hyaline, smooth, I–, oblong-ellipsoid, 4.5-7 $\times$ 2-3.5 μ. *K. grisea* appears to be a valid species.

2. GRAMMOTHELE MAPPA Berk. & Curt.

Jour. Linn. Soc. 10: 327. 1868.

Reviving once or twice, probably becoming widely effused; margin brownish, cartilaginous in appearance, appressed, smooth, up to 0.5 mm wide; pore surface cream, the tubes arising as wartlike protuberances which elongate and form extremely short dissepiments, a true poroid surface scarcely present, the openings daedaloid, about 3 per mm, edges thick, covered with small hydnoid processes; context scarcely present, the bases of the tubes becoming stuffed in development, forming a generally white tissue with yellowish resinous streaks where the hydnoid processes are present.

Tissue not markedly changing color in KOH, the context hyphae rarely branched, appearing solid, nonseptate, 1.5-2.5 μ in diam., the tramal tissue similar but thin-walled, the pore area filling with an alveolar tissue of extremely slender, much-branched hyphae about 0.5-1 μ in diam.; hymenium about 15 μ thick; hyphal fascicles present, 25-55 μ in diam.; cystidia none; basidia uncertain, apparently clavate, 16 × 4-5 μ; spores hyaline, smooth, I–, ellipsoid, 4-4.5 × 1.5-2 μ.

On angiosperms, associated with a white rot, known only from the type from Cuba. The species is doubtfully distinct from *G. lineata,* which appears to differ in having better formed pores, and perhaps in having larger spores. Spores were found only on the isotype in the Bresadola collection at Stockholm, and they may not represent the full size attained by this species.

Porogramme Pat.

Essai Tax., p. 63. 1900. *Poria* (*Porogramme*) Pat., Soc. Mycol. France Bull. 15: 199. 1899.

Sporophore resupinate, with a poroid hymenial surface; tubes well or imperfectly formed, with the hymenium restricted to the base of the tubes; hyphal fascicles absent; basidia nonseptate; spores hyaline.

Key to the Species

1a. Tissue brick red
- 2a. Tubes soon formed and pores of definite outline, 6-11 per mm — 6. *P. graphica*
- 2b. Tubes irregularly formed and pores of irregular shape, 10-20 per mm — 7. *P. lateritia*

1b. Tissue dark brown to black
- 3a. Tubes well formed, 4-20 per mm — 4. *P. fuligo*
- 3b. Tubes imperfectly formed, about 6 per mm — 5. *P. camptogramma*

1c. Tissue white to cream or brownish	
4a. Cystidia present	8. *P. delicata*
4b. Cystidia absent	
5a. Pores 10-15 per mm	1. *P. richeriae*
5b. Pores 2-8 per mm	
6a. Spores 5.5-6.5 × 4.5-5.5 μ; pores 4-8 per mm	2. *P. calcicolor*
6b. Spores 11-17 × 3.5-6 μ; pores 2-5 per mm	3. *P. delicatula*

1. Porogramme richeriae (Pat.) Pat.

Essai Tax., p. 64. 1900. *Poria richeriae* Pat., Soc. Mycol. France Bull. 15: 200. 1899.

Perennial, effused to 8 cm or more, up to 1 mm thick; margin white, minutely mycelioid, up to 2 mm wide; pore surface conspicuously areolate, glaucous-cream, the tubes quickly outlined on the marginal tissue and elongating to form fragile dissepiments up to 0.3 mm long, the pores (Pl. I, Fig. F) angular or very irregular in shape, about 10-15 per mm, the edges thin, entire; context if present less than 0.1 mm thick, mostly composed of tramal tissue in indistinct layers, each 0.2-0.3 mm thick, the whole firm, waxy, white.

Tissue not changing color in KOH, the tramal hyphae compactly arranged, with thin or rather thin walls, nodose-septate, about 2-3 μ in diam.; hymenium not seen; hyphal fascicles or cystidia not present; basidia not seen; spores hyaline, smooth, I–, ellipsoid to ovoid, 2.5-3.5 × 2-2.5 μ.

On angiosperms, associated with a white rot, in Guadeloupe, known only from the type collection and one other in the Patouillard Collection.

No other species here treated resembles this. Spores as described were found in considerable numbers along the hymenial surface, but none was seen attached, nor is there an evidence of a hymenium. These may be foreign spores, but they appeared to belong to this specimen.

There is another specimen in the Patouillard Herbarium, named and no locality given. The type selected is No. 587; the other specimen is No. 698.

2. ***Porogramme calcicolor*** (Sacc. & Syd.) Lowe, comb. nov.

Poria calcicolor Sacc. & Syd., Syll. Fung. 14; 192. 1899. *Polyporus calceus* Berk. & Br., Jour. Linn. Soc. 14: 55. 1875 (not *P. calceus* Schw., 1822). ?*Poria amaniensis* P. Henn., Engl. Bot. Jahrb. 38: 109. 1905 (type sterile).

Annual, becoming widely effused sometimes but usually of quite limited extent; margin white, arachnoid or matted and more or less thick-membranous and determinate, up to 2 mm wide; pore surface white, in old collections cream and usually with a yellowish or pinkish tint, the tubes arising by the downward growth of quickly outlined dissepiments on the marginal tissue, soft and usually quite fragile, up to 0.5 mm long, the pores angular, about 4-8 per mm, the edges thin, entire to eroded or lacerate; context below white to cream, soft, up to 0.3 mm thick, abruptly becoming translucent yellowish just below the tubes, this continuous with the pore wall tissue.

Tissue of the context unchanged in KOH, the hyphae compactly arranged, infrequently branched, thick-walled to apparently solid, nonincrusted, rarely and inconspicuously both simple-septate and nodose-septate, 0.5-1.5 (-2) μ in diam.; tramal hyphae similar except appearing more or less fused and more translucent than in the context; hymenium 19-25 μ thick; hyphal fascicles or cystidia none; basidia clavate, 17-20 $\times$ 6-8 μ; type sterile, Malayan specimen with spores hyaline, smooth, I–, broad-oblong to ovoid or oval, 5.5-6.5 $\times$ 4.5-5.5 μ.

On wood of angiosperms, associated with a white rot, in Ceylon (type), Malaya (collected by Weir and specimen at Beltsville), and the Philippines. The species is most similar to *P. delicatula*, which differs in having cylindric spores 11-17 $\times$ 3.5-6 μ, usually stouter context hyphae 2-3.5 (-5) μ in diam., and generally larger pores, 2-5 per mm.

3. ***Porogramme delicatula*** (P. Henn.) Lowe, comb. nov.

Poria delicatula P. Henn., Engl. Bot. Jahrb. 34: 44. 1904. *Grammothele hydnopora* var. *bahianensis* Bres., Ann. Mycol. 18: 43. 1920.

Annual, usually in limited areas but sometimes becoming widely effused; margin whitish to cream, velvety or cottony, sometimes fimbriate at the extreme margin, up to 0.5 mm wide; pore surface white to cream, the tubes arising by uniform downward growth of the marginal tissue, usually soft and fragile and crumbling on pressure, but in some rather tough, up to 0.2 mm long, the pores round to angular, about 2-5 per mm, edges thin to often somewhat thickened, usually granulose, entire; context white to cream, soft, up to 0.2 mm thick; tramal tissue continuing unchanged from context or becoming resinously indurated in the outer part.

Tissue not changing color in KOH, the context hyphae of two kinds, the principal hyphae compactly arranged, rarely branched, with thick to apparently solid walls, nonincrusted, hyphae under the tubes with extremely rare simple septa, hyphae in the marginal tissue with both clamp connections and simple septa but both very rare, mostly 2-3.5 μ in diam., occasionally apparently gelatinizing in KOH and up to 5 μ in diam.; hymenium 17-25 μ thick; hyphal fascicles and cystidia lacking; basidia broadly clavate, 14-19 $\times$ 8-12 μ; spores hyaline, smooth, I–, cylindric to subellipsoid or somewhat fusiform, 11-17 $\times$ 3.5-6 μ.

On angiosperms, associated with a white rot, in Brazil, Liberia, Sierra Leone, Tanganyika Territory, ?Queensland and ?New South Wales in Australia–specimens sterile, and Java. This species is most similar to *P. calcicolor* which differs in having spores 5.5-6.5 $\times$ 4.5-5.5 μ, usually more slender context hyphae 0.5-2 μ in diam., and generally smaller pores, about 4-8 per mm.

4. Porogramme fuligo (Berk. & Br.) Pat.

Essai Tax., p. 64. 1900. *Polyporus fuligo* Berk. & Br., Jour. Linn. Soc. 14: 53. 1875. *Polyporus ravenalae* Berk. & Br., Jour. Linn. Soc. 14: 53. 1875. *Polyporus niger* Berk. "var. *poris minoribus*" Berk. & Br., Jour. Linn. Soc. 14: 53. 1875. *Polyporus büttneri* P. Henn., Verh. Bot. Ver. Prov. Brandenburg 30: 129. 1889. *Poria glauca* Pat., Jour. de Bot. 5: 312. 1891. *Poria albocincta* Cooke & Massee, Grev. 20: 106. 1892. *Poria fuligo* (Berk. & Br.) Cooke var. *aurantiotingens* Ell. & Macbr., Bull. Lab. Nat. Hist. State Univ. Iowa 3: 191. 1896.

Annual or reviving once or twice, becoming widely effused, up to 0.5 mm thick; margin white or whitish, sometimes with a lavender tint, arachnoid, sometimes conspicuous and up to 6 mm wide; pore surface pale glaucous blue or bluish gray, darkening in age and on weathering becoming brownish black, tubes quickly outlined on the marginal tissue and elongated by the downward growth of the dissepiment tissue, varying from rigid and fragile to papery tough, up to 0.5 mm long, the pores angular, 4-20 per mm, usually varying within a narrow range but occasionally varying widely on the same sporophore (Pl. I, Figs. D, E), edges thin, entire; context brown-black, sometimes apparently lacking, firm, up to 0.1 mm thick, continuous with the tramal tissue or the tramal tissue apparently springing directly from the substratum.

Tissue darkening in KOH, the context hyphae very compactly arranged and difficult to see individually, rarely branched, with rather thin to apparently solid walls, nonincrusted, rarely septate, 1.5-3 μ in diam.; tramal hyphae parallel-arranged, thick-walled, 2.5-3.5 μ in diam.; hymenium 13-18 μ thick; hyphal fascicles or cystidia lacking; basidia clavate, 4-spored, 12-17 × 5-6 μ; spores hyaline, smooth, I –, oblong or oblong-ellipsoid to ellipsoid, 4-7.5 × 2.5-4 μ.

On angiosperms, usually monocots, apparently throughout the tropics. *P. camptogramma* resembles this in having similar dark tissue and similar spores, and that species may be based on an immature example of *P. fuligo*.

5. Porogramme camptogramma Pat.

Soc. Mycol. France Bull. 29: 208. 1913.

Annual, of very limited extent, the largest 2 × 1 cm; margin white, arachnoid, up to 0.5 mm wide; pore surface steel gray with a brownish tint, the tubes arising from linearly arranged brown spines which become coalesced by a translucent material, or as isolated spines that fuse laterally to form plates, both of which coalesce irregularly and form more or less daedaloid or rarely regular poroid areas with fragile tubes up to 0.2 mm long, the pores about 6 per mm, the edges slender, incised; context dark brown, less than 0.1 mm thick, continued into the tramal tissue.

Tissue dark brown in KOH, the context hyphae very compactly interwoven, scarcely discernible individually, probably same as the tramal hyphae which are parallel-arranged, rather thin-walled, probably septate (one simple septa seen), 2-3 μ in diam.; hymenium 12-14 μ thick; hyphal fascicles absent; cystidia lacking; basidia clavate, 12-17 × 5-7 μ; spores hyaline, smooth, I –, oblong to oblong-ellipsoid, 5.5-8 × 2.5-3.5 μ.

On bamboo in Indo-China, known only from the type collection, which may be of imperfectly developed specimens of *P. fuligo*.

That has a similar dark context but appears to differ in having a well-formed poroid hymenial surface.

6. Porogramme graphica (Bres.) Pat.

Essai Tax., p. 64. 1900. *Poria graphica* Bres., Hedwigia 35: 282. 1896.

Perennial (but type of one layer), becoming widely effused, up to 6 mm thick; margin uncertain (type without a margin), in one specimen white and extremely thin, arachnoid, narrow, in another about 0.5 mm wide, brownish-orange, matted, about 0.5 mm wide; pore surface more or less pinkish-gray, the tubes quickly and regularly outlined on the marginal tissue, firm and corky to soft and papery, in distinct to indistinct layers each 0.3-0.4 mm thick, the dissepiments formed of intersecting labyrinthiform ridges which soon produce regular, angular to daedaloid pores 6-11 per mm (Pl. I, Fig. B), the edges thin, entire; context usually apparently absent or extremely thin, the underlying tissue composed of old tubes filled with mycelium to form a compact brick red tissue.

Tissue intense red in KOH, the tramal hyphae nonseptate, rather thick-walled, infrequently branched, 2.5-4 (-5) μ in diam.; hyphal fascicles lacking; cystidia lacking; basidia and spores not found, according to Weir, " spores ovate to ellipsoid, smooth, hyaline, average (50) 2.4×4.6 μ."

On angiosperms, associated rot unknown, in Cuba, Dominica, Montserrat, Nicaragua, and Brazil. *P. lateritia* is the only other species with brick red tissue; it differs in having much less regularly formed pores which are about 10-15 (-20) per mm.

Weir (1926, p. 90) placed *P. lateritia* in synonymy with this species. Only one Brazilian collection made and identified by him has been seen; it is *P. graphica* as here defined.

7. Porogramme lateritia (Pat.) Pat.

Essai Tax., p. 64. 1900. *Poria lateritia* Pat., Soc. Mycol. France Bull. 15: 200. 1899.

Perennial, becoming widely effused, up to 3.5 mm thick; margin not present on specimens seen; pore surface reddish brown with a distinct grayish blue bloom, the tubes often arising from more or less parallel-arranged plates which are intersected irregularly by other plates to form dissepiments which are fragile, 0.1-0.2 mm long, the pores very irregular in shape, of variable size, about 10-15 (-20) per mm (Pl. I, Fig. C), the edges narrow, entire, often white and minutely granular; context apparently lacking or extremely thin, the supporting tissue made up of brick red, woody tramal tissue.

Tissue bright red in KOH, the tramal hyphae very closely interwoven, rarely branched, with rather thick walls, nonseptate, 1.5-3.5 μ in diam.; hymenium not present; hyphal fascicles absent; basidia and spores not found.

On angiosperms in Guadeloupe, known only from the type and one other specimen with locality label annotated simply as "Antilles." *P. graphica* is the only other species with similar red tissue; it differs in having larger, much more regularly formed pores 6-11 per mm.

8. ***Porogramme delicata*** (Bres.) Lowe, comb. nov.

Grammothele delicata Bres., Hedwigia 56: 299. 1915.

Annual, probably of limited extent (largest seen effused to 2 cm); margin white, thin, radially fimbriate, up to 0.5 mm wide; pore surface deep cream, "Cream Buff" of Ridgway, the tubes arising as isolated spines or plates which expand laterally and intersect to form an imperfectly poroid or somewhat hydnaceous surface, extremely soft and fragile, up to 0.1 mm long, the pores somewhat angular, averaging 5 per mm but varying from 4-7 per mm, the edges thin, dentate to incised; context whitish, soft-fibrous, up to 0.1 mm thick.

Tissue not changing color in KOH, the context hyphae loosely interwoven, rarely branched, rather thick-walled, occasionally incrusted, rarely nodose-septate, 1.5-2.5 μ in diam.; marginal tissue of two types, mostly as above but also some thin-walled, nodose-septate or simple-septate, 2-4 μ in diam.; tramal hyphae similar; hymenium indistinctly limited, about 20 μ thick; hyphal fascicles not present; cystidia abundant, heavily incrusted, cylindrical to broadly clavate, up to 50 μ long, 4-13 μ in diam. (outside dimension); basidia slender-clavate, with 2 sterigmata in the few seen, size somewhat uncertain but about 16-19 × 4-5 μ; type sterile, in the Japanese collection spores hyaline, smooth, I –, ovoid, 3-4 × 2.5-3 μ.

On angiosperms, associated with a white rot, in the Philippines and in Japan. The species is the only one in *Porogramme* with cystidia and seems unrelated even generically to the other species treated here.

Species Omitted

Porogramme dussii (Pat.) Pat. Type consists of *Poria borbonica* Pat. overrun by a white mycelium. The concept of the species seems to have been based on the pores of *P. borbonica* showing through the extremely thin cover of white mycelium. This species appears to have been the one chosen by Patouillard to serve as type for the genus, and that unfortunate choice has been followed by subsequent students (Donk 1960).

Porogramme lateritia or *P. richeriae* could also serve as type for the genus; the latter species would be the better choice.

LITERATURE CITED

Berkeley, M. J., and M. A. Curtis. 1868. Fungi Cubenses (*Hymenomycetes*). Jour. Linn. Soc. Bot. 10: 280-392.

Donk, M. A. 1960. The Generic Names Proposed for Polyporaceae. Persoonia 1: 173-302.

Killermann, S. 1928. *In* A. Engler und K. Prantl, Die Natürlichen Pflanzenfamilien, 6 Band, Abt. II. 290 pp.

Patouillard, N. 1899. Champignons de la Guadeloupe. Soc. Mycol. France Bull. 15: 191-210.

——. 1900. Essai taxonomique sur les familles et les genres des *Hyménomycètes*. Lons-Le-Saunier, France. 184 pp.

Weir, J. R. 1926. A Pathological Survey of the Para Rubber Tree (*Hevea brasiliensis*) in the Amazon Valley. U. S. D. A. Dept. Bull. 1380. 129 pp.

PAPERS OF THE MICHIGAN ACADEMY OF SCIENCE, ARTS, AND LETTERS
VOL. XLIX, 1964 (1963 MEETING)

THE SIGNIFICANCE OF THE ORGANIZATION OF THE INFLORESCENCE OF *AMBROSIA DIVARICATA* (COMPOSITAE)[1]

WILLARD W. PAYNE
The University of Michigan

IT IS commonly recognized that the floral head of the Compositae is a modified inflorescence which has been derived from more primitive types within the Angiosperms. By its nature as an inflorescence, the head terminates the branch upon which it is borne, and in species which bear more than one head, development of the heads is basipetal. Such a flowering plan, which might be equated with a cymose arrangement of individual flowers, is representative for the Compositae. Maturation progresses from the apical head of the main axis, or of a branch system, through the heads borne on secondary, tertiary, and lower orders of floriferous or capitulous branches (Fig. 1A).

The inflorescence of *Ambrosia*, however, when examined superficially, presents a strikingly different aspect from the typical condition within the Compositae. As it is expressed in the majority of North American species, the ragweed inflorescence consists of two kinds of capitulum groups: (*a*) acropetally maturing, spikelike racemes of staminate heads borne at the tips of the branches, and (*b*) sessile clusters of pistillate heads located in the axils of leaves and bracts below the staminate racemes (Fig. 1B). The acropetal development of the more obvious staminate raceme and the arrangement of the heads and other modifications of the floral organs which have come about through adaptation of the plants for anemophily indicate such a departure from the usual condition that several workers, including Rydberg (1922) and Small (1913), have placed the ragweeds and their relatives in a separate family, the Ambrosiaceae.

[1] Paper No. 37 on Atmospheric Pollution by Aeroallergens, supported by research grant No. AP 00008-02 from the Division of Air Pollution, Bureau of State Services, U. S. Public Health Service. I wish to express my appreciation to the curators of the Herbaria of the University of Arizona (hereafter ARIZ), The University of Michigan (MICH), and the New York Botanical Garden (NY) for making specimens available for this study.

A few months ago I completed a paper which dealt with the morphology of the ragweed inflorescence (Payne, 1963). In this paper I was able to demonstrate, primarily through the use of teratological forms of *Ambrosia artemisiifolia, A. hispida,* and *A. trifida,* that each of the stalked, staminate heads of the staminate raceme is actually a reduced, ebracteate branch system which bears only a single, modified head at its apex. The acropetal maturation of the staminate head cluster thus becomes an acropetal development of reduced branch systems, each of which bears but a single head. Acropetal development of the inflorescent branches is uncommon in the Compositae but does occur scattered through several genera, including *Artemisia, Aster, Cichorium, Prenanthes,* and *Solidago.* It was further suggested that the sessile cluster of pistillate heads represents a shortened branch, homologous with the reduced staminate branch, but which still bears several heads that mature in the usual, basipetal direction (centrifugal now by virtue of loss of the axis). In Fig. 2 the branch system modifications from normal (Fig. 2A) to ambrosioid (Fig. 2B) are shown diagrammatically. Until recently, however, I had not observed any species which showed an organization truly intermediate between that representative for the Compositae and that of the typical ragweed. While pursuing other taxonomic work, I encountered a relatively obscure species from central Baja California, Mexico, which is perfectly intermediate. This species is a ragweed, although it has been placed in the genus *Franseria,* a classification which cannot be maintained for reasons outlined by Shinners (1949), Payne (1962a, 1962b, and 1963) and Gebben, Payne, and Wagner (1962). Accordingly, the new combination is made below.

Ambrosia divaricata (Brand.) Payne, comb. nov. *Franseria divaricata* T. S. Brandegee. Proc. Calif. Acad. 2. 2: 171. 1889.

Type: T. S. Brandegee 1889. In the herbarium of the University of California; Sheet No. 88805. Photograph of type seen from the collections of the New York Botanical Garden (NY).

Type locality: San Gregorio, Lower California.

Distribution: Central Baja California, Mexico.

In *A. divaricata* the lowermost staminate branches often bear two or three heads or more, and several of the lower staminate branches may be subtended by bracts (Fig. 3A). Pistillate heads are borne on the elongate lowermost branches, which are clearly homologous with the staminate branches above (Fig. 3A). Thus, this

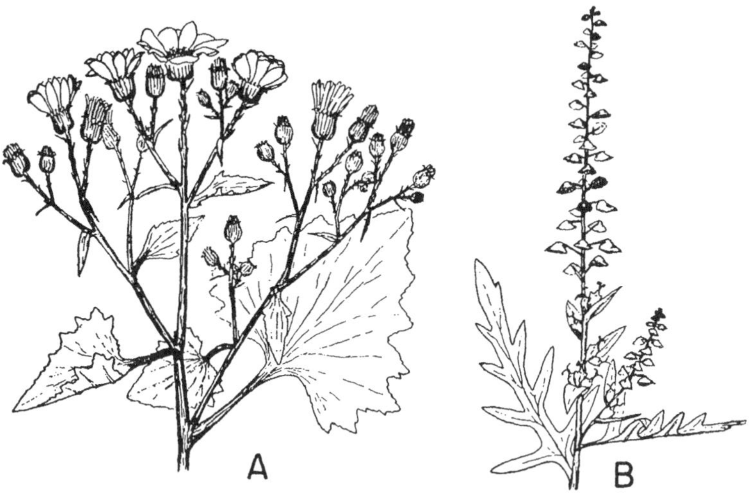

FIG. 1. Comparison of the inflorescence of a typical composite and typical ragweed. A. Terminal portion of the inflorescence of the florist's cineraria, *Senecio cruentus* DC., showing the basipetal head maturation characteristic of the Compositae. B. Inflorescence of *Ambrosia psilostachya* DC., showing the acropetal maturation of racemose clusters of staminate heads and the centrifugally developing clusters of pistillate heads in the axils of subtending leaves.

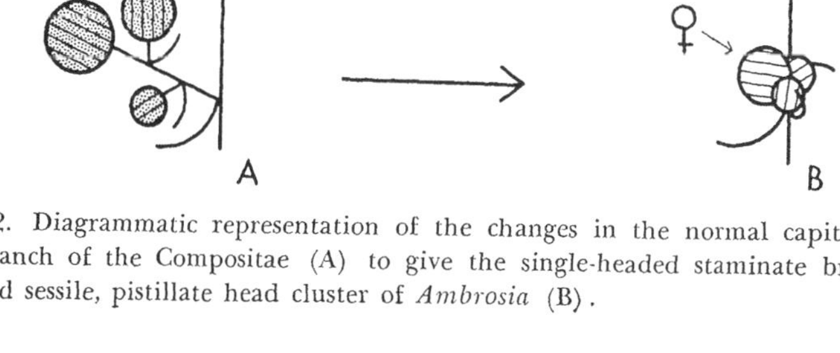

FIG. 2. Diagrammatic representation of the changes in the normal capitulous branch of the Compositae (A) to give the single-headed staminate branch and sessile, pistillate head cluster of *Ambrosia* (B).

species forms a connecting link, in regard to the organization of the inflorescence, between the usual condition in the Compositae and that typically encountered in *Ambrosia*. This character expression may be considered primitive for the genus.

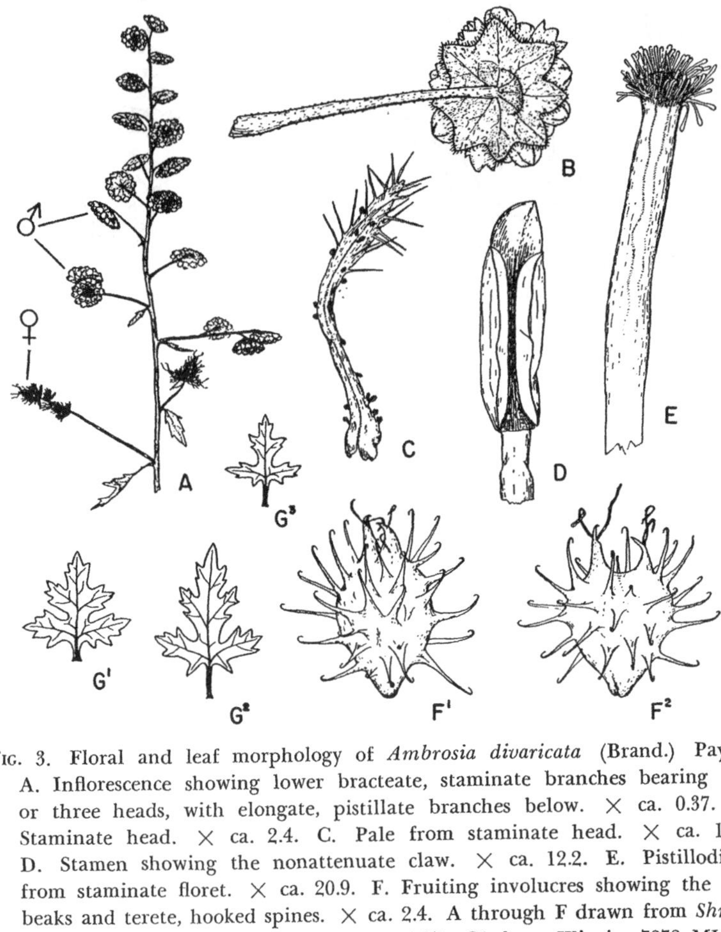

FIG. 3. Floral and leaf morphology of *Ambrosia divaricata* (Brand.) Payne. A. Inflorescence showing lower bracteate, staminate branches bearing two or three heads, with elongate, pistillate branches below. × ca. 0.37. B. Staminate head. × ca. 2.4. C. Pale from staminate head. × ca. 12.2. D. Stamen showing the nonattenuate claw. × ca. 12.2. E. Pistillodium from staminate floret. × ca. 20.9. F. Fruiting involucres showing the two beaks and terete, hooked spines. × ca. 2.4. A through F drawn from *Shreve* 7017 ARIZ. G. Cauline leaves; × ca. 0.37. G¹ from *Wiggins* 7878 MICH. G² from *Shreve* 7017 ARIZ. G³ from *Shreve* 6980 MICH.

Other presumably primitive character expressions for the inflorescence of *Ambrosia* (Payne 1962a and 1963) include tendencies to have: (*a*) the staminate involucre deeply toothed, (*b*) the pales which subtend the staminate florets considerably broadened at the apex, (*c*) the claws of the stamens triangular and not extended by attenuated tips, (*d*) several (three to seven) pistillate florets borne within each pistillate involucre, and (*e*) the spines of the pistillate involucre imbricated, flattened, and straight (not hooked at the apex). *A. divaricata* is more or less primitive or intermediate in regard to these characters and others. The staminate involucre (Fig. 3B) is not deeply toothed, and the pales of the staminate capitulum are linear to only slightly expanded distally (Fig. 3C). The stamen claws (Fig. 3D) are not distally prolonged, and the pistillodium (Fig. 3E) is of the usual type for *Ambrosia*. The fruiting involucres (Fig. 3F) commonly bear only two florets, and the imbricated spines are terete and hooked. The leaves of *A. divaricata* (Fig. 3G) are pinnately lobed, though sparsely, as is typical for the majority of species, and its shrubby perennial habit and alternate leaf arrangement are character states thought to be primitive.

In addition to *A. divaricata*, a few other species display similar intermediacy in their inflorescence organization. All these are presently classified in the genus *Franseria*, and, like *A. divaricata*, will be transferred to *Ambrosia* as soon as possible. These include the following species: *F. arborescens* Brand., which has an inflorescence similar to *A. divaricata* but more full and compact; *F. cordifolia* Gray and *F. sanctae-gertrudis* Rydb., which frequently bear the pistillate heads on elongate branches although the staminate heads are borne in conventional racemes. Superficial examination of *F. eriocentra* Gray indicates that the staminate " heads " of this species may, in some instances, actually consist of two or three capitula fused together at the tips of the capitulum stalks.

All the above species, including *A. divaricata*, are plants native to Baja California or to the adjacent Sonoran Desert region. In this area the ragweed flora abounds with species that display primitive character expressions. The relatively primitive inflorescence type of the species discussed above provides further evidence that this region may have been a center of specialization and diversification of the genus. The confirmation which the nature of the inflorescence organization of these species lends to the hypothesis of branch system homologies discussed earlier indicates that, while the ragweed inflorescence represents a departure from the typical condition, it is essentially only a modification of the basic plan of the Compositae. It also provides another illustration of the successful use of teratological morphology as a tool for the extrapolation of intermediate and primitive character expressions.

LITERATURE CITED

GEBBEN, A. I., W. W. PAYNE, AND W. H. WAGNER, JR. 1962. Botanical Phase, *in* J. M. Sheldon and E. W. Hewson. Atmospheric Pollution by Aeroallergens. Prog. Rept. 5. Ann Arbor: Univ. Mich.

PAYNE, W. W. 1962a. Biosystematic Studies of Four Widespread Weedy Species of Ragweeds (*Ambrosia*: Compositae). Ann Arbor (doctoral thesis).

——. 1962b. The Unique Morphology of the Spines of an Armed Ragweed, *Ambrosia bryantii* (Compositae). Madroño 16: 233-236.

——. 1963. The Morphology of the Inflorescence of Ragweeds (*Ambrosia-Franseria*: Compositae). Am. Jour. Bot. 50: 872-880.

RYDBERG, P. A. 1922. Carduales: Ambrosiaceae, Carduaceae; *in* North Am. Flora 33: 3-46.

SHINNERS, L. H. 1949. Notes on Texas Compositae III. Field and Lab. 17: 170-176.

SMALL, J. K. 1913. Flora of the Florida Keys. New York.

Papers of the Michigan Academy of Science, Arts, and Letters
Vol. XLIX, 1964 (1963 Meeting)

BIOTIC ASSEMBLAGE ASSOCIATED WITH A MASTODON SKULL FROM OAKLAND COUNTY, MICHIGAN

WARREN P. STOUTAMIRE
Cranbrook Institute of Science

WILLIAM S. BENNINGHOFF
The University of Michigan

In late September 1962, five students from Madison and St. Michael's junior high schools in Pontiac discovered a large skull and neck vertebra near the junction of Joslyn Road and the new I-75 expressway, north of the city. Terry Massey, Eddie Trippe, Gerald Henry, Mike Crawford, and Melvin Williams found the remains near the bottom of a clay embankment excavated from a nearby hillside during the construction of the expressway interchange with Joslyn Road. On the same hillside a bulldozer operator, Gene Karr, unearthed a tusk which probably was associated with the skull. Two visits have been made to the site by the senior author, but no other vertebrate remains have been found. The site is in the north-central part of Sec. 4, R. 10 E., T. 3 N. in Pontiac Township. The bones are those of the American mastodon, *Mammut americanum* Kerr.

A high morainal hill directly north of Pontiac projects westward from the Fort Wayne-Defiance morainal system (Martin 1955), and a large part of the southwest slope has been excavated in the course of expressway construction. The hill is composed of gray calcareous clays mantled by sandy and gravelly soils. The original location of the skull is not known because of the extensive earth-moving operations, so that it may have been moved several hundred feet to the embankment where it was found. The bones as found on the embankment were in a matrix of clay similar to that in the excavated slope of the distant hill. The transported clay is relatively rich in small macrofossils, but the material cannot be definitely associated with the skeletal remains because of the extensive disturbance of the site.

Persons living near the site say that a seepage area existed on the lower slopes of the hill, and this remained too wet to farm. Aerial survey photographs taken in 1957 before excavation began indicate gently sloping pasture with scattered brush and several shallow drainage channels. The skull probably was embedded in the wet clays associated with this seepage area.

The skull lacks the cranium and consists of the upper jaw and associated structures, the total 65 cm in length (Fig. 1). The left tusk is broken off at the opening of the tusk socket, and the right tusk is missing. A plug of fine sediment from the right tusk socket contained plant and animal material which will be discussed in the following sections. The associated tusk is 99 cm in length across the curve, and probably is the missing right tusk. The bones were treated with Alvar preservative, making carbon-14 dating unreliable. The skull, tusk, and vertebra are now in the collection of Cranbrook Institute of Science.

Analysis of Macrofossils

The discoverers removed a plug consisting mostly of clay (10 cm in diameter and 2-3 cm thick) from the right tusk socket. Quantities of plant and animal remains were visible on its surface. The possibility of separating enough organic material for carbon dating from the untreated clay, as well as speculation as to what remained hidden within, prompted the following treatment and examination. The clay was soaked in distilled water and gently shaken apart. It was then passed through graduated screens to separate the organic material, which consisted of pieces of bark, twigs, seeds, numerous needles and a cone of spruce (Fig. 2), mosses (Fig. 3), *Chara* fruiting bodies, and shells of several species of mollusks (Figs. 4-7). The screened clay was filtered, air-dried, and then subjected to pollen analysis (the results of which are discussed under that heading).

Representative twigs, needles, and bark samples were submitted to Francis Kukachka of the Forest Products Laboratory at Madison, Wisconsin, who identified them as twigs and wood of *Larix* sp.; twigs, wood, seeds, cone scales, and needles of *Picea* sp.; and twigs of *Salix* sp. The present southern limit of *L. laricina* (Du Roi) L. Koch in the Great Lakes area is in northern Ohio and Indiana, and it is common in a boggy area within a half mile of the site. Some of the spruce needles contain two large resin ducts and may be *P. mariana* (Mill.) B. S. P. (Little and Pauley 1958). They are

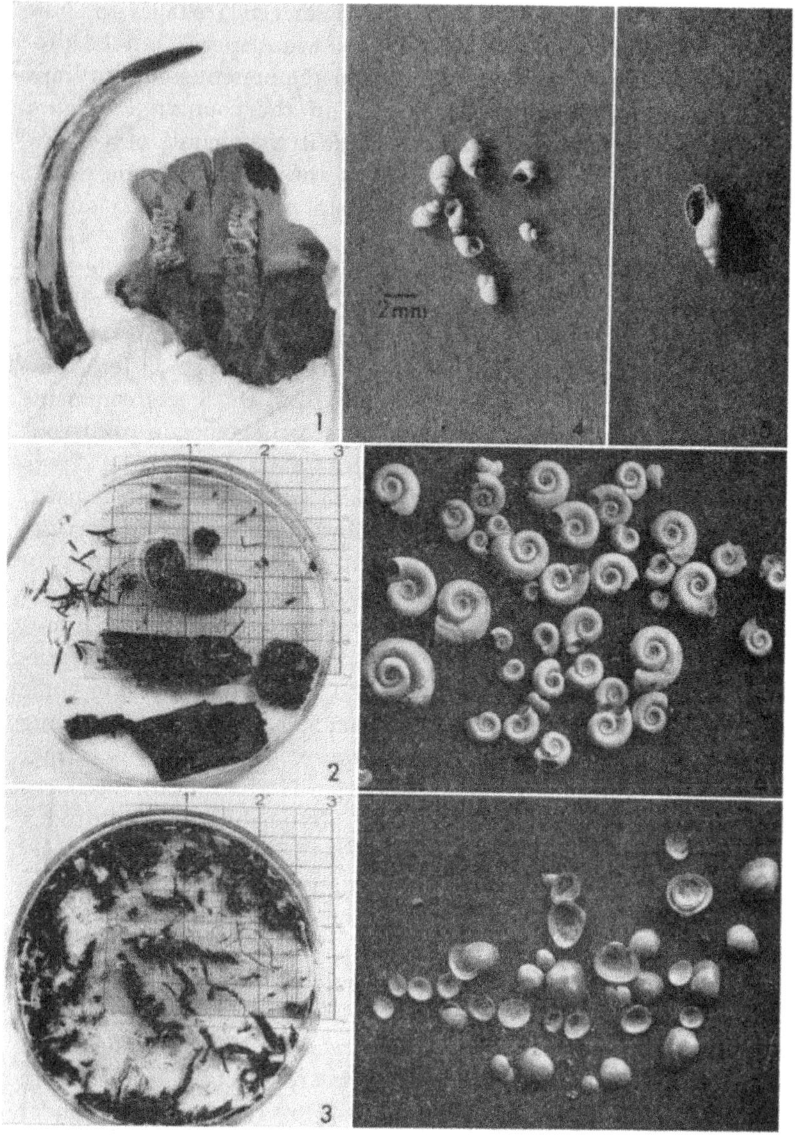

FIG. 1. Skull and right tusk of the Pontiac Mastodon.

FIG. 2. Fragments of bark, twigs, needles, and seeds, and a cone of *Picea glauca* (Moench) Voss, found in sediment removed from the right tusk cavity.

FIG. 3. Mosses found in sediment from the tusk cavity. Species include *Scorpidium scorpioides* (Hedw.) Limpr., *Campylium stellatum* (Hedw.) Lange & Jens., and *Drepanocladus revolvens* (C. Mull.) Warnst.

FIG. 4. *Lymnaea parva* Lea. All mollusks shown at same scale.

FIG. 5. *Lymnaea humilis* Say.

FIG. 6. *Gyraulus parvus* Say.

FIG. 7. *Pisidium nitidum* Jenyns.

too poorly preserved and too covered with clay particles to show the other features by which black and white spruce can be provisionally distinguished. The site is near the present southern limit of black spruce, which is infrequent in the county. *P. glauca* (Moench) Voss reaches its southern limit in the vicinity of Saginaw Bay, approximately 100 miles north of the site. The cone, seeds, and large number of needles lacking resin canals are clearly of this species.

Moss remains were identified by William C. Steere, of the New York Botanical Garden. They consist of abundant and well-preserved material of *Scorpidium scorpioides* (Hedw.) Limpr., with a few stems of *Campylium stellatum* (Hedw.) Lange & Jens. and *Drepanocladus revolvens* (C. Müll.) Warnst. Dr. Steere comments in his letters, "These same mosses still grow together in calcareous fens of northern Michigan and northward far beyond the Arctic Circle. The *Scorpidium* and *Drepanocladus* are both aquatic, emerging only if the water level goes down, whereas the *Campylium* grows in moist places, usually around the edges of fens, but not usually in the shade." A distribution map of the taxa in Michigan is given in Fig. 8. This was compiled from collections in herbaria at the University of Michigan and the Cranbrook Institute of Science. These collections also indicate general distributions of the three species in arctic, subarctic, and Canadian zones, with outlying colonies in the Rocky Mountains. *Campylium stellatum* is further represented in the University of Michigan Herbarium by specimens from Chiapas, Mexico, and Guatemala, presumably at high altitudes. Although these mosses occur at the present time in vegetation zones as diverse as temperate deciduous forest and arctic tundra, we believe their association in fens is most common in the Boreal conifer forest zone.

Molluscan shells were identified by Henry van der Schalie, of the University of Michigan, as *Lymnaea parva* Lea (Fig. 4), *L. humilis* Say (Fig. 5), and *Gyraulus parvus* Say (Fig. 6). These taxa occur in the area today, but our fossils are generally smaller than modern specimens. This suggests that the fossil mollusks lived under less than optimum conditions. The numerous bivalve shells were identified as *Pisidium nitidum* Jenyns by H. B. Herrington, of Westbrook, Ontario (Fig. 7). The Rev. Mr. Herrington comments that the shells are very fragile, suggesting that they came from quiet, shallow water, and that they probably came from a fairly soft but not muddy bottom. Several hundred *Chara* fruiting bodies were recovered and have been tentatively identified by R. B. Wood, of the University of Rhode Island, as *Chara vulgaris* L.

The remaining moss, wood fragments, twigs, needles, and cone scales were air-dried and weighed 26 gms. This aggregate sample of organic material was submitted to the Texas Bio-Nuclear radiocarbon dating laboratory, and the age was determined to be 11,900 ± 350 B. P. (Work Order # 801326-0000, sample No. 1).

Several mastodon finds from other parts of Lower Michigan have been dated by the University of Michigan radiocarbon dating laboratory. The dates are given here for comparison:

M-347	Lapeer Co.	Tusk	5950 ± 300 B. P. Crane & Griffin, 1959
M-67	Washtenaw Co.	Tusk	6100 ± 400 B. P. Crane, 1956 6300 ± 500 B. P. Crane, 1956
M-280	Lenawee Co.	Interior of tusk	7070 ± 240 B. P. Crane & Griffin, 1958
M-281	" "	Exterior of tusk	7820 ± 450 B. P. Crane & Griffin, 1958
M-282	" "	Wood associated with same tusk	9568 ± 1000 B. P. Crane, 1956
M-1254	Gratiot Co.	Tooth and jaw	10,700 ± 400 B. P. Oltz & Kapp, 1963

We have not made an exhaustive search of the literature, but the Pontiac specimen is the oldest dated mastodon for the state as far as we know.

Errors can arise in the carbon dating of bones through several factors, one of which is the replacement of original bone material by carbonates containing carbon of more recent origin. Skeels (1962) notes that the Washtenaw County find was in highly calcareous gravel and that the tusk was partly mineralized; therefore the radiocarbon date may be unreliable. The gravel is of Cary age, which, on the basis of other information, is considered to be approximately 13,000 to 15,000 B. P., suggesting that the tusk is much older than the C^{14} date. The Lapeer County specimen was buried under only 19 inches of muck in a 2-foot layer of marl, and the apparent youth of the find may have been influenced by penetrating organic matter of more recent origin. Our sample was treated to remove carbonates and humic acids of possible recent origin, and the organic carbon converted to benzene, the C^{14} content of which was then determined. This largely eliminates errors produced by carbonate replacement but does not rule out the possibility of slight contamination by rootlets of more recent plants. The plant material may have been washed into the skull shortly after death of the animal, or the skull may be much older than the plant debris. The age of the skull cannot be determined, now that the bone has been treated with organic preservatives. The date of 11,900 years must stand as a minimum date for the mastodon, and can be accepted as close to the actual date of deposition of the plant material. This fragile material could not have been redeposited into a more recent skull without being considerably more fragmented and eroded.

The date of extinction of the mastodon in the central U. S. is placed at about 7000 B. P. by Williams (1957) ; more recent evidence has led Skeels to suggest a date of 6000 B. P. She suggests several possible causes of extinction, among which is the vegetational change from a spruce-fir forest to which the browsing mastodons were apparently adapted, to a pine-dominated forest, the change being brought about by warming climate. It is also possible that mastodons were blocked from northward migration by the Great Lakes drainage system.

Pollen Analysis

Five samples were selected for pollen analysis as follows: (1) from the foramen of the atlas, (2) from a porous bone near the surface of the skull, (3) and (4) from the material filling the tusk cavity, and (5) from the sediment within the tusk cavity after screening. The dry weights of the sediment samples are given in Table I, along with the record of microfossils observed. To each sample 2 cc of *Cycas* pollen suspended in glycerin was added in order to provide a standard of exotic pollen against which the fossil pollen population in the sediment could be compared (Benninghoff 1962). The suspension was determined by 100 hemacytometer counts to contain 127,500 grains of *Cycas* pollen per cc, plus or minus 5%. The fossil pollen population per mg of dry sediment was calculated by the following proportion:

$$\frac{\text{no. of exotic grains counted}}{\text{no. of exotic grains/mg added}} = \frac{\text{no. of fossil grains counted}}{\text{no. of fossil grains/mg dry sediment}}$$

After addition of the exotic pollen, the glycerin was washed out with water, and all samples were subjected to the same chemical treatments. The samples were treated with 10% (by volume) HCl for 1 day, then with concentrated HF for 2 days, thus dissolving away the calcareous and siliceous mineral fractions. This was followed by the standard Erdtman acetolysis treatment, which, because of unusual amounts of strongly humified material, had to be repeated. After the final washing with water, the residues were transferred through 95% and absolute ethyl alcohol into silicone oil (Andersen 1960). Three slides were made of each preparation, with 22 × 40 mm cover glasses, and they were counted under a 40 × phase objective and 10 × oculars on a Zeiss microscope.

The results of the analysis are shown in Table I. Special notations are used to show the certainty of identification (Benninghoff and Kapp 1962). The suffix *id.* (for *idem,* identical) emphasizes that the identification is positive. *Comp.* (for *comparatus,* comparative) signifies that the fossil compares in all respects with reference preparations from the named taxon; and *sim.* (for *similis,* similar) indicates that the fossil resembles closely the reference material of the named taxon but the fossil may also resemble material from other, sometimes quite unrelated, taxa. Even in these twice-acetolyzed preparations the fossil pollen is in such low concentration that considerable effort at the microscope yielded relatively low total counts. The highest pollen concentration, that of sample 5, resulted from screening the coarse organic material from the tusk cavity sediment. This sample had 295 fossil grains per mg of sediment, less than one-tenth the concentration commonly found in younger postglacial peat deposits. The five samples show strong variations in the fossil pollen population densities, from a low value of 16.1 grains/mg to a high value of 294.5 grains/mg dry weight. No explanation of this variation is obvious, but it may be significant that the lowest values occur in unscreened sediment taken from skeletal cavities.

The five samples are relatively consistent with each other with respect to the various entities and their proportions, so that there is some justification for combining the counts to show the percentages in terms of the total tree pollen as 100%, the common practice. Bryophyte spores, fungal spores, unidentified pollen and spores, and the obviously rebedded spores were excluded from the total Nonarboreal Pollen and Spore (NAP) count, which, never-

TABLE I

Fossil Pollen Observed

Sample No.	1	2	3	4	5	Totals for All Samples	
Source (see text)	Vertebra, foramen	Skull, external	Tusk, cavity	Tusk, cavity	Tusk, screened		
Wt. of sample (mg)	12,400	2,500	1,200	800	900		
Cycas grains/mg added	28.6	142.0	295.8	443.8	394.4		
Cycas grains counted	70	142	457	508	101		
Fossil grains counted	39.5	60.5	55.5	23.5	75.5	Total	%
Fossil grains/mg sediment	16.1	60.5	35.9	20.5	294.5	No.	AP
	number counted						
Arboreal Pollen (AP)							
Picea	16.5	23.5	19.0	9.5	28.0	96.5	82.8
Pinus	2.0	1.0	0.5	1.0	4.5	9.0	7.7
Larix			tr				tr
Populus comp.		1			2	3	2.6
Quercus	2	1	1		1	5	4.3
Juglans cinerea			1			1	0.9
Ulmus				1		1	0.9
Acer saccharum		1					0.9
Total AP	20.5	27.5	21.5	11.5	35.5	116.5	100.0
Nonarboreal Pollen (and Spores) (NAP)							
Salix	1	3	1		1	6	5.2
Alnus					2	2	1.7
Corylus		1				1	0.9
Typha latifolia tetrad	1	1	1	1	2	6	5.2

Sparganium		1		1	1	3	2.6
Sagittaria	2	14	9	4	3	32	27.5
Potamogeton			1			1	0.9
Calla sim.		1				1	0.9
Utricularia	2				1	3	2.6
Cyperaceae	3	1	8	1	4	17	14.6
Gramineae	1	1			1	3	2.6
Chenopod-Amaranth.			2			2	1.7
Rosaceae				1		1	0.9
Rubus comp.				1		1	0.9
Saxifragaceae					1	1	0.9
Labiatae (6-colp.)		1				1	0.9
Elaeagnus id. *commutata* comp.			1			1	0.9
Lysimachia sim.				1		1	0.9
Plantago sim.					1	1	0.9
Compositae (Astereae)		3	1		2	6	5.2
Artemisia	1				3	4	3.4
Ambrosia	1				1	2	1.7
Polypodiaceae endospore					1	1	0.9
Equisetum sim.	1					1	0.9
Lycopodium complanatum			1			1	0.9
Total NAP	13	27	25	10	24	99	85.0
Bryophyte spores	2	1	11	29	16	59	50.6
Tasmanites sim. (rebedded)	5	1	1	1	4	12	10.3
Lepidodendron megaspore (rebedded)			1			1	0.9
Fungal spores	4		12	7	5	28	24.0
Unidentified pollen and spores	6	6	9	2	16	39	33.5

theless, amounts to 85% of the AP count. So high a proportion of NAP is usually considered to indicate incomplete forest cover on the landscape. In this material the total pollen population is low, and more than one-half of the NAP represents vegetation of shallow

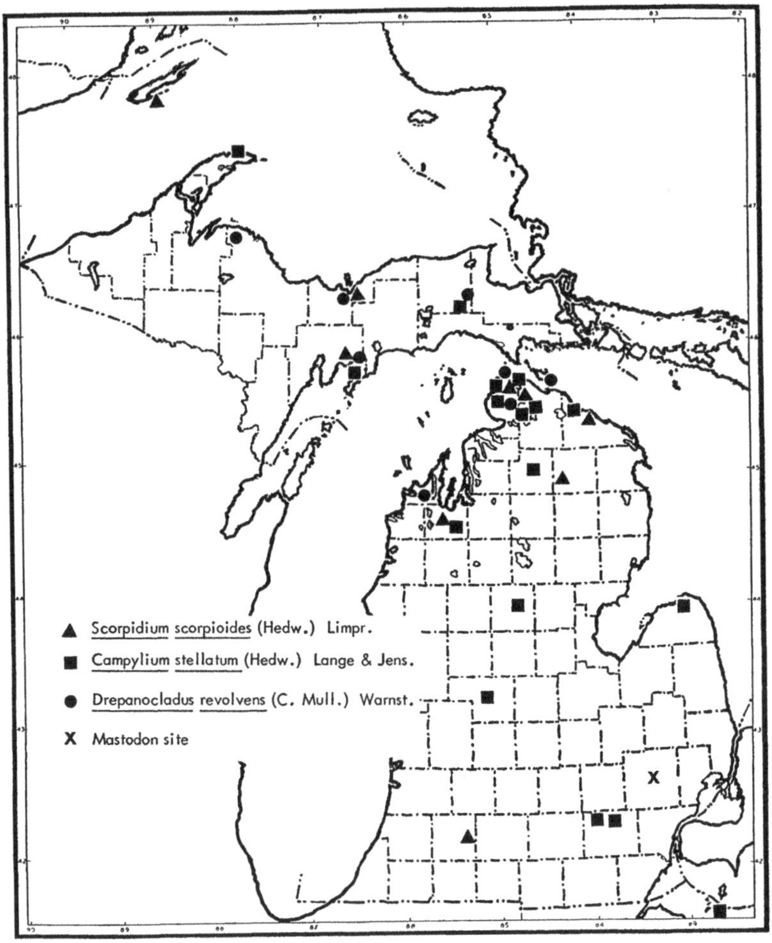

FIG. 8. Present distribution of the three identified moss taxa in Michigan and site of the fossil material.

ponds, shores, fens, marshes, or shrub swamps. It is likely that these characters of this pollen spectrum resulted from a general vegetation cover that was relatively poor in mesic site upland plants and richer in wet site and aquatic communities, while the surface drainage was still poorly integrated and the soils were still unmodified glacial deposits.

The dominance of *Picea* (spruce) pollen and presence of *Larix* (larch) pollen (*tr* indicates presence in frequency too low to appear in the counts) agree with the plant macrofossils. Although the *Picea* grains ranged in total size from 68 μ to 110 μ, a large proportion of them was broken or wrinkled and folded so that frequencies of size classes could not be used to confirm more than one species of *Picea*. The percentages of *Pinus* (pine) are as low or lower than those found in the recent pollen spectra of now pineless areas such as northeastern Indiana. Small percentages of *Quercus* (oak), *Ulmus* (elm), *Juglans cinerea* (butternut), and *Acer saccharum* (sugar maple) appear to be characteristic of Late-Glacial pollen spectra from the Erie Basin to eastern Minnesota (Benninghoff and Hibbard 1961; Fries 1962). *Carya* (hickory) and *Fraxinus* (ash), often present in similar amounts in Late-Glacial sediments, were absent here. It is very doubtful that these tree genera lived near the site at the time this material was deposited. Therefore, the pollen of these genera must represent either grains carried in the atmosphere as "long-range drift," or grains rebedded from fossil assemblages exhumed from older material that was incorporated in the glacial drift. The consistent values for *Quercus, Ulmus,* and *Juglans cinerea,* and complete absence of *Fagus, Tilia, Tsuga,* and other elements of the full nonglacial flora of this region, argue for the occurrence of the former group nearer than the others to the pioneer *Picea* forest.

In general the pollen spectrum shown in the right-hand column of Table I is strongly similar to that found in the material removed from the foramen of the woodland musk ox atlas found in a marl deposit near Kalamazoo and radiocarbon dated at 13,200 ± 600 B. P. (reported as 13,000 ± 600 B. P. in Benninghoff and Hibbard 1961). The spectrum from the Pontiac mastodon samples indicates that the nearest forest was almost exclusively of *Picea;* and the macrofossils indicate that *P. glauca* (white spruce) was present and that *P. mariana* (black spruce or bog spruce) may have been present. *Pinus* and the *Quercus, Ulmus,* and other broadleaved deciduous trees were at least several tens of miles distant.

Oltz and Kapp (1963) analyzed the fossil pollen of sediments associated with mastodon and mammoth remains in Gratiot County, central lower Michigan. The pollen spectrum associated with the mastodon tooth and jaw noted above (radiocarbon sample M 1254, 10,700 ± 400 B. P.) has 42.5% *Picea* and 37.2% *Pinus*. By comparison with the Pontiac site, we are led to the tentative deduction that the *Picea* dominance that existed near Pontiac about 12,000 years B. P. was beginning to give way to the " Pine Period " in an area about 90 miles to the northwest approximately 1,200 years later. That interval may represent approximately the duration of the pre-Pine Period, or Late-Glacial plus Spruce Period, of the pollen sequence in south-central Michigan.

The occurrence of *Elaeagnus commutata* (silverberry) is noteworthy, as it is a shrub of open river terraces and moist hillsides in the Boreal conifer forest. It was found by S. T. Andersen (1954) in the Late-Glacial sediments of Hidden Lake, southwest of Pinckney, Livingston County, Michigan. The earliest pollen spectra from Hidden Lake contain Cyperaceae, Gramineae, *Artemisia*, and *Ambrosia* in percentages almost identical with those in the Pontiac material. The Pontiac material, however, is richer in *Typha latifolia* (broad-leaved cattail), *Sparganium* (bur reed), *Sagittaria* (arrowleaf), and *Utricularia* (bladderwort), probably representing the shore and aquatic communities on the depositing area. The Bryophyte spores are believed to be identified as such with certainty as far as that group, but lack of adequate reference material prevented more precise determination. Spores of *Sphagnum* (peat moss) were not included in these Bryophyte spores. *Sphagnum* spores, as far as we know, appear well after the pioneer Late-Glacial vegetation phase in the Great Lakes region.

The strongly humified, resinous-appearing round bodies identified as Tasmanites range in size from 34 to 85 μ in diameter, and strongly resemble the 200 to 1000 μ microfossils of that form genus in Devonian rocks of the region. These microfossils are still being exhumed through erosion of glacial till and the native Devonian shales in some areas, but their occurrence appears to be characteristic of early Late-Glacial sediments, according to as yet unpublished findings of R. O. Kapp in central Indiana deposits and of W. S. Benninghoff in other deposits of southeastern Michigan. The one Lepidodendron megaspore found in the Pontiac material was strongly abraded in the equatorial flange area and was only 160 μ in diameter. This spore was undoubtedly rebedded from glacial drift, having been derived originally from Carboniferous age rocks of the region.

CONCLUSIONS

The different types of material associated with the Pontiac mastodon skull permit reconstruction of the environment at the site of deposition. Species of *Chara* grew in shallow, lime-rich water, and the *Limnaea, Pisidium,* and *Gyraulus* species lived in shallow, quiet water on a bottom which was soft but not muddy. The mosses *Scorpidium scorpioides, Drepanocladus revolvens,* and *Campylium stellatum* occurred on calcareous shoals and margins of the lake or pond. *Picea mariana* probably grew near the edges of the open water, and contributed needles. *P. glauca* did occur nearby, contributing the cone, seeds, and needles. Pollen of *Alnus, Typha, Sparganium, Sagittaria, Potamogeton, Calla,* and *Utricularia* species and of the family Cyperaceae suggest the immediate proximity of a sedge marsh or marshy shore vegetation. The presence of Tasmanites-like bodies and a Lepidodendron megaspore indicate that rock flour from Paleozoic sediments was being contributed to the deposit, a common occurrence on the fresh Late-Glacial surface. It has been reported in the literature that the mastodon browsed on conifer foliage. The landscape at the time of this deposit was mantled by stands of spruce, some closed, some parklike, with intervening stands of larch, meadows, marshes, and lakes. At undetermined greater distances away from the retreating ice, forests of pine and oak and other broad-leaved trees were becoming established. This reconstruction of a Late-Glacial environment in southern Michigan is consistent with evidence of other deposits of similar age in the region.

LITERATURE CITED

ANDERSEN, S. T. 1954. A Late-Glacial Pollen Diagram from Southern Michigan, U. S. A. Geol. Surv. Denmark, ser. 2, 80: 140-155.

——. 1960. Silicone Oil as a Mounting Medium for Pollen Grains. *Ibid.*, ser. 4 (1): 1-24.

BENNINGHOFF, W. S. 1962. Calculation of Pollen and Spore Density in Sediments by Addition of Exotic Pollen in Known Quantities (abstract). Pollen et Spores (Paris) 4 (2): 332-333.

—— AND C. W. HIBBARD. 1961. Fossil Pollen Associated with a Late Glacial Woodland Musk Ox in Michigan. Pap. Mich. Acad. 46: 155-159.

—— AND R. O. KAPP. 1962. Suggested Notations to Indicate Status of Fossil Pollen (abstract). Pollen et Spores (Paris) 4 (2): 332.

CRANE, H. R. 1956. University of Michigan Radiocarbon Dates I. Science 124: 664-672.

—— AND J. B. GRIFFIN. 1958. University of Michigan Radiocarbon Dates II. Science 127: 1098-1105.

——. 1959. University of Michigan Radiocarbon Dates IV. Amer. Jour. Sci. radiocarbon supplement. 1: 173-198.

FRIES, MAGNUS. 1962. Pollen Profiles of Late Pleistocene and Recent Sediments from Weber Lake, Minnesota. Ecol. 43 (2): 295-308 (4 diagrams in pocket).

LITTLE, E. L., JR., AND S. S. PAULEY. 1958. A Natural Hybrid between Black and White Spruce in Minnesota. Am. Mid. Nat. 60: 202-211.

MARTIN, HELEN M. 1955. Map of the Surface Formations of the Southern Peninsula of Michigan. Mich. Dept. Cons. Publ. 49.

OLTZ, DONALD F., AND RONALD O. KAPP, 1963. Plant Remains Associated with Mastodon and Mammoth Remains in Central Michigan. Amer. Mid. Nat. 70 (2): 339-346.

SKEELS, M. A. 1962. The Mastodons and Mammoths of Michigan. Pap. Mich. Acad., 47: 101-133. 3 figs., 3 pls.

WILLIAMS, S. 1957. The Island 35 Mastodon: Its Bearing on the Age of Archaic Cultures in the East. Amer. Antiq. 22: 359-372.

PAPERS OF THE MICHIGAN ACADEMY OF SCIENCE, ARTS, AND LETTERS
VOL. XLIX, 1964 (1963 MEETING)

FORESTRY

SOME THOUGHTS ON STAND VOLUME TABLES

JOHN CAROW
The University of Michigan

WE are all familiar with cycles of fashion and taste that are a continuing part of our everyday life. Many of us have had the experience of keeping some out-of-date article of clothing until it once again became fashionable. It is much less common for ideas and ways of doing things to come back into favor once they have been supplanted. While the revival of a once popular technique is usually a result of changed conditions, it also attests to the essential soundness of the original concept. Such a pendulum swing of popular usage appears to be occurring in the methodology of timber volume estimation.

According to early mensuration texts, a common way of calculating timber volume in Europe in the last quarter of the 19th century was by means of stand form factors. The cubic content of whole stands was found by multiplying the stand basal area by the average stand height and by the stand form factor. Thus the stand might be conceived as a forest of cylinders whose volume total when discounted by the average ratio of tree volume to cylinder volume gave the required stand volume estimate. Later, when yield tables became available, stand volume was directly estimated to be that shown by the tables for the appropriate age and site quality. However, as individual tree volume tables became common, inventory methods utilized them to an increasing extent, and the stand techniques fell into disuse.

In our country the early landlookers also used stand methods. From their experience with tracts that were estimated, cut, and scaled, early cruisers developed an eye for volume per acre by a sort of mental integration of tree size, density, and association. Theirs was a jealously guarded art that was difficult to describe or teach. As the forester brought scientific method into the woods and began training young men for the profession, tree volume methods proved superior for getting detailed information and for training inexperienced men to get satisfactory and consistent results. Cruising techniques using stand variables were almost forgotten.

Now there is renewed interest in stand volume tables. Probably the availability of aerial photography and the development of inventory techniques using photographic measurements have been primarily responsible. Foresters, eager to use this new tool, have found that the scales commonly available do not allow accurate counting of individual trees nor their classification into narrow size categories even when they are visible. However, stands and samples of stands can be measured for total height, average crown diameter, and crown closure; and these variables allow direct estimates of stand volume that are useful for many purposes. Aerial stand volume tables for a number of species and types are now appearing with increasing frequency in our literature (Avery et al. 1959, Moessner et al. 1951, and Pope 1962).

The Bitterlich method of angle-count sampling, called point-sampling by Grosenbaugh (1958), has also contributed in an important way to the current interest in the direct estimation of stand volume. Bitterlich's revolutionary technique has made basal area determination so simple that foresters can now make more use of this important measure of stand density. It can be a functional step in volume estimation rather than a laborious separate calculation.

The work of a few mensurationists has led to the revival of interest in stand volume tables. The most influential has been Spurr. His *Forest Inventory* (1952) detailed a systematic and thorough review of volume estimation which brought out basic principles of measurement and put volume table construction methods in common perspective. This monograph also led a current trend toward objective mathematical techniques in volume table construction and introduced the combined variable equation.

The combined variable formula $V = a + bD^2H$, where V is total cubic volume, D is diameter at breast height, H is total height, and a and b are constants determined for species, condition, and locality, uses individual tree measurements. Spurr recognized that when the above formula is changed to $V = a + bBH$ (by substituting tree basal area, B, for diameter with corresponding change in value of constant b), it was essentially the single tree form of a stand volume equation. Summing individual tree volumes on an acre give stand volumes per acre. Thus:

$$\sum_{i=1}^{n}(a + bBH) = na + b\sum BH$$

and if H is the stand average and thus constant, the stand form becomes $NV = na + bH\sum B$, or more simply, $V = K + bHB$, where V and B are stand volume and basal area per acre and K is a regression constant related to stocking. Often the regression constant a of the combined variable tree equation is zero or negligibly small, as in Gevorkiantz' (1955) equation for total cubic content, $V = .42BH$, so the stand formula sums no constant and also becomes $V = .42BH$, with changed meanings for V, B, and H. Buckman, in a recently published stand volume table for red pine (1962), built on Gevorkiantz' equation but measured only dominant and co-dominant trees for height, so he got an equation $V = .4085BH$. Buchman published an equally accurate alternate equation $V = 15.03 + .4062BH$, in a standard regression form and made available similar equations in merchantable cord and board foot units of volume.

There has been criticism of stand volume tables because the regression constant is theoretically related to stocking and, if positive, as it usually is in total cubic foot equations, it would underestimate dense and young stands and overestimate poorly stocked and older tracts. In practice the regression and slope constants are determined empirically from sample plot data used in preparing the tables, and they average such variables as stocking and tree form. Often the regression constant is small and relatively unimportant, as in Buckman's alternate form. When tested against their basic data, the stand equations have shown standard errors of from ± 5 to $\pm 10\%$, which compare very well with standard errors of tree equations for estimating individual tree volumes. It should also be remembered that any volume table gives average volumes of trees or stands of certain dimensions, and when measurements depart from the averages trees will not be properly estimated for volume. The weaknesses of stand volume tables are those of all efforts to use average values for individual estimates.

Just as tree volume tables based on dbh alone can be used locally where site and age are uniform, so local stand volume equations using basal area as the only independent variable can be employed with success. These are akin to the volume-basal area line equations of Hummel (1955). In 1956 I made two stand volume equations for aspen stands in Iron County, Michigan. Regressions of total cubic foot volume per acre were fitted to the combined variable (basal area per acre times average height of five trees per plot), and also to basal area alone, with the following results:

Equation	Standard Error of Estimate Cu. Ft.	Standard Error of Estimate % of Mean	Basis No. of Plots
V = .3717BH + 90.23	39.6	4.6	20
V = 22.09B — 25.62	78.3	9.1	20

As would be expected, the stand volume equation that considered both basal area and average height had a smaller error than the local equation which used only basal area. To test the formulas, 20 angle-count samples were taken about the same 1/5-acre plot centers from which the original data were collected. The mean volumes from the formulas did not differ significantly, nor did they differ from a third estimate that tallied each qualifying tree into 10-foot height classes and used the basic tree volume table converted to show cubic feet per square foot of basal area.

Stand volume tables can reduce the time needed to get volume estimated in the field by the Bitterlich method. In my test the time needed for plot taking and computation, when each qualifying tree was classified into 10-foot height classes, was 3 minutes per sample point. With the combined variable equation the time was 4 minutes when three trees per point were measured for total height. It would be less if fewer measured trees were considered adequate or if estimates of height were used. The local table using only basal area cut the time to 1 minute.

The use of stand volume equations and tables made from them appears to be most advantageous in connection with angle-count sampling and in situations where detailed information on diameter distribution is not important. They seem particularly well adapted for reconnaissance examinations and prescription surveys of tracts of immature timber such as dense pulpwood stands and plantations. For these the savings in field and office time is important and the accuracy secured is satisfactory.

Stand volume tables are also of real value in growth prediction. In *Forest Inventory*, Spurr gives a cogent argument for stand projection, using both basal area and height growth and shows that stand volume equations are the best way to integrate these variables in the growth estimate. Thus, a proportion of two equations representing the stand volume before and after the growth period would be:

$$\frac{V_a}{V_b} = \frac{b_a B_a H_a}{b_b B_b H_b}$$

Since over a short projection period b_a and b_b, representing stand form, would be little changed, future volume will be present volume multiplied by the ratio of the combined stand variables. Basal area growth has been found to be linear through the greater part of the life cycle of a species and can be determined by increment core samples from the whole stand rather than from individual diameter classes. Height growth can also be projected from samples, or it can be read from site index curves.

Interest in stand volume tables and expected increased application seem to be counter to the progression of mensuration technique toward greater detail and objectivity as noted by Grosenbaugh (1959). In connection with aerial photo interpretation and angle-count sampling, however, they do have a place in reconnaissance and prescription surveys and in short period growth predictions. Where high accuracy and detail are not required, the stand volume formulas are scientifically sound and useful rules of thumb which most foresters will appreciate.

LITERATURE CITED

Avery, Gene, and M. P. Meyer. 1959. Volume Tables for Aerial Timber Estimating in Northern Minnesota. U. S. For. Serv., Lake States For. Exp. Sta. Pap. 78. 21 pp.

Buckman, Robert E. 1962. Development and Use of Three Stand Volume Equations for Minnesota. Jour. For. 59: 573-575.

Gevorkiantz, S. R., and L. P. Olson. 1955. Composite Volume Tables for Timber and Their Application in the Lake States. U. S. D. A. Tech. Bull. 1104. 51 pp.

Grosenbaugh, L. R. 1958. Point-Sampling and Line-Sampling: Probability Theory, Geometric Implications, Synthesis. U. S. For. Serv., South. For. Exp. Sta. Occas. Pap. 160. 34 pp.

——. 1959. Quantification and Estimation in Future Forest Management. Proc. Soc. Am. For. San Francisco: 117-121.

Hummel, F. C. 1955. The Volume-Basal Area Line. For. Comm. Bull. 24. 84 pp.

Moessner, Karl E., D. E. Brunson, and C. E. Jensen. 1951. Aerial Volume Tables for Hardwood Stands in the Central States. U. S. For. Serv. Cent. States For. Exp. Sta. Tech. Pap. 122. 15 pp.

Pope, Robert B. 1962. Constructing Aerial Photo Volume Tables. U. S. For. Serv. Pac. NW For. and Range Exp. Sta. Res. Pap. 49. 25 pp.

Spurr, Stephen H. 1952. Forest Inventory. New York: Ronald Press. 476 pp.

PAPERS OF THE MICHIGAN ACADEMY OF SCIENCE, ARTS, AND LETTERS
VOL. XLIX, 1964 (1963 MEETING)

PRACTICAL AND ECONOMIC ASPECTS OF BOLTER OPERATIONS IN SOUTHERN MICHIGAN

JAMES DUNCAN AND EVERETT ELLIS
The University of Michigan

MUCH of the merchantable timber in southeastern Michigan's estimated one and a half million acres is in small unmanaged woodlots of mixed hardwood species. In general, this timber is of poor quality and small in size. Standard sawmills in this area regularly cut a small but unknown quantity of lumber for local and industrial markets. A few bolter mills are also found, but they saw only special products. The potential role of bolter saws, cutting bolts, or short logs up to six feet in length has been largely overlooked. The short-log bolter is capable of producing quality furniture flatstock and dimension squares. The availability of timber, plus a nearby furniture market, certainly enhances the future role of the bolter saw in southern Michigan.

The primary purpose of this paper is to explore marginal cost and marginal revenue of a complete bolter operation.[1] Such an operation including all phases of logging and milling from the stump to kiln-dried lumber and dimension was conducted. This study evolved as a result of three previous studies conducted at the University of Michigan. Johnson (1958) performed an exploratory study on the bolter, converting black oak and elm to furniture squares. He developed and substantiated the use of a clear face bolt grading system. Harold (1959) converted basswood bolts from a post-logging operation into furniture dimension. This study was also one of an exploratory nature. Misiak (1962) studied the evaluation of lumber and dimension quality. He developed several bucking criteria, for both logs and bolts, which were followed in this study.

There are a few additional studies from different geographic areas which are also worthy of mention. Calvert (1958) performed an extensive study of bolter operations in Canada. He found that bolter-produced dimension was gradually replacing traditional

[1] This paper is a condensed account of a study conducted at The University of Michigan in conjunction with the U. S. Forest Service.

methods of cutting long lumber into small cuttings. Calvert contended that it was logically more economical to manufacture stock approximating final size prior to transportation, kiln-drying, and other intermediate manufacturing steps than to perform these operations on common grades of lumber, of which up to 50% will subsequently be discarded. Redman and Willard (1952) conducted an extensive study in North Carolina which substantiated the results already mentioned concerning bolter-sawed furniture dimension. Their study also showed that a bolter operation was economically sound and practical in the South. These above-mentioned studies include the most significant works presently available concerning the bolter saw and its potentiality.

Several stands of timber were investigated before a low-quality hardwood stand that would be suitable for the study was found. The stand was selected on the basis of past history and present

TABLE I

STOCK TABLE PER ACRE BASED ON 2.9-ACRE STUDY AREA

Status of Stand	Red & Black Oak B. A.	Red & Black Oak Bd. ft.	White Oak B. A.	White Oak Bd. ft.	Misc. Species * B. A.	Misc. Species * Bd. ft.	All Species B. A.	All Species Bd. ft.
Before cut..	50.24	4244	11.21	519	18.19	955	79.64	5718
Cut	20.22	1973	3.42	187	9.63	664	33.27	2824
After cut...	30.02	2271	7.79	332	8.56	291	46.37	2894

* Miscellaneous species includes hickory and sassafras, which were considered unmerchantable species.

condition of the standing timber, with the objective of performing a comparative economic study. An area of 2.9 acres was selected and logged in the fall of 1962 on the Newcomb Tract owned by the University of Michigan. This logging site had been grazed as late as 1951 and high-graded an estimated 25 years ago. The stand had never been under any formal type of forest management. This stand's history was certainly typical of those found in southeastern Michigan. A stand improvement cut was made removing undesirable species, cull, and mature trees. Table I is a stock table including the basal area in square feet and board foot volume by species of the stand before cut, of the cut trees, and of the residual stand after cutting.

In logging the area, the trees were first felled and bucked. The logs were then graded. Forest Products Laboratory's log grades for standard lumber were used; the log grades were No. 1, No. 2, and No. 3; No. 1 is the top grade. That part of the tree yielding a No. 2 log or better was bucked as a sawlog; any part of the tree yielding a No. 3 log was considered for bolts. If one or more clear bolts could be cut from the No. 3 log, it was bucked as bolts to a minimum diameter of 6 inches d. i. b. If the No. 3 log was of such a nature that no clear bolts could be obtained, it was left as a sawlog. The bolt material was left in multiple bolt lengths to be later bucked into single bolts at the mill. Any large branches that would yield a straight clear bolt were also taken.

The logs and bolts were then skidded out together, but decked separately at the landing. Logs and bolts were loaded separately, hauled to the mill, and unloaded. The bolts were then bucked into 4-foot lengths considering such bucking criteria as crook, sweep, defects, and taper. A total volume of 1626 bd. ft. of bolts and 6010 bd. ft. of logs was delivered to the mill. The time required to perform each logging operation was recorded separately for logs and bolts.

At the mill, the bolts were scaled by the International $\frac{1}{4}$-inch rule, graded according to a method developed by Johnson (3), and sawed on a Penney Model E short-log bolter. The bolter was equipped with a 32-inch inserted tooth saw and powered by a 40 hp electric motor, which operated the saw at 1280 rpm. The bolts were taper-cut and sawn-around for maximum quality recovery. Each bolt was first slabbed on the lowest-grade face. This face was turned down on the saw table and the next face was slabbed. Four quarter (4/4) dimension flatstock was removed from this second face until a paint grade face appeared. Each subsequent face was cut, removing all higher grade dimension. Then the paint grade flatstock was removed until only a low grade cant remained. This cant was sawn into 8/4 flats. The 4/4 flatstock was edged on a small stripper to remove the bark and wane; the low grade 8/4 flats were edged and ripped into 2-inch dimension squares. The dimension was scaled, graded, and recorded by bolt grade, diameter, and species (Table II).

The sawlogs were milled on a small conventional Farquhar No. 9 sawmill. This mill includes a 48-inch circular, inserted-tooth saw, a pony edger, and a swing cutoff saw. The logs were also taper-cut and sawn-around for maximum quality recovery. The high grade parts of the logs were sawn into 4/4 lumber, and the lower, common grades were sawn into planks. The lumber was then edged and trimmed to obtain maximum lumber quality. It was finally tallied and graded separately by log grade, scale, and diameter. Table III gives the percent recovery of dimension and lumber grade for bolts and logs by species.

TABLE II

DIMENSION RECOVERY IN PERCENT BY BOLT GRADE FOR RED AND BLACK OAK

Bolt Grade	Bolt Scale Int. ¼″	Mill Tally	Dimension Grade Recovery (Percent) 2 Clear Faces	1 Clear Face	Paint	Core	Sound	Total
Number of Clear faces	Bd. ft.	Bd. ft.						
4	56	59	20.8	30.7	29.0	11.1	8.4	100.0
3	105	110	5.3	19.5	42.9	11.5	20.8	100.0
2	101	101	1.7	6.3	43.6	20.6	27.8	100.0
1	73	64	--	–	20.0	41.6	38.4	100.0

TABLE III

DIMENSION AND LUMBER RECOVERED FROM 2.9-ACRE STUDY AREA

Species	Short Dimension Cl. 2	Cl. 1	Paint	Core	Sound	Total	Lumber FAS	Sel.	#1C	#2C	#3C	Total
Blk. & red oak												
Bd. ft.	186	280	355	439	338	1598	990	963	1626	1436	1097	6106
% of total	11.6	17.5	22.2	27.5	21.2	100.0	16.2	15.8	26.6	23.4	18.0	100.0
White oak												
Bd. ft.	8	48	26	40	47	169	24	40	109	244	244	661
% of total	4.7	28.5	15.4	23.6	27.8	100.0	3.8	6.2	16.8	35.7	37.5	100.0
All species												
Bd. ft.	194	328	381	479	345	1767	1014	1003	1735	1674	1341	6767
% of total	11.0	18.7	21.5	27.0	21.8	100.0	15.2	14.8	25.6	24.7	19.7	100.0

SEASONING

In the bucking of bolts, an identification mark was placed on all branch wood bolts. In the milling of bolts, all branch wood dimension was sorted from the stem wood. The dimension flatstock (4/4 dimension) was air-seasoned for approximately two months. After that time, 320 bd. ft. of branch wood was randomly selected, 80 bd. ft. in each of the top four dimension grades. Likewise, 320 bd. ft. of stem wood and 320 bd. ft. of lumber were selected for kiln-drying. The lumber and dimension were kiln-dried on a mild kiln schedule for 20 days. In regrading the material, no significant differences were found in the amount of degrade due to kiln-drying of branch wood, stem wood, or lumber. The only degrading factor present was slight cupping in a very few boards. On the basis of these samples there is no apparent reason for discriminating against stock cut from branch wood.

Economic Analysis

Table IV is a summary of cost and incomes from the entire 2.9 acre study area.

TABLE IV

Cost and Income of Entire Operation
Based on Logs and Bolts from 2.9 Acres

Cost or Income	Dimension *	Lumber †	Combined
Cost			
Logging costs	$ 74.51	$377.67	$452.18
Milling costs	28.98	158.14	187.12
Total costs	103.49	535.81	639.30
Income			
Red and black oak value	197.50	797.93	995.43
White oak value	19.37	66.17	85.54
Total value ‡	216.87	864.10	1,080.97
Net income	$113.38	$328.29	$ 441.67

* Based on 1767 board feet mill tally dimension.
† Based on 6106 board feet mill tally lumber.
‡ Total value represents price per M FOB mill

Work was performed at a normal work speed; machinery breakdowns, rests, and miscellaneous delays were included in the time data. The labor costs represent averages for the Lake States and were furnished by the U. S. Forest Service; they include Social Security and Workmen's Compensation. A stumpage price of $20. was charged against sawlogs, assuming a lump-sum sale, a common method of selling timber in the Lake States. The cost of felling was also charged against sawlogs; it was assumed that all marked trees had to be felled, whether bolts were taken or not. Milling costs per M were considerably different for logs and bolts. The high cost of milling sawlogs was associated with the size of the small sawmill and the lack of automatic mill equipment. It takes five men to operate the sawmill, while only three are required for the bolter mill. The prices paid for oak lumber were quoted from the Hardwood Market Report. Prices for dimension stock were averages taken from several mills in the South and East. The quality index method of lumber and dimension price determination was used.

Conclusions

1. On 2.9 acres of a low quality hardwood stand in southeastern Michigan, 1626 board feet of bolts was available, in addition to 6010 board feet of logs. Bucking these low quality logs and larger branches into short lengths produced good quality bolts.

2. In the commercial species cut, including white oak, red oak, and black oak, 48% of the bolts taken were from the main stem, and 52% were from branches above the main stem or larger branches of the crown.

3. By bucking most No. 3 logs into bolts, the average log for the sawmill was of better quality and was considerably larger. The efficiency of the sawmill was therefore increased considerably, since the average log was of greater volume and higher quality.

4. The marginal costs were 26% less to log and mill a thousand board feet of bolts than it was for logs. This was partly explained by writing off the stumpage price and felling charges as a cost against sawlogs.

5. The marginal net revenue was 8.5% higher for dimension than it was for lumber production on a per M basis. In other words, the profit margin for dimension was 8.5% higher than for lumber.

6. When a bolter mill was established in conjunction with a small standard sawmill, total production was increased substantially (up to 50%) and at a reasonably small investment.

7. There was no significant difference in amount of degrade due to the kiln-drying of branch wood, stem wood dimension, or lumber. The only degrading defect present was slight cupping in a very few boards.

Summary

Substantially acreages of low quality hardwoods are presently unmanaged in southeastern Michigan. That this timber could be economically managed if an efficient logging and milling operation was established. Small trees of low quality are capable of yielding high grade lumber and dimension. This domestic lumber and dimension could be sold to the nearby furniture manufacturers at a price comparable to lumber now being purchased from other parts of Michigan and neighboring states. The products produced from these small trees possess mechanical and physical properties comparable to those of lumber recovered from larger trees.

This study has shown that the establishment of a bolter mill in conjunction with a standard sawmill could produce these same wood products. A small investment of $4500 to $5000 would completely install a bolter mill capable of producing 4 to 5 M board feet per day. The high quality dimension produced could be sold as furniture stock, while the low quality material could be sold as pallet stock. By bucking small low quality logs into bolts, the average log size and quality would be greatly increased, increasing the efficiency and productivity of the standard sawmill substantially. This type of approach to the low-quality forest problem in southeastern Michigan would open the door to this forested area of limited productivity.

LITERATURE CITED

Calvert, W. W. 1958. The Short-log Bolter. Tech. Note 10, Dept. Northern Affairs and Nat. Resources, Ottawa, Canada.

Harold, Monte R. 1959. Use of a Bolter Saw to Reduce Logging Residuals in Southeastern Michigan. Unpublished Master's Wood Technology thesis, Univ. Mich.

Johnson, Walter. 1958. Preliminary Study of a Bolter Saw Operation in Southeastern Michigan. Unpublished Master's Wood Technology thesis, Univ. Mich.

Misiak, Roger D. 1962. An Evaluation of Quality of Lumber Cut from a Typical Farm Woodlot in Southeastern Michigan. Unpublished Master's Wood Technology thesis, Univ. Mich.

Redman, George P., and Rudolph Willard. 1952. Short-log Bolter for Furniture Stock. Southern Furn. Mfrs. Assn., High Point, N. C., Feb. 1952.

PAPERS OF THE MICHIGAN ACADEMY OF SCIENCE, ARTS, AND LETTERS
VOL. XLIX, 1964 (1963 MEETING)

FROST DAMAGE IN A TULIP POPLAR PLANTATION AS RELATED TO FOLIAR POTASSIUM CONTENT[1]

DONALD P. WHITE
Michigan State University

RAYMOND F. FINN
Central States Forest Experiment Station

INTRODUCTION

A yellow poplar (*Liriodendron tulipifera* L.) plantation established in 1940 from locally collected seed on Michigan State University's Russ Forest in Cass County shows unsatisfactory development. The trees exhibit a wide range in heights and diameters as well as susceptibility to sunscald and dieback. The summer color of the leaves is usually greenish-yellow to yellow-green. The larger and taller trees generally have greener leaves.

The planting site is level farmland which was abandoned because of poor yields. The soils are Fox and Warsaw sandy loam. From the color of the leaves and general low vigor, a deficiency of soil nitrogen and perhaps potassium is probable.

In the spring of 1959 we planned a fertilizer experiment consisting of 36 $\frac{1}{40}$-acre plots within the plantation. Fertilizers containing nitrogen in several forms, potassium, and phosphorus were applied in four replications of nine treatments. Nitrogen, phosphorus and potassium were added in several combinations and at two levels of application. Subsequently, growth measurements and leaf sample collections were made annually (1959, 1960, 1961). Foliar analysis included several nutrient elements including potassium. Results of these experiments will be reported elsewhere.[2]

While the leaf collections were being made in mid-September 1960, we observed that the leaves on trees on some of the fertilized plots abscissed at a much later date than they did on the control

[1] Journal Article 3136, Michigan Agricultural Experiment Station. The work was supported in part by a grant-in-aid from the Allied Chemical Company.

[2] R. F. Finn, The Mineral Nutrition of Tulip Poplar (Ph. D. thesis in preparation, Michigan State University).

plots. Late abscission might, in effect, prolong the latter part of the growing season for trees so affected. Concomitantly, we also wanted to know if late abscissing trees leafed out in the spring before the others did. While we were making these observations in the spring we encountered a severe frost killing of yellow poplar leaves.[3]

Frost Damage

The drop in temperature that produced the killing frost of May 27, 1961 affected not only yellow poplar but other planted hardwood species on adjacent areas.

The degree of leaf mortality varied widely. All tulip poplar trees showed some damage. The amount of crown area on which the leaves had been killed by the frost was ocularly estimated as a percentage of the total crown area. As little as 5% of the leaves on some trees was killed; on other trees the kill was 100%.

Analyses

The leaves collected in the fall of 1960 were analyzed for total potassium.[4] While the absolute amounts of foliar potassium present in the leaves at the time of frost injury probably were a little less than the amounts present in the fall of 1960, it may be assumed that the relative concentrations remained essentially the same.

Correlation analysis was used to determine the significance of the relationship between foliar potassium and leaf mortality caused by low temperatures.

Results

Frost damage and potassium percentages for control trees and trees treated with the high level of potash fertilizer are shown in Fig. 1 and correlation coefficients in Table I. The extent of leaf mortality due to frost damage varied from 5% on some trees to as much as 100% on others. Potassium percentages were as low as 0.52 and as high as 1.42.

Correlation coefficients for leaf mortality and potassium content

[3] The frost damage occurred after a heavy freeze on the morning of May 27, 1961. John Bright, resident forester on the Russ Forest, estimated the percent of leaves killed on each of the trees in the test plots.

[4] Leaves were oven-dried, ground in a Wiley intermediate mill, and extracted with 1 *N* ammonium acetate; the amount of potassium present was determined on a Coleman flame photometer.

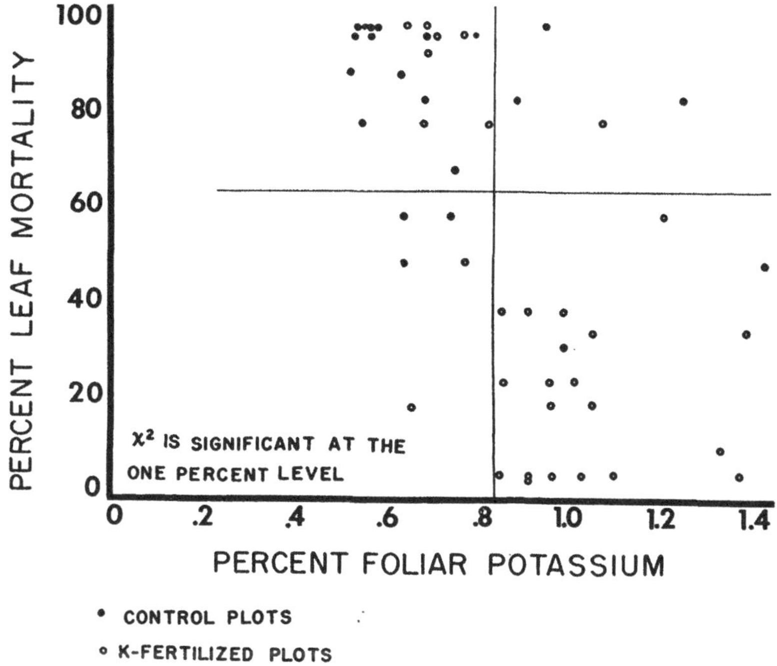

FIG. 1. Relation between foliar potassium content and percent of leaf mortality after a late spring frost.

TABLE I

CORRELATION COEFFICIENTS BETWEEN FOLIAR POTASSIUM AND LEAF MORTALITY DUE TO FROST

Group Analyzed	N	Correlation Coefficient	Statistical Significance	Percent of Leaf Mortality Accounted for by Foliar Potassium
Control	19	−.452	5% level	20
K-fertilized *	31	−.572	1% level	33
Control and K-fertilized	50	−.596	1% level	36

* Plots fertilized at rate of 276 lbs of KCl per acre.

were significant for the potash-fertilized trees, for the control group, and for both groups combined. The relationship was inverse in all these analyses; that is, the higher the foliar potassium percentage, the lower the percent mortality due to frost.

The frost mortality and foliar potassium data for the test trees shown in Fig. 1 clearly separate the least affected group (20% or less frost injury) as the group with the highest foliar potassium content. It is also significant to note that 18 out of the 19 untreated trees had more than 50% of the crown killed by frost.

The chi-square test applied to the quarters formed by equal separation of points in Fig. 1 shows a highly significant relationship between potash content and frost injury.

Foliar potassium levels account for about 36% of the observed leaf mortality caused by frost damage to the potash-fertilized trees. The reduction is about 20% for the control trees. The reduction in leaf mortality is statistically and practically significant.

The optimum foliar potassium concentration for yellow poplar is 2%.[5] In natural stands in the northeast part of the country, the percentage is about 1.6; it ranges from 1.3 to 2.0 (Mitchell and Chandler 1939). In the present study the highest average concentration (0.93%) occurred in the leaves of the potassium-fertilized trees. This is less than half the optimum concentration and only about 6/10 the average concentration found in leaves from stands in the northeastern U. S.

Discussion and Conclusions

The relation of potassium to frost hardiness was noted for strawberries by Davis et al. (1934), who found that the lack of frost hardiness was more markedly associated with potassium deficiency than with any other element. Jung and Smith (1959) found that cold resistance of alfalfa was enhanced by increasing potassium availability up to an amount equivalent to 200 lbs of potassium per acre when phosphorus was held constant at 80 lbs per acre.

It is generally known that increasing soil potassium availability tends to increase winter survival of forage crops. Bendel and Kresge (1962) present evidence to show that potassium increases the electrolyte level of root cells of alfalfa and orchard grass and suggest

[5] Unpublished data from a sand culture experiment by the junior author.

this as an important factor in reducing root injury caused by freezing.

Potassium is unique among other macrochemical elements required by plants in that it is not definitely known to be a part of any compound essential for the existence of the plant, although potassium salts of organic acids do occur in plant cells. Potassium probably exists for the most part as ions in the cells. It plays a part in the synthesis of carbohydrates and proteins, and living cell elasticity is increased by increasing the cell content of potassium.

Cell damage caused by freezing is ascribed to the formation of ice crystals which rupture membranes and the cell wall and to disorganization of the protoplasm caused by dehydration during ice formation.

The existence of potassium as ions in the cell sap coupled with the role of potassium in bringing about the synthesis of sugars and proteins would tend to increase the osmotic pressure of the cell, thus lowering the freezing point and increasing the leaf's resistance to frost damage. The lowering of the freezing point would also help to reduce cell damage caused by dehydration during ice formation, and the increased elasticity of the cell wall would decrease the damage due to ice formation, since the cell would be less apt to rupture.

The results reported here could have an important bearing on the protection of plantation hardwoods from late frosts. This would be of practical value in successfully establishing high value hardwoods.

The relation between leaf mortality and foliar potassium concentration should be studied under controlled environment conditions to remove the variance introduced by subjective observations and the variation resulting from a range of microenvironmental conditions in the plantation.

Summary

A hard late-spring frost killed yellow poplar leaves in a plantation on Michigan State University's Russ Forest. The leaves were fully or nearly fully developed at the time, and leaf mortality varied from 5 to 100% of the leaves on the individual trees. Leaf mortality was lowest on trees fertilized with potash fertilizer.

LITERATURE CITED

BUNDEL, V. A., AND C. B. KRESGE. 1962. Potassium Winterizes Your Forage Crop Roots. Crops and Soils 15 (2) : 12-13.

DAVIS, M. B. et al. 1934. Nutritional Studies with *Fragaria.* Sci. Agr. 14: 411-432.

JUNG, G. A., AND DALE SMITH. 1959. Influence of Soil Potassium and Phosphorus Content on Cold Resistance of Alfalfa. Agron. Jour. 51: 585-587.

MITCHELL, H. L., AND R. F. CHANDLER, JR. 1939. The Nitrogen Nutrition and Growth of Certain Deciduous Trees of the Northeastern United States. Black Rock For. Bull. 11. 94 pp.

PAPERS OF THE MICHIGAN ACADEMY OF SCIENCE, ARTS, AND LETTERS
VOL. XLIX, 1964 (1963 MEETING)

GEOLOGY

PRECAMBRIAN GEOLOGY OF THE PEMBINE AREA, NORTHEASTERN WISCONSIN [1]

J. ALLAN CAIN
Western Reserve University

INTRODUCTION

IN northeastern Wisconsin, there are extensive areas of granitic and metamorphic rocks which have been mapped only on a reconnaissance scale. In Michigan, however, due largely to the impetus of investigations for iron ore, the geology has been studied in considerable detail. The area described herein is adjacent and to the south of Precambrian exposures in Michigan which have been investigated recently by the U. S. Geological Survey (James 1958; Prinz 1958; Dutton 1960; James et al. 1961). In part my hope was to provide a small area of detailed knowledge from which further studies in northeastern Wisconsin might benefit.

The area described consists of part of the Pembine and Dunbar 15′ quadrangles of the U. S. G. S. (Fig. 1).

Previous work in northern Wisconsin is summarized by Cain (1963); a detailed historical review is, therefore, not given here. Mention should be made, however, of work by Lyons (1947), who described the geology around Niagara; and Prinz (1958) and Mancuso (1960), who studied areas north and south of the present region respectively.

GENERAL STATEMENT

As a result of detailed mapping, nine rock units were distinguished within the previously undifferentiated, largely granitic, area of northern Wisconsin. Several units established to the north of the present area by Prinz (1958) were traced south, and four newly named rock units were also recognized in the mapped area: Dunbar Gneiss, Twelve Foot Falls Quartz Diorite, Newingham Granodiorite,

[1] Contribution No. 6, Department of Geology, Western Reserve University.

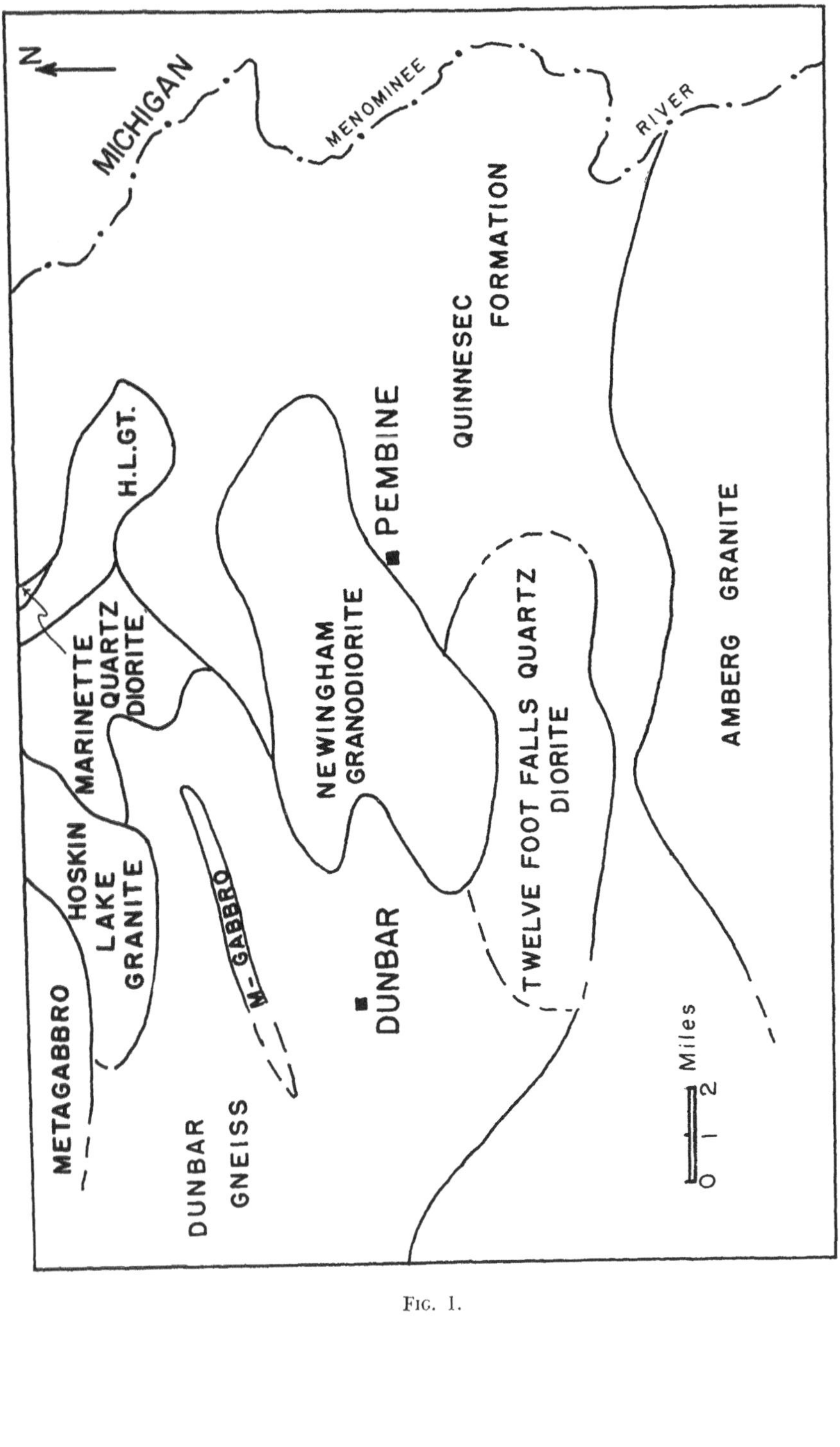
N
MICHIGAN
MENOMINEE
RIVER
QUINNESEC FORMATION
PEMBINE
H.L.GT.
NEWINGHAM GRANODIORITE
TWELVE FOOT FALLS QUARTZ DIORITE
AMBERG GRANITE
MARINETTE QUARTZ DIORITE
HOSKIN LAKE GRANITE
M-GABBRO
DUNBAR
METAGABBRO
DUNBAR GNEISS
0 1 2 Miles

FIG. 1.

and Amberg Granite (Fig. 1). Rock units recognized in this study and their possible age relationships are:

Youngest	Diabase dikes
	Amberg Granite
	Newingham Granodiorite
	Hoskin Lake Granite
	Metagabbro
	Twelve Foot Falls Quartz Diorite
	Marinette Quartz Diorite
	Dunbar Gneiss
Oldest	Quinnesec Formation

Age relationships.—Table I summarizes the evidence of age relationships obtained in this study. Ideally, such a table would have

TABLE I

AGE RELATIONSHIPS IN NORTHEASTERN WISCONSIN

	Diabase dikes	Amberg Granite	Newingham Granodiorite	Hoskin Lake Granite	Metagabbro	Twelve Foot Falls Quartz Diorite	Marinette Quartz Diorite	Dunbar Gneiss	Quinnesec Formation
Diabase dikes	///	X	X			X		X	X
Amberg Granite		///				X		X	X
Newingham Granodiorite			///	X		X	X	X	X
Hoskin Lake Granite...				///			X	X	X
Metagabbro					///			X	X
Twelve Foot Falls Quartz Diorite						///			X
Marinette Quartz Diorite							///		X
Dunbar Gneiss								///	X
Quinnesec Formation...									///

X means the rock unit on the left is demonstrably younger than the rock unit indicated on the top.

all spaces above the diagonal occupied by X. The tentative nature of this succession is thus indicated by the blank spaces above the diagonal.

Description of Rock Units

Quinnesec Formation

The age of the Quinnesec Formation has been variously interpreted as Keewatin to Keweenawan (Lower to Upper Precambrian) and is, as yet, not unequivocal. James (1958) considered it Lower Precambrian, while the State geologic map (1949) shows the formation as Upper Precambrian. The confusion arises from the absence of contact relationships between the Quinnesec Formation and the Middle Precambrian Animikie Series of the Menominee District to the north (Fig. 1). Dutton (1960 and personal communication, 1962) studied the relationships in the Florence area and found a greenstone unit (presumed to be the Quinnesec Formation) showing tops to the north but lying immediately south of Animikie strata having tops to the south. The implication, therefore, is of a possible fault contact between the units. Radiometric determinations (Abelson 1960, pp. 153-154) appear to correlate with degree of metamorphism rather than observed relative ages. No new direct evidence resulted from this study, but the Quinnesec Formation is older than all other units present. In view of the complex and extensive igneous and metamorphic rocks of post-Quinnesec age, however, it would seem very unlikely that the formation is Upper Precambrian in age, and I concur with James (1958) in assigning it, tentatively, to the Lower Precambrian.

The Quinnesec Formation consists of three main rock types: metavolcanics, metapyroclastics, and metasediments. The distribution of these three rock types appears random. Reconnaissance by William F. Read (personal communication, 1959), however, suggests that metavolcanics are dominant in the northeast part of the area and metasediments more common in the southwest. North of the present area, Prinz (1958) recognized metavolcanic and metasedimentary rocks but did not distinguish mappable lithologic units within the formation, except for the very localized Pier Gorge Schist Member.

Metavolcanic rocks are probably the most abundant rock type within the mapped unit and are typically represented by metamorphosed andesitic or basaltic flows, though rhyolitic varieties are dominant locally. Massive varieties showing pillow structures are present; and, locally, there are feldspathic hornblende schists. Xenoliths in the later intrusives are commonly feldspathic amphibolites showing hornfelsic texture.

Metamorphosed basalts probably make up most of the Quinnesec metavolcanics within the present area and constitute the typical greenstones. Characteristic mineralogy includes plagioclase and hornblende, together with chlorite, epidote, calcite, and quartz, in a relict diabasic texture. Closer to the granitic masses, the metavolcanics show localized alteration and recrystallization along minor shear zones. Modal compositions of metavolcanic, metapyroclastic, and metasedimentary rocks from the Quinnesec Formation are given in Table II.

TABLE II

AVERAGE COMPOSITION OF ROCK UNITS IN NORTHEASTERN WISCONSIN
(Each is the average of five specimens).

		1.		2.	3.		4.	5.		6.	7.	8.
Andesine					35	25	45		25			15
Oligoclase	25	30	35	30						20	45	
Labradorite								25				
Microcline				5		15				40	10	40
Epidote	25	15			5	5		25	5			
Quartz	tr	tr	7	20	15	20	25			20	25	20
Biotite		10	5	35	30	25			10	15	12	15
Hornblende	3	35	45	3	5	5	15	10	55		3	5
Actinolite	30							30				
Chlorite	10	5	5	5	3		10			3		
Calcite	5	3										
Iron ores and others	2	2	3	2	7	5	5	10	5	2	5	5

LEGEND: 1 = Quinnesec Formation; 2 = Dunbar Gneiss; 3 = Marinette Quartz Diorite; 4 = Twelve Foot Falls Quartz Diorite; 5 = Metagabbro; 6 = Hoskin Lake Granite; 7 = Newingham Granodiorite; 8 = Amberg Granite.

The metapyroclastic rocks consist of hornblende and plagioclase, with quartz and iron ores. Closer to the granitic bodies, microcline, albite, and muscovite are also present. The common pyroclastic rock appears to have been a lithic-crystal tuff which subsequent metamorphism has made difficult to distinguish from lava flows. The irregular outline of the larger crystals and the presence of large, limonite-stained, almost opaque fragments suggest that the original rock was pyroclastic in origin. Since more severe alteration is probably to be expected in tuffs than in lavas, the extensive alteration which makes difficult any precise petrographic determinations itself suggests the original character of the rock.

Within the mapped area, metasediments are apparently the least common rocks in the Quinnesec Formation. The metasedimentary rocks, in common with the rest of the Quinnesec Formation, show clear evidence of metamorphic changes: euhedral garnet crystals, fresh biotite, actinolitic hornblende, and sodic oligoclase. The effects of retrograde metamorphism are evidenced by the development of chlorite within biotite, orientation of chlorite grains perpendicular to the schistosity, and sericitization of the feldspars. Metasediments are typically present as schists containing quartz, biotite, and hornblende. Other constituents include microcline, sodic plagioclase, epidote, muscovite, sericite, chlorite, carbonates, and iron ores.

Regional metamorphism of the Quinnesec Formation has produced rocks of the greenschist facies, though amphibolites are developed marginal to, and xenolithic within, the granitic masses. Different metamorphic facies (as defined by Turner and Verhoogen 1960) are distinguished within the Quinnesec Formation as follows:

Greenschist facies.—Rocks of this facies are typically greenstones which consist of actinolite, albite, chlorite, epidote, and quartz, as well as minor amounts of sphene. The rocks could therefore belong to the Quartz-Albite-Muscovite-Chlorite Subfacies or the Quartz-Albite-Epidote-Biotite Subfacies.

Almandine-amphibolite facies.—Representative of this facies are amphibolites which contain hornblende, oligoclase, quartz, epidote, and biotite. Such a mineral assemblage would place the rocks in the Staurolite-Almandine Subfacies.

Although no attempt was made to map different metamorphic facies within the Quinnesec Formation, it was observed that amphibolites are only present close to the granitic masses. Similar observations were made to the north of the present area in a more detailed study of the Quinnesec Formation by Prinz (1958). It would thus appear reasonable to infer that the amphibolites represent the effect of contact metamorphism on regionally metamorphosed greenstones.

Petrogenesis.—Recent detailed studies by Shapiro (1952), Froelich (1953), Thompson (1955), Fulweiler (1957), and Prinz (1958) have demonstrated that the Quinnesec Formation is a highly variable sequence of volcanic and sedimentary rocks. Rocks of the Quinnesec Formation are present over many hundreds of square miles in northeastern Wisconsin and typically have fairly steep dips. Although folding and faulting probably have caused repetition of the formation, it still seems reasonable to assume a considerable thickness for this unit. It is thus suggested that the Quinnesec Formation represents interstratified sediments, flows, and pyroclastics characteristic of a geosynclinal succession (cf. Jones 1938).

DUNBAR GNEISS

The unit consists of two main types: medium-to-coarse-grained banded gneiss and fine-grained migmatitic gneiss.

The banded gneiss is commonly uncontorted and locally resembles foliated granitic rock. The alternating leucocratic and mafic (mainly biotite) bands range from ½ inch to 2 feet wide. The feldspar is subhedral, gray, and rarely shows twinning in hand specimen. Biotite is abundant and occurs in elongated aggregates. Locally, the gneiss exhibits augen texture with gray feldspar ovoids some 5 cm long. Typically, however, the gneiss is medium-grained and is characterized by regular bands up to 4 or 5 cm wide.

The migmatitic gneiss is fine-grained and appears, in hand specimen, to consist almost entirely of biotite. Leucocratic bands and stringers extend through the rock, locally forming agmatites. The migmatitic gneiss occurs most commonly as scattered xenoliths up to 15 feet long within the banded gneiss. However, west of Dunbar (Fig. 1) the migmatitic gneiss becomes increasingly abundant, locally to the exclusion of the banded gneiss.

Banded gneiss.—The gneiss has a granulated appearance in both hand specimen and thin section and a schistosity emphasized by elongated biotite clusters. Cataclastic texture is commonly developed, as indicated by the development of quartz and feldspar mosaic. Characteristic minerals are plagioclase, quartz, biotite, microcline, apatite, and sphene. Bands of biotite, sphene, and iron ores commonly alternate with leucocratic mosaic aggregates.

The composition of the banded gneiss is given in Table II.

Migmatitic gneiss.—Migmatitic gneiss is commonly fine-grained and shows leucocratic material present in thin stringers. These quartz-plagioclase bands are commonly connected by similar crosscutting veins and in thin section are seen to truncate constituents of the mafic gneiss. The migmatitic gneiss, which occurs as xenoliths within the banded gneiss, is typically contorted into asymmetrical folds of some two to three-foot amplitude. Insufficient outcrops were located from which fold axes orientations could be measured; thus, no structural interpretation is given. Near the granitic masses, megacrysts of potash feldspar up to 2 cm in length are present. Characteristic minerals of the fine-grained migmatitic gneiss are feldspar, quartz, biotite, epidote, apatite, sphene, and garnet. The feldspar is commonly sodic and typically oligoclase or albite.

Structure.—Structural data were obtained from only the banded gneiss. Poles to foliation in this unit (Fig. 2) plot as an incomplete girdle. There is, therefore, the suggestion that the gneiss is isoclinally folded, though the data are insufficient for detailed analysis. The dominance of southern dips shown in Fig. 2 suggests that the axial planes dip to the south. The proposed great circle indicates that the structure is possibly triclinic since the poles do not lie on one great circle. There is, thus, the possibility of at least two periods of deformation.

Petrogenesis.—The texture of the banded gneiss indicates that deformation and recrystallization have modified the original appearance of the rock. The quartz and feldspar mosaic commonly encloses biotite grains and suggests cataclastic deformation. Although the feldspar crystals are now composite and embayed, the original texture was probably granitic, and non-gneissic areas do have hypidiomorphic-granular texture. Impersistent areas containing microcline megacrysts are common throughout the gneiss, especially close to the Hoskin Lake Granite.

It is posulated in the sequel that metasomatism of a preexisting granitic mass might have produced the porphyritic Hoskin Lake Granite. There is thus the possibility that the banded gneiss, now granodioritic in composition, is part of that earlier granitic body. In support of this hypothesis are the following relationships: (*a*) foliation directions defined by mafics within the gneiss are continuous into the granite; (*b*) local areas of microcline megacrysts in the gneiss (especially close to the granite) give the gneiss an appearance almost identical to the granite; (*c*) the boundary between the two units, although not well exposed, appears gradational, and a sharp contact between gneiss and granite was not seen.

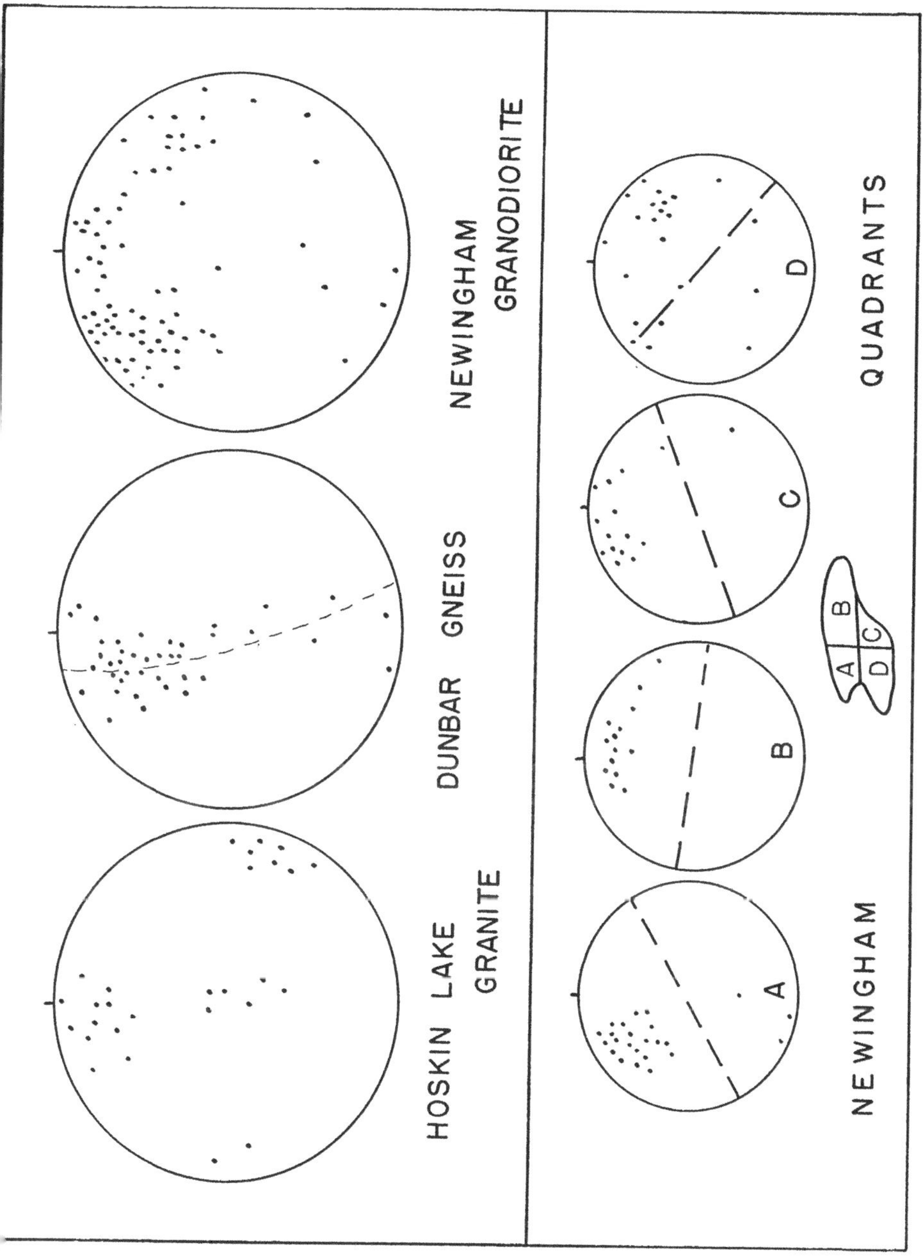
HOSKIN LAKE GRANITE
DUNBAR GNEISS
NEWINGHAM GRANODIORITE
A
B
C
D
NEWINGHAM QUADRANTS

FIG. 2.

The Dunbar Gneiss is apparently part of the " mafic quartz diorite " described by Lyons (1947) , a unit which he recognized as older than a " normal quartz diorite." The " normal quartz diorite " is probably part of the unit I mapped as Newingham Granodiorite.

Evidence for the relationship of the Dunbar Gneiss to the Marinette Quartz Diorite and the Twelve Foot Falls Quartz Diorite was not established because of lack of exposures. It is conceivable, however, that the Quinnesec Formation and migmatitic gneiss are penecontemporaneous and represent different lithologies within a geosynclinal sequence of sedimentary, volcanic, and associated intrusive rocks (banded gneiss, Marinette Quartz Diorite, and Twelve Foot Falls Quartz Diorite) .

Marinette Quartz Diorite

The Marinette Quartz Diorite was named by Prinz (1958) and is one of the most variable rock units in the area. It consists of three main types: coarse-grained schistose, medium-grained massive, and fine-grained massive. Locally, the quartz diorite is characterized by varying numbers of microcline megacrysts in a matrix of plagioclase, quartz, biotite, and minor hornblende. In more schistose phases, feldspar megacrysts are up to 10 cm long; and the rock, locally, is almost identical to the Hoskin Lake Granite. More commonly, feldspars are less than 1 cm in length, and the rock has a mottled black and white appearance. Also found are massive black varieties with a grain size less than 2 mm. Gradations among these three main types typically occur over distances of only a few feet. Xenoliths of Marinette Quartz Diorite up to 20 feet long are present within the Hoskin Lake Granite, and apophyses of granite cut the quartz diorite up to a mile each side of the boundary indicated in Fig. 1.

The massive fine-grained quartz diorite consists of plagioclase, biotite, quartz, hornblende, epidote, and iron ores. Plagioclase (An_{40}) is present as unoriented subhedral laths showing faint albite, Carlsbad, and pericline twinning.

Locally, where the quartz diorite is schistose, microcline crystals are commonly present. Associated with the potash feldspar are quartz, apatite, and some tourmaline. The remainder of the rock consists of subparallel bands of elongated biotite, hornblende, and andesine crystals emphasizing the schistosity. Augen of andesine are common, and microcline appears to have developed across planes of schistosity with the C crystallographic axis usually parallel to the foliation of the rock. Potash feldspar is also present in veinlets and commonly occurs as subhedral crystals up to 2 cm in length which embay quartz and biotite. There is little difference in mineralogy between the medium- and fine-grained quartz diorite, but the absence of potash feldspar readily distinguishes these from the coarse-grained schistose variety. Compositions of the fine- and coarser-grained quartz diorite are given in Table II.

Petrogenesis.—Apophyses of Marinette Quartz Diorite intrude the Quinnesec Formation (Prinz 1958) and the quartz diorite is interpreted as an igneous intrusion. However, local and impersistent zones of microcline megacrysts occur throughout the Marinette Quartz Diorite where the rock is schistose. The feldspar megacrysts contain inclusions of groundmass and cut across the foliation; veinlets of potash feldspar are also present. Biotite close to the microcline megacrysts is dark brown, somewhat poikilitic, and larger than biotite crystals elsewhere in the rock. Numerous veinlets of quartz mosaic are also common in the schistose rocks. It is therefore suggested that metasomatism developed feldspar porphyroblasts, quartz, and biotite where structural deformation had localized the metasomatic effects (see also the discussion of the Hoskin Lake Granite). The schistose parts of the unit contain too high a proportion of potash feldspar to be classified as quartz diorite. However, since the schistose areas are only local and most of the unit is almost devoid of potash feldspar, the name Marinette Quartz Diorite is retained to include both schistose and massive phases.

Twelve Foot Falls Quartz Diorite

This rock unit was mapped on reconnaissance scale by the author and was named and studied in detail by Wadsworth (1962). I am indebted to Wadsworth for much of the information included in this brief description.

To the northeast and south, the Twelve Foot Falls Quartz Diorite intrudes Quinnesec Formation greenstones and metasediments. This has resulted, in places, in recrystallization of the greenstones to hornblende schists. The eastern part of the quartz diorite is homogeneous with only weak foliation. Toward the west, the unit increasingly shows more pronounced gneissic texture, which is typical of the northwestern outcrops.

Macroscopically, the quartz diorite is characterized by subhedral, flesh-colored feldspar crystals up to 1 cm long, smaller subhedral hornblende crystals, and abundant anhedral blebs of blue quartz up to 1 cm in diameter. The texture of the rock is commonly gneissic with bands about 3 cm wide; but, locally, medium- to coarse-grained hypidiomorphic-granular texture is present.

Characteristic minerals of the Twelve Foot Falls Quartz Diorite include plagioclase, hornblende, quartz, and chlorite, together with apatite, magnetite, epidote, calcite, and veinlets of microcline. In thin section, plagioclase is partly altered to epidote, clinozoisite, sericite, and clear albite. Twinning is rarely seen but, where discernible, the composition of the plagioclase is sodic andesine. Inclusions of anhedral hornblende and chlorite are common in plagioclase crystals, and the plagioclase also typically embays hornblende crystals. Hornblende is present in subhedral to anhedral grains in which inclusions of apatite and magnetite are common, together with epidote and quartz. Alteration of hornblende to chlorite is extensive; and, locally, chlorite is present as pseudomorphs after hornblende. Quartz is present as veinlets which cut the rock and shows shadowy extinction. Occasional isolated pods of quartz occur, which exhibit sharp linear borders in contact with feldspar and hornblende crystals. The modal composition of the quartz diorite is given in Table II.

Petrogenesis.—The Twelve Foot Falls Quartz Diorite contains xenoliths of Quinnesec Formation and probably represents an igneous intrusion. Veinlets of quartz mosaic and locally gneissic texture give evidence of subsequent injection and deformation. The unit crops out between, and is older than, the Newingham Granodiorite and Amberg Granite (Fig. 1) and thus was probably subjected to the physical and chemical effects of the granitic masses. The absence of any relict pyroxene suggests that the Twelve Foot Falls Quartz Diorite was originally dioritic in composition.

Metagabbro

Metamorphosed diorites and gabbros crop out in the northwestern part of the mapped area. A field and petrographic comparison with metagabbro sills described to the northeast by Prinz (1958) suggests that the diorites and gabbros within the present area are a westward continuation of similar sills. An elongate area of similar rocks crops out within the Dunbar Gneiss (Fig. 1). The rocks, characterized by unoriented subhedral hornblende crystals up to 2 cm long, are typically massive, but locally they are schistose with alternating leucocratic and amphibole-rich bands. Occasionally, boudinage structure is evident in the thinner apophyses of the unit.

The massive metagabbro consists of plagioclase, hornblende, epidote, apatite, quartz, chlorite, and iron ores. Plagioclase is labradorite, which typically exhibits albite twinning and commonly shows relict diabasic texture with actinolitic hornblende. The labradorite occurs in laths up to 7 mm long and is partly altered to epidote, zoisite, and sericite. Hornblende is present as subhedral crystals to which actinolite commonly forms kelyphitic borders. (Actinolite is here distinguished from hornblende by its weak pleochroism and fibrous habit.) Pseudomorphs of hornblende after augite are also present. In the groundmass, minor pyrite and ilmenite show alteration to limonite, leucoxene, and sphene. Quartz is present in small veins throughout the rock.

Also present within the metagabbros are rocks of more dioritic composition, which may represent original magmatic differentiation within the gabbro. The diorites show similar textures, but with andesine instead of labradorite and no actinolitic borders to hornblende crystals. Other differences include the presence of biotite, the rarity of plagioclase twinning, and a marked reduction in the amount of epidote minerals. The differences between the gabbroic and dioritic types are shown in Table II.

Petrogenesis.—The metagabbros are thought to represent gabbroic, or locally dioritic, intrusions of sill-like form.

HOSKIN LAKE GRANITE

The Hoskin Lake Granite was named by Prinz (1958). The characteristic feature of this granite is the development of euhedral potash feldspar crystals which are locally up to 15 cm long. The feldspar crystals are gray or flesh-colored and commonly show Carlsbad twinning. Rapakivi texture is locally developed. Biotite is present as small crystals (less than 5 mm), and quartz occurs as anhedral blebs less than 1 cm in diameter. Locally, especially in the western part of the mass, the mafic constituents exhibit a strong foliation.

Although porphyritic granite is typical of the Hoskin Lake Granite, feldspar megacrysts vary in size. In general, coarse-grained, gneissic-textured granite dominates in the west; coarse-grained weakly foliated granite in the central area; and medium-grained leucocratic granite with feldspar crystals rarely over 2 cm long in the eastern exposures. Near the borders of the mass, finer-grained granite is common. Nonporphyritic granitic dikes, similar in composition and appearance to the Newingham Granodiorite, cut the Hoskin Lake Granite.

The granite mass has sharp boundaries with Quinnesec Formation to the north and east, but occurs in local and impersistent areas within biotite gneiss to the west and Marinette Quartz Diorite to the south. Xenoliths of amphibolite, quartz diorite, and schist up to 20 feet long are sparsely present throughout the unit. The xenoliths typically show sharp contacts with the enclosing granite and are more common near the southern and western margins of the mass.

The Hoskin Lake Granite consists of microcline, plagioclase, quartz, biotite, epidote, chlorite, tourmaline, apatite, muscovite, and iron ores. Feldspar is present as euhedral to subhedral microcline megacrysts which show inclusions of groundmass concentrated at the borders of crystals. Many microcline crystals are composite and consist of three or four large crystals. Commonly, subhedral microcline microperthite crystals have embayed and apparently replaced other constituents. Megacrysts of oligoclase are also present; and oligoclase, albite, and perthite are common in the groundmass. Clear albite not infrequently forms a rapakivi-like discontinuous border to microcline crystals. The modal composition of the Hoskin Lake Granite is given in Table II.

Structure.—Fig. 2 indicates three maxima of poles to foliations, mutually at right angles. One (foliation 080° at 70-80° S.) is defined by both mafics and feldspar megacrysts, whereas the other two maxima are defined by feldspar crystals only. The three foliation orientations tend to be mutually exclusive and to occur in definite geographic areas. The dominant foliation direction in the western part of the mass is N 80° E, which is the regional structural trend. In the central and eastern parts, the foliation strikes almost north-south. In several outcrops, however, both foliation directions are present and the north-south direction is generally better developed. The horizontal foliation was observed only in the central part of the area in a few outcrops of considerable relief. In other parts of the unit, foliation data were obtained from outcrops with little relief. The observed roughly horizontal foliation might thus be largely the result of different sampling procedures in vertical exposures, and insufficient data were obtained to permit detailed analysis. However, it seems clear that one generation of megacrysts is oriented parallel to the regional structure (James et al., 1961), and another generation is oriented perpendicular to that trend.

Petrogenesis.—A preferred orientation of the feldspar megacrysts is common, though on account of their large size, it is only recognized macroscopically. Although the feldspar crystals, inclusions, and matrix of the Hoskin Lake Granite commonly are aligned together, in the central part of the granite mass the megacrysts commonly cross the foliation of the groundmass at almost 90°. T. F. W. Barth (personal communication, 1960) pointed out that such relationships are common and suggested that a release of stress might have produced a low-pressure environment favorable to the growth of potash feldspar megacrysts. As mentioned in the description of the Marinette Quartz Diorite, coarse-grained varieties of the quartz diorite closely resemble Hoskin Lake Granite. Dunbar Gneiss to the west of the Hoskin Lake Granite also shows progressive increase in the number of feldspar megacrysts toward the granite.

One implication of such relationships is that metasomatism has produced rocks now of granitic composition in the central area, while to the south and west, the original rocks are still recognizable but with the development of megacrysts locally. Quinnesec Formation in contact with Hoskin Lake Granite shows only a minor increase in potash feldspar, and most changes in mineralogy may be directly attributed to thermal metamorphism. It could be argued, however, that the fine-grained massive nature of volcanic rocks would render them virtually impervious to the agents of metasomatism. Dikes of Hoskin Lake Granite cut the Quinnesec Formation; and, as previously mentioned, xenoliths of quartz diorite, amphibolite, and schist are common. The original rock, now represented by the Hoskin Lake Granite, was therefore presumably intrusive into the other rocks. If this rock were originally granitic in composition, potash metasomatism might initially merely increase the size of preexisting potash feldspars, whereas in rocks of nongranitic composition the development of potash megacrysts would be more local and dependent upon favorable environment. This hypothesis is somewhat borne out by considering changes within the Hoskin Lake Granite: The eastern outcrops of Hoskin Lake Granite are medium-grained, with feldspar megacrysts only 1-2 cm long, and the granite is probably closer in appearance to the original rock. Metasomatism of such a medium-grained granite would, in all probability, produce the coarse-grained variety common in the central area. The nature of the feldspars also tends to support this reasoning. Microcline in the groundmass contains inclusions throughout the crystals, whereas larger microcline megacrysts commonly have inclusions of groundmass concentrated near the margins of crystals. This might imply that some of the larger crystals represent further growth on preexisting microcline nucleii. On the other hand, many large microcline microperthite crystals show no evidence of a prior nucleus; but their interdigitating relationship with other constituents, plus small veinlets of potash feldspar, suggests introduction of potassic material. It is therefore suggested, on the basis of field and petrographic evidence, that the potash megacrysts formed, at least in part, as a result of metasomatism. As pointed out by Lyons (1947), feldspar megacrysts within the Marinette Quartz Diorite tend to occur where the quartz diorite is locally schistose. Thus, it seems likely that structural control (as suggested by Barth) was an important factor in localizing the effects of metasomatism.

Newingham Granodiorite

The Newingham Granodiorite is the only rock unit in the present area for which complete geologic boundaries were established. For this reason, this unit was studied in detail for specific gravity and modal data (Cain 1962, 1963a, 1964).

The granodiorite is fairly homogeneous in the eastern and central parts of the mapped area; but toward the west, it becomes increasingly xenolithic. This unit shows contact relationships and is well exposed in the neighborhood of the Newingham farm, in the northwestern part of the mass.

The typical Newingham Granodiorite is medium- to coarse-grained, leucocratic, gray, with rather weak foliation exhibited by biotite crystals. The central part of the mass exhibits little or no foliation; but toward the margins, planar banding of minerals becomes increasingly pronounced. Mafic schlieren and autoliths are common throughout the unit. A peripheral belt of gneissoid granodiorite up to $\frac{1}{2}$ mile wide is typical and shows granulation of constituent minerals. Throughout the mass, xenoliths of Dunbar Gneiss and Quinnesec Formation up to $\frac{1}{2}$ mile in length are common. The former are usually partly digested, but the latter typically show sharp contacts with the enclosing granodiorite.

Outlying plugs, up to 300 yards in diameter, resembling Newingham Granodiorite, intrude the Hoskin Lake Granite, Marinette Quartz Diorite, and Dunbar Gneiss, up to three miles from the main Newingham mass.

The characteristic mineralogy of the Newingham Granodiorite is feldspar, quartz, and biotite, together with hornblende, sphene, and epidote. Plagioclase is present as subhedral crystals of oligoclase. Commonly the plagioclase is zoned with a more calcic core and shows alteration of sericite, zoisite, and clinozoisite. Blebs of microcline are locally extremely numerous within plagioclase crystals, and subhedral microcline crystals are also present. Fine-grained aggregates of potash feldspar occur between larger lenticular feldspar crystals. Microcline is typically present as discontinuous borders to plagioclase crystals, as well as in thin veinlets. Myrmekite is present, and micrographic feldspar and quartz is commonly displayed. Hypidiomorphic-granular texture is typical of the granodiorite and readily distinguishes it from the Dunbar Gneiss, which exhibits shear zones and highly altered biotite and feldspar.

Within the Newingham mass, rocks of quartz dioritic composition are also present, generally as a finer-grained marginal facies of the pluton. The quartz dioritic rocks contain euhedral to subhedral plagioclase crystals, which are corroded at the periphery by quartz and occasional potash feldspar. Hornblende, which is subhedral in the granodiorite, is ragged and poikilitic in the quartz diorite. Biotite has apparently partly replaced hornblende. Except for differences in quartz and potash feldspar content, and the presence of greater amounts of hornblende, the quartz dioritic rocks are virtually identical to the granodiorite. Modal data from the Newingham Granodiorite are given in Table II.

Structure.—Fig. 2 shows the foliation data for both the Dunbar Gneiss and Newingham Granodiorite. It is clear from this figure that a greater range of foliation orientations is present in the granodiorite than in the gneiss. The dominant foliation direction in the Dunbar Gneiss coincides with the observed structural trend of the Lake Superior Region (N 70° E). The Newingham Granodiorite shows a similar concentration of foliation orientations but also many orientations divergent from the regional trend. The implications of the spread in foliation orientations is that a primary foliation pattern may be present in the granodiorite. To demonstrate this point, Fig. 2 shows the same foliation data for the Newingham

Granodiorite plotted for each quarter of the mass. Ideally, such a figure might be expected to emphasize any concentric, or primary, foliation pattern. It can be seen from Fig. 2 that such is, in fact, so. I thus believe that the foliation pattern shown by the granodiorite is sufficiently different from the poorly defined girdle exhibited by the Dunbar Gneiss to justify the acceptance of a primary, roughly concentric, foliation pattern within the Newingham Granodiorite.

Petrogenesis.—The granodiorite intrudes adjacent rock units and contains numerous xenoliths. In addition, the roughly concentric foliation directions and systematic zonal pattern of color index values also suggest an intrusive origin for the pluton. Preliminary results of a quantitative analysis of this unit (Cain 1962) indicate that microcline distribution does not correspond with that of other mineralogical variates. Veinlets of microcline are also present, and it is thus suggested that potash feldspar represents a late stage addition, rather than the product of normal magmatic crystallization. It is possible that autometasomatism produced the microcline, since it is local in distribution and nowhere abundant.

Amberg Granite

The Amberg Granite was studied only in reconnaissance, and the following description is of the northernmost exposures of the unit. A detailed study of this unit and its relationship to the High Falls Granite to the south (Mancuso 1960; Cain 1963) is currently in progress (Cain and Beckman 1964).

Marginally, a medium- to fine-grained gray granite crops out; but farther south, the typical medium- to coarse-grained pink Amberg Granite apparently intrudes the medium-grained gray border phase. Foliation is not well developed in the typical Amberg Granite. Dark, mafic inclusions are present but are not deformed into schlieren as in the Newingham Granodiorite.

Macroscopically, the granite is characterized by subhedral pink feldspars up to 3 cm long. The feldspar is typically Carlsbad-twinned and locally may have rapakivi texture. Quartz may be present in both large discrete blebs and fine-grained sugary aggregates. Biotite is commonly present in large flakes but locally may occur in aggregates of smaller grains. Minor shear zones are present locally, but the granite is characteristically fresh, undeformed, and has hypidiomorphic-granular texture.

The coarse-grained pinkish granite consists of microcline, plagioclase, quartz, and biotite, together with sericite and zoisite in hypidiomorphic-granular texture. Microcline is present as subhedral to anhedral grains showing microcline twinning. Peripheral to many potash crystals is a discontinuous border of plagioclase in a rapakivi texture. Microcline, commonly present as perthite, has ragged boundaries and contains inclusions of plagioclase, quartz, and biotite. Plagioclase is sodic andesine, which rarely shows twinning. In general appearance, the granite is less altered than other rock units in the area. Modal composition of the coarse-grained pink granite is given in Table II.

Petrogenesis.—In petrographic character, the Amberg Granite exhibits no unusual features to suggest that it originated by any other means than magmatic crystallization. Petrographic and hand specimen comparison of the Amberg Granite with the High Falls Granite (Cain 1963, fig. 2) suggests that these granites are part of one large granite body.

Diabase

Diabase dikes cut rocks as young as the Newingham Granodiorite. These dikes are fine-grained, show sharp linear contacts with the intruded rock, and are rarely over ten feet wide. Diabasic texture is evident in thin section, with laths of feldspar in a groundmass of augite and hornblende. Typical minerals include labradorite, augite, and hornblende, together with biotite, epidote, zoisite, sericite, and iron ores. Labradorite (An_{55}) is the dominant constituent, making up almost half the rock, and occurs as euhedral laths showing polysynthetic, Carlsbad, and pericline twinning. The laths show no preferred orientation and are commonly altered to sericite and epidote along twinning planes. Hornblende is the major constituent of the groundmass and commonly has kelyphitic borders of actinolite. Metamorphism is evidenced by the uralitization of poikilitic augite to fibrous aggregates of actinolite, associated with which are biotite, chlorite, and magnetite.

The almost vertical, uncontorted attitude of the diabase dikes indicates that they post-date major deformation in the area. The diabase dikes are thus considered the youngest igneous rock in the mapped area and, as indicated by James (1958), are probably Keweenawan in age.

Summary and Conclusions

It is commonly pointed out in the literature that igneous activity in orogenic areas typically conforms to a general pattern:

a) Eruption of basic lavas during the geosynclinal phase.

b) Injection of basic plutons during the early stages of folding.

c) Development of granodioritic and granitic batholiths during and following the main period of folding.

d) Eruption of lavas after the main phase of folding and plutonic activity.

Within the Precambrian of the Lake Superior Region, several orogenies have been recognized and named (Table III). Thus, the

TABLE III

Orogenies in the Lake Superior Region
(After Goldich et al 1961)

Precambrian Division	Orogeny	Approximate Age (in billion years)
Late	Grenville	1.1
Middle	Penokean	1.7
Early	Algoman	2.5
	Laurentian	2.6?

" typical " igneous cycle has probably been repeated many times in the Lake Superior Region. Interpretation of the geologic history of an area therefore requires that rocks associated with, or involved in, a particular orogeny can be distinguished from those of another. Sediments deposited after orogenic episodes provide marker horizons in many localities which aid interpretation of the geologic history. Unfortunately, within the present area, the Quinnesec Formation is the only extensive stratigraphic unit and its age is, as yet, not unequivocal. Structural analysis has aided the present study, but a dominant regional foliation direction (N 70° E.) probably obscures many differences in structural pattern that may have existed among the various rock units.

Radiometric dates in northeastern Wisconsin are, so far, scarce and somewhat inconclusive (Abelson 1960).

The Amberg Granite is undeformed and shows very weak foliation. It is, therefore, probably a post-orogenic body. The youngest major orogeny recognized in the Lake Superior Region is the Penokean orogeny (Goldich et al. 1961). Thus, it appears likely that the Amberg Granite (and possibly the High Falls and Belongia Granites, described by Mancuso 1960) is post-Penokean in age. Radiometric dates on similar pink rapakivi granite from central Wisconsin suggest a post-Penokean age (Goldich et al. 1961).

The Newingham Granodiorite does have a dominant foliation parallel to the regional trend, but structural analysis suggests that a roughly concentric foliation pattern is also present. It seems unlikely that the Newingham mass could be deformed during the Penokean orogeny and still retain an individual foliation pattern. It is therefore postulated that the Newingham mass was intruded after the climax of the Penokean orogeny but while regional stresses were still operating.

The Hoskin Lake Granite shows a fairly strong parallelism of feldspar porphyroblasts. This could reflect either a preexisting preferred orientation or orientation induced during development of the potash feldspars. Structural analysis suggests that stresses, which may have determined the orientation of feldspar porphyroblasts, acted in two or possibly three mutually perpendicular directions—possibly indicating the release of major stresses in the area. Age determinations of microclines from similar porphyritic granites in Michigan suggest they are younger than the main Penokean orogeny (Goldich et al. 1961, pl. 4 and p. 178).

The above three rock units are thus believed to be associated with the Penokean orogeny. The Quinnesec Formation, Dunbar Gneiss, Twelve Foot Falls Quartz Diorite, and metagabbro show the results of more intensive deformation than the above granitic bodies. However, since an orogeny commonly develops over a period of time with cumulative minor deformations, such differences in deformation do not necessarily prove the more deformed rocks were involved in an earlier orogeny (Algoman or Laurentian).

An interpretation of the geologic history of the present area is given in Table IV, although it should be remembered that the indicated age relationships are not completely established (Table I).

Comparison of Table IV with the typical igneous cycle indicates an incomplete record in northeastern Wisconsin, but the presence of two phases of igneous intrusion is suggested. The earlier phase represents a complete igneous cycle except for the absence of younger volcanic rocks, and the second phase has no sedimentary record.

The first phase of igneous activity is possibly related to the Laurentian or Algoman orogenies, since the rocks are probably pre-Animikie in age. The oldest representative of the second phase is considered to be the metagabbro sills, and these are thought to be post-Animikie in age (James 1958). The second phase is thus

TABLE IV

Geologic History of Part of Northeastern Wisconsin

	Geologic History	Name of Present Rock Unit
SECOND CYCLE?	Intrusion of diabase dikes	Diabase dikes
	Minor folding	
	Emplacement of granite (of batholithic dimensions)	Amberg Granite
	Deformation	
	Emplacement of granodiorite pluton	Newingham Granodiorite
	Development of feldspar porphyroblasts in preexisting granitic mass	Hoskin Lake Granite
	Deformation	
	Intrusion of gabbro and diorite sills	Metagabbro
	Penokean deformation (main)	
FIRST CYCLE?	Emplacement of granitic mass	Banded Dunbar Gneiss
	Intrusion of intermediate or basic plutons	Marinette Quartz Diorite Twelve Foot Falls Quartz Diorite
	Folding	
	Geosynclinal lavas, pyroclastics, and sediments	Migmatitic Dunbar Gneiss Quinnesec Formation

probably associated with the Penokean orogeny and is a complete cycle except for the absence of the sedimentary record, which is presumably represented partly by Middle Precambrian strata (Animikie Series) in the Menominee District. The break between the two igneous cycles therefore probably corresponds to the unconformity separating Lower and Middle Precambrian rocks in the Lake Superior Region.

Acknowledgments

I am grateful to the Research Corporation for a grant to carry out this investigation and to A. L. Howland, L. H. Nobles, W. F. Read, and E. H. T. Whitten for discussions of the problems involved.

LITERATURE CITED

ABELSON, P. H. 1960. Annual Report of the Director of the Geophysical Laboratory. Carnegie Inst. Wash. Y. B. 59: 153-154.

CAIN, J. A. 1962. Precambrian Granitic Complex of Northeastern Wisconsin. Ph. D. thesis, Northwestern Univ.

——. 1963. Some Problems of the Precambrian Geology of N. E. Wisconsin: A Review. Ohio Jour. Sci., 63: 7-15.

——. 1963a. Correlations among Mineralogical and Specific Gravity Data from a Granodiorite Pluton. Ohio Jour. Sci., 63: 210-214.

——. 1964. Areal Variability of Specific Gravity Data from a Granodiorite Pluton. Amer. Jour. Sci. (in press).

—— AND W. A. BECKMAN, JR. 1964. Preliminary Report on the Precambrian Geology of the Athelstane Area, Northeastern Wisconsin. Ohio Jour. Sci. 64: 57-60.

DUTTON, C. E. 1960. Precambrian Rocks of the Florence District, Wisconsin: Abstract. Bull. Inst. Lake Superior Geol., Madison, Wis., p. 19.

FROELICH, A. L. 1953. The Geology of a Part of the Wisconsin Granite-Quinnesec Greenstone Complex, Florence Co., Wisconsin. M. Sc. thesis, Ohio State Univ.

FULWEILER, R. E. 1957. The Geology of Part of the Igneous and Metamorphic Complex of Southeastern Florence County, Wisconsin. M. Sc. thesis, Ohio State Univ.

GOLDICH, S. S., A. O. NIER, H. BAADSGAARD, J. H. HOFFMAN, AND H. W. KRUEGER. 1961. The Precambrian Geology and Geochronology of Minnesota. Minn. Geol. Surv., Bull. 41: 193 pp.

JAMES, H. L. 1958. Stratigraphy of Pre-Keweenawan Rocks in Parts of Northern Michigan. U. S. Geol. Surv., Prof. Pap. 314-C: 27-44.

——, L. D. CLARK, C. A. LAMEY, AND F. J. PETTIJOHN in collaboration with J. FREEDMAN, J. TROW, AND K. L. WIER. 1961. Geology of Central Dickinson County, Michigan. *Ibid.* 310, 176 pp.

JONES, O. T. 1938. On the Evolution of a Geosyncline. Geol. Soc. London Quart. Jour. 94: lx-cx.

LYONS, E. J. 1947. Mafic and Porphyritic Rocks from the Niagara, Wisconsin Area. Ph. D. thesis, Univ. Wis.

MANCUSO, J. J. 1960. The Stratigraphy and Structure of the McCaslin District, Wisconsin. Ph. D. thesis, Mich. State Univ.

PRINZ, W. 1958. The Geology of Part of the Menominee District, Michigan and Wisconsin. Ph. D. thesis, Yale Univ.

SCHAPIRO, N. 1952. The Geology of a Part of the Quinnesec Greenstone Complex, Dickinson County, Michigan. M. Sc. thesis, Ohio State Univ.

THOMPSON, G. L. 1955. A Study of Part of the Precambrian Granite-Greenstone Complex in Southeast Florence County, Wisconsin. M. Sc. thesis, Ohio State Univ.

TURNER, F. J., AND J. VERHOOGEN. 1960. Igneous and Metamorphic Petrology. 2d ed. McGraw-Hill Inc. 602 pp.

WADSWORTH, W. B. 1962. Petrogenesis of a Quartz Diorite Pluton near Pembine, Wisconsin. M. S. thesis, Northwestern Univ.

PAPERS OF THE MICHIGAN ACADEMY OF SCIENCE, ARTS, AND LETTERS
VOL. XLIX, 1964 (1963 MEETING)

CTENACANTH SHARKS FROM THE MID-MISSISSIPPIAN OF MICHIGAN

JOHN A. DORR, JR., AND FRANK MOSER
The University of Michigan

REMAINS of pre-Pleistocene fossil vertebrates are sufficiently rare in Michigan that new discoveries, even of fragmentary remains, still are of local interest. This paper records the occurrence of several specimens of fossil shark remains all identifiable as belonging to the Class Chondrichthyes, Subclass Elasmobranchii, and to the extinct Order Cladoselachii and Family Ctenacanthidae, the so-called ctenacanth sharks. Ctenacanths were primitive types of sharks thought to have been close to the ancestry of all later sharks (Romer 1945, p. 64). Several genera and dozens of species have been described, most of them based upon isolated and often fragmentary spines or teeth. Unfortunately, the cartilaginous nature of the shark skeleton led to decay of most of the body during the fossilization process and commonly resulted in the dissociation of teeth and spines, normally the only preservable hard parts. No doubt many species based upon isolated spines are synonymous with other species based upon isolated teeth. In particular, teeth from the front of the mouth (classified in the genus *Cladodus*), teeth from the back of the mouth (classified as *Orodus*), and spines (classified as *Ctenacanthus*) may in many instances be parts of the same animal or at least parts of individuals of the same genus and species. Instances of fossil preservation where the several parts are associated are too few to allow the elimination of very many of the probable synonymies. Ctenacanth spines and cladodont and orodont teeth all occurring in the same sedimentary rock formation may be thought of as representing a primitive type of shark, a "ctenacanth" in common usage, although not necessarily a single species.

The two spines, anterior and posterior, of ctenacanths were placed along the dorsal midline of the body (Pl. I, Fig. 10). They apparently were inserted in the body for about one-fourth to one-third of their length, the exserted part being directed upward and backward at an angle that varied with the species. The anterior spine generally was more slender and the longer of the two. Accord-

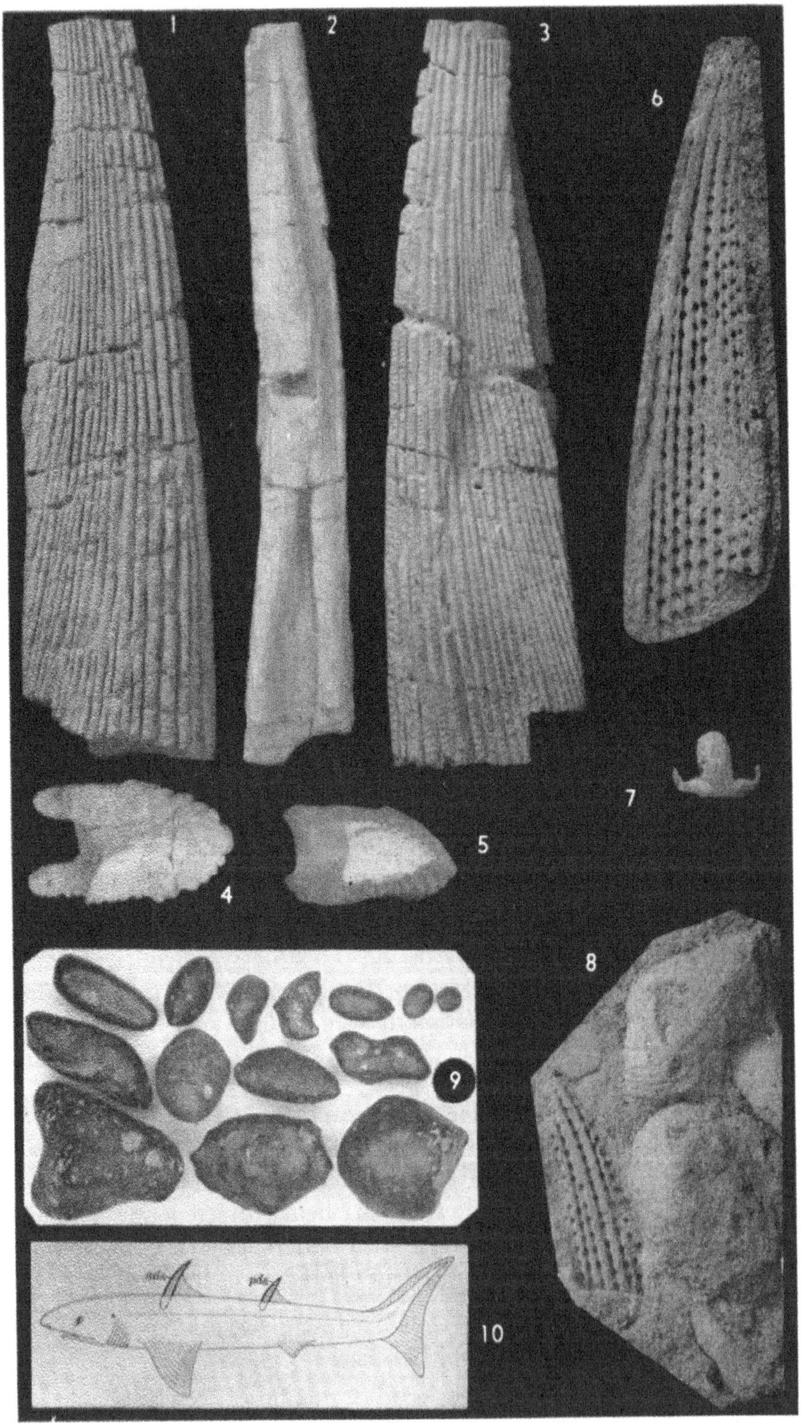
1
2
3
6
7
4
5
8
9
10

EXPLANATION OF PLATE I

FIGS. 1-5 AND 9. UMMP 45738. Partial spine of *Ctenacanthus* near *C. varians* St. John and Worthen. From mid-Mississippian Michigan formation at Grand Rapids, Michigan. All *ca.* × 1. 1—right lateral, 2—posterior, 3—left lateral, 4—transverse outline at natural break near basal insertion line, 5—transverse outline at natural break near distal tip, 9—coprolites found in association with specimen.

FIG. 6. UMMP 23845. *Ctenacanthus* sp. indet., spine impression, right lateral surface, in matrix of mid-Mississippian Marshall sandstone from Marshall, Michigan. *Ca.* × 1.

FIG. 7. UMMP 13613. "Cladodus" tooth from Mississippian age sandstone near Port Austin, Michigan. *Ca.* × 2.

FIG. 8. UMMP 45739. *Ctenacanthus* sp. indet., spine impression, left lateral surface, in association with pelecypods in block of Marshall sandstone from Stony Point Quarry near Hanover, Michigan. *Ca.* × 1.

FIG. 10. Drawing to show location of dorsal fin spines of ctenacanth sharks. Ads.—anterior dorsal spine. Pds.—posterior dorsal spine.

ing to Romer (1945, p. 63), the spines served as supports and "cutwaters" for the leading edges of a pair of dorsal fins. They grew progressively outward by additions near their bases. Their function as cutwaters is suggested by the fact that their surface ornamentation, consisting of many-beaded longitudinal ridges called costae, commonly is worn smooth at the distal (older) end of the spine.

The known stratigraphic range of ctenacanths is from the Middle Devonian through the Early Permian in North America and Middle Devonian through the Late Permian in Europe. They were particularly abundant during the Mississippian and Pennsylvanian periods. The time-stratigraphic ranges of spines or teeth assigned to particular genera or species are too poorly known to make such remains reliable age index fossils. The Mississippian age of the rocks in which they occur in Michigan is established by fossil invertebrates. The ctenacanths clearly were marine fishes.

Ctenacanth shark materials from Michigan, including those catalogued in the collections of the University of Michigan Museum of Paleontology (UMMP), are listed below.

UMMP 45738. *Ctenacanthus,* near *C. varians* St. John and Worthen, 1875, pp. 422-423, and plate 14, fig. 2 (Pl. I, Figs. 1-5, 9).

Eastman (1902, p. 99), in a general review of the taxonomy of ctenacanth spines, attached taxonomic significance, in decreasing order of importance, to spine curvature, cross-section shape, length of inserted portion, number, shape and direction of longitudinal ribs (costae), and the nature of the anterior and posterior faces. On these criteria he reduced *Ctenacanthus speciosus* St. John and Worthen, 1875, to synonymy with *C. varians,* thus including in the latter a suite of characters that come nearest to matching those of the specimen listed above. UMMP 45738 consists of a fragment of a posterior spine including a portion from just above the insertion line almost to the distal tip. Dimensions of the fragment are: maximum preserved length—9.5 cm; maximum anteroposterior diameter at base of fragment—2.6 cm; maximum transverse diameter at fragment base—9 mm. The spine is robust and tapers relatively rapidly toward the distal end, though not quite so rapidly as in the type of *C. varians.* The anterior margin is gently arched and the lateral faces are flattened. The maximum number of costae, at the base of the fragment, is 19 on the right side and 21 on the left. The costae commonly bifurcate toward the base, are broader than the narrow grooves separating them from one another, and decrease in breadth posteriorly, the anterior 4 to 6 being markedly broader than those behind. The anterior costae bear coalesced tubercles which form closely spaced, transverse ridges, whereas the tubercles on the posterior costae are single and more nearly circular. The posterior face of the spine is as in the type of *C. varians*, having " a strong median ridge slightly channeled on either side, and which becomes exceedingly prominent on reaching the middle of the spine " (St. John and Worthen, 1875, p. 423). Costae and tubercles on the distal third of the spine are worn smooth. Prominent denticles once were present along the posterolateral angles of the distal third of the spine but were broken off upon extraction from the matrix.

The type specimen of *C. varians* was collected from " the Kinderhook formation; Flint River, near Burlington, Iowa " (St. John and Worthen, 1875, p. 423). The specimen described here is from the mid-Mississippian Michigan formation. The proximal half of the specimen was collected in 1961 by Kenneth Rollins of the Best Wall Gypsum Company in that company's mine at Grand Rapids and was sent, via Jane Smith at Michigan State University, to A. S. Romer at the Harvard Museum of Comparative Zoology. The distal half was collected from the same locality and horizon in July 1962 by the junior author in company with Mr. Rollins. Professor Romer kindly donated Harvard's portion, which fit perfectly with the portion in our collection. The specimen was found in an eight-inch, sandy shale zone within the main gypsum bed at the Best Wall mine. X-ray analysis of the shale, by Warren Haney, shows quartz and illite as major constituents, with minor chlorite and anhydrite. The gypsum horizon at Grand Rapids can be traced into the subsurface where it changes into anhydrite; in this form it is known to petroleum geologists as the " triple gyp " (due to its characteristic form on electric logs). Toward the center of the Michigan basin there are anhydrites both above and below the " triple gyp," but at the Grand Rapids locality, which was not far from the distal edge of the sea during this part of Mississippian time, the " triple gyp " represents the first evaporite of the Michigan formation. Below the gypsum at Grand Rapids there is about 180 feet of sandstone belonging to the Marshall formation. A subsurface study currently being made by the junior author indicates that most, if not all, of the Marshall formation of southwestern Michigan interfingers and is time equivalent with the lower part of the Michigan formation. These considerations are important in an interpretation of the paleoecology of the Mississippian ctenacanth sharks described in this paper, as well as for interpretation of the environment of deposition of the mid-Mississippian sedimentary rocks of the state.

The specimen described here was partly imbedded in a small pocket of rather pure, medium-sized, well-rounded, and well-sorted quartz sand grains. Associated with the specimen and also dispersed beyond it in the matrix were abundant objects which appear to be coprolites (Pl. I, Fig. 9). These range from wheat grain size to slightly over 2 cm in maximum diameter, have smooth, shiny, striated surfaces, include what appear to be minute bone or shell fragments, and prove upon X-ray analysis to be composed almost entirely of carbonoapatite. If the coprolites were produced by the sharks of the time, they would indicate that the spine itself could have belonged to a creature that actually lived in the waters at that site rather than having been accidentally introduced by floating from another area after death.

UMMP 23845–*Ctenacanthus* species indeterminate (Pl. I, Fig. 6).

This specimen is simply an impression of another spine fragment, 6.9 cm in length. Its preserved features are similar to those of the specimen listed above, but they are insufficient for specific identification. The specimen was obtained by V. B. Monnett and presented to the University of Michigan in August 1946. Monnett acquired it from a quarry man said to have collected it from the mid-Mississippian "Lower Marshall" sandstone member of the Marshall formation from a now abandoned sandstone quarry on the south side of Marshall, Michigan. Although this quarry commonly is called the "type locality" of the Marshall sandstone, the sandstone exposures "approximately 100 feet upstream from the Emmett St. Bridge in Battle Creek" were originally so designated by Winchell (Martin 1956, p. 226). The brown sandstone matrix preserving the impression of this spine consists almost entirely of well-sorted and moderately well-rounded quartz grains with a few muscovite flakes. The stratigraphic relationship of this specimen to the one listed above cannot be determined directly by superposition nor from surface outcrop relations, but current subsurface studies by the junior author suggest that the two may be nearly or actually age equivalent, coming from two different lithofacies of the same time-rock unit, the interfingering Michigan and Marshall formations.

UMMP 45739–*Ctenacanthus*, species indeterminate (Pl. I, Fig. 8).

This impression, about 3.3 cm in length, of a fragmentary spine occurs on fine-grained, micaceous, quartzose sandstone in association in the matrix block with the marine pelecypods, *Palaeoneilo sulcatina* (Conrad) and *Solenomorpha scalpiformis* (Winchell). The specimen was collected May 5, 1962 by an introductory geology student, G. G. Baldrighi, while on a field trip in company with his instructor, David Darby. It was found in a loose sandstone fragment from the Lower Marshall sandstone member of the Marshall formation on the quarry floor at the Stony Point Quarry, NW $\frac{1}{4}$, Sec. 31, T4S, R2W., Jackson County, about $2\frac{1}{2}$ miles southwest of Hanover, Michigan. The exact level from which the specimen came in the quarry could not be determined. Other fossils common at this horizon and locality include nautiloid and ammonoid cephalopods, gastropods, and fossilized, water-worn driftwood fragments including *Calamites*. The specimen is too fragmentary for specific identification but evidently represents a ctenacanth shark from a nearshore marine sandstone.

UMMP 13613–*Cladodus*, species indeterminate (Pl. I, Fig. 7).

This specimen, labeled *Cladodus* sp. in the UMMP collection, is too fragmentary and poorly preserved to indicate more than that it is a cladodont type of shark tooth. As noted earlier, such teeth may have belonged to sharks of the ctenacanth type. The tooth is tricuspid, and the central cusp is highest; AP length is about 5 mm; maximum height is about 4 mm. This specimen was presented to the University in April 1931 by P. E. Fitzgerald, then of the Pure Oil Company. Accompanying notes state that it was collected from the " Lower Marshall " sandstone member of the Marshall formation in the " northern part of Sec. 2, T. 18N., R. 12E., (Hume Twp.) Huron County, Michigan." Although the notes do not specifically so state, this locality probably was one of the roadside sandstone quarries in the so-called Marshall sandstone along M-25 about 2 miles southwest of Port Austin in the " Thumb Area " of Michigan.

Other ctenacanth specimens (from Michigan but not in the University of Michigan collections). Newberry (1889, pp. 177-178 and pl. 27, fig. 10) redescribed and figured a 5-cusped cladodont tooth which he previously had named *Cladodus romingeri* (Newberry 1875, pp. 49-50). In 1875 he first listed this specimen as "found by Dr. C. Rominger, in a calcareous sandstone of Waverly age, at Battle Creek, Michigan." This probably would have been from Winchell's original type locality of the Marshall sandstone at Battle Creek (see above). In 1889 Newberry listed this same specimen as having come from the "Waverly sandstone; Battle Creek, Michigan. Collected by Dr. Charles Rominger, to whom it is dedicated." Both reports may have been in error, however, because he indicated (1889) on plate 27 of his later monograph that the same specimen was from the "Waverly, Grindstone City, Michigan." And on page 121 of the same monograph the specimen is listed as from the "Berea grit, Michigan." Possibly Dr. Rominger sent the specimen to Newberry from Battle Creek, although it might have been found at Grindstone City. The specimen could have come from either place and the question cannot now be resolved.

Another tooth, likewise listed by Newberry (1889, pp. 205-206 and pl. 27, fig. 29) as from the "Waverly sandstone at Grindstone City, Michigan" was figured and identified by Newberry as *Orodus ramosus* Agassiz. This specimen also was presented to Newberry by Rominger. "*Orodus*" teeth are flattened teeth probably from the posterior region of the tooth series of sharks of the same general type that bore "ctenacanth" spines on their backs and cuspidate, "cladodont" teeth in the front of their mouths.

Both of the specimens just listed may now be in the U.S. National Museum collections, although Newberry does not so state.

Conclusions

In the discussion of specimens that may have come from the "Thumb area" of Michigan near Grindstone City or Port Austin, we refer to the specimens as being from the "so-called" Marshall formation. We do so because current subsurface investigations of the junior author now indicate that the "Marshall formation" (or Waverly, or Berea, of Newberry and Rominger) in the Thumb area may be an older (though nevertheless Mississippian) horizon *not* lithologically continuous with or age equivalent to the sandstones of the "true" Marshall formation of southern and southwestern Michigan.

The presence of fossil ctenacanth shark remains in association with marine invertebrates and driftwood in the nearshore marine sandstones of the Marshall formation in southern Michigan, and the presence of ctenacanth shark remains in association with coprolites in a thin, sandy shale parting within what may be the near age equivalent portion of the evaporite sequence in the Michigan formation at Grand Rapids, is significant. We suggest that the sharks and associated organisms lived in the more normal saline waters freshened by entering streams from the land along the margins of the Mississippian Age "Michigan Sea" at the same time that evaporites were accumulating in offshore areas of deposition. The salt-saturated offshore waters ordinarily would have been intolerable for such creatures except during such times, as represented by the sandy shale partings in the Michigan formation at Grand Rapids, when current changes or increased influx of fresh water from the land freshened the sea for a greater than normal distance from shore. The sharks of that time may have skirted around the margins of the sedimentary basin, avoiding the central portions.

LITERATURE CITED

EASTMAN, C. R. 1902. Some Carboniferous Cestraciont and Acanthodian Sharks. Harvard Mus. Compar. Zool., Bull., 29 (3): 53-99, 7 pls.

MARTIN, H. M., AND M. T. STRAIGHT. 1956. An Index of Michigan Geology, 1823-1955. Mich. Dept. Cons., Geol. Surv. Div., Publ. 50: x + 461, 13 figs., 2 unbound charts.

NEWBERRY, J. S. 1875. The Structure and Relations of *Dinichthys* with Descriptions of Some Other New Fossil Fishes. Rep. Geol. Surv. Ohio, 2 (pt. 2): Paleontology: separate pages 1-64, pls. 54-59.

——. 1889. The Paleozoic Fishes of North America. U. S. Geol. Surv., Mon. 16: 340, 53 pls.

ROMER, ALFRED S. 1945. Vertebrate Paleontology, 2nd ed. Univ. Chicago Press. ix + 687 pp.

ST. JOHN, O., AND A. H. WORTHEN. 1875. Geology of Illinois. 6—Geology and Paleontology, Part I, Sec. 1, Descriptions of Fossil Fishes: 245-488, pls. 1-22.

PAPERS OF THE MICHIGAN ACADEMY OF SCIENCE, ARTS, AND LETTERS
VOL. XLIX, 1964 (1963 MEETING)

A CONTRIBUTION TO THE SAW ROCK CANYON LOCAL FAUNA OF KANSAS

CLAUDE W. HIBBARD
The University of Michigan

INTRODUCTION

THE recovery of the lower dentition of a large marmot and remains of a new *Peromyscus,* in association with the Saw Rock Canyon local fauna, during the summer of 1962, warrants the placing on record these new faunal elements.

Since the discovery of this local fauna in June 1943, by Thad McLaughlin and myself, numerous accounts have been published on various members of the fauna. I described (1944 and 1944a) the first vertebrates from this local fauna. At that time, I assigned the age of the fauna to the Middle Pliocene because of the presence of *Osteoborus* and *Dipoides.* A later study of the area and fauna (Hibbard 1949) showed that the deposits (XI member of the Rexroad Formation) contained a fauna intermediate between the Edson Quarry and Rexroad faunas. Because of the close relationship of the fauna to the overlying Rexroad fauna, I stated that it could be assigned either to the upper part of the Middle Pliocene (Hemphillian) or the lower part of the Upper Pliocene.

The fauna has been assigned various ages during the past ten years. Leonard (1952) and Frye and Leonard (1952) assigned the molluscan faunule from the Saw Rock Canyon local fauna to the Pleistocene (Nebraskan age). I reported (1953) on the fauna as known at that time and stressed the fact that it helped to bridge the gap between the Hemphillian (Middle Pliocene) and the lower Blancan (Upper Pliocene) and placed the fauna in the lower Upper Pliocene. Reed (1962) described a mole, *Hesperoscalops sewardensis,* from the Saw Rock Canyon fauna and assigned a late Pliocene or early Pleistocene age to the specimen. She gives no reason for the possible Pleistocene age assignment.

Unfortunately, no horses are known from the fauna, but from the evidence of the microvertebrates one would expect the horses which lived at the time of the Saw Rock Canyon local fauna to be more like those in the Buis Ranch local fauna of late Hemphillian age than like the horses in the Rexroad fauna. *Dipoides*, an " index fossil " for the Hemphillian, occurs in the Rexroad fauna, so it is of no great concern whether the fauna is placed in the late Hemphillian or early Blancan, as long as it is understood that early Blancan implies early Upper Pliocene.

The Saw Rock Canyon Local Fauna

Following is a list of the vertebrates, a systematic account of two new members of the fauna with comments on certain groups, and the bibliography on the fauna.

Phylum Mollusca

The mollusks are not listed. They were recently studied by Dwight W. Taylor (1960, p. 23). He reported 2 species of pelecypods and 23 species of gastropods.

Phylum Vertebrata

- Class Osteichthyes
  - Order Cypriniformes
    - Family Ictaluridae
      - *Ictalurus sawrockensis* Smith
  - Order Cyprinodontiformes
    - Family Cyprinodontidae
      - *Fundulus* sp.[1]
- Class Amphibia
  - Order Caudata
    - Family Ambystomatidae, indet.
  - Order Salientia
    - Frog and toad bones
- Class Reptilia
  - Order Chelonia
    - *Geochelone* sp.
    - *Terrapene* sp.
      - (Pond and river turtles not studied)

[1] I have been requested by C. L. Smith to correct the following identifications: *Aplodinotus grunnieus* C. L. Smith 1954, p. 286; and ?*Aplodinotus* sp., 1962, pp. 510, 515, 516; and Hibbard and Taylor 1960, p. 57, are species of the genus *Fundulus*. Reidentified by Gerald R. Smith.

Order Squamata
Family Iguanidae
Phrynosoma cf. *P. cornutum* Harlan
Family Scincidae
Eumeces cf. *E. striatulus* Taylor
The snakes are being studied by Bayard H. Brattstrom.

Class Aves
The bird remains are being studied by William A. Lunk.

Class Mammalia
Order Insectivora
Family Talpidae
Hesperoscalops sewardensis Reed
Family Soricidae, indet.
Order Rodentia
Family Sciuridae
Marmota sawrockensis sp. nov.
Citellus sp.
Family Geomyidae
Pliogeomys sp.
Family Heteromyidae
Perognathus mclaughlini Hibbard
Prodipodomys sp.
Family Castoroidae
Dipoides wilsoni Hibbard
Family Cricetidae
Onychomys larrabeei Hibbard
Baiomys sawrockensis Hibbard
Peromyscus sawrockensis sp. nov.
Cimarronomys stirtoni Hibbard
Ogmodontomys sawrockensis Hibbard
Order Carnivora
Family Canidae
Osteoborus progressus Hibbard
Family Mustelidae
Buisnictis cf. *B. schoffi* Hibbard
Order Lagomorpha
Leporid sp.

Order Artiodactyla
Family Camelidae
Gigantocamelus cf. *G. spatulus* Cope
Camel, small sp.

SYSTEMATIC DISCUSSION

CLASS MAMMALIA

ORDER RODENTIA

FAMILY SCIURIDAE

Marmota sawrockensis sp. nov.

(FIG. 1)

Marmota? sp. Hibbard, 1953, p. 392.

Holotype.—No. 45775, University of Michigan, Museum of Paleontology, part of the right incisor and P_4-M_3 of a young adult. Collected by Claude W. Hibbard and party in the summer of 1962.

Paratypes.—UMMP 29227, a right P^3 and part of the right upper incisor.

Horizon and type locality.—Lower upper Pliocene, XI member of the Rexroad Formation, Saw Rock Canyon, near the center of the west section line of Sec. 36, T. 34S., R. 31W., XIT Ranch (west part of the old XI Ranch), Seward County, Kansas. Taken from stream-laid sandy silts directly across the canyon from the type locality of *Osteoborus progressus* and up the canyon a short distance from the main quarry of the Saw Rock Canyon local fauna.

Diagnosis.—A marmot the size of *Marmota nevadensis* (Kellogg) from the Thousand Creek local fauna of Nevada. It differs from the latter by the absence of an anteroconid (protoconulid of Bryant 1945) on P_4. A distinct basin occurs on the metalophid of P_4-M_3. The basin occurs between the metalophulid I and metalophulid II of Wood and Wilson (1936). *M. sawrockensis* is larger than *M. oregonensis* Shotwell.

Description of holotype.—The P_4 is broken and lacks the hypoconid, but the tooth was not as large as M_1 (Fig. 1). The metaconid of P_4 is low and not recurved as in *Marmota nevadensis* (Kellogg

1910, p. 424) and Recent species of *Marmota*. The metaconid is slightly higher than the protoconid. The talonid floors of P_4, M_1, M_2 and M_3 are rugose.

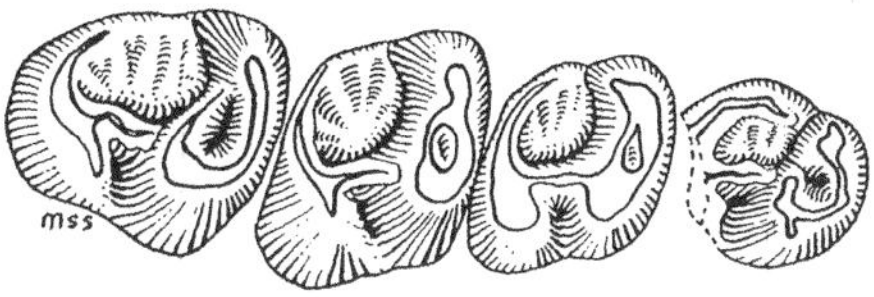

FIG. 1. *Marmota sawrockensis* sp. nov. holotype, UMMP 45775, right P_4-M_3. Occlusal view. × 2.

The anteroposterior length of M_1, M_2 and M_3 is: 6.25 mm; 6.75 mm; and 8.5 mm. The maximum width of M_1, M_2 and M_3 is: 6.4 mm; 8.0 mm; and 7.7 mm.

The presence of a basin on the metalophid of M_2 and M_3 in the ground squirrels has only been observed in specimens of the genus *Cynomys*. *Marmota sawrockensis* appears to represent a specialized sideline of the Pliocene marmots.

The lower incisor is broken. The enamel surface is smooth. One of the fragments has a depth of 6.0 mm and a width of 4.0 mm.

The paratypes were recovered from the main quarry in the summer of 1951, in association with the holotype of *Cimarronomys stirtoni*. The enamel of the fragment of the right upper incisor is rugose and striated. It has an anteroposterior depth of 6.65 mm and a transverse width of 4.3 mm. The right P^3 has an anteroposterior length of 4.9 mm and a transverse width of 4.0 mm.

Remarks.—In the summer of 1950, the carapace and plastron of *Terrapene* sp., No. 44648, were recovered at the locality where the holotype of *Marmota sawrockensis* was taken. Part of a plastron, UMMP 37186, of another individual of this box turtle was taken at the main quarry where matrix was removed for washing to recover microvertebrates.

FAMILY CRICETIDAE

Peromyscus sawrockensis sp. nov.

(FIG. 2)

Holotype.—UMMP 45779, right lower jaw with incisor, M_1-M_3. Taken by Claude W. Hibbard and party in the summer of 1962.

Paratypes.—UMMP 41394, part of right jaw with incisor and M_3; No. 46175, part of left jaw with M_2 and M_3; and No. 46176, part of left jaw, with incisor and M_1. All were taken in the summer of 1953. No. 45777, part of a right jaw with M_2 and M_3, was taken in the summer of 1962.

Horizon and type locality.—Lower upper Pliocene, XI member of the Rexroad Formation, Saw Rock Canyon, near the center of the west section line of Sec. 36, T. 34S., R. 31W., XIT Ranch (west part of the old XI Ranch), Seward County, Kansas, Saw Rock Canyon local fauna.

Diagnosis.—Size smaller than *Peromyscus antiquus* Kellogg, *P. nesodytes* Wilson, and *P. pliocenicus* Wilson. It is larger than *P. kansasensis* Hibbard. It is near the size of *P. dentalis* Hall but is distinguished from the latter by the shorter anteroposterior length of M_2 and the placement of the mental foramen on the dorsal surface of the ramus instead of on the lateral surface as in *P. dentalis.*

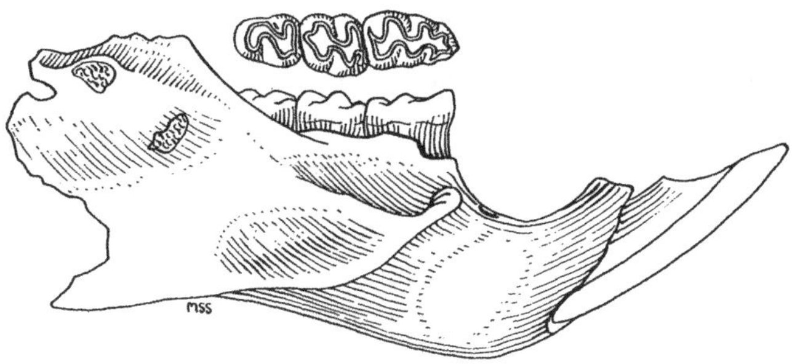

FIG. 2. *Peromyscus sawrockensis* sp. nov. holotype, UMMP 45779, part of right lower jaw, with incisor, M_1-M_3. Labial and occlusal views. $\times$ 5½.

Description of holotype.—The lower jaw lacks the articular condyle (Fig. 2). The jaw is large for the size of the molar teeth. The stage of wear in the molars is that of an adult. The masseteric ridge is well formed and ends opposite the anterior root of M_1. The mental foramen is located on the dorsal surface of the broad diastema just anterior to the root of M_1. The molar teeth lack stylids and lophids. The re-entrant valleys also lack the distinct shelf which closes some of the openings of the re-entrant valleys in *Cimarronomys.*

The base of the incisor ends in a well-developed capsular process. The anteroconid of M_1 appears to have been divided by a shallow groove. M_3 is large for the size of M_1 and M_2 when compared with the M_3 of Recent specimens of *Peromyscus leucopus* (Rafinesque) and *P. californicus* (Gambel).

There is no fossa between the M_3 and the base of the coronoid process as occurs in *Peromyscus dentalis* (Hall 1930). A slight depression does occur posterior to M_3 on the posterior edge of the horizontal ramus. The development of the fossa between M_3 and the coronoid process in *P. dentalis* and the fact that M_2 is nearly as long as M_1 in the species seem to indicate that *P. dentalis* is closely related to the ancestral stock of the subgenus *Megadontomys.*

TABLE I

MEASUREMENTS IN MILLIMETERS OF *Peromyscus sawrockensis* sp. nov.

	UMMP 45779	UMMP 45777	UMMP 46175	UMMP 41394
M_1-M_3, alveolar length	4.50	–	–	–
M_1-M_3, occlusal length	4.20	–	–	–
M_1, length	1.70	–	–	–
M_1, breadth	1.05	–	–	–
M_2, length	1.20	1.30	1.30	–
M_2, breadth	1.15	1.15	1.15	–
M_3, length	1.30	1.30	1.40	1.45
M_3, breadth	1.05	1.05	1.10	1.10
Depth of ramus below M_1 labial side	4.00	–	–	–
Depth of ramus below M_2 labial side	3.50	–	–	3.50

Peromyscus sawrockensis is distinguished from *P. kansasensis* by the more massive jaw, broader incisor, a dorsally placed mental foramen, and the presence of the capsular process for the base of the incisor.

The dental pattern and the characters of the lower jaws of the paratypes agree with those of the holotype. The dentitions are those of adults or old adults. Paratype No. 45777, a fragment of a jaw with M_2 and M_3, possesses teeth with the least wear. The M_1 of paratype No. 46176 has an anteroposterior length of 1.7 mm and the greatest width of 1.0 mm. The tooth appears slightly eroded and the width was probably greater (see table of measurements for those of the holotype and other paratypes).

Ogmodontomys sawrockensis Hibbard

(Fig. 3B)

A nearly perfect right lower jaw, UMMP 45754, an adult of the above vole was recovered with the holotype of *Peromyscus sawrockensis*. The mental foramen is located on the labial side anterior to the root of M_1 just anterior to the masseteric crest (Fig. 3B) as in other jaws of this vole previously recovered. This specimen provides the first information concerning the posterior part of the lower jaw. There is only a slight enlargement on the side of the coronoid process where the base of the incisor ends. This is in contrast to the condition observed in *Ogmodontomys poaphagus* from the Rexroad fauna. In *O. poaphagus* the base of the incisor either ends in a pronounced swelling or in a capsular process separated by a shallow sulcus from the coronoid process as shown in specimen UMMP 28212 (Fig. 3A) from the Fox Canyon locality of the Rexroad fauna. The lower jaw, UMMP 28212, is that of a very old adult, which may account for the development of a larger process than noted in some specimens of *O. poaphagus*. The definite increase in size of the area of the jaw where the base of the incisor ends in the specimens from the Rexroad fauna is one of the evolutionary trends observed in the microtines.

There are a number of separate trends in the evolution of the teeth of microtines. One or more of the following trends may be observed in a given group, through time: (1) the increased length of the lower incisor, which results in the development of a swelling to that of a pronounced capsular process for the reception of the base of the incisor on the coronoid process; (2) development of higher crowned teeth; (3) the reduction of root size and the number of roots of some teeth; (4) delayed development of roots on the teeth until individuals have reached adult or old adult age; (5) the loss of roots and the development of ever-growing cheek teeth; (6) the development of cement in the reentrant angles of the teeth of certain late groups; (7) the development of dentine tracts along the sides of the teeth which produce an interrupted enamel pattern in some with rooted teeth and in some ever-growing teeth; and (8) an increase in the number of alternating triangles of M_1 and M^3.

Much is still to be learned of the early history of these rodents. The known fossil record indicates that *Prosomys mimus* Shotwell

(1956) may be the ancestral stock that gave rise to *Nebraskomys mcgrewi* Hibbard. More complete material is needed of both forms before their true relationship is known.

Ogmodontomys sawrockensis or a closely related species is considered as the ancestral stock of *O. poaphagus*. *O. poaphagus* occurs

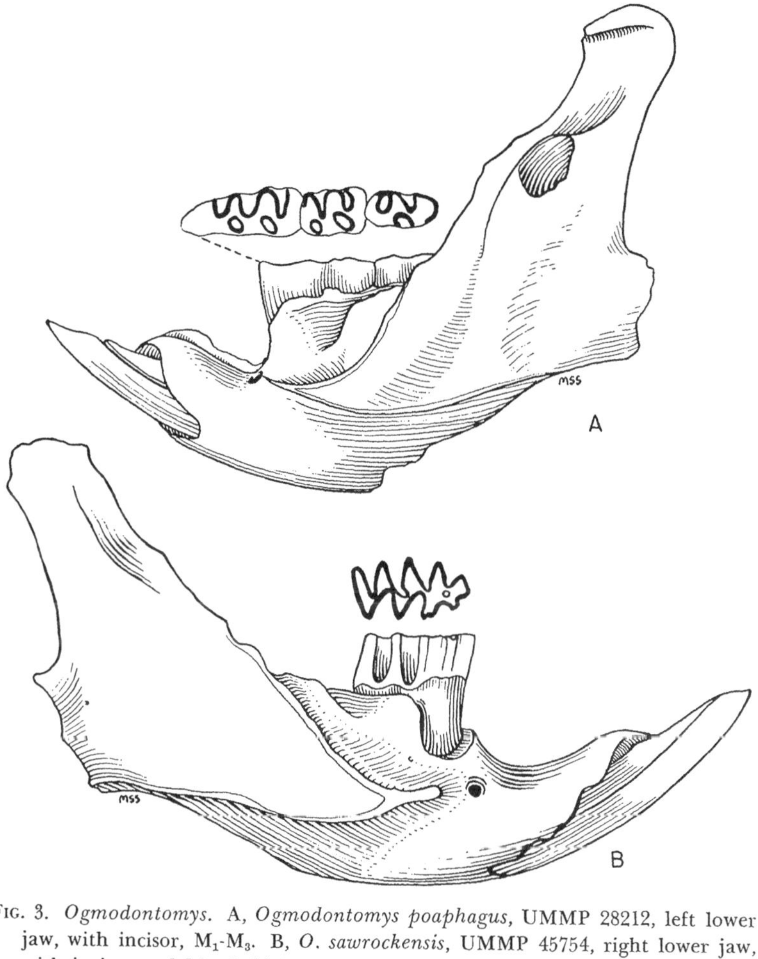

FIG. 3. *Ogmodontomys*. A, *Ogmodontomys poaphagus*, UMMP 28212, left lower jaw, with incisor, M_1-M_3. B, *O. sawrockensis*, UMMP 45754, right lower jaw, with incisor and M_1. Labial and occlusal views. All × 5.

early enough to have given rise to *Cosomys primus* Wilson of the Hagerman fauna. *C. primus* from the Hagerman fauna shows a definite root reduction in M^3 and has a better developed capsular process for the base of the incisor than occurs in *O. poaphagus*. After a study of specimens of *Mimomys pliocaenicus* Forsyth Major in Europe, I consider *Cosomys* a distinct genus. *M. pliocaenicus* possesses more hypsodont teeth, which develop roots later in the life of the individual than in *Cosomys*. Furthermore, the reentrant angles of the teeth of *M. pliocaenicus* are filled with cement, which is lacking in the genus *Cosomys*.

Ogmodontomys appeared early enough in the Pliocene to have been ancestral to the genus *Mimomys*, though I know of no reason why *Cosomys* could not be the ancestral stock of *M. pliocaenicus*.

Relationship of the Saw Rock Canyon local fauna to the Buis Ranch and Rexroad local faunas.—Unfortunately, most of the vertebrates of the Buis Ranch local fauna (Hazard 1961; Hibbard 1954; and Tihen 1955) are mainly upland forms when compared chiefly to stream, marshland, and lowland valley inhabitants known from the Saw Rock Canyon fauna. The presence of *Buisnictis* cf. *B. schoffi* and *Pliogeomys* cf. *P. buisi* in the Saw Rock Canyon fauna, as well as the presence of *Perognathus* cf. *mclaughlini* in the Buis Ranch local fauna, indicates that these two faunas are nearly the same age. How much time elapsed between the time the Saw Rock Canyon fauna inhabited the region and the time it was inhabited by the later Rexroad fauna is not known, but it was long enough for the gophers to develop ever-growing cheek teeth. There is no evidence of a climatic change between the time the Saw Rock Canyon fauna inhabited the region and the time the Rexroad fauna lived. Both faunas are considered as having occupied the region in a uniform subtropical (frost-free) climate, with more effective moisture than at present (Hibbard 1960).

Many members of the Saw Rock Canyon fauna carried on into the Rexroad fauna unchanged, such as 22 species of mollusks. It was for this reason that Frye and Leonard, 1952, considered the two faunas to be of the same age.

Some of the vertebrates in the Saw Rock Canyon fauna appear to be the ancestral stocks from which members of the Rexroad fauna were derived, such as:

Saw Rock Canyon Local Fauna		Rexroad Local Fauna
Ictalurus sawrockensis Smith	→	*I. benderensis* Smith
Phrynosoma cf. *P. cornutum* Harlan	→	*P. cornutum*

Eumeces cf. *E. striatulus* Taylor	→	*E. striatulus*
Hesperoscalops sewardensis Reed	→	*H. rexroadi* Hibbard
Pliogeomys	→	*Geomys*
Perognathus mclaughlini Hibbard	→	*P. rexroadensis* Hibbard
Onychomys larrabeei Hibbard	→	*O. gidleyi* Hibbard
Ogmodontomys sawrockensis Hibbard	→	*O. poaphagus* Hibbard
Osteoborus progressus Hibbard	→	*Borophagus*
Buisnictis cf. *B. schoffi* Hibbard	→	*B. breviramus* Hibbard
Gigantocamelus cf. *spatulus* Cope	→	*G. spatulus*

More intensive collecting of the Buis Ranch and Saw Rock Canyon faunas should furnish better evidence for some of the ancestral stocks that gave rise to the numerous carnivores and cricetine rodents of the Rexroad fauna.

Acknowledgments.—I am indebted to the following members of the 1962 University of Michigan field party who helped collect the fossils: James B. Stevens and Margaret Skeels Stevens of the University of Michigan; G. Nelson Greene and Jerry G. Smith of Alma College; John R. Bolt, Michigan State University; and Richard Zakrzewski, Wayne State University. We are grateful to David C. Coleman, chief of the Game Division of the Kansas State Forestry, Fish and Game Commission; and Harry Smith, superintendent of Meade County State Park, who gave us permission to live in the park and to wash the fossil-bearing matrix there. I am also indebted to David Adams, XIT Ranch, who allowed collecting and geologic work on the ranch.

This study was aided by permission to examine the specimens under the care of William A. Clemens, Museum of Natural History, University of Kansas.

Financial support to me for the fieldwork in Meade County during the summer of 1962 and for the line drawings by Margaret Skeels Stevens (MSS) was provided by the National Science Foundation (Project G-19458).

LITERATURE CITED

BRYANT, MONROE D. 1945. Phylogeny of Nearctic Sciuridae. Am. Midland Nat. 33 (2) : 257-390, 8 pls., 48 figs.

* ETHERIDGE, RICHARD. 1960. The Pliocene Lizard Genus *Eumecoides* Taylor. Bull. So. Calif. Acad. Sci. 59 (Pt. 2) : 62-69.

* FRANZEN, DOROTHEA S., AND A. B. LEONARD. 1947. Fossil and Living Pupillidae (Gastropoda = Pulmonata) in Kansas. Univ. Kans. Sci. Bull. 31 (Pt. II, No. 15) : 311-411, 6 pls., 15 figs.

* FRYE, JOHN C., AND A. B. LEONARD. 1952. Pleistocene Geology of Kansas. Kans. Geol. Surv. Bull. 99: 1-230, 19 pls., 17 figs.

HALL, E. R. 1930. Rodents and Lagomorphs from the Later Tertiary of Fish Lake Valley, Nevada. Univ. Calif. Publ. Bull. Dept. Geol. Sci. 19 (12) : 295-312, 1 pl., 29 figs.

HAZARD, EVAN B. 1961. The Subgeneric Status and Distribution in Time of *Citellus rexroadensis*. Jour. Mammal. 42 (4) : 477-483, 2 figs.

* HERRINGTON, H. B., AND D. W. TAYLOR. 1958. Pliocene and Pleistocene Sphaeriidae (Pelecypoda) from the Central United States. Univ. Mich. Mus. Zool., Occ. Pap. No. 596: 1-28, 1 pl.

* HIBBARD, CLAUDE W. 1944. A New Land Tortoise, *Testudo riggsi*, from the Middle Pliocene of Seward County, Kansas. Univ. Kans. Sci. Bull. 30 (Pt. 1, No. 7) : 71-76, 2 figs.

* —— 1944a. Two New Mammals from the Middle Pliocene of Seward County, Kansas. Univ. Kans. Sci. Bull. 30 (Pt. 1, No. 10) : 107-114, 2 pls.

* —— 1949. Pliocene Saw Rock Canyon Fauna in Kansas. Univ. Mich. Mus. Paleon., Contrib. 7 (5) : 91-105, 2 figs.

* —— 1953. The Saw Rock Canyon Fauna and Its Stratigraphic Significance. Pap. Mich. Acad. 38 (1952) : 387-411, illus.

* —— 1954. A New Pliocene Vertebrate Fauna from Oklahoma. *Ibid.* 39 (1953) : 339-359, 5 figs.

* —— 1957. Two New Cenozoic Microtine Rodents. Jour. Mammal. 38 (1) : 39-44, 2 figs.

—— AND D. W. TAYLOR. 1960. Two Late Pleistocene Faunas from Southwestern Kansas. Univ. Mich. Mus. Paleon., Contrib. 16 (1) : 1-223, 16 pls., 18 figs.

KELLOGG, LOUISE. 1910. Rodent Fauna of the Late Tertiary Beds at Virgin Valley and Thousand Creek, Nevada. Univ. Calif. Publ. Bull. Geol. Sci. 5 (29) : 421-437, 20 figs.

* LEONARD, A. BYRON. 1952. New Gastropods from the Blanco Formation (Nebraskan age) Pleistocene in Kansas. Nautilus 66 (2) : 37-45, 1 pl.

* PACKARD, ROBERT L. 1960. Speciation and Evolution of the Pygmy Mice, Genus *Baiomys*. Univ. Kans. Publ. Mus. Nat. Hist. 9 (23) : 579-670, 4 pls., 12 figs.

* REED, KATHERINE M. 1962. Two New Species of Fossil Talpid Insectivores. Breviora, Mus. Comp. Zool. 168: 1-6, 1 pl.

* SHOTWELL, J. ARNOLD. 1955. Review of the Pliocene Beaver *Dipoides*. Jour. Paleon. 29 (1) : 129-144, 6 figs.

—— 1956. Hemphillian Mammalian Assemblage from Northeastern Oregon. Bull. Geol. Soc. Am. 67: 717-738, 7 figs.

SMITH, C. LAVETT. 1954. Pleistocene Fishes of the Berends Fauna of Beaver County, Oklahoma. Copeia 1954, No. 4: 282-289, illus.

* —— 1962. Some Pliocene Fishes from Kansas, Oklahoma, and Nebraska. Copeia 1962, No. 3: 505-520, 9 figs.

* TAYLOR, DWIGHT W. 1960. Late Cenozoic Molluscan Faunas from the High Plains. U. S. Geol. Surv. Prof. Pap. 337: 1-94, 4 pls., 2 figs.

* TIHEN, JOE A. 1955. A New Pliocene Species of *Ambystoma*, with Remarks on Other Fossil Ambystomids. Univ. Mich. Mus. Paleon., Contrib. 12 (11) : 229-244, 1 pl., 1 fig.

* TWENTE, JOHN W., JR. 1952. Pliocene Lizards from Kansas. Copeia 1952, No. 2: 70-73, 4 figs.

WOOD, A. E., AND R. W. WILSON. 1936. A Suggested Nomenclature for the Cusps of the Cheek Teeth of Rodents. Jour. Paleon. 10 (5) : 388-391, 1 pl.

* Bibliography on the Saw Rock Canyon local fauna.

PAPERS OF THE MICHIGAN ACADEMY OF SCIENCE, ARTS, AND LETTERS
VOL. XLIX, 1964 (1963 MEETING)

GLYPTODON FREDERICENSIS (MEADE) FROM THE SEYMOUR FORMATION OF KNOX COUNTY, TEXAS

WILLIAM G. MELTON, JR.
The University of Michigan

THE occurrence of *Glyptodon fredericensis* (Meade) is reported for the first time from the Seymour Formation of Pleistocene age in Knox County, Texas. Many previously unknown parts of the carapace and skeleton of this glyptodon are described. Evidence is presented for the assignment of these remains to the genus *Glyptodon* Owen.

Walter W. Dalquest of the Midwestern University, Wichita Falls, Texas, in 1956 gave many of the specimens here reported to the University of Michigan Museum of Paleontology. Additional fieldwork was conducted briefly by Claude W. Hibbard and me in Knox County in the summer of 1957, and we did more intensive work in the spring of 1958 and 1962. We were able to recover additional fossil vertebrates. All of the material here described comes from the Seymour Formation. The late Kansan age of this fauna has already been discussed by Hibbard (1960), Hibbard and Dalquest (1960 and 1962), and Stricklin (1961).

The remains of glyptodons are second only to those of the horses in the numbers of individuals represented in the Gilliland faunule. Parts of 117 specimens of glyptodons are cataloged in the collection. No complete individual was found, nor was a complete carapace recovered. The largest parts of a carapace contain 384 scutes. Some of the remains may be widely scattered parts of the same individual. However, parts of 80 carapaces have been found over an area of more than 130 square miles, and it is more probable that these carapace fragments represent nearly as large a number of individuals.

The presence of so many of these large and probably warmth-loving mammals adds further evidence for the conclusions drawn by Hibbard (1960) concerning the climatic conditions during late Kansan time for this part of Texas. The glyptodons show the most southern distribution of the different groups of edentates found

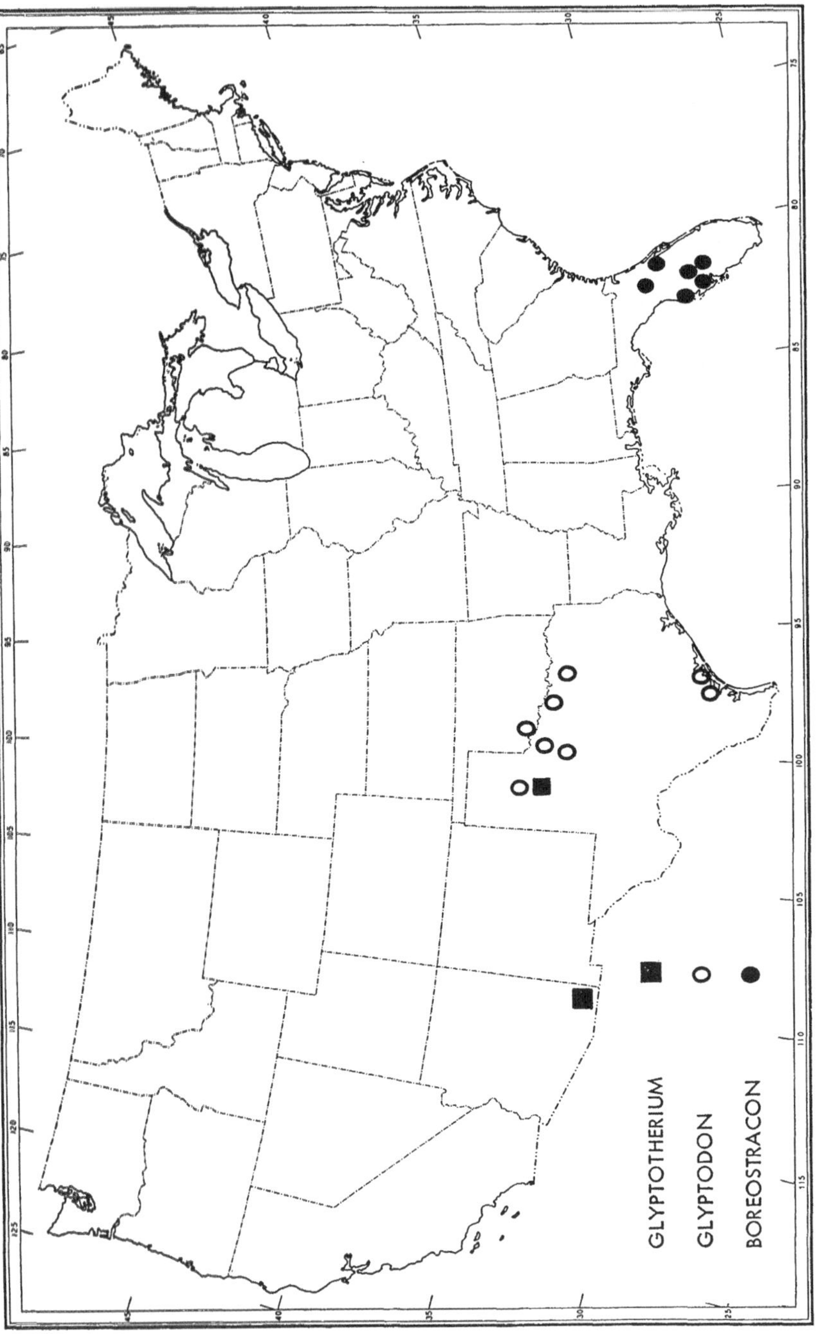

Fig. 1. Distribution of glyptodons in the United States.

in North America. Holmes and Simpson in 1931 published a map on the then known distribution of recent and fossil edentates in the U. S. Fig. 1 of this paper shows the localities where remains have been recovered of the three genera of glyptodons known from the U. S. References citing new localities shown in Fig. 1 are included in the literature cited (Brodkorb 1957; Hay 1916, 1926, and 1927).

CLASS MAMMALIA

ORDER EDENTATA

FAMILY GLYPTODONTIDAE

Genus *Glyptodon* Owen, 1838

Glyptodon fredericensis (Meade)

(Pls. I-III; Figs. 1-3)

Glyptodon petaliferous Cope, Hay and Cook, 1930, p. 10.

Xenoglyptodon fredericensis Meade, 1953, p. 455.

DESCRIPTION

Skull.—The skull of the glyptodon is characterized by the tremendous descending zygomatic processes which end well below the crowns of some of the anterior lower teeth (Pl. I, Fig. 3). The elements of the skull are fused posteriorly and there is a distinct occipital crest below the head shield posteriorly. The external occipital proturbance sits well forward for the attachment of heavy neck muscles. There are numerous foramina in the occipital-parietal area of the skull beneath the head shield. The frontals appear to lap or fold down over the maxilla in the center of the orbital area. The anterior opening of the infraorbital foramen is a wide vertical slit, and to the rear it is divided by a septa into a dorsal foramen and a ventral foramen. The front part of the frontals and the nasals are absent on both parts of the skulls recovered. Specimen No. 34826 UMMP is the most complete skull of the North American glyptodons so far reported. However, the nasals must have been very wide and short. A complex of turbinate bones are exposed

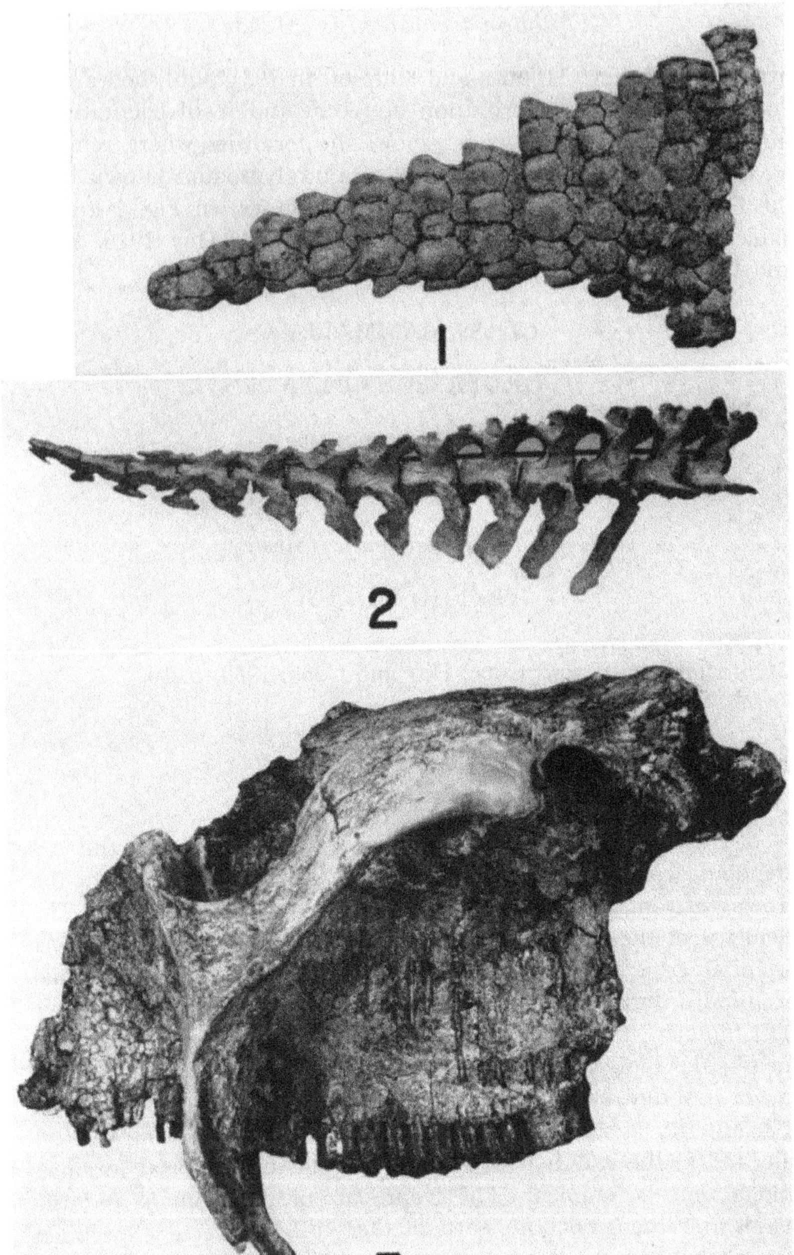

EXPLANATION OF PLATE I

FIG. 1. *Glyptodon fredericensis* tail armor (UMMP 34826) about $\frac{1}{12}$.
FIG. 2. *Glyptodon fredericensis* tail vertebrae (UMMP 34826) about $\frac{1}{12}$.
FIG. 3. *Glyptodon fredericensis* skull (UMMP 34826) about $\frac{1}{3}$.

between the maxillaries, which are fused, and no midline suture is apparent. The premaxillaries are either fused or absent as in other genera of glyptodons.

The upper teeth are described from specimen UMMP 34826. The occlusal length of the eight teeth in the upper tooth row is 203.3 mm on the right side and 200 mm on the left (see Table I for measurements of teeth). The teeth are here designated N^1, N^2-N^8, as suggested by Hoffstetter (1956). N^1 (the first upper tooth) is, more or less, a long irregular oval with an osteodentine pattern of a reversed sigma and a faint transverse line across the

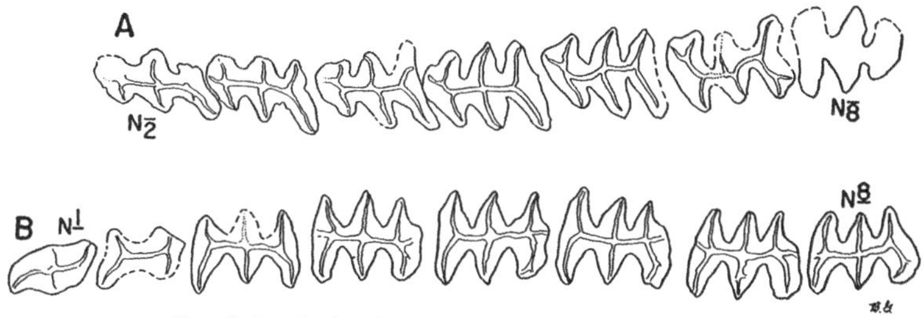

FIG. 2-A. Occlusal pattern of lower teeth. × ½.
B. Occlusal pattern of upper teeth. × ½.

center (Fig. 1B). N^2 contains a distinct front and rear lobe, but no center lobe. N^3 is the first trilobed tooth, and the remainder are trilobed through N^8. The osteodentine pattern consists of a central line nearly parallel to the long axis of the tooth with a branch extending at nearly right angles to the tips of each lobe. The central line extends from the anterior to the posterior margin of the tooth, or terminates at the point where the anterior or posterior branch enters the end lobes (Figs. 1A, B). This pattern, as well as the shape of the teeth, is quite variable, as shown by the teeth in specimens UMMP 34826 and UMMP 38761.

Lower jaw.—The most complete lower jaw, UMMP 38761, consists of the left half, which was broken just posterior to the first tooth (Fig. 3). It contains seven teeth, N_2-N_8 (Fig. 2A), see Table I for measurements. The right side lacks the anterior part and was broken between the second and third lobe of N_4. It contains the posterior part of N_4-N_8. The following measurements were taken from the left side of the jaw. The maximum height of the jaw is 238.0 mm (Fig. 3). The length of jaw from N_2 to the posterior

TABLE I

MEASUREMENTS (IN MM) OF THE OCCLUSAL SURFACE OF UPPER TEETH, UMMP 34826, AND LOWER TEETH, UMMP 38761, OF *Glyptodon fredericensis*

Tooth	Length	Width	Tooth	Length	Width
N^1	18.4	9.6	N_1	–	–
N^2	17.0	11.0	N_2	23.0	14.0
N^3	27.6	16.0 *	N_3	23.0	16.0
N^4	24.9	18.0	N_4	26.0	19.0
N^5	26.0	20.0	N_5	26.0	18.0
N^6	26.2	22.0	N_6	22.0	19.0
N^7	25.8	19.0	N_7	22.0	17.0
N^8	24.1	17.0	N_8 †	19.0	19.0

* Minimum width; medial lobe broken on crown of labial side.
† Crown missing; measurements approximate.

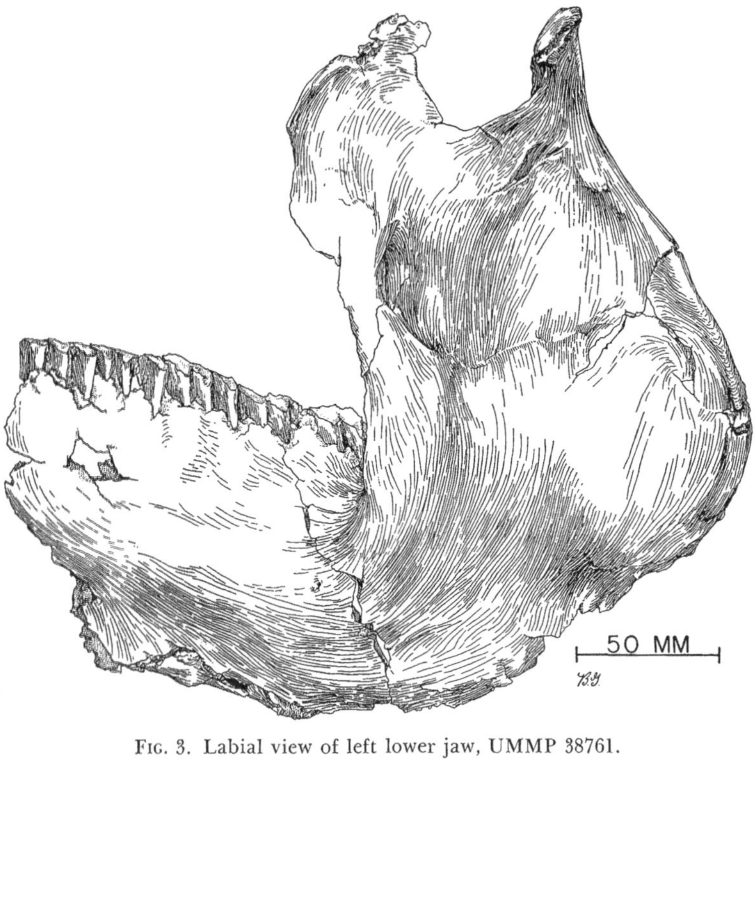

FIG. 3. Labial view of left lower jaw, UMMP 38761.

edge of the ascending ramus is 263.0 mm. The depth of the jaw below the fourth tooth (N_4) is 104.0 mm. There are parts of four other jaws without teeth in the collection, representing three other individuals. Tooth fragments of three other individuals are in the collection.

The lower teeth, like the uppers, are evergrowing. The front of the lower teeth are convex except N_8, which is concave. The posterior face in all the teeth but N_8 is concave. N_8 is the reverse, concave to the front and lobate posteriorly. The osteodentine pattern extends to the apexes of the lobes of the teeth. The anterior lobe of N_2-N_7 has a triradiate pattern which connects with the + pattern of the median and the posterior lobes of N_2-N_7. The crown of the N_8 is broken and missing, so it is not known if the osteodentine pattern of N_8 is reversed. The angle of the occlusal surface of the teeth with the posterior edge of the ascending ramus is 55°. The angle measured between the occlusal surface and the anterior edge of the ascending ramus is 53°. The anterior edge of the ascending ramus extends further forward in specimen UMMP 38761, than in two other rami of specimen UMMP 46237, and in the type of *Xenoglyptodon fredericensis* Meade. Specimen UMMP 38761 is a slightly larger and probably an older individual than the others.

Vertebra.—The atlas is free and is very short. The posterior dorsal parts of the alar processes extend upward rather than outward. The dorsal arch is convex, with a rough dorsal tuberosity. The alar foramina have distinct grooves running mesially to the dorsal tuberosity. The ventral arch is slightly convex. The intervertebral foramina pass through the alar processes just above the well-defined articular surfaces for the axis. The ventral tubercle is broadly concave, with a well-defined articular surface for the odontoid process. One complete specimen, UMMP 34826, and two fragmentary atlases are in the collection.

The axis, the 3rd, 4th, 5th, and 6th cervical are fused; and the neural canal forms a continuous tube. They are shorter in length than in width. There is a ventral and anterior articular surface on the odontoid process. No clearly defined separation occurs between the vertebrae other than the nerve openings. The neural spines are fused into a single ridge of bone. On either side of the neural tube at the rear are two articular facets for the large and complicated 7th cervical. The posterior mesial facets of the 6th vertebrae are oval, and the lateral facets are greatly elongated horizontally.

On the anterior ends of the 7th cervical, short oval articular processes extend forward mesially, as do lateral articular processes, which are considerably longer. The 7th cervical is fused with the first thoracic vertebra. They are so broken that it is not possible to determine whether the 2d thoracic was also fused with them. Fused to the 1st thoracic vertebra is the large first rib, which expands ventrally and mesially into a broad plate. The head of the rib appears to be fused in part to the 7th cervical vertebra.

No remains of other thoracic and lumbar vertebrae, pectoral girdle, humerus, ulna, and pelvis were found.

Manus.—A nearly complete left forefoot (manus) consisting of digits II, III, IV, and V and a left hind foot (pes) consisting of digits I to V are found with specimen UMMP 38761. The principal bones recovered for the forefoot are essentially the same as those described by Burmeister (1870, p. 338). The number of sesmoids is uncertain. There are seven carpals; the four bones of the proximal row are present with articular facets for the missing pisiforme. The second digit is missing. Metacarpals III, IV, and V and nine phalanges are present; however, the articular facets are present for metacarpal II. The distal end of the radius was associated with the above foot.

Femur.—Parts of three femori were found, one being a part of an articulated leg lacking only the patella. The femur is longer than any previously described from North American glyptodons.

The length of the complete femur, UMMP 46231, is 488.0 mm, measured parallel to the long axis. Other measurements are given in Table II. The femur has a large trochanter major and a tremendous lateral epicondyle (Sisson 1938). The medial epicondyle curves upward.

Tibia-fibula.—The tibia-fibula is solidly ankylosed, as in other glyptodons. Two complete bones and one broken about halfway down the tibial side are in the collection. The shortest tibia-fibula, UMMP 33524, is 253.0 mm, and the longest is 283.0 mm. UMMP 46232, which is 262 mm long, would be longer if all of the distal surface was present. The tibia-fibula are longer and average about 12% to 30% longer than the one described from *Glyptotherium arizonae* Gidley (1926). A summary of the measurements is given for comparison in Table III.

Pes.—The bones of the hind foot of the glyptodons are quite distinct from the other major groups of edentates.

Two completely articulated left rear feet were collected by

TABLE II

MEASUREMENTS (IN MM) OF *Glyptodon* AND *Glyptotherium* FEMURS

	Glyptodon fredericensis No. 46231	*Glyptotherium arizonae* *
Length of internal side............	488	456
Width at greater and internal trochanters	253	–
Width of lower condyles..........	135	130
Narrowest width of shaft..........	89	100

* Gidley, 1926 (No. 10536, U. S. National Museum).

TABLE III

MEASUREMENTS (IN MM) OF THE TIBIA-FIBULA OF *Glyptodon* AND *Glyptotherium*

	Glyptodon fredericensis No. 46231	No. 33524	*Glyptotherium arizonae* *
Total length	253	283	220
Proximal width	157	163	–
Width of tibia	147	132	–
Width of fibula	102	112	–
Anterior length of the tibia-fibula opening	123	160	–
Width of center opening...	64	64	–
Distal width	143	134	130
Width of astragulus articular facet	63	63	–

* Gidley, 1926 (No. 10536, U. S. National Museum).

Walter Dalquest, one a part of the articulated leg, UMMP 46231, and the other UMMP 38761. The tip of the calcanium is missing on both specimens. Several other isolated foot and toe bones also are present in the collection.

There are 39 bones in each of the complete left feet: 7 tarsals,

5 metatarsals, 15 phalanges, and 12 sesmoids. Burmeister lists 14 phalanges and 10 sesmoids for *Glyptodon asper*. He found no 2d phalanx and no distal sesmoids on digit I. In both of the feet recovered, digit I had 2 phalanges and 2 distal sesmoids. The tarsals are greatly compressed and flattened to bear the weight of the heavy bony carapace. The metatarsals and the proximal toe bones are flattened. The distal toe bones are short and wedge-shaped. The ungals are blunt and broad (Pl. II, 1 and 2).

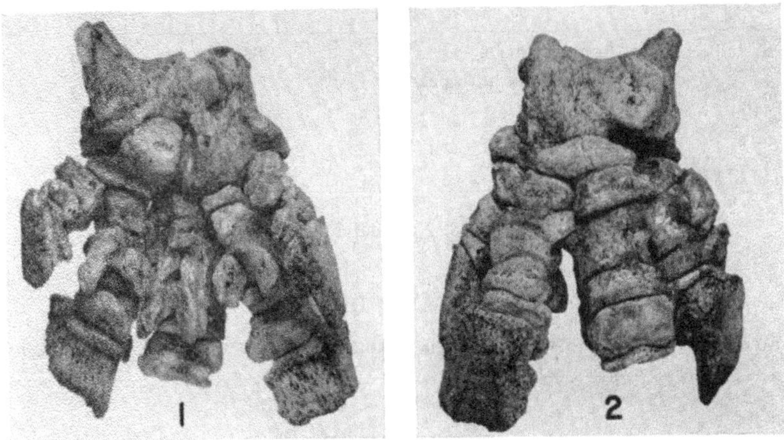

EXPLANATION OF PLATE II

FIG. 1. *Glyptodon fredericensis* rear foot (UMMP 38761) plantar (ventral) view. × ⅓.

FIG. 2. *Glyptodon fredericensis* rear foot (UMMP 38761) dorsal view. × ⅓.

Carapace and scutes.—The central carapace scutes are more or less hexagonal, with sturdy interlocking borders. The scutes along the midline of the carapace are fused. They come apart readily a few scutes from the midline, except in the sacral region. There is no evidence that the vertebra were fused to the carapace, except in the sacral region.

In general, the central figure of a scute is round to hexagonal, being larger along the margins and growing relatively smaller toward the center of the carapace. The central figure is deeply depressed. The depression is often below the radial grooves that delimit it. In some cases the central figure is no larger than the peripheral figures. In outline the scutes are generally hexagonal, but they frequently depart from this pattern, some even approaching a four-sided condition, or others a highly irregular shape. Occasionally corners of four adjoining scutes are contiguous. No two scutes present the same appearance.

On many of the body scutes, the peripheral figures are a part of two scutes, and the peripheral number may vary from six to as many as 15 in the sacral region of the carapace. Near the margin of the carapace where the scutes are most consistently alike, there are seven to nine peripheral figures which are entirely contained on the scute. Nine is the most consistent number of peripheral figures.

The sutures, while highly irregular, are generally interlocked firmly. There are no depressions or radiating grooves along the margins of the scutes as in *Boreostracon* (Holmes and Simpson, 1931, pp. 582-583). Castellanos (1953, pp. 397-398) states that the sculpture of the scutes of the genus *Boreostracon* differs from the sculpture of the scutes of *Glyptodon* especially in the dorsal regions of the carapace. In *Boreostracon*, the central figures and the slightly smaller peripheral figures are not very rough, although they are very punctate.

Near the anterior margin of the carapace of *Glyptodon fredericensis* there are generally only a few large pores on each scute. About the 4th row of scutes posterior to the anterior margin there are four to seven pores in the anterior part of the groove around the central figure. Toward the center of the carapace, they become fewer, or more obscure.

The surface texture of most of the scutes are extremely rough, particularly near the front margin. The first row of nuchal scutes and those near the anterior margin have no recognizable central figure or peripheral figurines, although they generally contain a deep central depression. The surface of the marginal scutes in the nuchal area are slightly convex to deeply depressed, varying with the individual. These scutes are relatively small with four suture surfaces, with a suture surface laterally and two posteriorly. Quite frequently a very small irregularly shaped scute fills in a gap between two normal-sized scutes.

The anterior lateral border scutes are elliptical. Their ventral edge is straight along the anterior lateral margins and become more pointed until they reach a nearly triangular shape with the apex directed rearward (Pl. III, 2). The size and shape varies somewhat with the individual. The elliptical scutes are generally depressed in the center. The scutes along the side are arranged in fairly

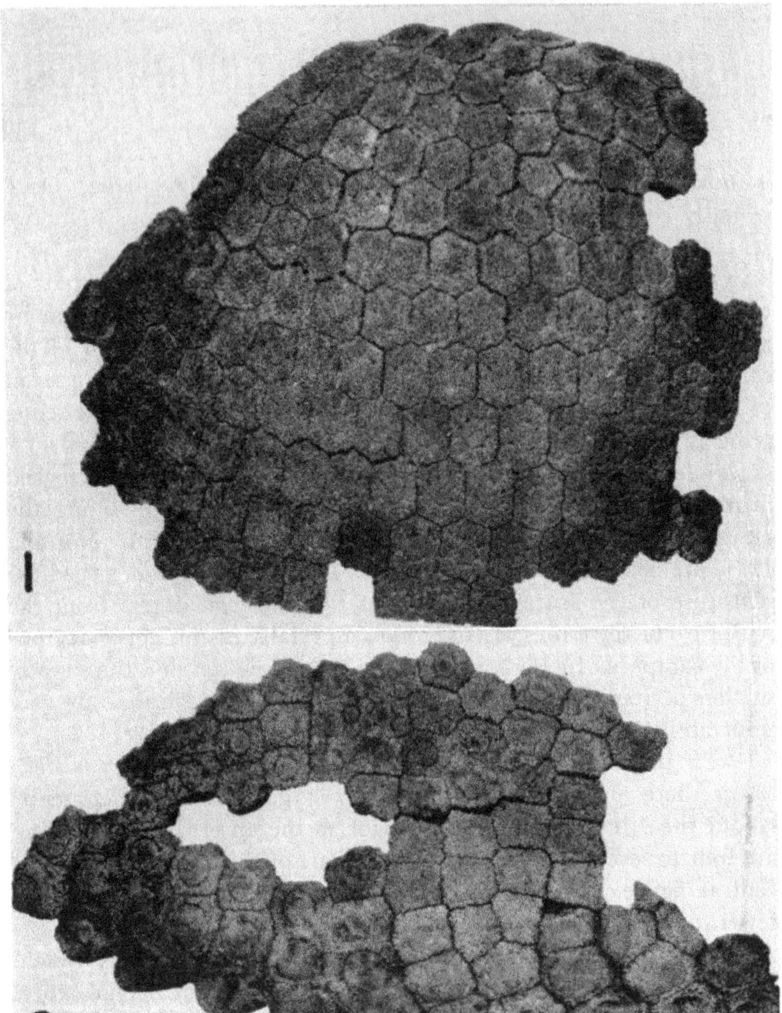

EXPLANATION OF PLATE III

FIG. 1. Front of carapace. Nuchal border at top. (UMMP 46241).

FIG. 2. Lateral border of carapace. Caudal opening towards left. (UMMP 46230).

FIG. 3. Posterior border of carapace. Caudal opening towards left. (UMMP 38761). All × 1/6.

straight vertical rows starting from the marginal scutes. The vertical rows divide where necessary to provide the curvature of the carapace.

The posterior marginal scutes are very large and thick. They have a large circular or oval central figure of up to 63 mm in diameter, although the average size is around 50 mm. The greatest width of the scute is 86 mm, and the greatest thickness is 53 mm. The central figure may be convex but never approaches a peaked or pointed outline. Other scutes are depressed deeply.

The exact shape of the carapace can only be inferred from the larger fragments. The front opening is lower than the rear. The rear of the carapace slopes downward and turns upward in the caudal region.

The maximum width of carapace UMMP 46320 exceeds 2140 mm, measured across the curvature of the carapace just anterior to the sacral region.

Scutes of five young individuals are present in the collection. These scutes are small, their greatest width being less than 35 mm, they contain no central figure and no peripheral figures. They possess a very rugose surface and a deep central depression on the exterior of the scute. One specimen, UMMP 46297, consisting of five scutes, has a distinct groove on the ventral side of the scutes which extends around the interior margin. This groove may have been for the attachment of the scute to the dermis. These scutes are closely sutured on the dorsal surface, but on the ventral surface they are separated by a distance of two to four mm. The dorsal surface has numerous large pores around the central depression. Not enough material is present to tell which part of the carapace these scutes are from.

The scutes that form the head shield are highly irregular in size and shape, varying from sesmoid-shaped drops, long cylinders, to a variety of flat or flattened irregularly shaped scutes. The anterior left half of the head shield was removed intact from skull UMMP 34826. The scutes are not as rough or rugose on the surface as the body scutes but are more punctate. The scutes are closely sutured but are thinner and do not interlock very well.

Tail.—The tail is composed of 13 caudal vertebrae and 12 rings of bony scutes. The last two vertebrae are slightly fused at the center of the joining faces of the centra. On caudal vertebrae seven and nine, a large foramen opens on the right side of the neural arch near the posterior end of these vertebrae for a branch from the spinal cord. The metapophyses extend upward and laterally above the short neural spine on the 2d to the 11th caudal vertebrae to support the caudal rings. The first three caudal vertebrae have long transverse processes, and the dorsal surfaces of these possess articular facets which apparently articulated with the posterior part of the pelvis. The transverse processes of the 4th through the 11th caudal vertebrae have a sharply descending projection that articulates with the encircling ring of bony plates (Pl. I, Fig. 2). The longest downward process is on the 4th caudal vertebrae and they

TABLE IV

MEASUREMENTS (IN MM) OF CAUDAL VERTEBRAE AND CHEVRON BONES OF UMMP 34826

	VERTEBRA				
	Length of	Greatest Antero-		Transverse-width	
No.	Centra	Posterior Width	Length	Top	Bottom
1	73	–	–	–	–
2	68	49	162	66	36
3	73	44	137	64	60
4	84	54	116	53	51
5	88	59	91	49	43
6	91	64	81	40	52
7	89	55	67	38	60
8	87	56	48	32	52
9	83	57	32	23	48
10	78	55	28	25	44
11	79	32 *	19	26	29
12	74	42	18	20	–
13	84	–	9	fused	13

* Approximate.

gradually diminish to the 11th. The transverse process is poorly developed on the 12th and 13th caudal vertebrae. The chevron bones are fused on the 12th and 13th vertebrae and vary considerably in size and shape from anterior to posterior (Table IV and Pl. I, Fig. 2).

All of the caudal vertebrae but the first two have a complete circle of bony armor (Pl. I, Fig. 1). No armor was recovered for the first vertebra. Dorsal to the second vertebra is a double row of scutes that descended to about the midline of the tail. The armor of the 2d and 3d caudal vertebrae must have been under the margin of the carapace, since they are rather smooth and less punctate than the last ten bony rings. All of the rings are composed of two rows of scutes. The proximal row of scutes are smoother, smaller, and flatter than those of the distal row. The dorsal and lateral scutes of the distal row in the bony rings three to ten are more or less tubercular. The tubercular projections on the two dorsal scutes, as well as the scutes along the lateral margins of the tail of the distal row of each ring, are better developed than on the other scutes. In the bony rings three to ten, the two ventral scutes are flat and smooth, and they join so that there is a noticeable V notch toward the rear where the chevron bone on the next anterior vertebrae articulated. The scutes of the last two rings are more irregular in shape and are all convex. The end of the caudal tube is bluntly rounded.

The total length of the caudal vertebrae is 42 inches or 1102 mm. The length of the caudal vertebrae in *Glyptotherium texanum* is 620 mm, and in *G. arizonae* 895 mm.

Discussion

Castellanos (1953) revised the subfamily Glyptodontinae and the genus *Glyptodon* Owen. He considers that the sculpture of the external face of the scutes of the carapace distinguishes the species of glyptodons. Therefore, he is able to establish some limits of generic characters which agree with other skeletal characteristics. Castellanos (p. 391) lists the following characters of the scutes as distinctive for the genus *Glyptodon*: (1) the central figure of each scute is pentagonal to round or elliptical; (2) the outer surface is generally very rough or rugose but may have some smooth-surfaced scutes; (3) the outline is flat, slightly convex, or deeply depressed; (4) the surrounding peripheral figures are small, generally smaller than the central figure, but rarely equal; (5) the number of peripheral figures on each plate varies from 5 to 9, with 9 being exceptional; (6) the outline is polygonal or angular. Species with 9 peripheral figures never have all the typical features of the genus *Glyptodon*.

Castellanos distinguishes the genus *Glyptodon* from *Glyptotherium* Osborn because *Glyptotherium* has a very large central figure on the scutes, which always represents more than half of each scute. The central figure is surrounded by 7 to 12 peripheral figures, which are separated by wide deep furrows. The tail of *Glypto-*

therium is different from any of the species of the genus *Glyptodon*, because it has seven movable rings and a short caudal tip. The tail of *Glyptodon* consists of 10 to 13 caudal vertebrae and 8 or more caudal rings and the caudal tip.

Simpson (1929, pp. 582-583) noted that specimens assigned to *Glyptodon petaliferus* by Hay from Hunt and San Patricio counties, Texas, differs from *Boreostracon floridanus* because the sculpture of the plates from the homologous regions of the carapace is different in that the marginal areas of the plates assigned to *G. petaliferus* are strongly punctate. In *Boreostracon*, these areas of the carapace are marked by prominent, irregular, radial grooves. Three distinct species of the genus *Boreostracon* have since been recognized in South America.

Simpson (1929) also stated, "Copes specimen (*Glyptodon petaliferus*) is of historical interest as being the earliest indication of a glyptodon north of Mexico, but his name cannot now be used for specimens from other localities . . . I therefore have rather reluctantly taken the step of basing a new name on the incomparably better Seminole materials and of considering both *G. petaliferus* and *G. rivipacis* as *nomina nuda* [*nomina vanem*]."

From a study of the glyptodon material recovered from the Seymour formation, it is evident that these specimens belong to the genus *Glyptodon* Owen.

Meade (1953) described *Xenoglyptodon fredericensis*, a member of the Frederick local fauna recovered from the Holloman gravel pit a short distance north of the town of Frederick in Tillman County, Oklahoma. The holotype of *X. fredericensis* is in the University of Texas collections and has been compared with the material from Knox County, Texas. These specimens have been found to be conspecific. Therefore the name *Glyptodon fredericensis* is used for the specimens from Knox County, Texas, and Tillman County, Oklahoma.

The deposits at the Holloman gravel pit and the vertebrates from there have been studied. The deposits are considered to be of the same age as the sands and gravels of the Seymour formation (Late Kansan).

Acknowledgments

I wish to express my appreciation to Claude W. Hibbard, curator of vertebrates at the University of Michigan Museum of Paleontology, for permission to publish the description of these specimens and for the many helpful suggestions in preparing this manuscript and to Walter W. Dalquest of Midwestern University, who collected nearly all of the specimens described.

I also wish to thank John A. Wilson of the Department of Geology, University of Texas, for permission to examine the type of *Glyptodon fredericensis* and other glyptodon specimens in his care at Austin, Texas.

Permission to examine the formation from which the specimens were collected was kindly granted by the ranchers in the area, and I wish especially to thank O. L. Patterson and Bruce Burnett.

Financial support for the fieldwork was accorded by the Museum of Paleontology of the University of Michigan and by the National Science Foundation, Project G-19458.

The photographs for the plates were taken by Herbert Weinert and Karoly Kutasi. The figures were drawn by Barbara Gagnon (Bilge). The translations of Spanish, French, and Portuguese were done by Veronica Melton.

LITERATURE CITED

Brodkorb, Pierce. 1957. New Passerine Birds from the Pleistocene of Reddick, Florida. Jour. Paleon. 31 (1): 129-138.

Burmeister, German. 1870-1874. Monographica de los Glyptodontes en el Museo Publica de Buenos Aires, II (6): 1-395.

Castellanos, Alfredo. 1953. Anotacoes e Retificacoes ao Genero "Glyptodon" Owen e A Familia "Glyptodontinae". An. Acad. Brasil. Cien. 25 (4): 391-410.

Gidley, James W. 1926. Fossil Proboscidea and Edentata of the San Pedro Valley, Arizona. U.S. Geol. Surv., Prof. Pap. 140B: 83-95, 13 pls., 4 figs.

Hay, O. P. 1916. Description of Two Extinct Mammals of the Order Xenarthra from the Pleistocene of Texas. Proc. U.S. Nat. Mus. 51: 107-123, 5 pls.

—— 1926. A Collection of Pleistocene Vertebrates from Southwestern Texas. *Ibid.* 68 (24): 1-18, 8 pls.

—— 1927. The Pleistocene of the Western Region of North America and Its Vertebrated Animals. Carnegie Inst. Wash. Pub. 322B: 346 pp., 12 pls., 19 figs.

—— and Harold J. Cook. 1930. Fossil Vertebrates Collected Near, or in Association with, Human Artifacts at Localities near Colorado, Texas; Frederick, Oklahoma; and Folsom, New Mexico. Proc. Colo. Mus. Nat. Hist. 9 (2): 4-40, 14 pls.

Hibbard, Claude W. 1960. An Interpretation of Pliocene and Pleistocene Climates in North America. Mich. Acad. 62nd Ann. Rep. 5-30.

—— AND WALTER W. DALQUEST. 1960. A New Antilocaprid from the Pleistocene of Knox County, Texas. Jour. Mammal. 41 (1) : 20-23, 1 fig.

—— 1962. Artiodactyls from the Seymour Formation of Knox County, Texas. Pap. Mich. Acad. 47: 83-99, 4 figs.

HOFFSTETTER, ROBERT. 1956. Edentata: Traite de Paleontologie 6 (2) : 535-636, 64 figs.

HOLMES, W. W., AND G. G. SIMPSON. 1931. Pleistocene Exploration and Fossil Edentates in Florida. Bull. Am. Mus. Nat. Hist. 59 (7) : 383-418, 21 figs.

MEADE, GRAYSON E. 1953. An Early Pleistocene Vertebrate Fauna from Frederick, Oklahoma. Jour. Geol. 61 (5) : 452-460, 1 pl.

OSBORN, H. F. 1903. *Glyptotherium texanum*, a New Glyptodont from the Lower Pleistocene of Texas. Bull. Am. Mus. Nat. Hist. 19 (17) : 491-494, 1 pl.

OWEN, R. 1838. In Parish's "Buenos Ayres and La Plata", p. 178b (*Fide* Lydekker, 1887, p. 114). Cat. Fossil Mammalia Brit. Mus., pt. 5: 1-345.

SIMPSON, G. G. 1929. Pleistocene Mammalian Fauna of the Seminole Field, Pinellas County, Florida. Bull. Am. Mus. Nat. Hist. 56 (8) : 561-599, 22 figs.

SISSON, S., AND J. D. GROSSMAN. 1938. The Anatomy of the Domestic Animals. W. B. Saunders Co. 972 pp., 770 illus.

STRICKLIN, JR., FRED L. 1961. Degradational Stream Deposits of the Brazos River, Central Texas. Bull. Geol. Soc. Am. 72 (1) : 19-35, 9 figs.

PAPERS OF THE MICHIGAN ACADEMY OF SCIENCE, ARTS, AND LETTERS
VOL. XLIX, 1964 (1963 MEETING)

ISOPOSITIONAL SUBSTITUTION IN CRYSTALLINE SUBSTANCES

R. M. MILLER AND H. B. STONEHOUSE
Michigan State University

INTRODUCTION

THE classical investigation of the structure and properties of crystals by V. M. Goldschmidt (1926) directed attention to the principles governing distribution of trace elements among the magmatic minerals. Considerable interest was developed in the phenomenon in which atoms or ions of particular elements occupy the structural positions of atoms or ions of other elements in a mineral, and Strunz (1937) introduced the term "diadochie" in discussing this structural relationship.

Making the assumptions that ionic bonding prevails and the ion making the larger contribution to the energy of the crystal structure is preferred, Goldschmidt (1937) proposed a set of rules governing diadochie substitution of ions in magmatic minerals:

a) For two ions to substitute diadochically one for another in a crystal structure, the ionic radii must not differ by more than 15%.

b) When two ions possessing the same charge, but different radii, compete for a position in the crystal structure, the ion with the smaller radius is preferred.

c) Ions having similar radii but different charges (of the same sign) may substitute diadochically in a crystal, in which case the ion having the larger charge has preference over the ion with the smaller charge.

The discovery that ionic size of trace elements govern their substitutional behavior to a greater extent than their position in the periodic table was accepted as a fundamental relationship. However, in recent years, workers in the field have expressed the need for revision of the rules governing trace element distribution among common rocks and minerals. Fyfe (1951) and Ramberg (1952) called attention to necessary modifications. Shaw (1953) questioned the validity of the rules and listed exceptions to them. Ahrens (1952) pointed to the chemical bond as the answer to much of the questionable geochemical phenomena, with ionization potential as the clue to be investigated. Ringwood (1955) proposed to modify the Goldschmidt rules dealing with replacement of trace elements in magmatic minerals to allow for differing electronegativities of elements which display diadochy. This investigation is directed toward the replacement process in synthetic crystals. When the crystallization process is standardized, the manner in which the replacement ions enter the structure may be dependent upon their radii, ionization potential, electronegativity, electron configuration, or a combination of these. The number of impurity ions replacing the metallic cations can be determined by spectrographic analysis. The effect of isopositional substitution of foreign ions on the lattice parameters of the aluminum phosphate crystal may be followed by X-ray diffraction techniques. Measurement of these two variables, when correlated with data pertaining to the particular ion being substituted, bears directly on the Goldschmidt rules and the revisions being proposed by current investigators.

Procedure

The production of silica tetrahedral structures requires pressures and temperatures not ordinarily available in mineralogy laboratories. Goldschmidt (1932) pointed to the possibility of YPO_4, $YAsO_4$, and $CaCrO_4$ forming isomorphous structures with quartz; and in the search for a substitute for crystalline quartz, in short supply during World War II, Gruner (1946) investigated synthetic aluminum phosphate.

A similarity in crystalline structure between aluminum phosphate and α-quartz was reported by Brill and deBretteville (1948) as part of a Signal Corps project. On the basis of similarity (Table I) it seems logical that, in an investigation of Goldschmidt's rules for diadochic replacement of ions, synthetic aluminum phosphate might conveniently be used to provide a replica of the silicon tetrahedral structure so common in the upper lithosphere, and trace element substitution made for the aluminum cation. To limit the number of variables the ions selected for substitution need to be trivalent and within the limitation of 15% of the size of aluminum.

The Signal Corps investigation of aluminum phosphate (Stanley 1954) proved good quality crystals could be grown from a solution of 1.45 moles aluminum phosphate per liter of 6.5 M phosphoric acid when held in a glass-lined stainless steel autoclave for 48 hours at 167° C. Such a method, with slight modification, was used successfully to produce the crystals for this investigation.

When separated from the mother liquor and washed and dried, the crystals appear as in Fig. 1. They are uniformly small, clear, well-formed, and up to 1 mm in length. The scalenohedron form is noticed and fully developed rhombohedral faces are the rule.

TABLE I

A COMPARISON OF PHYSICAL PROPERTIES ASSIGNED α-QUARTZ AND ALUMINUM PHOSPHATE

Item	$SiSiO_4$	$AlPO_4$
lattice	hex-R	hex-R
crystal class	32	32
a	4.90 Å	4.93 Å
c	5.39 Å	2 × 5.47 Å
c/a	1.10	2 × 1.11
n_e	1.553	1.524
n_w	1.544	1.524
optic sign	plus	plus
transmitted light	colorless	transparent to translucent
bi-ref.	0.009	0.006
cleavage	not distinct	none
fracture	conchoidal	conchoidal
hardness	7	6½
sp. g.	2.65	2.56
modifications	alpha & beta	alpha & beta
tem. of inv.	573° C	585° C

Comparison of the weaker lines in the back reflection region of an X-ray diffraction pattern with those of the original work (Huttenlocher 1935) confirm the crystals as $AlPO_4$. Addition of impurity elements to the reactants in amounts of 2.5, 5, 7.5, 10, 15, 20, and 25%, by mol weight, was used as a first approximation toward determining the quantity of impurity needed to produce a detectable structural change. Ga (metal), $FePO_4 \cdot 2H_2O$, $CrPO_4 \cdot 3H_2O$, Co_2O_3, Co (metal) and V (metal) were selected to supply ions for substitution in the crystalline framework.

Emission spectroscopy provided information as to percent metal impurity in a sample of the order of 0.01%. Spectral analyses were made of a sample of uncontaminated $AlPO_4$ and on five series of

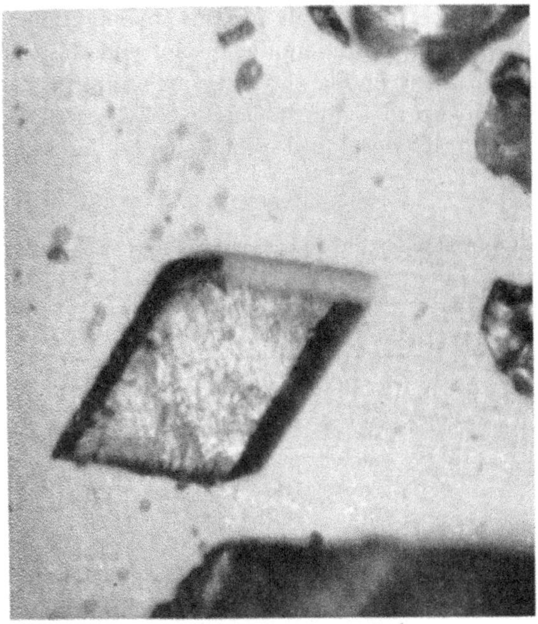

FIG. 1. A crystal of $AlPO_4$ approximately 1 mm in length.

TABLE II

THE RESULTS OF SPECTROSCOPIC ANALYSIS OF $AlPO_4$ CRYSTALS IN WHICH SPECIFIC IMPURITIES WERE INCORPORATED

Sample	Percent Ga	Percent Fe	Percent Cr	Percent Co	Percent V
uncontaminated	0.0035	0.0125	0.0006	0.001	0.001
A	2.0	0.15	0.10	0.01	0.01
B	3.0	0.16	0.11	0.02	0.03
C	4.0	0.23	0.13	0.03	0.035
D	5.6	0.25	0.17	0.04	0.04
E	10.6	0.32	0.18	0.05	0.05
F	18.4	0.57	0.19	0.33 *	–

* Considered in error due to inclusions of metallic cobalt mixed with sample.

mixed crystals, each of which contained a specific impurity ion (Table II). An X-ray diffractometer survey of the 2θ region between 130° and 150° produced six prominent, well-defined peaks on the strip-recorder chart (Fig. 2).

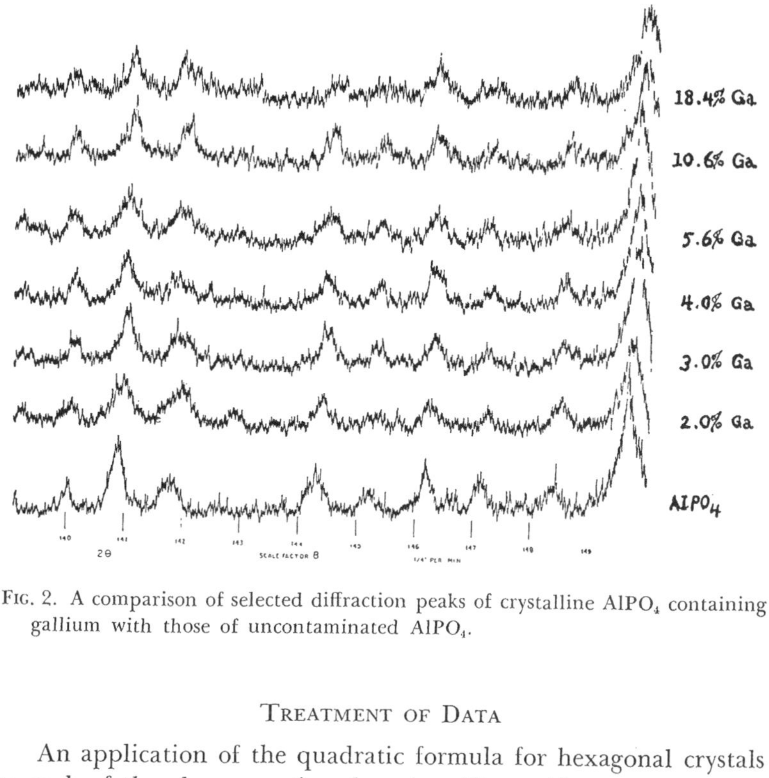

FIG. 2. A comparison of selected diffraction peaks of crystalline $AlPO_4$ containing gallium with those of uncontaminated $AlPO_4$.

TREATMENT OF DATA

An application of the quadratic formula for hexagonal crystals to each of the above-mentioned peaks will provide an equation in two unknown lattice parameters *a* and *c*.

$$1/d^2 = (4/3a^2)\,(h^2 + hk + k^2) + l^2/c^2$$

By solving these equations in combination, the best values for *a* and *c* may be determined (15 possible combinations). To check the assumption that all planes are of equal value in such an operation, a graphic examination of the situation was made (Fig. 3). Such an examination is facilitated if the following substitutions are made: A for $1/a^2$, C for $1/c^2$, and D for $1/d^2$. The equation becomes:

$$A = -\,3/4\,(h^2 + hk + k^2)\,l^2C + 3/4\,(h^2 + hk + k^2)\,D$$

It is evident that the Miller indices control the slope, while any experimental error in interplanar-spacing affects the intercept. Thus, each equation may be represented by a line broadened to a zone by the experimental error (Fig. 3-a). The intersection of two such zones provides an area in which solutions for a and c exist. It is apparent that a set of two equations with similar slopes (planes 421-331) produces a zone of intersection entirely different from a set with dissimilar slopes (planes 225-331). Lines running parallel have no common intersection (indeterminate solution) while the lines meeting at right angles have the most clearly defined intersection.

The angle at which lines meet (Fig. 3-b) determines the area and shape of the parallelogram of intersection containing a and c. Since this angle changes with the slopes, the variation created in a and c follows some trigonometric relationship. Consequently every pair of quadratic equations containing measured 2θ values need not be of equal importance when solving for a and c. The ability of the diffractometer to record to the nearest 0.01° and of the experimenter to interpret the strip-record with comparable accuracy control the variation in the intercept on A (Fig. 3-a). In each instance the Miller indices assigned to the lattice plane determine the angle at which the line representing the plane meets the C axis.

In order to determine the significance factor to be assigned to each such pair, the range of the resulting slope and intercept variation is derived for the general case. The possible values for A and C exist within the parallelogram of intersection, and the ranges are denoted by ΔA and ΔC. In Fig. 3-b the slope of line 1 is represented by the tangent of α_1; this angle also occurs in triangle NHL as angle NHL.

From the definition of tangent function:

$$\frac{\mathrm{NM} + \mathrm{ML}}{\mathrm{HL}} = \tan \alpha_1$$

by substitution,

$$\frac{\Delta \mathrm{A} + \mathrm{x}}{\Delta \mathrm{C}} = \tan \alpha_1.$$

In triangle MGL the angle MGL is equal to α_2. By definition:

$$\frac{\mathrm{ML}}{\mathrm{GH} + \mathrm{HL}} = \tan \alpha_2$$

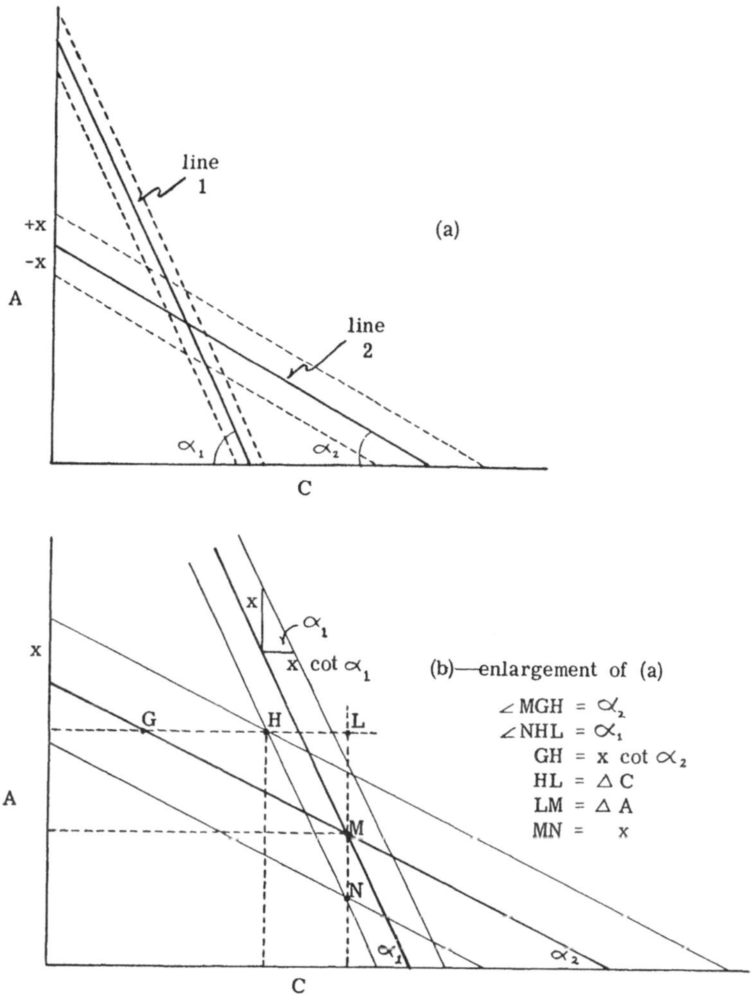

FIG. 3. A graphic solution of the quadratic formula as applied to two lattice planes of a hexagonal crystal and solved simultaneously for *a* and *c*.

by substitution,

$$\frac{\Delta A}{\Delta C + x \cot \alpha_2} = \tan \alpha_2.$$

In order to solve for ΔA and ΔC, the expression for α_1 and α_2 are combined to produce:

$$\Delta A = \frac{x \tan \alpha_2 \ (\cot \alpha_1 + \cot \alpha_2)}{1 - \tan \alpha_2 \cot \alpha_2}, \ \Delta C = \frac{-x \ (\tan \alpha_2 \cot \alpha_2 + 1)}{\tan \alpha_2 - \tan \alpha_1}.$$

Since α_1 and α_2 are controlled by hkl values and x is assumed to be a constant, ΔA and ΔC may be evaluated for any specific set of lines. Each lattice plane represented by a line is assigned a value for α_1. Holding α_1 and x fixed, ΔA and ΔC are determined by using any other line producing an angle α_2 with the axis and intersecting this reference line. Repeating the procedure, in turn, for each of the six reflection peaks provides data used in preparing the graphs required to weight the relative importance of each intersection.

TABLE III

A TABLE OF WEIGHTING COEFFICIENTS ASSIGNED TO VALUES OF *a* AND *c*

Intersecting hkl Planes	Weighting Factor *a*	Weighting Factor *c*
225-404	× 1	× 2.8
413-404	× 1.7	× 1.5
413-225	× 2.5	× 2.8
413-331	× 3.6	× 1
413-421	× 3.6	× 1
404-331	× 3.8	× 2.4
404-421	× 3.8	× 2.4
225-331	× 4.2	× 4.1
225-421	× 4.2	× 4.1

The 420 plane fails to fit the graphic scheme for testing the solution and is omitted from further consideration. The intersection of lines representing 421-331 planes is considered indeterminate and this pair is rejected. Each of the remaining intersections is significant, and a coefficient assigned to the importance of its solution. Note that an intersection of two given lines need not provide solutions of equal significance for both *a* and *c* (Table III).

The X-ray measurements were reproducible and any variations in the results appear to be due to instrumental and propagation

TABLE IV

EXPERIMENTAL 2θ VALUES FOR SELECTED hkl PLANES OF CRYSTALLINE $AlPO_4$ CONTAINING INCREASING AMOUNTS OF THE IMPURITIES Ga, Fe, Cr, Co, AND V

	A	B	C	D	E	F	
Ga	149.75°	149.79°	149.75°	149.72°	149.84°	149.82°	hkl
Fe	149.67	149.65	149.69	149.63	149.69	149.64	421
Cr	149.66	149.63	149.64	149.63	149.64	149.59	
Co	149.73	149.65	149.65	149.67	149.62	149.68	
V	149.67	149.71	149.69	149.72	149.67	–	
Ga	146.24	146.29	146.19	146.29	146.23	146.24	hkl
Fe	146.23	146.22	146.22	146.20	146.19	146.21	420
Cr	146.16	146.14	–	146.14	146.13	146.18	
Co	144.26	146.18	146.18	146.20	146.17	146.17	
V	146.21	146.31	146.28	146.21	146.16	–	
Ga	144.39	144.37	144.37	144.42	144.46	144.51	hkl
Fe	144.31	–	144.30	144.30	–	144.25	331
Cr	144.33	144.26	144.27	144.26	144.26	144.31	
Co	144.37	144.28	144.30	144.37	144.34	144.32	
V	144.28	144.38	144.34	144.34	144.40	–	
Ga	140.95	141.03	140.96	141.01	141.06	141.03	hkl
Fe	140.94	140.85	140.85	140.90	140.99	140.87	225
Cr	140.88	140.87	140.83	140.87	140.91	140.80	
Co	140.89	140.90	140.88	140.89	140.90	140.91	
V	140.92	140.98	140.90	140.98	140.92	–	
Ga	135.72	135.75	135.76	135.77	135.75	135.80	hkl
Fe	135.68	135.64	135.67	135.64	135.67	135.65	413
Cr	135.62	135.68	135.67	135.68	135.67	135.67	
Co	135.73	135.67	135.63	135.63	135.63	135.67	
V	135.64	135.68	135.70	135.72	135.67	–	
Ga	133.15	133.18	133.20	133.15	133.12	133.12	hkl
Fe	133.15	133.11	133.13	133.15	133.18	133.10	404
Cr	133.11	133.11	133.13	133.11	133.11	133.09	
Co	133.19	133.17	133.10	133.14	133.11	133.17	
V	133.13	133.17	133.17	133.20	133.14	–	

errors. The diffraction angles (2θ) are presumed to be correct to 0.01° (Parrish 1956). The hkl planes, 413-225, were selected to provide numbers tending to create a maximum propagation error

TABLE V

THE BEST EXPERIMENTAL *a* AND *c*/2 VALUES FOR $AlPO_4$ CRYSTALS CONTAINING GALLIUM, IRON, CHROMIUM, COBALT, AND VANADIUM

	Percent Impurity	*a* (Å)	*c*/2 (Å)
Gallium	0.0	4.9292	5.4633
	2.0	4.9285	5.4629
	3.0	4.9282	5.4615
	4.0	4.9278	5.4633
	5.6	4.9278	5.4615
	10.6	4.9271	5.4615
	18.4	4.9268	5.4629
Iron	0.0	4.9292	5.4633
	0.15	4.9354	5.4536
	0.16	4.9289	5.4543
	0.23	4.9292	5.4652
	0.25	4.9289	5.4566
	0.32	4.9352	5.4505
	0.57	4.9299	5.4642
Chromium	0.0	4.9292	5.4633
	0.10	4.9289	5.4651
	0.11	4.9292	5.4624
	0.13	4.9292	5.4647
	0.17	4.9296	5.4624
	0.18	4.9299	5.4633
	0.19	4.9292	5.4661
Cobalt	0.0	4.9292	5.4633
	0.01	4.9278	5.4647
	0.02	4.9292	5.4629
	0.03	4.9292	5.4638
	0.04	4.9289	5.4647
	0.05	4.9292	5.4647
	0.33	4.9289	5.4629
Vanadium	0.0	4.9292	5.4633
	0.01	4.9292	5.4638
	0.03	4.9257	5.4629
	0.035	4.9285	5.4647
	0.04	4.9285	5.4615
	0.05	4.9285	5.4647

Error considered maximum for each value in column 2 is ± 0.0004; error considered maximum for each value in column 3 is ± 0.0024.

when used with the quadratic formula. A value 0.01° above, or below, the true 2θ value causes $1/d^2$ to vary ± 0.001 Å. Solving the quadratic formula for *a* produces an error of ± 0.0004 Å for the hkl combination chosen. However, when solved for *c* by substitution, this same hkl combination provided the limiting error of ± 0.0024 Å.

Discussion of Results

Spectroscopic analysis of the pure crystalline aluminum phosphate indicated the inherent trace element concentrations were 100-fold less than the amount of added impurity and may be considered

TABLE VI

Values for the Radius, Electron Configuration, Electronegativity, and Ionization Potential Assigned the Impurities Linking the PO_4 Tetrahedra in the Crystal Structure of $AlPO_4$

	Radius of Trivalent Ion (Å)	Electron Configuration							Electro-negativity	Total Ionization Pot. (0° K)
				M			N			
		K	L	(3s)	(3p)	(3d)	(4s)	(4p)		
Al	.57	2	8	2	1				1.5	1228 kcal
Ga	.62	2	8	2	6	10	2	1	1.6	1310
Fe	.67	2	8	2	6	6	2		1.8	1261
Cr	.64	2	8	2	6	5	1		1.6	1259
Co	.65	2	8	2	6	7	2		1.7	1365
V	.65	2	8	2	6	3	2		1.35	1094

negligible. The maximum inherent impurity was 125 ppm of iron in the uncontaminated crystal. The data indicate gallium entered the crystalline structure of aluminum phosphate in amounts near those predicted by stoichiometric calculations, whereas iron, chromium, cobalt, and vanadium were not accepted as anticipated.

The cobalt and vanadium ions were the least acceptable, with a range of substitutions from 0.0 to 0.05%. The iron and chromium ions entered the structure sufficiently to alter the color and occurred in amounts ranging approximately 0.1% for chromium and 0.5% for iron, a factor of ten more than cobalt and vanadium.

Based on Goldschmidt's rules for diadochic substitution and values for ionic radii (Table VI), gallium should be most successful in substituting for aluminum in the crystal framework, as is the case. Iron has the largest ionic radius of any of the cations presented for substitution and should be least successful in substituting for aluminum. This is not true, as iron ranked second to gallium in the ability to substitute for aluminum and created the greatest structural alteration of any of the five impurities.

Electronegativity is an ionic property considered to be a principal factor in determining the structure of crystals. Electronegativity is a combination of ionization potential and electron affinity, and extensive calculations were performed by Pauling and Haissinsky (Fyfe 1951) to provide the coefficients listed in Table VI. Fineman and Daignault (1959) question Pauling's relationships for bonds with a great deal of ionic character and present different coefficients from those of Pauling and Haissinsky. Ringwood (1955) attributed a major significance to the role played by electronegativity in elements which display diadochy. His paper indicated an element displaying higher electronegativity than the host forms a weaker bond and will not be accepted readily, while an element of lower electronegativity than the host will be preferentially incorporated. Such a relationship is reported to apply when the electronegativity difference between the host and the trace element is greater than 0.1 (approximately).

Using this criterion, vanadium should be accepted by the crystalline aluminum phosphate in the largest concentration, and in turn provide the greatest change in lattice parameters, while iron will be least acceptable and provide the smallest alteration in the crystal structure. The present investigation does not support this view.

Ionization potential in its own right may be considered to play a fundamental part in the ability of an ion to substitute in a structure. Ionization potential is a measure of the energy required to move a single electron from a free atom or ion to infinity, or conversely, as a measure of the energy lost by a recombination of the resulting ion and the electron. Ahrens and Morris (1956) employed the ionization potential of the cation as an approximate comparative measure of its power of attraction for the most loosely bound electrons of the neighboring anions when considering the formation of crystals from different cations of the same charge and ionic radii with a given anionic constituent. According to their argument, the greater the ionization potential of the cation, the greater the deviation from ionic bond character, the weaker the bond, and the less desirable the cation to the structure. Recent values for ionization potentials assembled by Basolo, Pearson, and others (1958) are listed in Table VI.

Following the ideas presented by Ahrens and Morris, vanadium should be preferred by the crystalline structure over the gallium due to the vanadium's lower ionization potential. This was found to be experimentally not true.

The chemical nature of the trace ions entering a crystalline structure affects the bonding and is necessarily a prime factor in establishing any substitutional pattern. The electron configurations of the metallic elements involved in this study are listed in Table VI (Rankama and Sahama 1950, p. 796). It is evident that the 8-electron outer shell of trivalent aluminum has reached a degree of stability. The 18-electron cation of gallium also contains the maximum number of electrons in its M shell. This similarity, along with the physical properties previously listed, permit gallium to proxy for aluminum in a predictable manner. X-ray evidence shows a reduction in the *a* axis length with addition of gallium and, as the amount of substituted gallium increases, the axial contraction is more pronounced. Pauling (1929) noticed this type of deformation and wrote:

> Deforming action of an eighteen-shell ion is much greater than the deforming action of an eight-shell ion of the same size and valence. This is due to large effective nuclear charge.

Iron, chromium, cobalt, and vanadium are transition metals and exist with their M shells containing less than a stable 18-electron configuration. As these atoms oxidize to the trivalent state, it is assumed the N shell loses its electrons first, and then the 5 d-orbitals of the M shell may alter. The exact manner of alteration and any simple prediction for the relative stability of a resulting ion is a matter of conjecture (Gould 1955, p. 21). Corwin (1958), in studying reactions under controlled conditions between salts of alkaline earth metals and silica, found the formation of crystalline structure is correlated with the initial pH of the solution as well as the relative dimension of the ions involved.

The iron and chromium originally were put in the crystallizing solution as hydrated ferric and chromic phosphate. Spectroscopic evidence indicates only a few tenths of a percent of either the Fe (III) or the Cr (III) ions succeeded in proxying for aluminum. In view of the specific crystallizing temperature, pressure, and concentration, this failure to substitute may be due to either the inability to become simple, mobile ions or to the energy demands of the lattice sites. The latter possibility appears the more probable.

The insolubility of the cobaltic oxide, in addition to its electron configuration, may be the cause for its lack of success in substitution. From the appearance of the reaction where the cobalt metal was added to produce cobaltic cations, the cobaltous cation formed was not suited (smaller charge, larger size) to the diadochic process. The vanadium metal was soluble and the presence of large numbers of V (III) ions was indicated by the intensely green crystallizing solution. Again, like the cobaltic cations, the vanadium was unable to occupy the structural aluminum sites, and spectroscopic analysis showed less than 0.05% concentration of either ion under conditions which allowed gallium to substitute up to 18%.

Iron was the most successful of the transition metals in proxying for aluminum and provided a unique change in the structural patterns of the crystal. Whereas the gallium impurity contracted the *a* axis by 0.003 Å at the maximum substitution of 18%, iron expanded the *a* axis by 0.006 Å with the minimum substitution of 0.15%. It is significant that (1) iron altered the axis in a manner opposite from gallium, (2) caused a relatively large axial change with trace substitution, (3) shifted the axial lengths for discrete percentages of impurity, and (4) allowed the lattice to remain relatively unchanged for intermediate values. One explanation for this may be the electron configuration.

Conclusions

The tetrahedral structures AlO_4 and PO_4 are replicas of the SiO_4 tetrahedral unit, and crystalline $AlPO_4$ may be produced under ordinary laboratory conditions. Impurities of Ga, Fe, Cr, Co, and V may be introduced into crystallizing $AlPO_4$ and produce lattice constant deviations without destroying the crystal's gross morphology or lattice structure. The electron configuration of the cation attempting to proxy for aluminum in the $AlPO_4$ structure appears to be the prime factor in its occupying a given structural site. Until the present uncertainty concerning atomic and molecular constants is reduced, a precise explanation of structural alterations in crystalline substance caused by isopositional substitution of foreign ions is impossible.

ACKNOWLEDGMENTS

The authors are grateful to the Department of Soil Science, Michigan State University, for the use of laboratory facilities and to John Ligtvoet, Henry Ford Community College, for help in preparing the manuscript. Financial assistance for this work was kindly provided by the Petroleum Research Fund administered by the American Chemical Society.

LITERATURE CITED

AHRENS, L. H. 1952. The Use of Ionization Potentials. Part 1: Ionic Radii of Elements, Geochimi. et Cosmochimi. Acta, 2: 155.

— AND D. F. C. MORRIS. 1956. Ionization Potentials and the Chemical Binding and Structure of Simple Inorganic Crystalline Compounds I-II. Jour. Inorg. Chem. 3: 263.

BASOLO, F., AND R. G. PEARSON. 1958. Mechanisms of Inorganic Reactions. New York: John Wiley & Sons, N.Y. 426 pp.

BRILL, R., AND A. DEBRETTEVILLE. 1948. On the Crystal Structure of $AlPO_4$. Am. Mineralogist 33: 750.

CORWIN, J. F. 1958. Hydrothermal Reactions under Super Critical Conditions. Jour. Phys. Chem. 62: 1086.

FINEMAN, M. A., AND R. DAIGNAULT. 1959. Electronegativities Calculated Using a Modification of Pauling's Equation. Jour. Inorg. and Nucl. Chem. 10: 205.

FYFE, W. S. 1951. Isomorphism and Bond Type. Am. Mineralogist 36: 538.

GOLDSCHMIDT, V. M. 1926. Geochemische Verteilungsgesetze der Elemente, VIII. Skrifter Norske Videnskaps-Akad. Oslo, I, Mat.-Naturv. Kl., pp. 156.

— 1929. Crystal Structure and Chemical Composition. Trans. Faraday Soc. 25: 253.

— 1932. Kristallographie und Stereochemic Anorganischer Verbindungen. Stereochemie, verlag von Frank Deuticke in Leipzig (reprint).

— 1937. The Principles of Distribution of Chemical Elements in Minerals and Rocks. Jour. Chem. Soc. London, p. 655.

GOULD, E. S. 1955. Inorganic Reactions and Structure. New York: Henry Holt & Co. 470 pp.

GRUNER, J. W. 1946. Isostructural Relationship of $AlPO_4$ and $SiSio_4$. Am. Mineralogist 31: 196.

HUTTENLOCHER, H. F. 1935. Kristallstruktur des Aluminiumorthophosphates $AlPO_4$. Zeit. Krist. 89-90: 508.

PARRISH, W. 1956. X-ray Intensity Measurements with Counter Tubes. Phillips Tech. Rev. 17: 206.

PAULING, L. 1929. Structure of Complex Ionic Crystals. Jour. Am. Chem. Soc. 51: 1012.

RAMBERG, H. 1952. Chemical Bonds and the Distribution of Cations in Silicates. Jour. Geol. 60: 331.

RANKAMA, K., AND TH. G. SAHAMA. 1950. Geochemistry. Chicago: Univ. Chicago Press. 912 pp.

RINGWOOD, A. E. 1955. The Principles Governing Trace Element Distribution During Magmatic Crystallization. Part I. Geochimi. et Cosmochimi. Acta 7: 189.

SHAW, D. M. 1953. The Camouflage Principle and Trace Element Distribution in Magmatic Minerals. Jour. Geol. 61: 141.

STANLEY, J. M. 1954. Hydrothermal Synthesis of Large Aluminum Phosphate Crystals. Ind. and Eng. Chem. 46: 1684.

STRUNZ, H. 1937. Titanit and Tilasit. Uber die Verwandtschaft der Silikate mit den Phosphaten und Arsenaten. Zeit. Krist. 96: 7.

Papers of the Michigan Academy of Science, Arts, and Letters
Vol. XLIX, 1964 (1963 Meeting)

THORACIC ARMOR OF A NEW ARTHRODIRE (*HOLONEMA*) FROM THE DEVONIAN OF PRESQUE ISLE COUNTY, MICHIGAN

MARGARET SKEELS STEVENS
The University of Michigan

Introduction

During the early part of August 1962, five ornamented plates of a new Devonian fish were found by a workman, Chad Chapman, in the Onaway Stone Company quarry. This quarry is located about one third of a mile north of Onaway in the SE$\frac{1}{4}$, NE$\frac{1}{4}$ Sec. 6, T. 34 N., R. 2 E., Presque Isle County, Michigan. The fish remains were recovered from the lowest part of an approximate 3$\frac{1}{2}$- to 5-foot bed of commercial, thin-bedded limestone exposed in the quarry. Very few invertebrates occur in this unit. This bed is in all probability the Koehler limestone, which lies just above the Genshaw Formation, and possibly an equivalent of the Newton Creek limestone, a part of the Traverse Group (Selley and Smith 1947). Above this unit less dense fossiliferous (invertebrate) limestone occurs; below it massive noncommercial limestone is found. This massive limestone forms the floor of the quarry.

Four of the five plates are associated and comprise most of the ventral armor of an arthrodiran fish. These plates are the right and left antero-ventralaterals and the right and left postero- ventralaterals. These four plates are complete and include most of the original bony material, as well as impressions in the supplementary piece of limestone. The fifth plate is almost complete and is the median-dorsal from a different individual.

ORDER ARTHRODIRA

Family Holonemidae

Holonema farrowi, sp. nov.

(Pls. I, II, Figs. 1-2; 1C, F)

Holotype.—UMMP 46647; four ventral plates collected by Chad Chapman. August 1962, and donated by Mrs. Jean Adrian.

Paratype.—UMMP 46648; an almost complete median-dorsal plate.

Horizon and type locality.—Middle Devonian Koehler limestone of the Traverse Group, Onaway Stone Company quarry, SE¼, NE¼ Sec. 6, T. 34 N., R. 2 E., Presque Isle County, Michigan.

Diagnosis.—A large arthrodire, the ventral shield of which is composed of rather thin, simple, oval antero-ventralateral plates and small, thin, subtriangular postero-ventralateral plates. Ornamentation can be distinguished from all other species of *Holonema* in that it consists of nonstellate tubercles arranged in beadlike rows, and these tuberculate rows never fuse to form a ridge. These rows are either concentric, radial, or longitudinal, depending on the position they occupy on the plates. The paratype, a median-dorsal plate, is also characteristically ornamented, and thin.

Description of holotype.—The ventral armor of *Holonema farrowi* (UMMP 46647) consists of the right and left antero-ventralateral plates, (right and left AVL), and the right and left postero-ventralateral plates, (right and left PVL) (Pl. I). The antero-ventral median is represented only as a slight and incomplete impression and little detail can be observed. The postero-ventral median plate cannot be located. Either it is not preserved, or it is so deeply buried beneath the other plates that trying to expose it would destroy the medial margins of these plates.

The antero-ventralateral plates are oval, longer than broad, the width being approximately one-half of the length. The anterior margins are rounded, and the posterior margins come to a slight apex externally (Pl. I). The maximum breadth is, approximately, across the center of each plate. The greatest length of the right plate is 201.3 mm., and the greatest width is 100.3 mm. The greatest measurements attained for the length and width for the left plate are 204.9 mm and 98.7 mm, respectively.

Upon the external surface of both of these plates (AVL) there is a sensory canal (Pl. I). This canal crosses from the antero-internal margin toward the center of each plate. The canal is 63.4 mm long on the right plate and 59.3 mm long on the left plate. These measurements were obtained from the block containing the impressions of these plates, since all of the details are preserved (Pl. II, Fig. 2). Part of the surfaces of the plates themselves are denuded. Both canals are straight and very distinct. A continuation of the

PLATE I

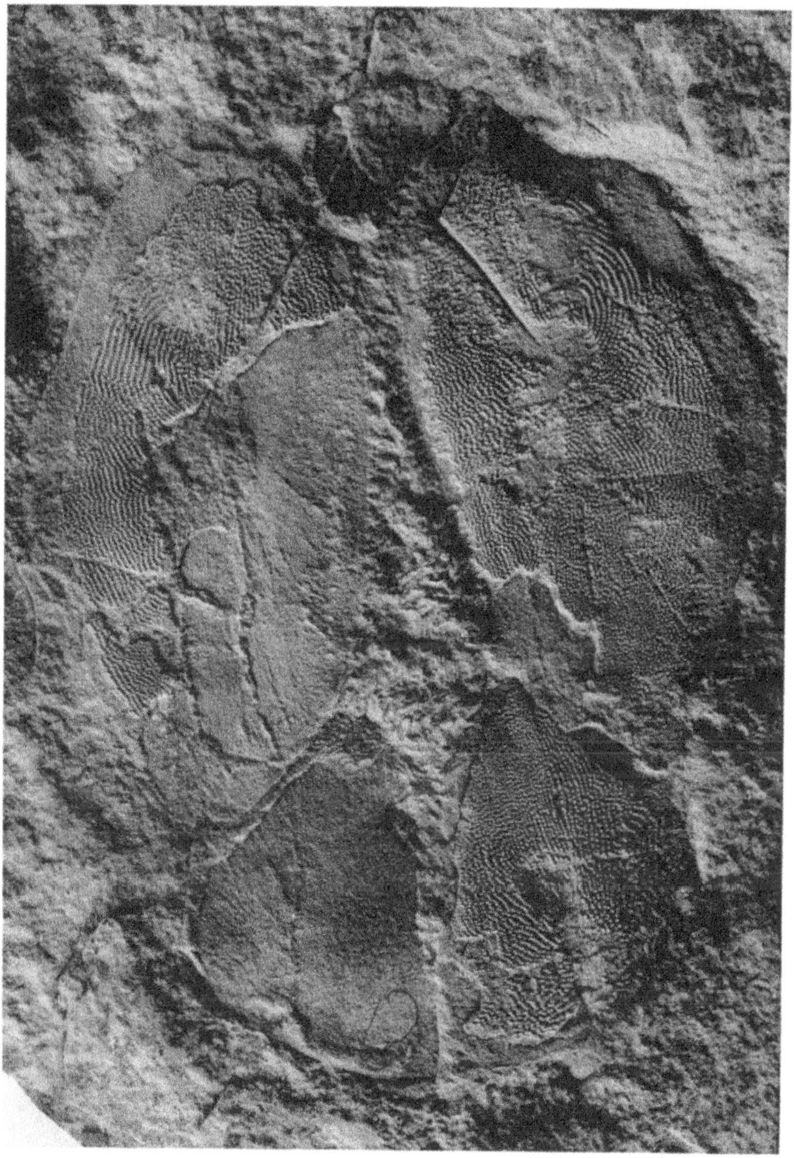

The ventral armor of *Holonema farrowi*, sp. nov., holotype, UMMP 46647, showing the right and left antero-ventralateral plates and the right and left postero-ventralateral plates. Also preserved is part of the antero-ventral median plate with its sensory canal and the continuation of this canal on the antero-ventralateral plates. × approx. ½.

canal is shown on the impression of the antero-ventral median plate (Pl. I).

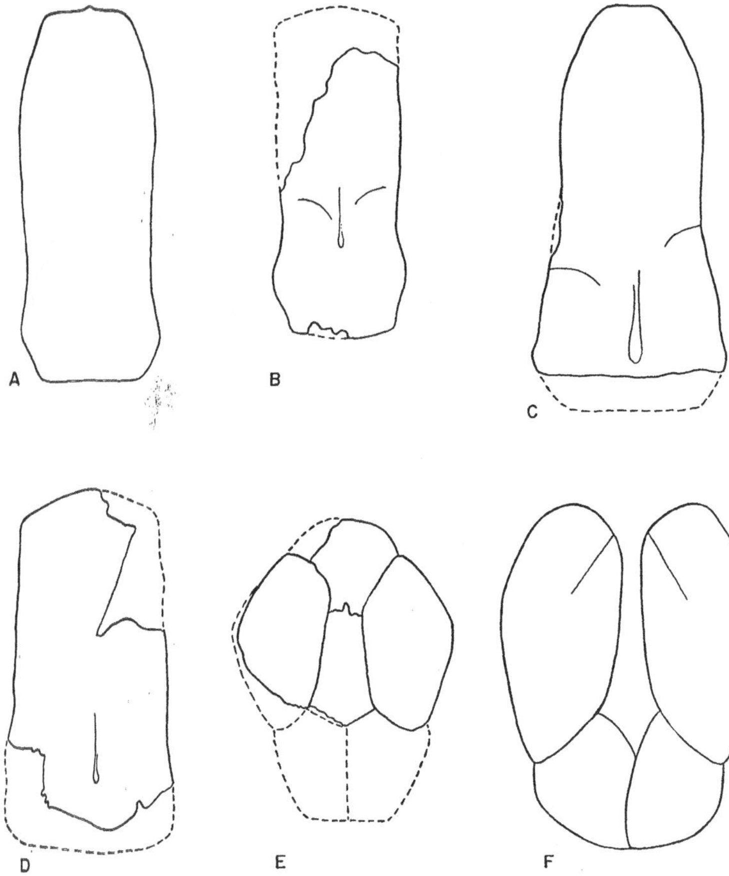

FIG. 1. Outline of the median-dorsal (MD) plate in various species of *Holonema*, A-D. (A) MD of *H. radiatum*, after Obrutschew, 1933. × 1/6. (B) MD of *H. rugosum*, after Wells (1943), showing center of ossification and nerve canals. × approx. 1/4. (C) MD of *H. farrowi*. × 1/2. (D) MD of *H. abbreviatum*, after Wells (1942). × approx. 1/10. (E) The ventral armor of *H. rugosum*, showing arrangement and outline of the plates, after Williams (1893). × approx. 1/11. (F) The ventral shield of *H. farrowi*. × 1/4.

The surface sculpture of both antero-ventralateral plates consists of rounded, nonstellate tubercles arranged in definite beadlike rows, with the exception of a few randomly placed tubercles. On both

PLATE II

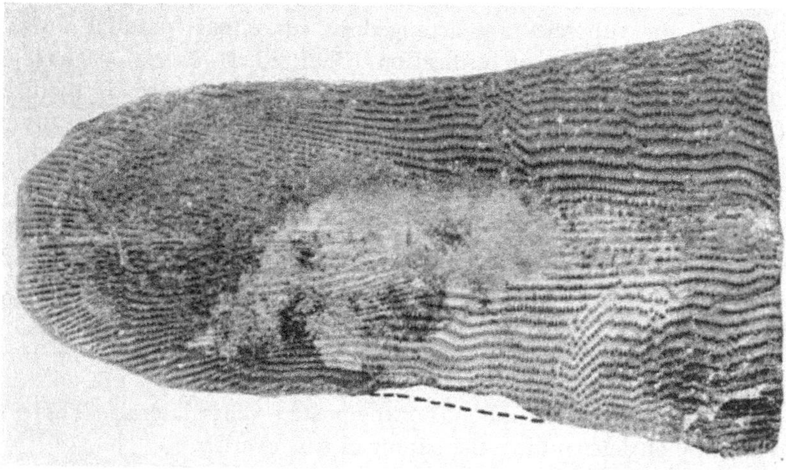

FIG. 1. The median-dorsal plate of *Holonema farrowi* showing style of ornamentation, nerve canals, and posterior "club." × approx. ½.

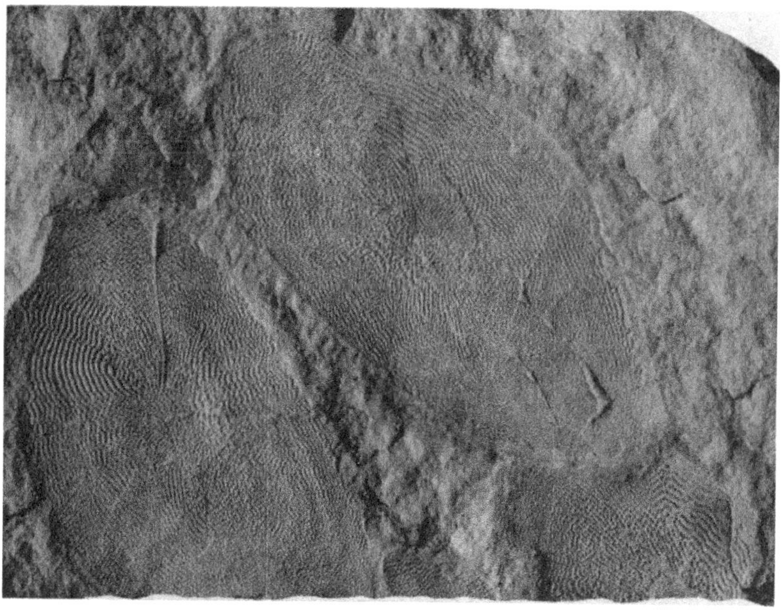

FIG. 2. Impression of the ventral shield of holotype, UMMP 46647, showing ornamentation. × ⅜.

plates the tubercles are either arranged radially, concentrically, or longitudinally, on different parts of the plate. On the antero-lateral portions, the tubercles are arranged in concentric, parallel rows anterior to the center of ossification (Pl. I; Pl. II, Fig. 2). I have followed the use of the term center of ossification as used by Wells (1942, p. 652). These rows often have interspaces wider than the rows themselves. On the antero-internal and posterior half of each of these plates the tubercle alignment is more radial, being directed outward from the center of ossification and more parallel to the long axis of the plate.

There is no evidence that these two anterior plates ever made contact with each other, as there is no indication of a squamate articulation between them. Their position in the matrix appears natural. However, the postero-internal ends, medial to the apex, overlie the antero-external ends of the postero-ventralaterals. There is no way of determining the extent of this overlay.

The postero-ventralateral plates are situated posterior to the antero-ventralaterals, but are partly overlain by them. The right plate (PVL) is approximately 109.4 mm long. The width of this plate cannot be determined, since it is overlain by the left plate (PVL). The greatest width observed is 64.5 mm. The length and width of the left plate (PVL) are 111.8 mm and 85.1 mm, respectively.

No sensory canals have been found on either postero-ventralateral plate. The surface sculpture displayed by the left postero-ventralateral is similar to that seen on the anterior pair of plates, though the rows of tubercles are more radial (Pl. I). The tubercle rows on the external-lateral side of the left plate (PVL) are concentric, and radial from the center of ossification throughout the remainder of the plate. The center of ossification is located along the postero-central margin. Ornamentation on the right plate (PVL) cannot be determined, as its surface is denuded and the block showing the complete impression of this plate was not recovered.

Description of paratype.—The paratype (UMMP 46648), a median-dorsal plate, is almost complete. This specimen was not found in association with the holotype. However, it was also found in the lower part of the $3\frac{1}{2}$-5 foot Koehler limestone, which is the commercial stone of the quarry. This plate agrees very well with similar plates of *Holonema* and especially with the characteristics of the holotype, though it is probably from a larger individual.

Therefore, it is considered to be the same species. There is no evidence indicating that it should be excluded. Further evidence concerning the median-dorsal plate in *Holonema farrowi* may alter this opinion.

The median-dorsal plate is long and narrow. It is rounded on its anterior margin. The posterior portion of this plate is not preserved, but if it follows the condition found in other holonemids, the margin would be fairly straight. The sides are slightly concave (Pl. II, Fig. 1), and this plate is not as rectangular as the median-dorsal plate of most species of this genus. The median-dorsal plate of *Holonema farrowi* is flat (it is possible that some flattening has occurred) and shows a slight medial thickening, ventrally. No keel has been observed, as the antero-ventral surface of the plate is buried in matrix. The length of the portion preserved is 197.3 mm, as measured along the axis. The actual length may have extended only about 2 cm beyond the broken posterior margin (Fig. 1C). The maximum breadth was ascertained as 109.4 cm. The broadest part is located very near the posterior border.

Two rather faint sensory canals can be observed on this plate. These canals lack absolute symmetry, but both originate near the postero-central margins and curve gently backward for approximately 30 mm. They are not deeply set into the plate but are expressed more as regular interruptions of the tubercle rows. These rows, directly behind each canal, make a sharp inward turn, then once more become more longitudinally aligned (Pl. II, Fig. 1).

Ornamentation is extremely well developed on this plate. The tubercles are aligned in strong beadlike rows and, for the most part, follow the long axis of the plate. Near the postero-center is located the center of ossification. This center was formed in a club-like manner, the "club" of J. W. Wells (1942, p. 652). The club effect is produced by rows of tubercles, parallel to the long axis of the plate, curving around and closing in posteriorly behind it. This produces a more or less imperfect half circle (Pl. II, Fig. 1). Here again, as with the holotype, the rows are formed by rounded, nonstellate tubercles. The rows of tubercles on the median-dorsal are straight, parallel, or gently curving, with interspaces between them often broader than the rows themselves. Tuberculation is coarsest toward the posterior one-half of the plate.

An interesting observation regarding the tubercular rows in both holotype and paratype is that when the tubercles are more definitely aligned, the tubercles are broader than long as they lie in the row. That is, the long axis of each tubercle is perpendicular to the long axis of the row.

Discussion.—Remains of species belonging to *Holonema* are rare. *H. eifeliense* (Kayser) is not a well-known form. It was originally described in 1880 under the name *Dinichthys eifeliense*, from specimens collected from the Middle Devonian in and around Germany. This species was later put into the genus *Holonema* by Gross (1933). An antero-ventralateral plate is known for this species. It is similar in most respects to the corresponding plate of *H. rugosum* (Claypole). Eastman (1897, pl. 5, fig. 4), figures the AVL of *H. eifeliense* but presents no description. His photograph is vague, but it appears in most respects also similar to the AVL plates of *H. farrowi.* A sensory canal is present on the AVL plate of *H. eifeliense* and is very similar to those seen on the AVL plates of *H. farrowi.*

Tuberculation, in *Holonema eifeliense,* consists of radial, concentric, or irregular rows of tubercles (Gross 1933). If these tubercles are arranged in single rows they have wide interspaces, a condition *H. farrowi* displays. However, if they are arranged in parallel series of one to four rows, the interspaces are very narrow, a condition not observed in the latter species. Separating the one- to four-row series is a wider, deeper interspace. Gross (1933) says, "*Holonema eifeliense* Kayser is a large Middle Devonian species of *Holonema* which is recognized by the type of sculpture which is one to four fold rows of small tubercles which are fused into walls and convolutions. It is distinct from *H. rugosum* and is closer to the Upper Devonian species *H. radiatum* Obr."

An unusual median-dorsal plate has also been assigned to this species (Gross 1937). It is approximately oval, dorsally keeled medially on its anterior half (posterior of Gross 1937) and strongly ornamented with broad, generally concentric, convolutions. The median-dorsal plate of *Holonema farrowi* shows no dorsal keel whatever. The convolutions on the posterior margin (anterior of Gross) of the median-dorsal plate of *H. eifeliense* become so closely set that the interspaces between them become interrupted, producing a pitted surface. This condition is not observed in *H. farrowi.* The convolutions on the median-dorsal plate of the latter species are more nearly parallel with the long axis of the plate and close in behind producing the "club." This club is not seen on the median-dorsal plate of *H. eifeliense.*

Holonema rugosum (Claypole) is much better known than *H. eifeliense*. *H. rugosum* was originally described as *Pterichthys? rugosus* (Claypole 1883) from the Upper Chemung (Upper Devonian) beds of Pennsylvania. This species was later made the type of a new genus, *Holonema*, by Newberry (1890).

The armor of *Holonema rugosum* is characterized by being rather thin, flat, and ornamented with raised enamel ridges, most of them parallel but, on occasion, broken, winding, and anastamosing. These ridges radiate from the center of the plates (Newberry 1890). Williams (1893) gives a reconstruction of the ventral shield based on photographs of a slab containing most of the plates (Fig. 1, E). No description of the ornamentation is given. The general shape of these plates is similar to those of *H. farrowi*; however, differences can be noted. Specifically, the right and left antero-ventralateral plates of Williams' figure are more pointed anteriorly than the anterior ends of these plates in *H. farrowi*. Conversely, the posterior margins are less pointed in *H. rugosum* than the same margins on these plates in *H. farrowi* (Fig. 1F). The postero-ventral median (the "lozenge" plate of early authors) is broader in *H. rugosum* than this plate could have been in *H. farrowi*. Although the postero-ventral median plate was not located in *H. farrowi*, its basic shape is indicated by the natural position of the remaining plates (Pl. I).

In addition to the ventral shield, a median-dorsal plate is known (Wells 1943). He described an impression of this plate found as a float block derived from the Upper Devonian Chemung or Enfield Formation. The impression is almost complete; only part of the anterior margin is lacking. It is long, narrow, and rectangular, with nearly straight sides and a slight bulge near the posterior end (Fig. 1B). The posterior edge of this plate, as with holonemids in general, is nearly straight across. The center of ossification is located near the postero-median line; a condition observed in *Holonema farrowi*. The ornamentation consists of enamel ridges. These ridges run from the center of ossification forward, following the long axis of the plate. Toward the posterior end, however, these ridges become undulating and close in behind this center, forming the "club." This condition is also found on the median-dorsal of *H. farrowi*. However, near the margins of the median-dorsal plate of *H. rugosum* the axial alignment of the ridges changes and becomes perpendicular to the edge, a condition not found on this plate in *H. farrowi*. Also, in contrast with *H. farrowi* the ridges of *H.*

rugosum have interspaces that are narrower than the rows themselves (Wells 1943). Two sensory canals are present on the median-dorsal plate of *H. rugosum* and are located in asymmetrical positions similar to those on the median-dorsal plate of *H. farrowi*.

Holonema horridum Cope (1892) was also described from the Chemung beds of Pennsylvania and was based on a postero-ventralateral plate. According to Cope (1892), this differed from similar plates of *H. rugosum* mainly in its extremely coarse ornamentation. As stated above, *H. rugosum* is characterized by radial sculpture, diverging from a central point. In *H. horridum* the ridges are longitudinal following the main axis of the plate (Cope 1892). According to him, these ridges are interrupted as the edge of the plate is neared, but less so toward the center. This condition is not found in *H. farrowi*. He also states that the roughness of the ornamentation exceeded that of all other armored fishes known at that time.

Eastman (1907) described *Glyptaspis abbreviata*. This species was later put into the genus *Holonema* by Hussakof and Bryant (1918). Still later this form was made the type of a new genus, *Deirosteus*, by Wells (1942). A postero-ventralateral (antero-ventralateral of Eastman 1907) is known for *H. abbreviatum*. It is short, broad and subtriangular. According to Eastman this plate is ornamented only on the central portion. Sculpture consists of raised lines or broken rows of tubercles. The style of ornamentation on *H. farrowi* consists of tubercles closely aligned in rows; however, these tubercles never fuse to form a ridge and cover the entire surface of the plate.

Wells (1942) referred a median-dorsal plate, again from the Upper Devonian Enfield Formation, to *Holonema abbreviatum*, but while doing so set it up for the type of the new genus *Deirosteus*. This median-dorsal plate, as is the condition of all holonemids, is long, narrow, and quite rectangular with nearly straight sides and slightly arched posteriorly (Fig. 1D). There also are indications of a slight median keel on the ventral side of this specimen (Wells 1942). As indicated by the foregoing description of the median-dorsal plate of *H. farrowi*, this plate is not rectangular and shows only a slight medial thickening posteriorly. Probably an antero-ventral keel exists. There is no indication of any great deformation in the plates from the Onaway quarries, though some flattening has undoubtedly occurred.

Surface sculpture of *Holonema abbreviatum* consists of sharp ridges formed by small, fused tubercles or rows of tubercles. These tubercles are definitely stellate. The tubercles on *H. farrowi* are not stellate. Toward the posterior end of the median-dorsal plate of *H. abbreviatum* the tuberculate ridges are scalloped and close in behind to form the " club " around the center of ossification. Tuberculation also becomes finer and simpler toward the margins, a condition seen in *H. farrowi*.

I am not convinced that the new genus *Deirosteus* is valid. Wells (1942) based his distinction on two points: (1) the ornamentation of *Deirosteus* is distinct from all true holonemids by being essentially concentric, and (2) by possessing interspaces wider than the ridges. In this respect, *Holonema farrowi* is intermediate; possessing either radial, concentric, or longitudinal rows and also possessing interspaces that are wider or narrower than these rows, depending on the position of the rows on the plate. This is interesting, as *H. farrowi* is a Middle Devonian form and *H. abbreviatum* is an Upper Devonian form.

Holonema radiatum Obrutschew is a very large Lower Upper Devonian Arthrodire from Russia, and the only other species of this genus that is fairly well known. Obrutschew (1933) has described various plates of this species, among them a left antero-ventralateral plate and a median-dorsal plate. The antero-ventralateral is ornamented with definite convolutions which radiate outward from the center of ossification. These ridges occasionally anastamose, a feature not seen on plates of *H. farrowi*. A sensory canal is present on this plate and runs forward, from the center of ossification, toward the antero-median margin. This same canal continues forward onto the antero-ventral median plate. The canal is very similarly located to those on the antero-ventralateral plates of *H. farrowi*. Also, there is a vague indication of this canal continuing on the antero-ventral median of *H. farrowi*. According to Obrutschew (1933), the antero-ventralateral is 18 cm long. His specimen, however, is not complete, and he estimates the total length as 30 cm.

The median-dorsal of *Holonema radiatum* is elongate, 29.5 cm long and 11.5 cm wide (after Obrutschew 1933). This plate is slightly arched, both transversely as well as longitudinally, and rectangular (Fig. 1A). The ornamentation of the median-dorsal, as in *H. rugosum*, radiates from the center of ossification, and it also displays the " club." The interspaces on this median-dorsal plate are narrow as in *H. rugosum,* and the ventral surface shows only a very slight keel.

Holonema farrowi is not a member of any of the foregoing species. This form shows some characteristics of each but not all of any one of them. These differences seem too great to be variations among any one species. The presence of nerve canals, simple non-stellate tubercles aligned in rows, the lack of tubercles fused into ridges, and either radial, concentric, or longitudinal arrangement of these rows seem to indicate a generalized stock among the holonemids. Its Middle Devonian age also supports this opinion.

The Middle Devonian *Holonema farrowi* seems closer to the Upper Devonian *H. rugosum, H. radiatum, H. abbreviatum,* and *H. horridum* species than to the Middle Devonian form *H. eifeliense.* This latter species seems divergent in regard to its ornamentation and very unusual median-dorsal plate.

Acknowledgments

The University of Michigan Museum of Paleontology is deeply indebted to the kindness of Mrs. Jean Adrian, co-owner of the quarry, who donated the specimens to the Museum. She is the daughter of the late Dan Farrow, the founder of the quarry and the man for whom this species is named. Thanks are also given to G. M. Ehlers and E. C. Stumm, of the Museum of Paleontology, who first reported the specimens, for aiding in the stratigraphy, C. W. Hibbard for helpful criticism, and J. B. Stevens for aiding in translating the German.

LITERATURE CITED

Claypole, E. W. 1883. Note on a Large Fish-Plate from the Upper Chemung (?) Beds of Northern Pennsylvania. Proc. Am. Philos. Soc. 20 (113) : 664-667, 1 fig.

Cope, E. D. 1892. On Some New and Little Known Paleozoic Vertebrates. *Ibid.* 30: 221-229, 2 pls.

Eastman, C. R. 1897. On the Relationships of Certain Plates in the Dinichthyids with Descriptions of New Species. Bull. Mus. Comp. Zool. 31 (2) : 19-44, 5 pls.

—— 1907. Devonic Fishes of the New York Formations. N.Y. State Mus. Mem. 10, Albany: 1-235, 15 pls., 34 figs.

Gross, W. 1933. Die Wirbeltiere des rheinischen Devons. Abh. Preuss. Geol. Landesanst. 154: 1-84, 11 pls., 20 figs.

—— 1937. Die Wirbeltiere des rheinischen Devons. Part 2. Abh. Preuss. Geol. Landesanst. 176: 1-83, 10 pls., 29 figs.

HUSSAKOF, L., AND W. L. BRYANT. 1918. Catalog of the Fossil Fishes in the Museum of the Buffalo Society of Natural Sciences. Bull. Buffalo Soc. Nat. Sci. 12: 1-346, 70 pls., 64 figs.

KELLY, W. A., AND G. W. SMITH. 1947. Stratigraphy and Structure of Traverse Group in Afton-Onaway Area, Michigan. Am. Assoc. Pet. Geol. Bull. 31 (3) : 447-469.

NEWBERRY, J. S. 1890. The Paleozoic Fishes of North America. Mem. U. S. Geol. Surv. 16: 1-340, 53 pls., 3 figs.

OBRUTSCHEW, D. 1933. Holonemidae des Russischen Devons. Trav. Inst. Pal. Acad. Sci. USSR 2: 97-115, 4 pls., 27 figs.

WELLS, J. W. 1942. Arthrodiran Fish Plates from the Enfield Formation (Upper Devonian) of New York. Jour. Paleon. 16 (5) : 651-656, 3 pls.

—— 1943. A Median Dorsal Plate of *Holonema* from the Upper Devonian of New York. Bull. Am. Paleon. 27 (107) : 3-10, 1 pl., 1 fig.

WILLIAMS, H. S. 1893. On the Ventral Plates of the Carapace of the Genus *Holonema* of Newberry. Am. Jour. Sci. 46 (whole no. 146) : 285-288, 5 figs.

PAPERS OF THE MICHIGAN ACADEMY OF SCIENCE, ARTS, AND LETTERS
VOL. XLIX, 1964 (1963 MEETING)

ZOOLOGY

VALVE ACTIVITY IN FRESH-WATER PELECYPODS [1]

JAMES A. SALBENBLATT [2] AND ARLAN L. EDGAR
Alma College

INTRODUCTION

FRESH-WATER pelecypods in their natural environment exhibit varying amounts of movement. Some species of clams which characteristically inhabit gravelly riffles of streams probably travel little or not at all once they are established. Others, in different water and bottom conditions, respond to seasonal and other fluctuations by burrowing or moving from one location to another (Baker 1928; Pennak 1953). Pelecypods move typically by extension and retraction of the muscular foot, with attendant opening and closing of the valves. This valvular activity was chosen as the criterion of clam movement partly for convenience, but also because it was surmised that this activity was an indication of much of the dynamic organization of the animal. Feeding, respiration, excretion, locomotion, and other functions may be carried out somewhat independently in the clam; however, all these probably influence, to some degree, the position and movement of the valves.

The main purpose of this study, then, was to describe the valvular activity of certain fresh-water clams. Especially sought was evidence of biological rhythmicity as manifested by the abduction and adduction of the valves. The effects of temperature and light were observed.

Three species of clams in the family Unionidae were investigated: *Anodonta grandis, Lampsilis siliquoidea,* and *Anodontoides ferussacianus.* The first two species were collected from Silver and Windover lakes respectively and the third from the south branch

[1] This study received support from the Michigan Academy of Science, Arts, and Letters and from the National Science Foundation, Grant No. G-21933. Grateful acknowledgment is made to Alma College for the use of laboratory facilities.

[2] Present address: The University of Michigan Medical School, Ann Arbor, Michigan.

of the Tobacco River. All were collected in Clare County in Lower Michigan during June 1962.

Materials and Methods

Each clam was supported by a stationary mount clamped to one valve and was suspended in a three-gallon aquarium. A wire was cemented to the other valve and this articulated with an ink-writing lever. The latter recorded valve movement on a standard kymograph drum which was geared to a speed of one revolution every six days. 12 such setups were put into operation. Four specimens of each of the three species were used. Strained beef and beef heart were fed at irregular intervals and apparently maintained the animals in a healthy condition.

One clam of each species was placed in constant dark (DD), one in constant light (LL), and one in 12 hours of light beginning at 6 A.M. followed by a 12-hour period of dark (LD). With one exception, each clam remained in the same light regime throughout the entire experiment. However, each spent approximately 10 days in each of three different temperatures: $11 \pm 1°$ C, $22 \pm 1°$ C and $29 \pm 1°$ C. At any one time at a given temperature the valve activity of all three specimens of the same species were recording on the same drum. All temperatures given are water-temperature readings. A fourth specimen of each species was placed under conditions of natural light in the laboratory at a room temperature of $23 \pm 2°$ C. With the exception of the LD chamber at 29° C, where it was necessary to use a $7\frac{1}{2}$ watt bulb to prevent an excessive temperature increase, illumination for the other LD animals came from 15-watt bulbs at a distance of approximately one foot. Laboratory ceiling lights were used for LL animals at all three temperatures. Natural light entered through north windows; the clams never received direct sunlight.

To each record of activity was added a time base which was divided into three-hour intervals. The actual length of the activity trace in excess of the baseline was measured by the use of a small calibrated wheel. These values were used to compute the mean activity curves for each of the eight three-hour periods in a day. Actual distances that the valves moved were small; therefore, these were amplified by the lever system for greater accuracy in measurement. Although the lever arrangement for each of the setups was identical, a small difference in the position of the lever fulcrum may have caused the amplitude to vary slightly. The recordings were made between June 12 and August 23, 1962.

We decided that valve opening after several hours of closure constituted the best measure of the onset of activity. We recognized that other parts of the recorded activity might contain significant information, but we felt unable to identify them with confidence.

Results and Discussion

It was found, after several days of recording, that characteristic elements in the tracings could be recognized. Some patterns were peculiar to the individual clam, and others seemed characteristic of the species. For example, all animals showed periods of complete closure. The clam may remain closed a majority of the time, as with *Anodonta,* or only for brief periods, as shown by *Lampsilis.* Valve closure was usually preceded and/or followed by vigorous activity which suggested effort on the part of the animal to flush away debris or excrement or to effect increased gaseous exchange across gill surfaces.

The thin-shelled *Anodonta* opened and closed rapidly at more or less regular intervals. *Anodontoides* exhibited a characteristic pattern in which the valves open wide just after a closure. During the rather long periods where the valves were open, there was a two-to-three-hour cycle in which the extent of gap changed. On this shorter cycle were five to seven smaller changes in valve position; this gave the appearance of waves upon waves. The valves of *Lampsilis* typically remain closed for much shorter periods than those of *Anodonta* and *Anodontoides.* Also typical of *Lampsilis* were vigorous flushing or ventilating movements accompanying the opening and closing of the valves. This species has a curious steplike rather than continuous movement when changing the position of the valves.

The animal whose activity was recorded for the longest period was a specimen of *Anodonta* (Fig. 1). This animal experienced two different lighting conditions and three temperature changes. During the early part of this record there was indication of rhythmic activity (Fig. 1). The mean period from one valve opening to the next during the time from June 12 to July 2 was very close to 24 hours. The vigorous activity usually occurred during the night. Unfortunately, there was an interruption in the recording from

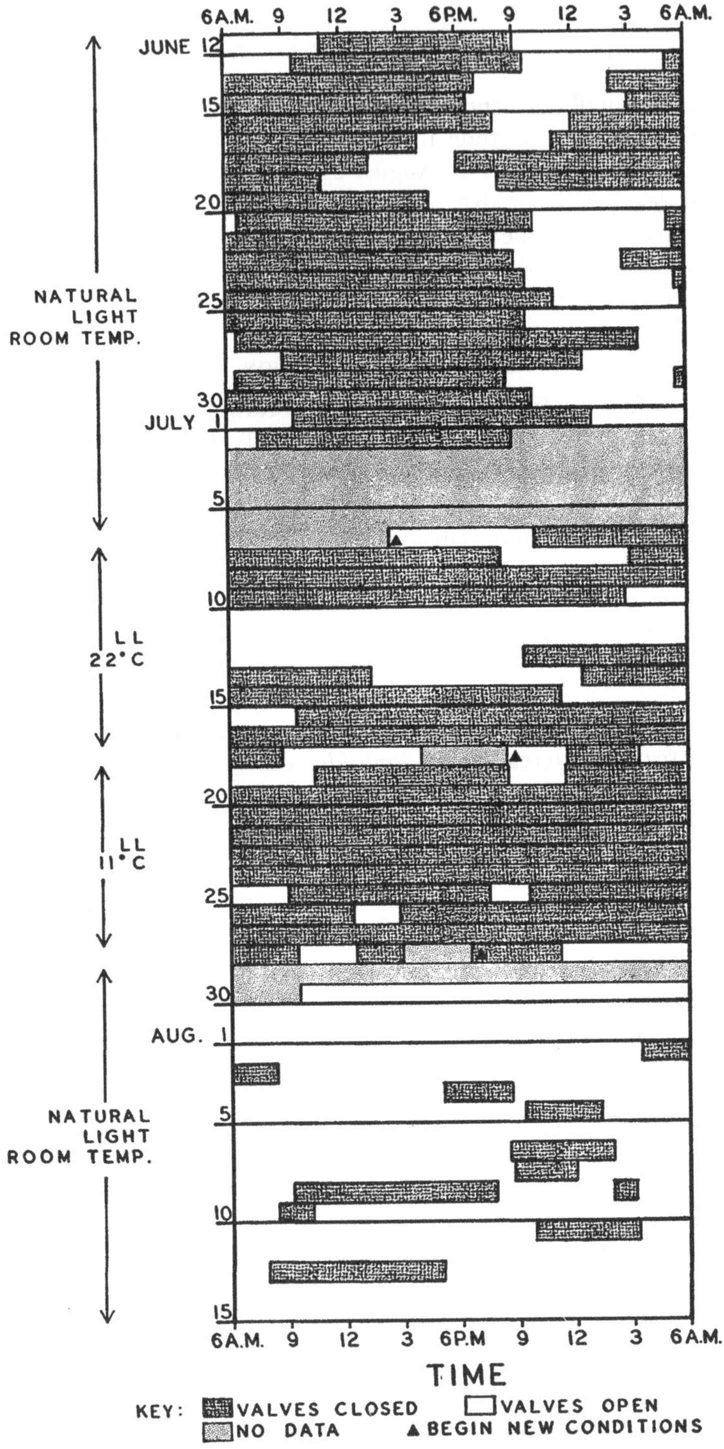

FIG. 1. Valvular activity of *Anodonta* from June 12 to August 15.

July 2, until the animal was placed in LL at 22° C on July 7. From that time until the record stopped on August 15, the clam did not exhibit any consistent repeated activity. Fig. 1 shows that the clam did not respond to daylight as it had in the early recordings, after its exposure to LL.

Another specimen of *Anodonta* exhibited an 8-hour cycle followed by a 24-hour cycle in DD at 22° C. From these records and personal impressions gained from the total record, it is suggested that this species possesses a period of valve opening of approximately 24 hours, with occasional shorter cycles of 8 hours. On several occasions regular patterns of valve closing and opening were observed, but these would disappear for no detected reason. The most regular recordings were obtained from animals in natural conditions or LD, but the period was not always 24 hours. A mean period of 39 hours over 16 consecutive valve openings was shown by *Lampsilis* in natural conditions. On another occasion a different individual of the same species in LD averaged 42 hours over four consecutive openings.

In an early record of *Anodontoides* in natural light, periodic peak activity of approximately 25 to 26 hours appeared. This suggests the possibility of a lunar influence. The method by Brown (1955) was used to analyze the data. Peak activity was exhibited later and later in the dark portion of consecutive days until the peak occurred near dawn. The interval to the next peak was shortened so that it appeared again during the early part of the next dark period. This process was repeated for three cycles of approximately 6 days each. Later records showed a gradual lengthening between distinct closing, and finally only variation in extent of opening without complete closure was shown.

In all three species these periodic movements appear to be easily disrupted. Change from one temperature or light condition to another, as well as unknown influences during apparently constant conditions, seemed to repress periodicity. Perhaps vibration from opening and closing of the heavy doors of the constant temperature rooms or fluctuation in atmospheric pressure were factors in this regard. No record of barometric pressure was kept, and the authors view this as an unfortunate omission.

The data in Figs. 2, 3, and 4 show total mean values of valve movement for the three species investigated. Values were given for each of the eight 3-hour periods into which the day was divided.

Generally, there was only one period of relatively high activity per day. A specimen of *Anodontoides* in LD was the most active animal of those tested. In general, the gross activity of this species exceeded that of the other two tested; the preponderance of the movement occurred either during the subjective day (LL and DD) or during the dark period (natural light and LD) of the alternating regimes. *Anodonta* and *Anodontoides* in natural light and LD were most active from 6 P. M. to 6 A. M. *Anodonta* and *Lampsilis* in LL and DD generally showed less total activity and no obvious preference for any particular period. *Anodontoides*, in LL and DD, exhibited more valvular activity during the subjective day. The typical peak of activity for almost all clams in LD and natural light occurred during the night.

A rationale for the timing of peak activity of *Anodontoides* in LL and DD conditions is suggested. Two " cues " to the animal may operate to cause peaks of activity; the first is endogenous, initiating the gradual accumulation of energy and is triggered by the second, an exogenous cue. According to Bethe (1946) the slow accumulation of energy will discharge when it reaches a maximum. It can also discharge earlier if stimulated by an external excitatory process. Without the latter, a greater period is necessary for sufficient energy to accumulate before discharge. In our case of *Anodontoides* in LL and DD (Fig. 3) there is no external light cue, and a period of approximately 12 hours may be required before spontaneous discharge level occurs. With *Anodonta* and *Lampsilis*, the endogenous chronometer-like cue may exist but never reaches a threshold level where spontaneous discharge occurs in the absence of light change; hence, reduced and arhythmic movement results. Bunning's remarks (1960) concerning the possible existence of a relaxation oscillation may be applicable to the findings from these latter genera. A single stimulus was necessary to initiate the oscillation. In LL there was no dark period or relaxation phase, and in DD no tension phase induced by light. Therefore, repeated high activity periods were not initiated.

The influence of the three temperatures upon the period from valve opening to valve opening is shown in Table I. For each temperature-light combination, two numbers may be seen. The one in parentheses indicates the number of valve opening cycles included, while the other is the mean value of these cycles, in hours. There are certain obvious shortcomings in treating data in this

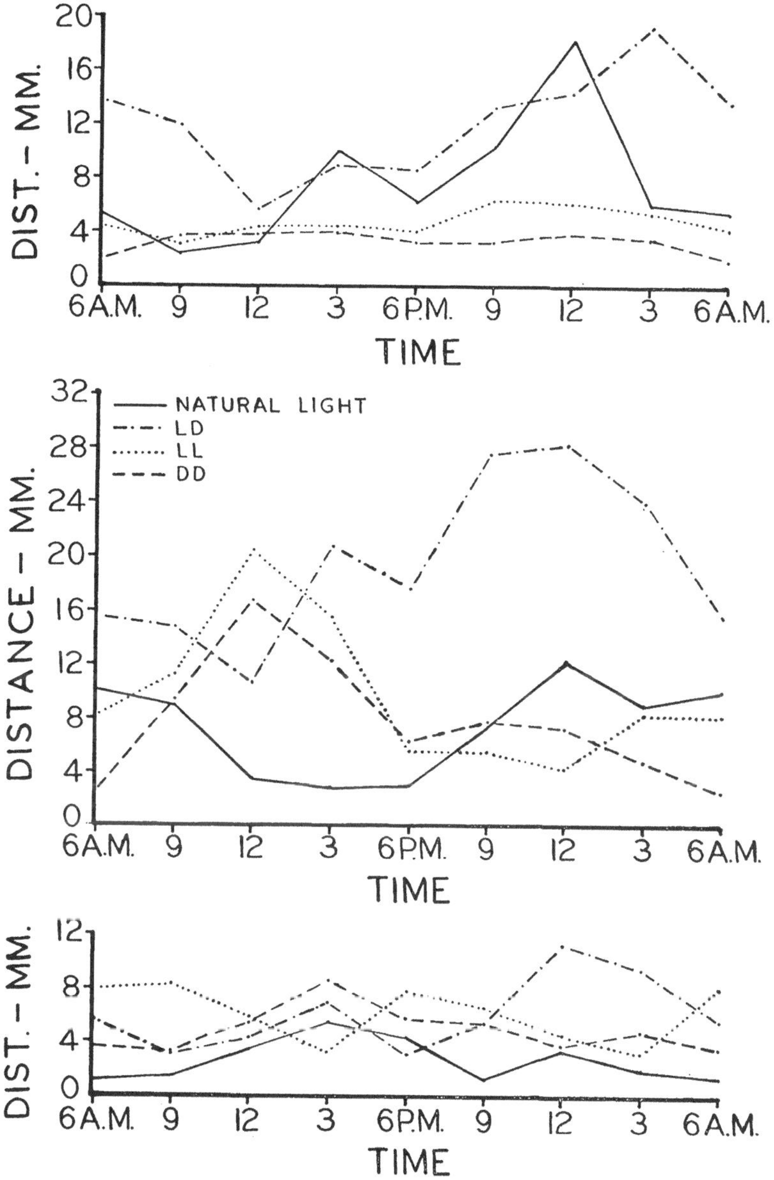

FIGS. 2-4. Mean distances of recorded valve movement over the entire period of investigation. FIG. 2 (top) *Anodonta*, FIG. 3 (middle) *Anodontoides*, FIG. 4 (bottom) *Lampsilis*.

manner: the amount of actual valve movement is not shown; the valves might remain completely closed with sporadic openings, or they might exhibit vigorous movement with short closures. Further, mean values do not reveal sequences where regular openings oc-

TABLE I

MEAN PERIODS (IN HOURS) FROM ONE VALVE OPENING TO THE NEXT

Species	Temp.	Light: LL	Light: DD	Light: LD	Light: Daylight (23 ± 2° C)
Anodonta	11° C	44 (5)	22 (4)	55 (3)	
	22° C	58 (4)	13 (19) a	32 (5)	24 (20) *
					18 (47) *
	29° C	12 (4)	7 (6)	27 (19) b	
				23 (18) †	
Anodontoides	11° C	23 (10)	remained closed	38 (4)	
	22° C	26 (17)	49 (8)	61 (8)	31 (22) c
					25 (18) †
	29° C	remained open until death			
Lampsilis	11° C	74 (7)	26 (4)	46 (4)	
	22° C	37 (4)	31 (7)	42 (4)	41 (21) d
	29° C	22 (8)	31 (7)	23 (10)	

a-d lists (below) actual consecutive periods, in hours, from opening to opening.

a. 10.5, 16, 7, 9.5, 9, 9, 7, 6.5, 8, 8, 7, 8, 8, 8, 13.5, 15.5, 48, 26, 27.5.

b. 13, 12, 15.5, [92], 10.5, 16, 15.5, 34, 17, 29, 26, 23, 19, 29, 24.5, 24.5, 39, 31.5, 35.5

c. 26.5, 23.5, 25, 29, 17, 21, 28, 24.5, 26, 31, [87], 26, 26, 22.5, 26, 21.5, [67], 27, [253], 30, [50], 19

d. 42, 29, 24, 18, 31.5, 74, 50, 33, 24, 37, 39.5, 31, 69.5, 51, 48, 40, 39, 40, 91.5, 30, 26

* Different animals.

† The data immediately above this number were recalculated without including the atypically high valves shown in brackets at the bottom of the table.

curred. Nevertheless, the fact that a closure occurred where the valve muscles showed no overt activity for a period is thought to be of significance. Parts of the record made just after moving the animal and during other major disturbances were not included.

Anodonta and *Lampsilis* generally opened and closed their valves more frequently with increase in temperature. The reaction of

Anodontoides to temperatures revealed the upper and lower limited tolerance. All specimens of *Anodontoides* tested in 29 ± 1° C succumbed within 3 days. The animal in DD at 11° C failed to open during the period lasting 11 days. *Anodonta* exhibited a thermal-death temperature intermediate to *Anodontoides* and *Lampsilis,* in that two animals in a preliminary test lived only 5 and 7 days at a temperature of 29 ± 1° C. A single specimen was still living after 23 days in a temperature of 27 ± 0.5° C.

The valve-opening periods for the specimen of *Anodontoides* in daylight showed a mean of 31 hours; however, if one were to discount the four longest periods (50, 67, 87, 253 hours) the remaining ones fall close to an average of 25 hours. This is also true for the specimen of *Anodonta* in LD. If the 92-hour period is discounted, a 23-hour periodicity remains. These atypically long valve periods, 253 hours for example, suggest that the laboratory environment may be deficient or abnormal in some important way. The *Anodonta* in DD at 22° C showed a very rhythmic period of 8 hours for 4 days and then changed suddenly. Throughout the remainder of the record it never returned to this rhythmicity. This leads to a recommendation that, in further studies, effort should be made to recognize basic natural habitat requirements and to provide for them in the laboratory situation. Another parameter (previously mentioned) which might be helpful in the interpretation of the results is barometric pressure. A lever, recording activity of the foot, would be desirable to show periodicities in locomotory activity. Possibly this would shed light upon certain characteristics of valve movement which have remained uninterpreted to this point.

Conclusions

Where there is a light cue as in LD and normal daylight, the valvular activity is greatest during the dark period. *Anodonta* and *Lampsilis* showed no peak activity when placed in constant light or dark; *Anodontoides* exhibited a peak during the subjective day for these constant conditions. *Anodonta* and *Anodontoides* did not survive exposure to 29° C. *Anodonta* and *Lampsilis* generally showed valve openings to increase in frequency with a rise in temperature, while the reverse was indicated in *Anodontoides.* On several occasions cycles of valve openings approximated periods of 8 hours or multiples thereof. 8-hour periods were shown by *Anodonta* in DD at 22° C. *Anodontoides* exhibited 24-hour periods in LL at 11° C and 22° C and also in daylight. *Anodonta* also showed 24-hour periodicity in LD at 27° C and in daylight. A 40-hour period was shown by *Lampsilis* in LD at 22° C and in daylight.

In all three species valve movement occurred when approximately the same period was repeated for several cycles. In all cases these patterns were disrupted by a change of light or temperature, and often they would disappear for no apparent reason. The sporadic occurrence of prolonged activity and valve closure suggested that more attention should be given to the simulation of natural conditions of the clam's environment.

LITERATURE CITED

BAKER, F. C. 1928. The Fresh Water Mollusca of Wisconsin. Pt. II. Bull. Univ. Wis. 70. 495 pp.

BETHE, A. 1946. Irritabilitat, Rhythmik und Periodik. Naturwiss 33: 86-91.

BROWN, F. A., JR., R. O. FREELAND, AND C. L. RALPH. 1955. Persistent Rhythms of O_2-Consumption in Potatoes, Carrots and the Seaweed, *Fucus*. Plant. Physiol. 30: 280-292.

BUNNING, E. 1960. Opening Address: Biological Clocks. Cold Spring Harbor Symp. Quant. Biol. 25: 1-9.

PENNAK, R. W. 1953. Fresh-Water Invertebrates of the United States. New York: Ronald Press Co. 769 pp.

Papers of the Michigan Academy of Science, Arts, and Letters
Vol. XLIX, 1964 (1963 Meeting)

ENTOMOLOGY

LEPERISINUS ACULEATUS AND *PHLOEOTRIBUS LIMINARIS*, TWO BARK BEETLES OF MICHIGAN HARDWOODS

BRUCE H. BAKER
The University of Michigan

Introduction

The subjects of this study are both members of the family Scolytidae (Coleoptera). *Leperisinus aculeatus* (Say) is commonly known as the ash bark beetle, its hosts being the various species of ash: *Fraxinus oregona, F. acuminata,* and *F. americana* (Chamberlin 1939). *Phloeotribus liminaris* (Harris) is referred to as either the peach bark beetle or the peach-tree bark beetle. Hosts include peach (*Prunus persica*) and other cultivated fruit trees such as plum (*P. domestica*) and cherry (*P. avium*). In addition, this insect lives in black cherry (*P. serotina*) and mountain ash (*Sorbus americana*) (Gossard 1913). The activities of these two species have aroused relatively little attention regarding damage to forest trees. The peach-tree bark beetle has, however, created some concern to orchard growers.

Objective

The objective of this study was to analyze some biological aspects of *L. aculeatus* and *P. liminaris.* These are their life cycles, their population densities at the branch level, their larval survival, and their egg-laying preferences relative to branch surface and branch position.

These two insect species were selected for study because of the similarities in their habits. The entrance tunnel of each species connects with a forked egg gallery which is transverse with respect to the branch axis (Blackman 1922; Chamberlin 1939). The larval galleries extend out from the parent egg galleries. The larval galleries of *L. aculeatus* are quite straight and uniform, whereas those of *P. liminaris* are extremely winding and less orderly. Their geographic distributions overlap to the extent that the ranges of their hosts do. The two species may be distinguished from each other by appearance as well as by host specificity. *L. aculeatus* is dark brown with a dorsal marking formed by light-colored scales (Blackman 1922). *P. liminaris* lacks the mottled dorsum and has feebly rugose elytra (Chamberlin 1939).

Location of the Study Area

The area selected for a study of the two insect species is located within 25 miles of Ann Arbor, Michigan. Two hardwood lots were selected within Stinchfield Woods, a forest research area maintained by the School of Natural Resources of The University of Michigan (Univ. Mich. 1960). The selected lots typify the mixed hardwood forests found in southeastern Michigan.

In each of these lots, the harvested trees have been utilized to approximately a four-inch diameter limit. Therefore, the potential breeding sites in deadwood are in the stumps, the branches less than four inches in diameter, whatever dead branches might appear in living trees, and the stacked fuel wood which is left in the stand to season for one summer.

Sampling Techniques—Summer 1962

In lot 59, one large white ash (27.5-inch stump diameter) was harvested in a 1961-62 partial cutting. This provided an abundance of slash resulting from the extraction of sawlogs and fuel wood. Various branch samples were collected from this tree. After a commercial cutting in lot 11, the slash of one black cherry (15-inch stump diameter) was included in the post-logging residue. This, combined with a storm-stricken tree of the same species, provided an adequate number of branches from which to sample beetle activities.

Sampling of branches for beetle development and activity was conducted at approximately two-week intervals. A branch sample is defined as a 36-square-inch sample of a branch measuring from two to ten inches small end diameter. These samples are categorized in one of the following four classes with respect to position and orientation.

Class 1 Sample taken from the upper surface of a branch lying on the surface of the forest litter.

Class 2 Sample taken from the lower surface of a branch lying on the surface of the forest litter.

Class 3 Sample taken from the upper surface of a branch positioned aerially from 5 to 20 inches above the surface of the forest litter.

Class 4 Sample taken from the lower surface of a branch positioned aerially from 5 to 20 inches above the surface of the forest litter.

In each instance the two samples from a given vertical position (on litter surface or aerial) were paired as they were from the same section of the same branch, thus eliminating the variability between branch segments. Insect activities were then compared with respect to these various positions. In order to dissect the branch samples under laboratory conditions, each piece was marked and then preserved with the bark intact under freezing temperatures until a detailed analysis could be made. At such time, a specimen was thawed in a tray of water to loosen the bark. Upon removal of the bark, measurements were made regarding egg gallery lengths, larval gallery lengths, and the ratio of pupal chambers to egg niches. Discretion was displayed in the selection of branch samples in that signs of beetle activity were sought. There were, however, a number of collected branch samples which had the outward appearance of containing either the ash or the peach-tree bark beetle but which when further analyzed showed no evidence of their presence.

Life Cycles

The developmental rates of the two species of beetles are compared in Fig. 1. The duration of each stage is an estimate based upon actual observations made on the indicated dates. The taper of each lens is an estimate of the relative rates of increase or decrease in numbers of individuals in each stage.

Leperisinus aculeatus.—The eggs began to hatch during the latter half of May, and the first pupae appeared in early July. Emergence of the new adults was completed in late September. In Mississippi, Blackman (1922) recalled one year in which the larvae were from half to full grown and some pupae had formed by April 24. He noted that the new adults emerged between May 16 and the last of June. As may be expected, the developmental sequence begins sooner and is more accelerated in Mississippi than in southeastern Michigan.

Phloeotribus liminaris.—The beetles observed in Stinchfield

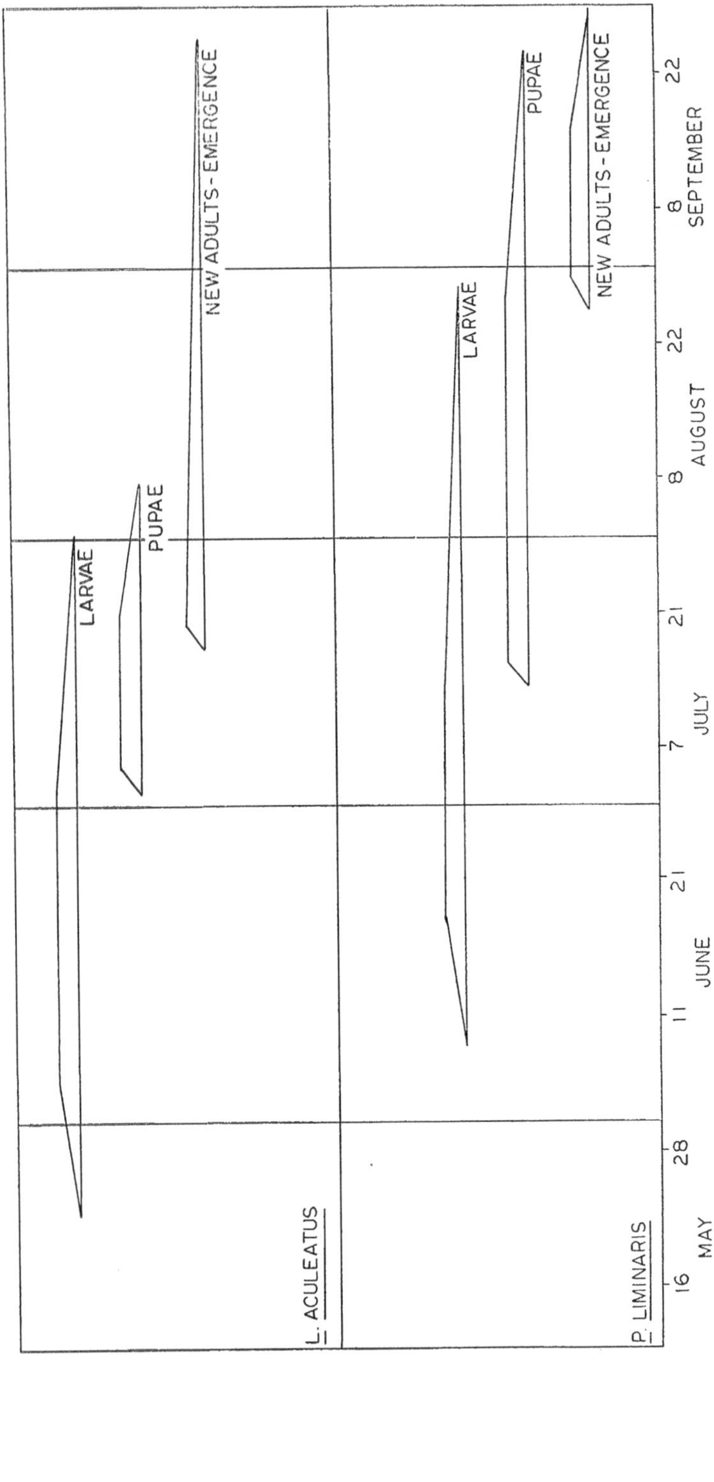

Fig. 1. Seasonal development of *Leperisinus aculeatus* and *Phloeotrib us liminaris*, 1962, Stinchfield Woods, Washtenaw County, Michigan.

Woods began reaching maturity in late August. According to Gossard (1913), in Ohio the summer brood of peach-tree bark beetles matures and emerges from about July 20 to early October. Although the Michigan beetles reached maturity earlier than those in Ohio, the last Michigan beetles emerged at approximately the same time as those in Ohio (early October).

Population Characteristics

Larval gallery characteristics.—The larval gallery lengths for each species are as follows:

	Mean length (cm) of completed larval galleries
L. aculeatus	1.98 ± 0.22 at 95% probability
P. liminaris	3.46 ± 0.32 at 95% probability

Gossard's values (1913) for the larval gallery length of the peach-tree bark beetle are from 1.5 to 2.9 in. (3.8 to 7.4 cm).

Larval mortality.—As an index of egg failure and larval mortality in the ash bark beetle, adult galleries were examined; and a ratio of egg niches per egg gallery to pupal chambers per egg gallery was established. An average of 46% of the egg niches developed into pupal chambers. This percentage would vary depending on parasitism and other environmental pressures. A similar value could not be calculated for the peach-tree bark beetle in black cherry because the egg niches are minute and often difficult to distinguish. In addition, the pupal chambers are difficult to associate positively with the egg galleries.

Blackman (1922) states that the average adult gallery of the ash bark beetle has from 40 to 60 egg niches.

Insect Preference Relative to Branch Surface and Branch Position

Blackman (1922) observed that no ash bark beetle galleries had been started upon the uppermost side of one felled tree, but the sides and underside were "peppered" with entrance holes. Using the four classes of branch samples mentioned earlier, an effort was made to compare the preference of adult beetles for branch surface, upper or lower. An additional comparison involved branches positioned on the ground and those positioned aerially.

These comparisons were made for both species. The results appear in Tables I and II. These data compare the relative numbers of galleries as well as the average gallery lengths. The significance of

TABLE I

PREFERENCE OF *Leperisinus aculeatus* RELATIVE TO BRANCH POSITION AND BRANCH SURFACE IN POST-LOGGING SLASH OF WHITE ASH

	MEAN GALLERY LENGTH Aerial position			Ground position		
	Upper	Lower	Total	Upper	Lower	Total
Total combined length of egg galleries in cm....	196.6	615.9	812.5	98.4	441.9	540.3
Total number of egg galleries	61	167*	228	25	94*	119*
Mean egg gallery length in cm	3.2	3.7	3.6	3.9	4.7	4.5

* Estimated to be a significant difference.

TABLE II

PREFERENCE OF *Phloeotribus liminaris* RELATIVE TO BRANCH POSITION AND BRANCH SURFACE IN POST-LOGGING SLASH OF BLACK CHERRY

	MEAN GALLERY LENGTH Aerial position			Ground position		
	Upper	Lower	Total	Upper	Lower	Total
Total combined length of galleries in cm........	24.7	129.4	154.1	44.1	180.0	224.1
Total number of egg galleries	10	42*	52	18	50*	68*
Mean egg gallery length in cm	2.5	3.1	3.0	2.4	3.6	3.3

* Estimated to be a significant difference.

the differences has been estimated. Due to the striking differences between comparable values, no statistical significance tests were applied.

Summary and Conclusions

The ash bark beetle (*Leperisinus aculeatus*) and the peach-tree bark beetle (*Phloeotribus liminaris*) are members of the family Scolytidae. The following conclusions may be drawn from a study of the two species.

1. The two species are quite similar in size. The most distinctive feature separating the adults in appearance is the mottled pattern on the dorsum of *L. aculeatus*.

2. The egg gallery patterns of the two species are similar in that they both form two transverse burrows, one on each side of the entrance hole and nuptial chamber. In each case, the eggs are deposited in niches along the margins of the egg gallery. The average length of *P. liminaris* larval galleries is about 3.5 cm, whereas the average for *L. aculeatus* is only 2 cm. The larval galleries of *L. aculeatus* are quite parallel and uniform, whereas those of *P. liminaris* are meandering and nonuniform in length.

3. Both species are attracted to dead, dying, or weakened wood of their particular host tree species. Therefore, the economic importance of these insects under present standards of utilization is low.

4. Both species tend to enter the branches of post-logging slash via the lower surfaces. This tendency is probably due to an attraction to cool and shaded surfaces where conditions are favorable for moisture.

A significant difference apparently exists between the number of *P. liminaris* egg galleries on the branches positioned aerially and the ones positioned on the ground. In the case of *L. aculeatus*, 228 egg galleries were started on the aerially positioned branch samples. The total number on those branch samples located on the ground was only 119. It would be interesting to determine the air temperatures and relative humidity levels at various elevations above the forest floor. Granted that temperature and moisture conditions inside the bark may regulate copulation, gallery construction, and egg deposition, it is nevertheless the conditions prevailing outside the bark which attract the adult insects. It is possible that the aerially positioned branch is cooler than the one on the ground. Thus, one would expect the highest count of galleries to be on the lower surface of the aerially positioned branch. This was the case with *L. aculeatus*.

5. The geographic ranges of the insects overlap to the extent that their host tree ranges overlap.

6. *L. aculeatus* entered each of its developmental stages earlier in the summer season than did *P. liminaris.* Despite the earlier development of the ash bark beetle, the populations of both insects completed their emergence at nearly the same time, late September.

There was ample time left in the summer season for at least one additional generation of ash bark beetles to develop. Blackman (1922) concludes that the insect has three generations per season in Mississippi.

Gossard (1913) maintains that two broods of *P. liminaris* develop per year in northern Ohio. One might conclude that the first adults to emerge in Stinchfield Woods could form new egg galleries which would produce mature adults before the end of a single growing season. Further, a more extensive survey might indicate the presence of a second brood resulting from the early developing adults.

ACKNOWLEDGMENTS

I am especially grateful to Fred B. Knight, School of Natural Resources, The University of Michigan, for his interest and assistance throughout the fieldwork and the preparation of the manuscript. I wish also to thank W. H. Anderson of the U. S. Department of Agriculture for his verification of the identities of the two bark beetles studied.

LITERATURE CITED

BLACKMAN, M. W. 1922. Mississippi Bark Beetles. Miss. Agr. Exp. Sta. Tech. Bull. 11.

CHAMBERLAIN, W. J. 1939. The Bark and Timber Beetles of North America. OSC Cooperative Association, Corvallis, Oregon. Litho. U. S. A. 513 pp. illus.

GOSSARD, H. A. 1913. Orchard Bark Beetles and Pin Hole Borers. Ohio Agr. Exp. Sta. Bull. 264.

University of Michigan School of Natural Resources. 1960. A Guide to Stinchfield Woods.

PAPERS OF THE MICHIGAN ACADEMY OF SCIENCE, ARTS, AND LETTERS
VOL. XLIX, 1964 (1963 MEETING)

TWO CASES OF ABNORMAL DEVELOPMENT IN MALE POSTABDOMENS OF FLIES OF THE FAMILY SCIOMYZIDAE (DIPTERA)

GEORGE C. STEYSKAL
Entomology Research Division, U. S. Department of Agriculture, Washington, D. C.

DURING the course of taxonomic studies, one specimen each of *Sepedon haplobasis* Steyskal and *Limnia ottawensis* Melander was encountered with interesting and apparently hitherto unrecorded abnormalities in the development of the male postabdomen. The specimens are in the collections of the U. S. National Museum.

Sepedon haplobasis Steyskal

(FIG. 1)

1960. Proc. Ent. Soc. Wash. 62: 42.

One male specimen of the original series collected by Neff and Matthews in Mexico and reared in the Cornell University laboratories was found to have the postabdomen only partially circumverted and to have retained the dorsal tracheal trunks of the larval stage. The circumversion of the postabdomen apparently takes place between the 7th and 8th somites (cf. Steyskal 1957), the protandrium therefore being only secondarily affected by the rotation of the andrium. The retention in this specimen of the larval tracheal trunks (tt) has apparently contributed to the failure of the andrium to circumvert and to bring the hypandrium (hyp) and aedeagus (ae) into a ventral position. The 6th and 7th spiracles of the adult (6, 7 sp) are apparently not furnished with tracheae, although they are approximately in their final position. The location of the tracheal trunks just posterad of the anterior margin of the 6th tergite is interesting, since they are usually considered in the larva to be on the 8th or 9th abdominal segment (Hennig 1948, p. 41; Snodgrass 1924, p. 4; Snodgrass 1935, p. 438).

Limnia ottawensis Melander, new status

(Fig. 2)

1920. *Limnia saratogensis,* var. *ottawensis.* Ann. Ent. Soc. Am. 13: 324.

Studies now in progress indicate that this form is entitled to specific rank. A specimen from Zanesville, Ohio, June 11, 1954 (D. L. Haynes), represents a far more complete development than the foregoing one. The structures are almost completely and characteristically developed; only the andrium has failed to make the

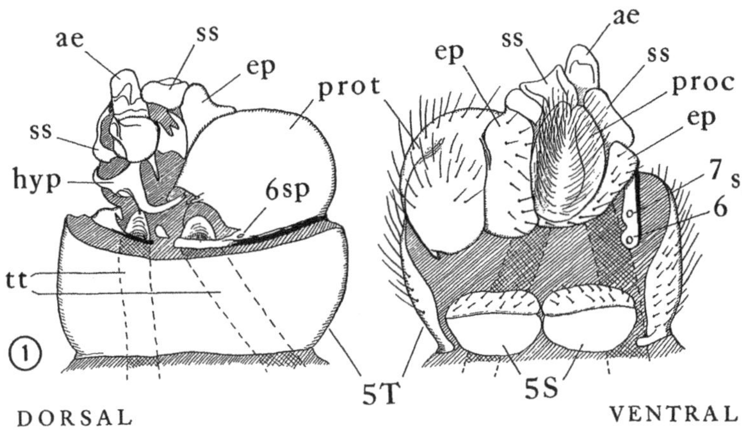

Fig. 1. *Sepedon haplobasis* Stey., abnormal male postabdomen.

last 90° of its rotation. In the figure, points A_1 and A_2 should be adjacent to each other, the hypandrium (hyp) should underlie the flap (7S, probably 7th sternite), and the aedeagus (ae) should be ventral rather than lateral in position. It was at first thought that the specimen had been pried open and the andrium displaced while the insect was still soft, but when the postabdomen was macerated the membranes were found intact and the true condition was discovered.

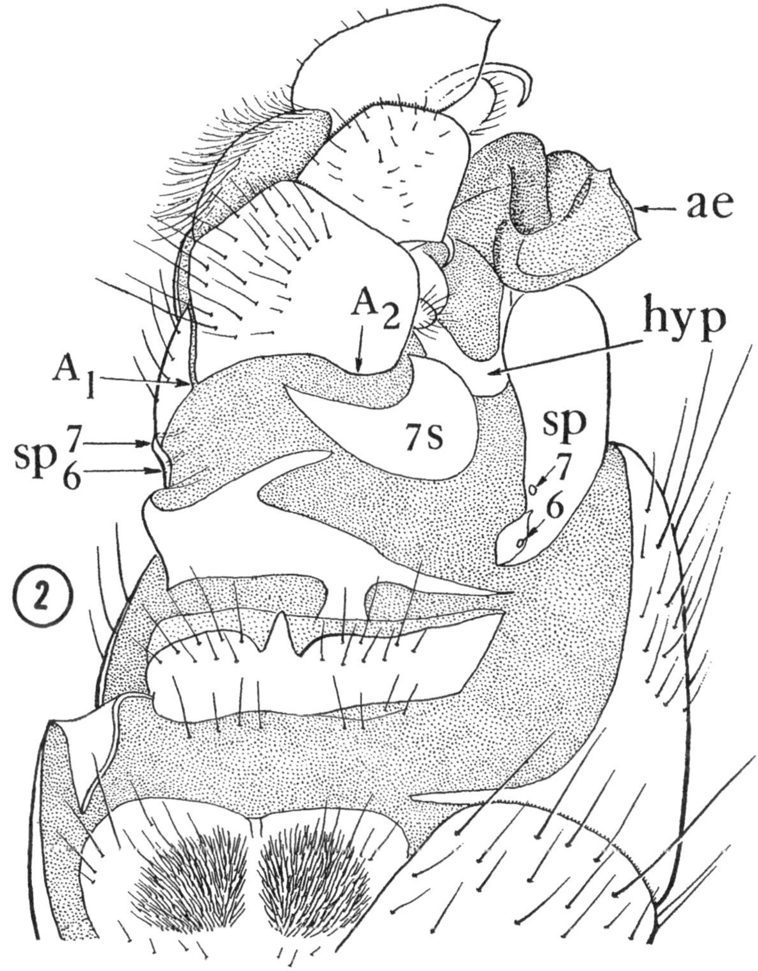

FIG. 2. *Limnia ottawensis* Mel., abnormal male postabdomen, ventral view. LEGEND: A_1, A_2—reference points (see text); ae—aedeagus; ep—epandrium; hyp—hypandrium; proc—proctiger; prot—protandrium; sp—spiracle; ss—surstylus; S—sternite; T—tergite; tt—tracheal trunks.

LITERATURE CITED

HENNIG, W. 1948. Die Larvenformen der Dipteren. 1. Teil. Akademie Verlag, Berlin, pp. 1-(185).

SNODGRASS, R. E. 1924. Anatomy and Metamorphosis of the Apple Maggot, *Rhagoletis pomonella* Walsh. Jour. Agr. Res. 28: 1-36, pls. 1-6.

—— 1935. Principles of Insect Morphology. New York and London: McGraw-Hill Book Co., 667 pp.

STEYSKAL, GEORGE C. 1957. The Postabdomen of Male Acalyptrate Diptera. Ann. Ent. Soc. Am. 50: 66-73.

Papers of the Michigan Academy of Science, Arts, and Letters
Vol. XLIX, 1964 (1963 Meeting)

PERITHOUS DIVINATOR ROSSI AND ITS HOST, *PEMPHREDON LETHIFER* SHUCKARD

SYLVAN J. THOMAS
St. Clair Shores, Michigan

For the past several years, I have been rearing individual specimens of the twig-nesting wasps and their parasites collected in Michigan. Several specimens have been exposed to varying environmental conditions resulting in both normal and abnormal developmental periods. Two such rearings during the post-diapause period produced specimens of *Perithous divinator* Rossi, an ichneumon parasite, and *Pemphredon lethifer* Schuckard, its host.

The literature relating to the host and the parasite is scanty. H. and M. Townes (1960) have published on *P. divinator*, which includes the collection dates, rearing records and host species. They recorded *P. divinator* from the locality of East Lansing, Michigan. The only host species listed for this parasite is *Pemphredon unicolor* Latreille in France. Krombein (1959) has published notes recording the distribution of *P. lethifer* in the U. S. Chandler (1960) has published some biological notes relating to the summer broods of *P. lethifer*.

The purposes of this paper are (1) to record *P. lethifer* as a host of the ichneumon *P. divinator* in Michigan; (2) to present some data on the synchronized developmental periods of this host and its parasite; and (3) to describe the developmental periods of one specimen of *Perithous spp.* reared under refrigeration, delaying post-diapause development.[1]

Observations of Normal Post-Diapause Development

Two infested nests of *P. lethifer* (numbers 611 and 612) were collected in Ogemaw County, Michigan, in twigs of sumac on April 30, 1961. Nest 611 contained five cells separated by a thin partition. Cells one, two, four and five contained yellow diapause larvae char-

[1] I am indebted to Dr. Karl V. Krombein of the USNM who identified the host sphecid, *P. lethifer*. I am also indebted to Miss Luella Walkley of the U. S. National Museum and H. Townes, of the University of Michigan, who identified the ichneumons and provided several biological references cited.

acteristic of the *Pemphredons*. Cell three contained a white diapause larva of *P. divinator* measuring 8-9 mm in length. No other distinguishing structural characteristics were noted. During the post-diapause developmental period, the larvae were kept in temperatures above 60° F. No apparent structural characteristics developed to indicate post-diapause development until May 7, when eye pigmentation appeared in the *Perithous* larva of cell three. This larva pupated during the following five days, between May 9 and May 12. Pigmentation was apparent in the exoskeleton by June 2, and the adults emerged June 9.

Three of the host *Pemphredon* larvae in the nest not parasitized by *P. divinator* pupated between May 9 (cells two and four) and May 12 (cell one). The larva of cell five died without pupating. The host pupae were pigmented in the exoskeleton by June 2 and the adults emerged about June 9.

Nest 612 contained six cells. Cells one to five contained the typical white diapause larvae of the parasite *P. divinator*. Cell six, nearest the tunnel opening, contained a typical diapause larva of *P. lethifer*. Eye pigmentation appeared in the *Perithous* larvae about May 3. By May 7, larvae in cells four and five had pupated and by May 9, the remaining parasitic larvae pupated. An adult emerged from cell two on May 30, and an adult emerged from cell four on June 1. The remaining specimens were in adult form by June 6.

The single host in this nest pupated by May 12, and the adult emerged by June 1.

These nests were kept under normal seasonal conditions, except that they were transported from Ogemaw County to Macomb County for the remainder of their developmental periods.

Observations of Retarded Post-Diapause Development

An unusual set of circumstances resulted in the rearing of a *Perithous* specimen during the 1962 season. I had collected a nest of *Pemphredon* containing diapause larvae in Macomb County, Michigan on March 3. The larvae were kept indoors for five days and then transferred to capsules and placed in refrigeration at about 43° F. to retard development until later in the season. The specimens were checked on July 10, and at this time I noted that the eyes had become pigmented in the *Perithous* larvae, a condition which was not in evidence before refrigeration began. The second observation made on July 30 confirmed the observation that the parasite had continued development at this temperature. It was now in the pupal form. I removed the capsule on November 11. Three days later pigmentation appeared in the thoracic region. Pigmentation of the integument continued and on November 23, the specimen was in the adult form. It had reached maturity but was defective in several areas. The wings, which normally should have been extended during the last ecdysis, were in the pads. However, the abdomen was pulsating, and the appendages were free and flexing as the insect attempted to move the antennae and legs. The insect survived until November 29, about six days after reaching the imago. The elapsed time for the developmental period from collection date was 260 days.

LITERATURE CITED

CHANDLER, L., N. WORK, AND F. SHEWMAN. 1960. The Life History of *Pemphredon lethifer lethifer* Shuckard (Hymenoptera: Sphecidae). Proc. Ind. Acad. Sci. 69 (1959): 148.

KROMBEIN, KARL V. 1959. Two Additional Adventive European Wasps in the United States. (Hymenoptera: Sphecidae, Chrysididae). Bull. Brooklyn Entomol. Soc. No. 3, 54: 95-96.

TOWNES, HENRY AND MARJORIE. 1960. Ichneumon Flies of America North of Mexico. U.S. Natl. Mus. Bull. 216 (2): 215-216.

Papers of the Michigan Academy of Science, Arts, and Letters
Vol. XLIX, 1964 (1963 Meeting)

A SIMPLIFIED TECHNIQUE FOR REARING AND MAINTAINING A COLONY OF HOUSE FLIES (*MUSCA DOMESTICA* L.)

G. W. OZBURN
Northern Michigan University

Introduction

The following technique has been modified after that originally described by Fisher and Morrison (1949). It is an extremely easy, inexpensive, and relatively clean method of culturing house flies and, in addition, eliminates the necessity of keeping materials in the laboratory which may be subject to insect and mite infestation.

Rearing the Maggot

The method consists of using 6 six-gram lots of shredded cellu-cotton compressed into pads in 6 sterilized pint milk bottles. 360 ml of homogenized milk containing approximately 6 mg of cholesterol and 3 gm of brewer's yeast are mixed in a Waring Blender. This mixture is added to the pads, 60 ml per pad, prior to seeding with eggs.

Eggs are collected in the morning from collecting pads placed in the cages the previous evening. These pads, consisting of two pieces of absorbent cotton fitted into a petri dish, are soaked with a 1:3 milk-distilled water mixture and are arranged to form a central groove in which the flies tend to concentrate their eggs during oviposition. Skim milk powder and sugar are sprinkled on the pads, as well as a few drops of ammonia to act as an attractant to ovipositing flies.

The masses of eggs are gently lifted off the pad with slide forceps and placed in distilled water in a 50 ml beaker. When necessary, to ensure a sufficient number of eggs, some of those adhering to the cotton pad are washed off with a jet of water from a wash bottle. By means of squeezing the rubber bulb of a modified glass pipette, the water containing the eggs is agitated and the eggs separated from one another. Floating nonviable eggs are decanted off and discarded. Viable eggs from the bottom of the beaker are transferred to a two-inch graduated section of a 10 ml pipette which has a piece of fine copper gauze soldered to one end. Enough eggs are transferred to give 0.3 cc when compacted.

The 0.3 cc of eggs are then transferred from the two-inch section of pipette by inverting it over a test tube and washing from the bottom with distilled water. The test tube is then filled with water to a level of 9 ml and the eggs dispersed by agitation. Uniformity of egg distribution is ensured by agitation of the suspension. Eggs are pipetted from this suspension with a 5 ml graduated pipette marked off at 1.5 ml, and 1.5 ml or about 250 eggs are pipetted into each of six culture jars. These jars are then loosely stoppered with cotton wool and placed in a darkened cabinet at 27 ± 1° C.

On the third day, when the larvae are approaching maturity, a piece of dry crumpled cheesecloth is placed on the top of the culture for the larvae to crawl into and pupate. On the fourth day the cotton plug is removed to allow aeration and thus encourage drying of the culture.

After pupation has taken place, the wad of cheesecloth containing pupae is removed. On the eighth day after seeding the culture, the pupae are separated from the cheesecloth by running it between the forefinger and thumb, thus dislodging the pupae into a large plastic basin. Before reuse, the milk bottles and cheesecloth are autoclaved, the latter being cleaned of any larval medium adhering from the previous use.

Care of the Adults

Adult flies kept for maintaining the colony are held in conventional 12″ × 12″ × 12″ cages. Two sides and the top of these cages are covered with copper screening. The bottom is made of galvanized iron. The front and back are made of masonite, constructed to fit tightly into slots, and are removable. The front panel contains a tube-like cotton sleeve fastened around a circular hole of 7″ diameter by a piece of masonite. This cloth sleeve is approximately 20 inches in length, permitting it to be knotted to prevent the escape of flies. These cages are readily cleaned after use by removing the front and back panels; yet they are rigid due to the solid metal floor. When in use the cages are kept in a rearing room at a constant temperature of 27 ± 1° C and a relative humidity ranging from 40-60%.

Two methods of rearing adults are used, one to ensure a daily supply of newly emerging flies, the other to keep the colony in a good state of production for periodic use.

The 11-day plan has evolved for daily supply and may be outlined as follows:

Sun	Mon	Tu	Wed	Th	Fri	Sat	Sun	Mon	Tu	Wed	Th	Fri
A	A	A	*A*	*A*	*A*	A						
	B	B	B	*B*	*B*	*B*	B					
		C	C	C	*C*	*C*	*C*	C				
			D	D	D	*D*	*D*	*D*	D			
				E	E	E	*E*	*E*	*E*	E		
					F	F	F	*F*	*F*	*F*	F	
						A	A	A	*A*	*A*	*A*	A

The letters A, B, . . . F designate breeding cages, and the underlined letters refer to the presence of an oviposition dish on a specific day. The procedure is to stock cage A on Sunday with newly emerged flies. These are fed daily by means of a cotton pad soaked in fresh homogenized milk diluted 1:3 with distilled water, and with about a teaspoonful of skim milk powder and granulated sugar 1:1 sprinkled over the top. This pad is replaced Monday and Tuesday evening, but removed on Wednesday evening and replaced with an oviposition pad. On Saturday the flies are cleared from cage A, the cage cleaned, restocked, and the cycle begun again. Extra cages are useful so that the necessity of cleaning and drying a cage immediately is eliminated. This procedure is used for each cage.

Old flies are killed by placing the entire cage in a sterilizing oven prior to washing.

For the purpose of keeping the colony in reasonable numbers for periodic use, a modification of the above rearing program has been devised. This consists of starting new cultures every four days and thereby maintaining only three cages of flies. The oldest cage from which eggs are collected in the morning is replaced by a cage of fresh flies. The following illustration serves to describe this cycle, which may go up to 13 days for any one cage.

Sun	Mon	Tu	Wed	Th	Fri	Sat	Sun	Mon	Tu	Wed	Th	Fri	Sat	Sun
A	*A*													
B	*B*	B	B	B	*B*									
C	*C*	C	C	C	*C*	C	C	C	*C*					
	D	D	D	D	*D*	D	D	D	*D*	D	D	D	*D*	
					A	A	A	A	*A*	A	A	A	*A*	A
									B	B	B	B	*B*	B
													C	C

It was found that the flies maintained their egg-laying capabilities for this long a period provided they were fed ample food and the relative humidity was not allowed to drop to the point where feeding pads dried out before being replaced.

LITERATURE CITED

FISHER, R. W., AND F. O. MORRISON. 1949. Methods in Rearing and Sexing *Musca domestica* L. 80th Ann. Rep. Ont. Entom. Soc., pp. 41-45.

PAPERS OF THE MICHIGAN ACADEMY OF SCIENCE, ARTS, AND LETTERS
VOL. XLIX, 1964 (1963 MEETING)

FISHERIES AND WILDLIFE MANAGEMENT

CONSUMPTION OF NATURAL FOODS BY CAPTIVE CANVASBACK AND LESSER SCAUP

JERRY R. LONGCORE
The University of Michigan

GEORGE W. CORNWELL
Virginia Polytechnic Institute

INTRODUCTION

WATERFOWL food habits have been extensively studied, but measurements of the daily consumption of natural foods have not been made. This paper reports the daily intake of natural foods necessary to maintain captive diving ducks of two species in good physical condition. 37 juvenile canvasbacks, *Aythya valisineria*, and lesser scaup, *Aythya affinis*, were fed aquatic plants and invertebrate animals collected from the lower Detroit River. Feeding was carried out for 72 days, from October 2 to December 13, 1961.

PROCEDURES

Food collection sites were located on the lower Detroit River. They extended in a north-south direction from Grassy Island to Sturgeon Bar, and from the Gibraltar channels on the west to Stoney Island on the east side of the river. Most of the collections were made along the west shores of Grosse Ile and Celeron Island. This portion of the Detroit River area has been well described by Hunt (1957).

Food materials were raked or dredged from the river bottom and deposited in ten-gallon milk cans. Vegetation was washed and screened to concentrate invertebrate animals, mainly mollusks. A total of 28¼ cans of food materials was gathered in this manner. Field records were kept of all food items collected.

The contents of six ten-gallon milk cans were analyzed in the laboratory. They were emptied onto a stainless steel table and the various species of aquatic plants were sorted and identified. Invertebrate animals were sorted and weighed on a torsion balance.

The various plant species were washed free of mud and debris with a water spray, allowed to drain on paper toweling, and weighed on a torsion balance. All weights were recorded to the nearest tenth of a gram. Subsamples of this vegetation were set aside for

TABLE I

SUMMATION OF SAMPLE FOODS ANALYZED

Kind of Food	Weight in Pounds Wet	Oven-dried	Percent of Total Wet Weight
Vallisneria americana...........	165.9	9.95	76.2
Elodea canadensis..............	35.8	2.15	16.4
Chara sp......................	6.8	0.41	3.1
Heteranthera dubia............	3.2	0.19	1.5
Ceratophyllum dermersum......	2.4	0.14	1.1
Myriophyllum sp...............	0.3	0.02	0.1
Invertebrate animals...........	3.4	0.20	1.6
Totals	217.8	13.06	100.0

TABLE II

TOTAL FOODS COLLECTED DURING FEEDING PERIOD

Food Item	No. Units *	Weight in Pounds Wet	Oven-dried	Percent of Total Wet Weight
Vallisneria americana.....	18.00	653.4	40.46	63.7
Elodea canadensis........	4.25	154.3	9.55	15.1
Lemna sp................	1.00	36.3	2.25	3.5
Heteranthera dubia......	0.63	22.7	1.41	2.2
Myriophyllum sp.........	0.37	13.6	.84	1.3
Ceratophyllum dermersum	0.25	9.1	.56	0.9
Invertebrate animals......	3.75	136.1	8.43	13.3
Totals	28.25	1,025.5	63.50	100.0

* Ten-gallon milk cans.

dry weight determinations. These were dried at 37° C in an electric oven, with circulating air forced through the oven by a compressor. This decreased the drying time so that only a day and a half was required to dry each sample. A summation of sample foods analyzed and total foods collected is presented in Tables I and II.

The birds were confined in a pen described by Cornwell and Hartung (article in press). Captive birds adjusted well to the type of housing used. They spent a great deal of time on the water and did not exhibit signs of stress commonly associated with crowding.

The two species of ducks were fed in their respective compartments, and both preferred diving for food in the water rather than feeding from pans. Food was placed in the water tank whenever raking the bottom suggested depletion. An excess of food was kept in the tank at all times.

The tank was cleaned periodically to facilitate feeding by the waterfowl and for reasons of sanitation. Unconsumed food materials were removed at this time (Table III).

TABLE III

AMOUNT OF UNCONSUMED FOOD (POUNDS, WET WEIGHT)

Feeding Period	Canvasback	Lesser scaup
Oct. 2 to Oct. 24	7.3	7.3
Oct. 24 to Nov. 12	19.4	58.1
Nov. 12 to Dec. 13	00.0	43.6
Totals	26.7	109.0

Salmonellosis, a bacterial disease, was present in the captive flock and necessitated medication and supplemental feeding of shelled corn and wheat from October 7 to 31.

To determine if the criterion of adequate nutrition had been fulfilled, duck body condition was evaluated by three methods: (1) total body weight, (2) classification on the basis of fat deposits, and (3) measurement of emaciation. An Emaciation Index (Cornwell and Cowan, in press) was applicable.

RESULTS

Twelve trips were made to the Detroit River for food collection between October 1 and December 6. A total of 1,025 lb (wet weight) of plant and animal foods were collected. (All food weights are wet weights unless otherwise stated.) Of this material, wild celery (*Vallisneria americana*) comprised 63.7%; waterweed (*Elodea canadensis*), 15.1%; and invertebrate animals, 13.3%. The invertebrate animals were primarily mollusks (85.0%), but included aquatic insects, leeches, and freshwater scuds. Mollusks were well represented by the genera *Lymnaea, Campeloma, Amnicola, Physa, Valvata,* and *Helisoma.*

Salmonellosis was detected in the birds at the onset of the experiment. These ducks were hatched and reared at the Delta Waterfowl Research Station, and we believe the disease was contracted

TABLE IV

MEAN FOOD CONSUMPTION PER BIRD PER DAY

Feeding Period	Amount of Food in Pounds Wet	Oven-dried *	Waterfowl Consumption Days	Pounds per Bird per Day Wet	Oven-dried *
		NATURAL FOOD			
Oct. 2 to Nov. 12 (*41 days*)					
Canvasback	318.4	19.11	479	0.664	0.040
Lesser scaup	279.7	16.79	654	0.444	0.026
Nov. 12 to Dec. 13 (*31 days*)					
Canvasback	167.9	10.07	145	1.158	0.069
Lesser scaup	124.3	7.45	171	0.726	0.044
Oct. 2 to Dec. 13 (*72 days*)					
Canvasback	486.3	29.18	624	0.780	0.046
Lesser scaup	404.0	24.24	825	0.488	0.029
		MIXED GRAINS			
Oct. 7 to Oct. 31 (*24 days*)					
Canvasback	–	20.00 †	283	–	0.071 †
Lesser scaup	–	20.00 †	408	–	0.049 †

* Oven-dried weights were obtained by drying 21 wet vegetation samples and computing a conversion factor of 1 lb, wet weight, equals 0.06 lb, dry weight.

† Commercial weight.

there. Salmonellosis commonly occurs following the stress of changes involving transportation, food, or housing. Supplemental feeding of grains and treatment with antibiotic drugs suppressed the disease after a temporary disruption of feeding and a 10% mortality. After the second week of treatment, the ducks regained the weight lost, were again active, and no longer exhibited symptoms associated with salmonellosis. Although it cannot be concluded that the birds were free of disease, the effects of the disease upon food consumption were thought to be minimal.

The total consumption of food by the canvasback and lesser

scaup for the entire feeding period was 486.3 lb and 404.0 lb respectively (Table V). (Food consumption rates are based on waterfowl consumption days, WCD's. A WCD equals one duck feeding for one day.)

The mean WCD for the first 41 days was 0.66 lb for the canvasback and 0.44 lb for the lesser scaup. The remaining 31 days showed an increase in the mean WCD of 0.49 lb for the canvasback and 0.28 lb for the lesser scaup. The canvasback consumed 1.16 lb per WCD, and the lesser scaup consumed 0.73 lb over the remaining

TABLE V

Body Condition as Rated by Fat Deposits

Species	Rating			
	Excellent	Good	Fair	Poor
Canvasback	8	1	2	5
Lesser scaup	11	5	2	3

31-day period. For the entire 72-day period, canvasbacks consumed 0.29 lb more food per WCD than did the lesser scaups.

Grain consumption from October 7 to 31, 1961 was 0.07 lb for the canvasbacks and 0.05 lb for the lesser scaups. These amounts of grain are thought to be insignificant in relation to the total foods consumed.

Mean body weights of experimental canvasbacks were 1.97 lb for males and 2.03 lb for females. Male and female lesser scaup had mean weights of 1.23 and 1.21 lb respectively.

Classifications of body condition based on subcutaneous, inguinal, visceral, and cardiac fat deposits are presented in Table V.

Discussion

Moyle (1961) observed that a 3-lb mallard would have to consume approximately 1.5 lb of aquatic plants per day to obtain its protein requirement of 2 gm per lb of body weight. Our results show comparable consumption rates of 0.78 lb per day for a 2-lb canvasback and 0.49 lb per day for a 1-lb lesser scaup. Sincock (1962) estimated daily consumption of natural vegetation to be 0.27 lb, dry weight, for the canvasback and 0.20 lb for the lesser scaup. These calculations are based upon a consumption rate of

10% of the bird's body weight. Our ducks, however, consumed approximately 2 to 3% of their body weight daily in oven-dried foods. Sincock assumed that dry-weight determinations by various workers were equivalent, which may explain why our oven-dried weights of 0.05 lb for the canvasback and 0.03 lb for the lesser scaup do not agree with his estimates. Other studies of grain consumption by waterfowl have been made (Jordan 1953 [the pioneer work]; Holm and Scott 1954; and Bossenmaier and Marshall 1958).

Individual canvasbacks ate 0.49 lb more food per day in the period from November 12 to December 13 than during the period from October 2 to November 12. Lesser scaup consumed 0.29 lb more during the same interval. This greater food consumption probably reflects the increase in energy required to maintain body temperatures with declining air temperatures. The mean maximum (45.2° F.) and mean minimum (29.5° F.) temperatures of the last 31 days were considerably lower than the mean maximum (62.1° F.) and mean minimum (41.1° F.) temperatures of the first 41 days. These findings agree with Jordan (1953), who found that the amount of food eaten by mallards varied inversely with the seasonal air temperature.

The relationship between body weight and food intake parallels the observations of Jordan (1953). Mean canvasback food intake was 0.29 lb more than the mean lesser scaup food intake for the entire feeding period. Our canvasbacks were ½ to 1 lb heavier than the lesser scaups.

Mean body weights of experimental birds were comparable with weights found by other workers (Bellrose and Hawkins 1947; Dzubin 1959; Town 1961). The emaciation index obtained for canvasbacks, 66.8 compared closely with the 65.0 index obtained by Cornwell and Cowan (in press) for canvasbacks on the lower Detroit River.

The body condition based on fat deposits (Table V) was probably the most reliable indicator of condition in our study. Since the breast muscles were slightly atrophied because of disuse, the degree of emaciation as a criterion for rating birds in good to excellent condition was limited. Likewise, many of the ducks which had comparatively low body weights were small-bodied and actually were in excellent condition. Thus, evaluation by total body weight has shortcomings in respect to overall condition.

In the final six weeks of the experiment, the ducks maintained nearly constant body weights and were in excellent condition. The increased food consumption with decreasing air temperatures did not result in weight gains, suggesting that the birds were at a maximum weight under the experimental conditions and had adequate food supplies.

Application to Detroit River Mortality

A canvasback wintering on the Detroit River, on the basis of our data, requires a minimum of nearly 1 lb of aquatic vegetation and invertebrate animal foods per day to maintain body weight. As the duck expends additional energy to overcome environmental stresses, the daily food consumption becomes significantly greater than that of the relatively inactive and protected birds used in our study.

The role of starvation in winter waterfowl mortality on the Detroit River is unclear (Hunt 1957; Hunt and Cowan, in press). Aside from the spectacular die-offs associated with pollution, daily mortality occurs from January through late March. The dead or diseased ducks, when found, are usually severely emaciated. It is not known if starvation is the direct cause of their poor condition.

It was our experience that the most productive food collection sites became ice-covered. We then attempted to collect at several locations in which waterfowl were concentrated. The available food was sparsely distributed and difficult to locate. A food shortage does exist, in our opinion, during the later winter months on the Detroit River.

The carrying capacity of the Detroit River for wintering waterfowl could be determined by surveys of the critical food-producing areas and the availability of these areas during periods of ice cover. With knowledge of the amount of available food and food consumption rates, the carrying capacity, or WCD's, could be estimated. In the event of extensive ice formation and large wintering populations, or both, well-planned and concisely executed artificial feeding programs may be justified. Feeding programs, *if necessary*, could be initiated after the fall migration, thereby not attracting birds that would not normally winter in the area. Grains could be placed in areas where diving ducks normally congregate and thus not render them any more susceptible to pollution.

Artificial feeding of wildlife, like any other management technique, must be evaluated on its own merits within the context of each unique situation. If feeding 500 bushels of corn and wheat results in 500 more adult canvasback hens returning to the nesting grounds, the increase in waterfowl production may justify the expenditure. This is especially true if species preservation is a critical and primary management objective, as perhaps it should be in the case of the canvasback.

SUMMARY

Captive juvenile canvasback and lesser scaup were fed aquatic plants and invertebrate animals from October 2 to December 13, 1961. These food materials were collected on the lower Detroit River. Mean daily food consumption over this period was 0.78 lb for the canvasback and 0.49 lb for the lesser scaup. Both species consumed more food as air temperatures dropped during the winter. The amount of natural foods needed to maintain body weight may be helpful in determining the waterfowl carrying capacity of a wintering area. Observations made of diving-duck food scarcity during the winter months on the lower Detroit River suggests that a winter-food availability study of these waters is warranted. The results of such surveys would indicate if an artificial feeding program is necessary.

ACKNOWLEDGMENTS

For financial support we express our appreciation to the Wildlife Management Institute, the Delta Waterfowl Research Station, the Frank M. Chapman Fund of the American Museum of Natural History, the National Wildlife Federation, and the Game Division, Michigan Department of Conservation. George S. Hunt, A. B. Cowan, H. Albert Hochbaum, Peter Ward, and James Foote made valuable contributions to this study. To these men we extend our sincere thanks.

LITERATURE CITED

BELLROSE, F. C., AND A. S. HAWKINS. 1947. Duck Weights in Illinois. Auk 64: 422-430.

BOSSENMAIER, EUGENE F., AND W. A. MARSHALL. 1958. Field-Feeding by Waterfowl in Southwestern Manitoba. Wildl. Mono. 1, 32 pp.

CORNWELL, GEORGE W., AND R. HARTUNG. 1963. An Inexpensive Holding Pen for Diving Ducks. Jour. Wildl. Mgt. (in press).

—— AND A. B. COWAN. 1963. Helminth Populations of the Canvasback (*Aythya valisineria*) and Host-Parasite-Environment Interrelationships. Trans. 28th N. Am. Wildl. Conf. (in press).

DZUBIN, ALEX. 1959. Growth and Plumage Development of Wild-Trapped

Juvenile Canvasback (*Aythya valisineria*). Jour. Wildl. Mgt. 23 (3): 279-290.

HOLM, E. R., AND M. L. SCOTT. 1954. Studies on the Nutrition of Wild Waterfowl. N. Y. Fish and Game Jour. 1 (2): 171-187.

HUNT, GEORGE S. 1957. Causes of Mortality among Ducks Wintering on the Lower Detroit River. Ph. D. thesis, Univ. Mich. x + 296 pp.

—— AND A. B. COWAN. 1963. Causes of Deaths of Waterfowl on the Lower Detroit River—Winter 1960. Trans. 28th N. Am. Wildl. Conf. (in press).

JORDAN, JAMES S. 1953. Consumption of Cereal Grains by Migratory Waterfowl. Jour. Wildl. Mgt. 17 (2): 120-123.

MOYLE, JOHN B. 1961. Aquatic Invertebrates as Related to Larger Water Plants and Waterfowl. Minn. Dept. Cons., Inv. Rep. 233, 24 pp. mimeo.

SINCOCK, JOHN L. 1962. Estimating Consumption of Food by Wintering Waterfowl Populations. Proc. 16th Ann. Conf., S. E. Assn. Game and Fish Comm. (in press).

TOWN, RALPH H. 1960. A Survey of Helminth Parasites in Diving Ducks Found Dead on the Lower Detroit River. M. Wildl. Mgt. thesis, Univ. Mich., 61 pp.

PAPERS OF THE MICHIGAN ACADEMY OF SCIENCE, ARTS, AND LETTERS
VOL. XLIX, 1964 (1963 MEETING)

THE EFFECT OF RAISING THE WATER LEVEL ON THE PRODUCTIVITY OF A MARL LAKE

JAMES W. MERNA
Michigan Department of Conservation

INTRODUCTION

THIS study is an attempt to determine the effects of an increase in water level upon the productivity of a marl lake. Marl lakes are common in southern Michigan and tend to produce few fish or other organisms. Precipitation of calcium carbonate leads to the development of bottom soils which contain much calcium and little organic matter. The effects of change in water level upon ecosystems of this type have not been studied intensively.

Roelofs (1944), Rawson (1951), Moyle (1956), and others have discussed the relationship between soil fertility and production in aquatic habitats. Kadlec (1960) and Cook and Powers (1958) found that the productivity of impoundments can be increased by manipulation of the water level. Reflooding, following the drawdown of an impoundment, was believed to result in release of nutrients from soils that had been aerated and dried during the drawdown.

It was reasoned that the flooding of about 100 acres of "new" soil would result in a release of nutrients such as occurs upon the reflooding of an impoundment after a drawdown. The aerated soils surrounding a marl lake have a higher organic content than the marl deposits of the lake basin. Decomposition of this organic material could bring about release of nutrients. The solubility of phosphorus may be increased by soil submergence (Mortimer 1941). Under anerobic conditions, when high concentrations of carbon dioxide are present (as occurs in submerged soils), iron, calcium, and magnesium go into solution as bicarbonates, and phosphorus is released in a soluble form (Cook and Powers 1958).

STUDY AREA

Big Portage Lake is in Jackson County, Michigan, T 1S, R 2E, Sec. 31 and T 2S, R 2E, Sec. 6. In February 1957, the Conservation Department installed a dam which raised the water level 3 feet.

The area of the lake was increased from 335 acres to approximately 435 acres. The lake basin now has three depressions with maximum depths of 28, 34, and 43 feet. These depressions are separated by shoal areas where the water is less than 6 feet deep. Water depth is less than 5 feet in 34% of the basin and less than 20 feet in 85% of the basin. Beyond the 10-foot contour, there is a steep drop-off into the depressions.

In 1959 the aquatic vegetation on shoal areas consisted of *Scirpus* sp. and scattered beds of *Nymphaea odorata*. There were few weeds at depths greater than 5 feet; this is surprising since in 1946 a survey party reported that 20 species of higher aquatics were present. Of the 20 species, 10 were typically submergent forms and included *Potamogeton gramineus, P. pectinatus, P. praelongus, P. zosteriformis, Ceratophyllum demersum*, and *Myriophyllum heterophyllum*. These species were reported to be common. Kadlec (1960) found that relatively minor differences in water level will cause adjustments and changes in vegetation. Thus, the elevated water level may account for the present scarcity of vegetation.

Methods

In 1953 and in the spring of 1954, water samples were collected once a month. From February 1957 through February 1960, samples were collected at irregular intervals. All collections were made at the deepest part of the lake. The depths at which water samples were collected were varied, depending on the stratification of the lake. During summer stratification three samples were taken within the epilimnion, two in the thermocline, and two in the hypolimnion. At times of homothermal conditions, water samples were usually collected at 10-foot intervals.

Bottom fauna samples were collected at monthly intervals between May and October of 1953. Sampling stations were located on 2 transects at depths of 3, 15, and 25 feet. Stations 1, 2, and 3 were established in the east bay and 1A, 2A, and 3A in the west bay (Fig. 1). In 1959, after the water level was raised, samples were collected each month at these same stations, where water depths were 3 feet greater than in 1953. The recently flooded lake bottom of both the east and west bays was sampled in 1959 (Fig. 1, stations 4 and 4A). In both 1953 and 1959, four bottom samples were collected each month at each station with a 6-inch Ekman dredge.

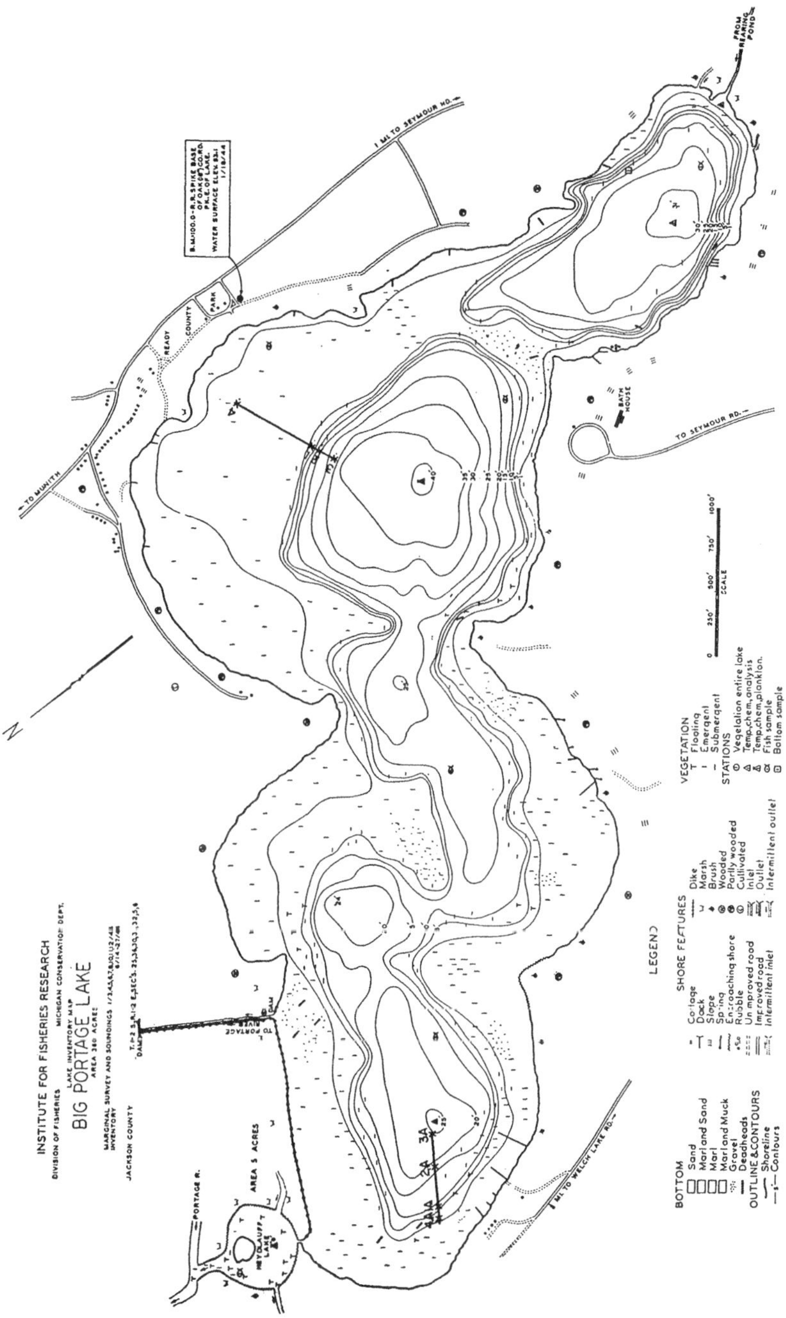

Fig. 1. Location of stations where bottom samples were collected in Big Portage Lake in 1953 and in 1959.

Fish were collected with trap nets in the late fall of 1953 and in the spring of 1954. Since there is virtually no growth over the winter, these fish were considered as one collection. The growth of the fish collected in 1953-54 was compared with fish collected with trap nets in the spring of 1958. The fish collected in 1958 had spent one growing season in the lake after the water level was raised.

Results

The mineral concentration of lake water is subject to large variations within a short period of time. This is especially true of phosphorus content, which may vary with depth and with stratification of the lake. Fig. 2 shows the phosphorus and alkalinity content from February 1953 to February 1960. The figures used on the graph represent average values for samples in the epilimnion during stratification, or average values for samples at all depths when the lake was not stratified.

In 1953 and 1954 total phosphorus ranged from a low of 6 ppb (May 1953) to a high of 25 ppb (January 1954), but it rarely exceeded 10 ppb.

In 1957 and 1958, after the water level was raised, total phosphorus ranged from a low of 23 ppb (September 1958) to a high of 41 ppb (January 1958). In 1959 and in February 1960 there was some indication of a decline in total phosphorus; such a decline was expected since the newly flooded bottom would soon release its available nutrients and begin to accumulate marl.

Total phosphorus values before and after the increase in water level were compared by means of a sign test (Dixon and Mood 1946). Samples collected in 1953 and 1954 during spring and fall circulation, and during summer and winter stratification, were compared with samples collected during similar periods after the water level was raised. Total phosphorus content was higher after the level was raised; the difference was statistically significant at the 5% level.

An increase in the soluble phosphorus after flooding was noted also; however, this difference was not significant at the 5% level. With few exceptions, soluble phosphorus in the samples collected in 1953 and 1954 ranged between 1 ppb and 3 ppb; only once during these two years was there a sample with a value greater than 3 ppb. In March, May, October, and December of 1953 the average amount of soluble phosphorus was less than 0.1 ppb. During 1957 and 1958,

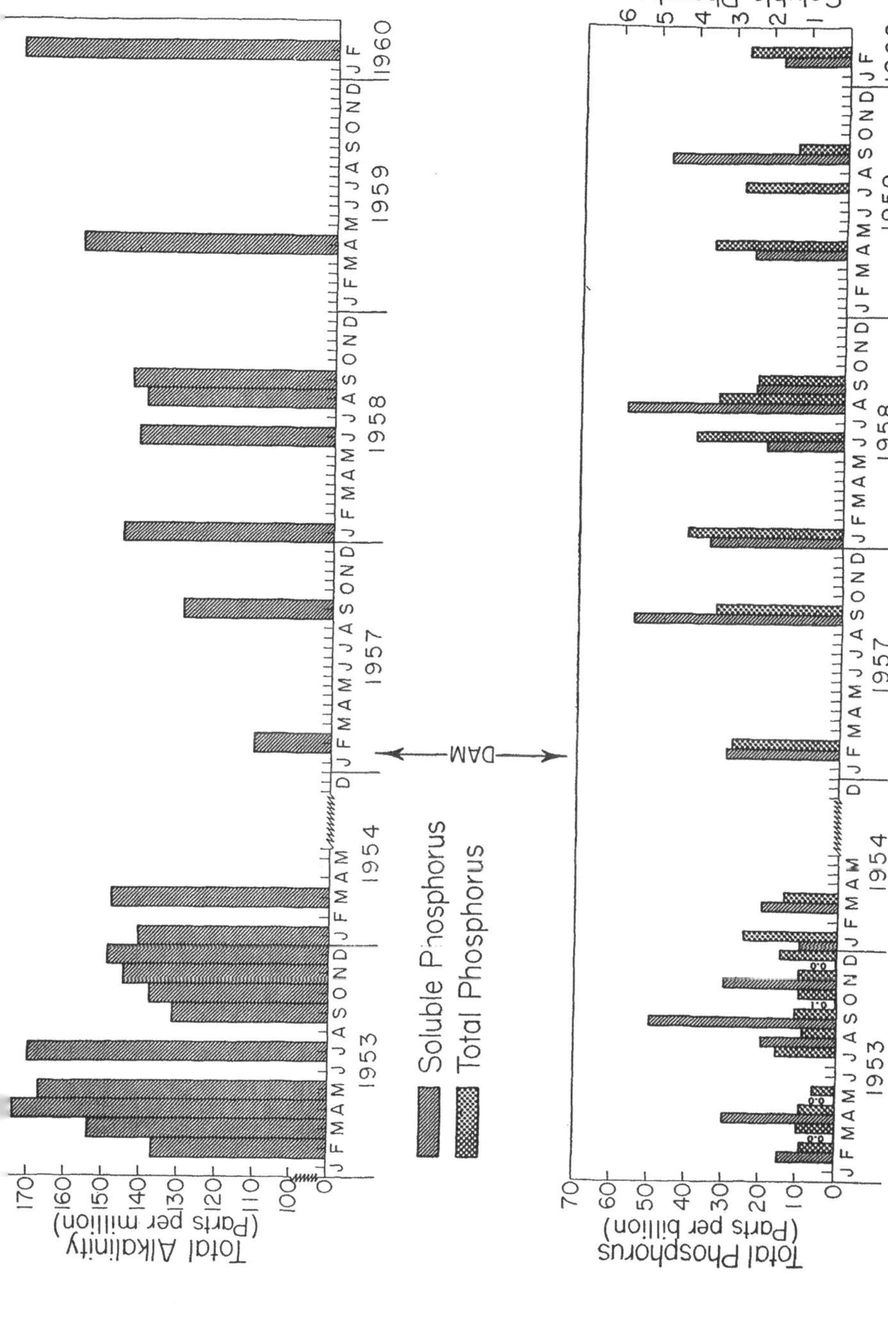

FIG. 2. Total phosphorus, soluble phosphorus, and total alkalinity concentrations of Big Portage Lake from February 1953 to February 1960.

soluble phosphorus values of less than 3 ppb were recorded only twice; all other values during these years ranged between 3 ppb and 6 ppb. In February 1960 the average soluble phosphorus content was less than 2 ppb. This was the lowest value recorded since 1954.

In 1953 and 1954 the total alkalinity ranged from 132 ppm (September 1953) to 174 ppm (April 1953); most of the values exceeded 145 ppm. In 1957 and 1958 the highest total alkalinity recorded was 146 ppm; the majority of the samples ranged between 130 ppm and 140 ppm. The lowest alkalinity recorded (110 ppm) was in February 1957 immediately after the dam was installed. Although total alkalinity appeared to decrease after the water level was raised, the sign test failed to show that the difference was statistically significant.

Ball and Tanner (1951) noted a decrease in alkalinity when phosphorus was added to unproductive trout lakes. This inverse relationship between phosphorus and alkalinity was also observed here. However, the increase in phosphorus and the decrease in alkalinity were conspicuous for only two or three years. By 1959 and February of 1960 there was an indication that the bicarbonate content was again approaching the level present in 1953 and 1954. The February 1960 value of 173 ppm was almost equal to the highest value recorded in 1953.

Bottom Fauna

The standing crop of bottom fauna was lower in 1959 than in 1953. In 1953 the average standing crop ranged between 25 and 45 organisms per $\frac{1}{4}$ square foot at all stations, except at station 1 in the east bay (Fig. 3). Here there were fewer than 5 organisms per $\frac{1}{4}$ square foot. In 1959 the average standing crop did not exceed 10 organisms per $\frac{1}{4}$ square foot at any station; at station 1 the crop of bottom fauna showed a modest increase over 1953 but did not exceed 10 organisms per $\frac{1}{4}$ square foot. Samples taken in 1959 from stations in "new" lake bottom (4 and 4A) contained very few organisms. They were poorer than samples taken from "old" bottom in the same general areas in 1953. For comparison, samples collected before and after the rise in water level were paired by months. A sign test indicated that the difference was significant at the 1% level.

Qualitatively, samples collected in 1953 were not greatly different from those collected in 1959. In 1953 the samples contained caddisfly larvae (*Polycentropus*), the phantom midge, *Chaoborus*, and seven species of midge larvae (Chironomidae). The midges identified were: *Procladius* sp., *Clinotanypus* sp., *Pseudochironomus* sp., *Calopsectra* sp., *Cryptochironomus* sp., *Tendipes tendipes decorus*, and *T. tendipes tentans*. The samples collected in 1959 contained the same species except for *Polycentropus* (they were not abundant in 1953) and the midge *Pseudochironomus*.

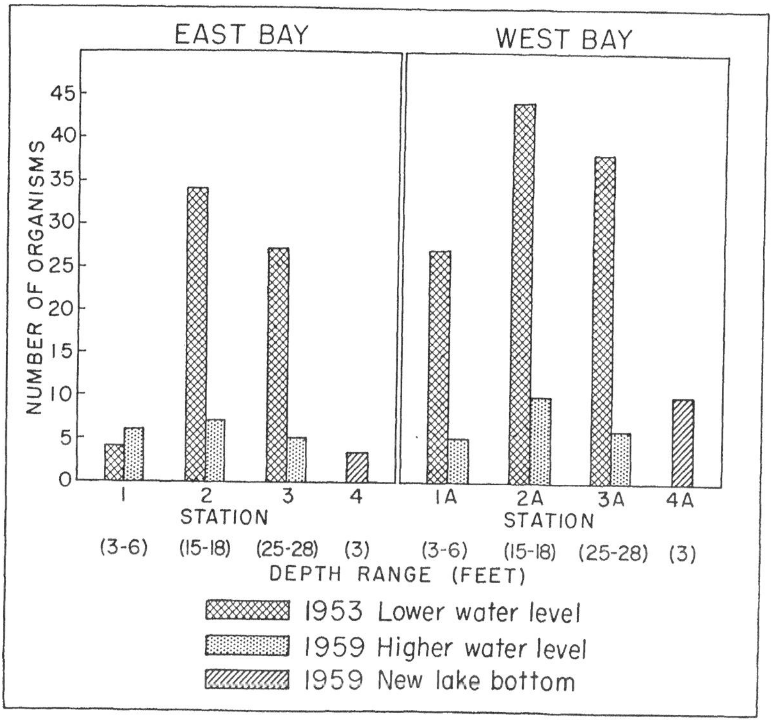

FIG. 3. Average standing crop of bottom fauna per ¼ square foot at stations sampled in Big Portage Lake in the years 1953 and 1959.

In both 1953 and 1959 *Chaoborus* and chironomids comprised nearly 100% of the population at the deep-water stations. The major difference between the samples was the great reduction in the number of *Chaoborus* in 1959. *Chaoborus* made up 87% of the population in 1953 and 11% in 1959. Except in the shallow-water samples from the east bay there was no marked reduction of chironomids in 1959. Hence it seems that the contribution to the food chain by the midges did not change appreciably.

Fish

Wide variations in the growth rate of fish may result from changes such as increased food supplies or decreased competition for food. At Big Portage Lake it was assumed that any increase in productivity resulting from the raised water level would be reflected in the growth rate of the fish. The acceleration (if any) in growth would be most pronounced in the growing season of 1957.

Bluegills (*Lepomis macrochirus*), black crappies (*Pomoxis nigromaculatus*), and largemouth bass (*Micropterus salmoides*) had

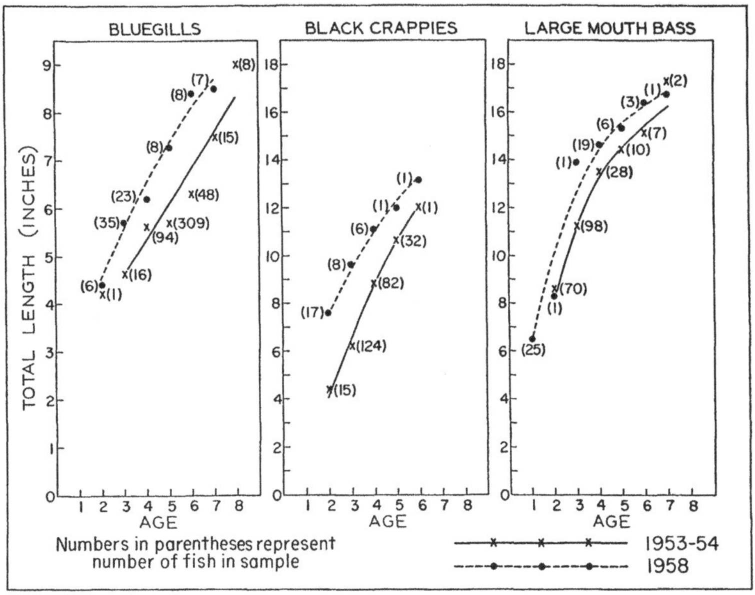

Fig. 4. Growth curves for fish collected from Big Portage Lake in the years 1953, 1954, and 1958.

a greater average total length after the water level was raised (Fig. 4). The average length of bluegills of each age group collected in 1958 was about 1 inch greater than that of fish of the same age collected in 1953-54; average lengths of 2- and 3-year-old black crappies collected in 1958 exceeded by almost 3 inches the average lengths of fish of the same ages collected in 1953-54, and for other age groups the improvement amounted to about 1 inch; all age groups of largemouth bass showed an increase of about 1 inch in 1958 over 1953-54.

To assess the effect of the raised water level, the amounts of growth made by fish during 1953 and 1957 were compared. Bluegills and largemouth bass which had completed their fourth growing season and black crappies which had completed their second growing season during these two years were selected for this comparison because these age groups were represented in sufficient quantity to permit reliable statistical comparison. The average growth increment of bluegills in their fourth growing season was 1.4 inches (growth during 1953) for those collected in 1953-54, and 2.4 inches (growth during 1957) for those collected in 1958. The individual growth increments during the fourth growing season were more variable for bluegills collected in 1958 than for those collected in 1953-54. For 1958, increments ranged from 0.3 inch to 3.7 inches (standard deviation 1.18); for 1953-54, the increments ranged from 0.7 inch to 2.8 inches (standard deviation 0.35). Because the variances of the increments were unequal, the Welch procedure (Brownlee 1960) was used to test whether the difference in increments, between 1953 and 1957, was statistically significant. The test showed the difference to be significant at the 0.01 probability level. The greater variation in growth after the water level was raised indicates that all bluegills did not respond equally. The larger fourth-year growth increment of the bluegills collected in 1958 accounted for their greater total length; there was not a significant difference in length between the two groups at the end of the third growing season.

The growth increment of black crappies in their second growing season was 2.9 inches in 1953 and 6.2 inches in 1957. This difference was significant at the 0.01 probability level. There was no significant difference between the first year growth of black crappies in the two collections; therefore the greater total length in 1958 was due only to greater growth during 1957.

The average growth increment of largemouth bass in their fourth growing season was 2.5 inches in 1953 and 3.0 inches in 1957. The bass taken in 1958, however, had been growing more rapidly during each year of life than those collected in 1953-54. The change in water level therefore did not account for the total increase in the length of largemouth bass collected in 1958 over those taken in 1953-54.

Discussion

There is evidence that the elevation of the water level of Big Portage Lake resulted in changes in the level of nutrients and therefore may have increased basic productivity. A higher phosphorus and lower alkalinity content of the water persisted for two years after the change in water level. During the third season these elements returned nearly to their former concentrations.

The bottom fauna did not respond to the increased phosphorus content of the water. There was a significant decrease in the average standing crop of benthos at the stations sampled between 1953 and 1959. It is possible that this decrease in bottom fauna took place between 1953 and the time the water level was raised (1957). It is also possible that the poor crop of benthos in 1959 was independent of the change in water level. *Chaoborus* exhibited the greatest decrease in numbers; however, there was little change in the number of chironomid larvae. Consequently the contribution of the bottom fauna to the food chain may not have changed appreciably. Chironomids and *Chaoborus* were the only abundant bottom organisms. The reduction in bottom fauna may have been related to the scarcity of aquatic weeds after the water level was raised.

Although an increase in bottom fauna did not occur, the growth rate of bluegills, largemouth bass, and black crappies increased after the water level was raised. Possibly the increased growth resulted from a greater production of zooplankton. The increase in phosphorus and the reduction in higher aquatic plants may have resulted in a greater production of phytoplankton, periphyton, and zooplankton.

Acknowledgments

Frank F. Hooper supervised the project and reviewed the manuscript. Richard O. Anderson, Kenneth E. Christensen, and Donald F. Thomas helped to collect samples. James T. McFadden gave advice on statistical procedures.

LITERATURE CITED

Ball, R. C., and H. A. Tanner. 1951. The Biological Effects of Fertilizer on a Warm-Water Lake. Mich. State Coll. Agr. Exp. Sta., Tech. Bull. 223. 32 pp.

BROWNLEE, K. A. 1960. Statistical Theory and Methodology in Science and Engineering. New York: John Wiley & Sons, Inc. 569 pp.

COOK, A. H., AND C. F. POWERS. 1958. Early Biochemical Changes in the Soils and Waters of Artificially Created Marshes in New York. N.Y. Fish and Game Jour. 5 (1): 9-65.

DIXON, W. J., AND A. M. MOOD. 1946. The Statistical Sign Test. Jour. Am. Stat. Assn. 41: 556-566.

KADLEC, J. A. 1960. The Effect of a Drawdown on the Ecology of a Waterfowl Impoundment. Mich. Dept. Cons. Game Div., Rep. 2276. 181 pp.

MORTIMER, C. H. 1941. The Exchange of Dissolved Substances between Mud and Water in Lakes. Jour. Ecol. 29: 280-329.

MOYLE, J. B. 1956. Relationships between the Chemistry of Minnesota Surface Waters and Wildlife Management. Jour. Wildl. Mgt. 20 (3): 303-320.

RAWSON, D. S. 1951. The Total Mineral Content of Lake Water. Jour. Ecol. 32: 669-672.

ROELOFS, E. W. 1944. Water Soils in Relation to Lake Productivity. Mich. State Coll. Agr. Exp. Sta., Tech. Bull. 190. 31 pp.

PAPERS OF THE MICHIGAN ACADEMY OF SCIENCE, ARTS, AND LETTERS

VOLUME XLIX

(1963 Meeting)

PART II. SOCIAL SCIENCE

1964

VOLUME XLIX CONSISTS OF THREE PARTS:

I. *Natural Science*

II. *Social Science*

III. *Literature and Art*

Papers of the
MICHIGAN ACADEMY
OF SCIENCE, ARTS, AND LETTERS

VOLUME XLIX

(1963 Meeting)

Editor

RALPH A. LOOMIS

PART II. SOCIAL SCIENCE

Published for the Michigan Academy of Science, Arts, and Letters by

THE UNIVERSITY OF MICHIGAN PRESS

ANN ARBOR

1964

MANUFACTURED IN UNITED STATES OF AMERICA

CONTENTS

Part II. Social Science

ANTHROPOLOGY

ASIAN STUDIES

A. Southeast Asia

B. Japan

C. Southern and Eastern Asia

ECONOMICS

GEOGRAPHY

A. Regionalism—Concepts, Methods, and Examples

PAPERS OF THE MICHIGAN ACADEMY OF SCIENCE, ARTS, AND LETTERS
VOL. XLIX, 1964 (1963 MEETING)

ANTHROPOLOGY

SOME CHARACTERISTICS OF PROJECTILE POINT BASES FROM THE HOLCOMBE SITE, MACOMB COUNTY, MICHIGAN *

JAMES E. FITTING
The University of Michigan

THE HOLCOMBE SITE

THE Holcombe site is a Late Paleo-Indian site in Macomb County, Michigan. The numerical designation is 20 MB 30, although D 36 and MB 150 have been used in the past. It is situated in Sterling Township in the SW$\frac{1}{4}$ SW$\frac{1}{4}$ Sec. 23.

The site was located by Jerry DeVisscher of Mount Clemens, Michigan. Mr. DeVisscher notified Wayne State University of his find, and they suggested notifying the University of Michigan. In the spring of 1961 a crew from the University of Michigan tested the site. The results were promising, and in the summer of 1961 a party of four graduate students under the direction of Richard O. Keslin spent seven weeks at the site working under a National Science Foundation grant. Later that year, and in the summer of 1962, Mr. DeVisscher returned to the site for further excavation. The material described and tested includes Mr. DeVisscher's surface collection and some excavated material, as well as the material from the University of Michigan excavations.

The purpose of this paper is to examine the applicability of quantitative methods to a sample of Paleo-Indian artifacts; hence only a brief description of these artifacts is given. The full descriptive report on the Holcombe site is being prepared by Dr. Keslin.

The most important artifact class in the present study is one composed of projectile points and bases. The Holcombe site points are lanceolate with concave bases. Complete points from the Holcombe site are extremely rare, but complete points with bases similar

* I would like to thank Richard O. Keslin and Arthur J. Jelinek, both of the University of Michigan Museum of Anthropology, for reading and commenting on this paper. I, however, assume full responsibility for the interpretation of the material from the Holcombe site here presented.

to the typical forms from the Holcombe site are found on a closely related site, 20 MB 171, located three-fourths of a mile east of Holcombe. They ranged from 5.9 to 6.8 cm in length. The maximum thickness of the points ranged from .3 to .6 cm. The sides were straight to convex, but most of the Holcombe sample consisted of basal fragments, so the degree of expansion from base to mid-section could not be determined for most of the specimens.

The basal treatment requires special explanation; it varies a great deal, ranging from completely unaltered bases to bases with an actual flake removed from one face. There were no bases with a single broad flake removed from both faces. Most of the material could be classified as basally thinned, that is, one to six long narrow flakes were removed from one or both faces. Since it has been postulated that the removal of such flakes is an intermediate stage in fluting (Mason 1958: 12; Witthoft 1952: 475), and there are finished "fluted" points exhibiting multiple "flutes" (Wormington 1957: 54, 75), it would appear that the removal of one or several channel flakes would represent one type of cultural activity. This group could be referred to as "basally modified," could include both basally thinned and "fluted" forms, and can be compared to that part of the sample which is unaltered. The Holcombe sample contains both basally modified and unmodified forms. Some of the projectile point fragments from this site, including most of the bases used in this study, are shown in Fig. 1.

There were a number of other artifacts from the site, some of which are shown in Fig. 2. In the upper row is a series of finely worked blades with rounded tips, as thin as the point fragments and exhibiting the same type of flaking. These are fragments of thin knives. There was a large series of artifacts which have, in many instances, the same configurations as the thin knives but lack the fine flaking and thinness. These will be referred to as thick knives although some undoubtedly are blanks for projectile points, and others appear to have functioned as sidescrapers.

The scraper types present an unusual distribution for a Late Paleo-Indian site in this area. Sidescrapers predominate; several spokeshave scrapers are present in the sample. One snub-nosed end scraper has been reported (Roosa 1962: 265); it is illustrated in Fig. 2, middle row, second from the left. No gravers were found, but one object that might have served as a graver or perforator and one similar drill-like form were included in the sample.

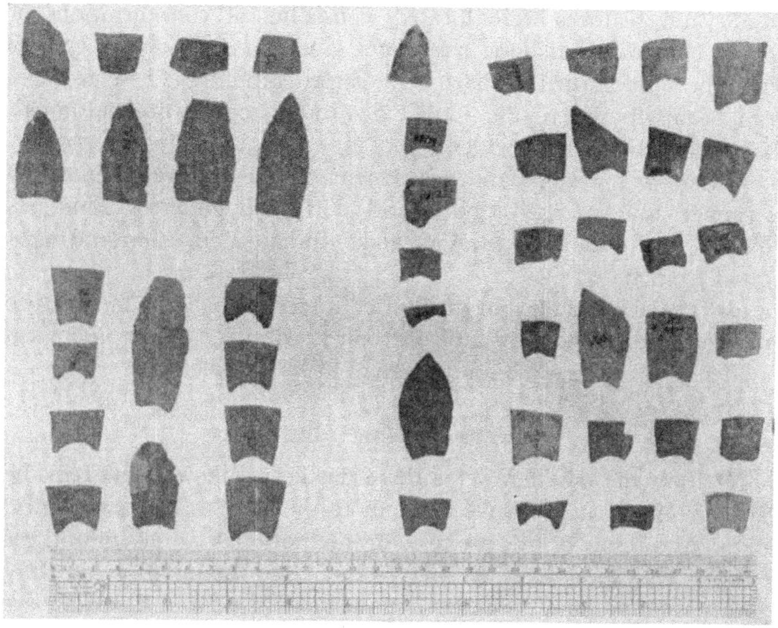

FIG. 1. Some of the projectile point bases and fragments from the Holcombe site.

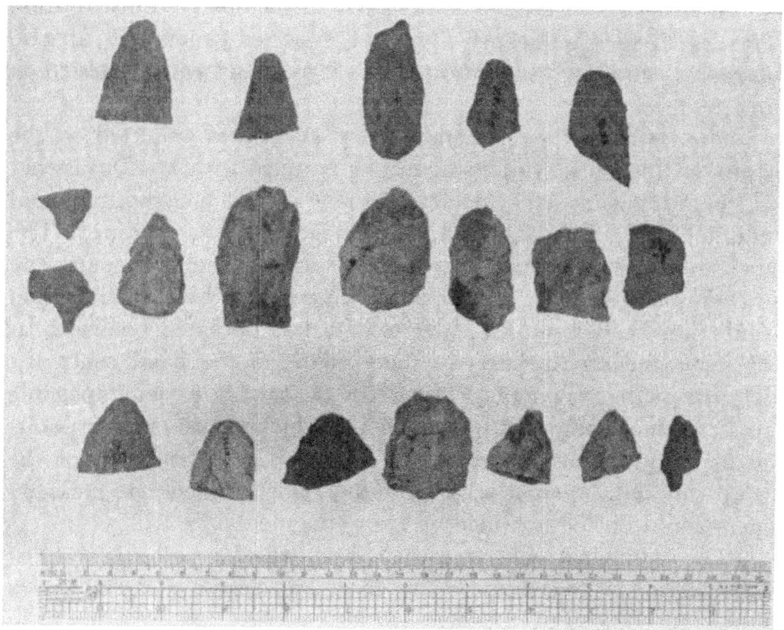

FIG. 2. Artifacts from the Holcombe site, including thin knives, thick knives, scrapers, and perforators.

Several features were located, consisting of concentrations of chips, charcoal, and bone fragments. Some of the bone fragments have been identified as part of a barren ground. These features can be attributed to the Late Paleo-Indian occupation, although some disturbance was evident.

There was a woodland component on the site, represented by stemmed and notched points and pottery fragments. This was located to the south of the Late Paleo-Indian occupation and separated from it.

While lack of sufficient carbonized material from any one feature precludes the possibility of Carbon-14 dating, geochronological evidence indicates a maximum age of 9000 to 10,000 B. C.

Analysis of Holcombe Bases

Within the past few years there has been a new interest in the application of quantitative methods to the analysis of archeological data. As a result of this interest, an international symposium on the subject was sponsored by the Werner-Gren Foundation at Burg Wartenstein in 1959. The methods of analysis used in this study are based, primarily, on a paper (1960) delivered by Albert C. Spaulding at that meeting.

In all tests, our null hypothesis is that the observed distribution does not differ significantly from the expected random or, in two instances, normal distribution. This hypothesis will be tested at the .05 level.

The data on the Holcombe point bases was collected in the spring of 1962, after discussing fluting patterns with Mr. DeVisscher at the Michigan Archeological Society spring meeting. A total sample of 44 points, bases, and basal fragments was examined. The total in any series of comparisons may vary, as some fragments were broken so that the width, thickness, or pattern of basal modification could not be determined. Metric data was obtained on width at the base and on thickness at the extreme of the basal concavity. Because of the fragmentary nature of the sample it was impossible to record maximum width or maximum thickness on enough points to make quantitative comparison worthwhile. The depth of the basal concavity was not recorded either, since an accurate measurement could not be obtained on many fragments.

The nonmetric data recorded included the pattern of basal modification, a description of the chert, and degree of lateral grinding. The information on lateral grinding proved too subjective for consistent description and was dropped from the test. Therefore, four attributes, two metric and two nonmetric, were recorded and compared with each other.

The first thing checked was the nature of the distributions of the two metric attributes. For the thickness at the base the following results were obtained:

Sample size = 44 Sum of values = 14.3 cm
Mean = .325 cm Variance = .0075
Standard deviation = .0866 cm

Class	Observed	Expected
-.24 cm	8	8.7
.25-.34 cm	21	18.5
.35-.44 cm	11	13.6
.45 plus cm	4	3.2

Our chi-square value was 1.0911 with one degree of freedom, so our probability factor is greater than .2. The distribution of basal thickness does not differ significantly from a normal distribution.

For the width of the points and bases at the base, the following data were recorded:

Sample size = 41 Sum of values = 75.8 cm
Mean = 1.85 cm Variance = .031
Standard deviation = .18 cm

Class	Observed	Expected
1.35-1.54 cm	1	1.9
1.55-1.74 cm	12	9.9
1.75-1.94 cm	16	17.1
1.95-2.14 cm	11	9.9
2.15-2.34 cm	1	1.9

The resultant chi-square equals 1.4909 with two degrees of freedom, and our probability factor is greater than .3. The distribution of base widths does not differ significantly from the expected normal distribution, so this looks like one population with respect to this attribute.

In testing attributes, our four attributes allow six associations. The first of these was the test of width against the pattern of basal modification. The width categories were 1.5-1.6 cm, 1.7-1.9 cm, and 2.0-2.3 cm. The patterns of modification were: no modification (No No) ; modification on one face only (No Plus) ; and modification on both faces (Plus Plus). The sample size was 39, and a chi square value of 4.8261 with four degrees of freedom was obtained. In this instance the probability factor was greater than .30, so the deviation from the chance association of these attributes was not significant.

We plotted thickness against fluting, using the same modification categories, and under .24 cm, .25-.34 and .35 plus cm as thickness categories. The sample size was 43, and a chi-square value of 3.0506 with four degrees of freedom was obtained. In this instance the probability factor was greater than .5. The deviation from the chance association was not at all significant.

In testing for an association between fluting and chert types, a problem was encountered. Since a wide range of chert types was present, a great deal of grouping had to be done. The final result was a white chert category, a light to dark gray chert category (the largest category), and a gray-brown to brown chert category. The sample size was 43, and a chi-square value of 2.1000 with four degrees of freedom was obtained. Here the probability factor is greater than .70, so there is very little deviation from the expected chance association. Keslin (in a personal communication) has suggested that the gray-brown to brown chert category is composed of patinated examples of the white and gray chert categories. If this is true, then this test demonstrates that patination, as well as chert type, is randomly associated with basal modification.

We used the same categories in testing the association of width and thickness. The sample size was 40, and a chi-square value of 6.0475 with four degrees of freedom was obtained. Here the probability factor was greater than .1, so there was no significant deviation from the expected random distribution.

For width and chert association the sample size was 41 with four degrees of freedom, and the chi-square value was 3.3670. The probability of this high a value was greater than .3 and not significant.

For thickness and chert type the sample size was 43. A chi-square value of 6.3335 with four degrees of freedom was obtained. The probability of this large a value is greater than .1 and, therefore, is not significant.

It can be noted that in the above three by three tests no significant differences were found. This, coupled with the normal distribution of the sample, suggests that the Holcombe sample is fairly homogeneous. It was decided to retest some of the associations with two by two tables. With one exception, no significant differences from the expected random distributions were noted. The one exception was the association between the pattern of basal modification and the width at the base.

In order to convert the three by three table into a two by two table, those bases which were modified on one or both faces were combined. The premise here is that basal modification is a culturally significant trait as opposed to nonmodification. The grouping of widths was accomplished by grouping the two narrow categories, 1.5-1.6 cm and 1.7-1.9 cm, into 1.5-1.9 cm, since this approximated the range of the third category, 2.0-2.3 cm. The following results were obtained.

	1.5-1.9 cm		2.0-2.3 cm	
	Observed	Expected	Observed	Expected
Unmodified	5	8	6	3
Modified	22	19	6	9

We obtained a chi-square value of 5.4989 with one degree of freedom. The probability of this occurring is less than .02, and considerably less than .05, which is required for the rejection of our null hypothesis.

Here then is a basis for the establishment of possible point classes. An examination of the attribute clustering shows that we have more than the expected frequencies in the narrow modified class and the wide unmodified class. We can form the hypothesis that the wider bases tend to be unmodified and the narrower modified. This would, at first glance, suggest two classes, but we are left wondering what to do with the two remaining groups. Do we have a narrow unmodified variant of the narrow modified class, or do we have a wide modified variant of this class? Since there were no objective criteria for making this type of decision, it was decided to consider all four groups as possible classes, with the realization that only two actually indicate attribute clustering.

Then we have four classes which may be defined as follows:

1. A wide modified form
2. A wide unmodified form
3. A narrow modified form
4. A narrow unmodified form

We had provenience data on 17 bases from the University of Michigan excavations. This sample size is, however, too small and the number of categories too large to check for deviation from chance association, but it does look as though there might be a gradation from area to area.

Conclusions

In a study such as this it is always possible that the sample is not representative of the population from which it is drawn. The fact that the conclusions are stated in terms of probability takes this factor, as well as others, into consideration. It should be noted that in no instance was a probability estimate in an intermediate range; all were either clearly significant or not significant.

The possible classes in this preliminary examination are, at best, suggestive. What this paper has attempted to demonstrate is the need for quantitative examination of artifacts from sites such as the Holcombe site. Traditional techniques for classification, which rely on qualities such as fluted and nonfluted, may be misleading unless other attributes can be associated with these characteristics.

For the most part, the four attributes studied are randomly distributed throughout the sample. The only significant association is between width and pattern of basal modification. The point bases from the Holcombe site might well be placed in subgroups based on these attributes when the final study of all artifacts from the site is completed.

LITERATURE CITED

Mason, Ronald J. 1958. Late Pleistocene Geochronology and the Paleo-Indian Penetration into the Lower Michigan Peninsula. Univ. Mich. Mus. Anthrop. Pap. 11.

Roosa, William B. 1962. Comment on "The Paleo-Indian Tradition in Eastern North America," by Ronald J. Mason. Current Anthrop. 3 (3): 263-265.

Spaulding, Albert C. 1960. Statistical Description and Comparison of Artifact Assemblages. *In* The Application of Quantitative Methods in Archeology, ed. by Robert F. Heizer and Sherburne F. Cook. Chicago: Viking Fund Publ. Anthrop., No. 28, pp. 60-83.

Witthoft, John. 1952. A Paleo-Indian Site in Eastern Pennsylvania: An Early Hunting Culture. Proc. Am. Philos. Soc., 96 (4): 464-495.

Wormington, H. M. 1957. Ancient Man in North America. Denver Mus. Nat. Hist., Pop. Ser. 4.

PAPERS OF THE MICHIGAN ACADEMY OF SCIENCE, ARTS, AND LETTERS
VOL. XLIX, 1964 (1963 MEETING)

GLUCOSE-6-PHOSPHATE DEHYDROGENASE ENZYME DEFICIENCY ON THE EAST COAST OF NICARAGUA

MARY W. HELMS
The University of Michigan

IN the past few years, an increasing body of data has shown significant correlations between the frequency and distribution of various polymorphic traits found within some human populations and the distribution of particular diseases. More specifically, attention has been focused on the influence of malaria as a factor in human mortality. Malaria is widespread in tropical, subtropical, and temperate zones and is said to kill 2 million children every year (Motulsky 1960: 42). Most fatalities are attributable to falciparum malaria, which is found most frequently in tropical and subtropical areas. From an evolutionary point of view, such a disease should be an important selective agent for or against the maintenance of particular protective traits within a population.

The various polymorphisms thought to be associated with malaria are found in the red blood cells. The malaria parasite is dependent upon some of the enzymes and metabolites of the blood cells for development, and the number of parasites in the red blood cells appears to be related to malarial mortality. Thus an abnormality in the workings of the cells which would be detrimental to optimal parasite development would be advantageous to the individual in that the number of parasites in the blood cells would be reduced. Sickle-cell anemia and thalassemia are two types of red blood cell abnormality in which individuals who are homozygous for the trait suffer from an often fatal anemia, but those who are heterozygous seem to have a resistance to malaria. This selective advantage against malaria accorded to heterozygote " carriers " would serve to maintain high percentages of the abnormal hemoglobins in the population in spite of the continual loss of genes due to fatal anemia. The geographical distribution of both sickle-cell anemia and thalassemia corresponds with the distribution of falciparum malaria.

In a somewhat similar way, although the evidence is not yet as clearcut, individuals with another red blood cell abnormality glucose-6-phosphate dehydrogenase (G-6-PD) enzyme deficiency—also appear to have a selective advantage against malaria. A. G. Motulsky has recently summarized the results of studies showing the occurrence of G-6-PD to be limited to a wide belt of tropical Africa, the Mediterranean area, the Near East, Indian Southeast Asia, and the Philippines—all malarial areas.

This blood cell abnormality was first discovered when it was noted that about 10% of American Negro soldiers suffered severe blood destruction when given primaquine, a new antimalarial drug. Subsequent investigations showed that susceptibility to drug-induced hemolysis was due to a deficiency of G-6-PD. Investigators in Italy and Israel have demonstrated that favism—hemolytic anemia caused by eating raw fava beans or from inhaling flowering fava bean pollen—occurs only in individuals with the same enzyme deficiency.

Motulsky noted that the lower glutathione level in enzyme-deficient cells would create suboptimal conditions for growth of the malarial parasites, which require glutathione for normal development. Since the number of parasites in the blood cells appears to be related to malarial mortality, individuals with enzyme-deficient cells would be expected to have a lower malaria mortality rate. G-6-PD deficiency is also known to affect older red blood cells with greater severity than younger, significantly shortening the life span of the cells even when they are not stressed by the presence of drugs. Thus the selective advantage probably relates to falciparum malaria and malaria caused by *Plasmodium malariae*, both of which preferentially parasitize older blood cells, rather than to vivax malaria, which more readily parasitizes young blood cells. This fact is consistent with the previous observation that falciparum malaria is the lethal malaria in most cases. The protective trait that would be selected for would be that one which protects against the factor responsible for greatest mortality.

Since the trait has always appeared as a polymorphism so far, there must be some disadvantage associated with it to counterbalance its protective effect. The lowered fitness of those susceptible to favism would be one such disadvantage, and hemolytic anemia induced by virus or bacterial infection has been suggested as another. Probably more harmful than either of these is still a third defect recently traced to G-6-PD deficiency, i. e., severe hemolytic disease of the newborn not associated with blood group incompatibility (Motulsky 1961: 179).

The defect appears to be inherited as a sex-linked character with full expression in hemizygous males and homozygous females and partial expression in heterozygous females. For purposes of determining gene frequency, population surveys on males will give clear-cut bimodal results. The frequency of the abnormal gene equals the frequency of the affected hemizygote males in the population. Since heterozygote females may have normal or low enzyme levels, frequency data collected on females may be misleading.

The occurrence of G-6-PD enzyme deficiency is not limited to the Old World. 2% incidence of the deficiency has been found among Carib populations, and 16% incidence has been reported among the South American Oyana Indians, who have lived in malarial areas for many generations (Motulsky 1960: 50). This report describes the results of a survey conducted among 487 Mosquito Indian and Creole males located on the east coast of Nicaragua (Mosquito Coast), a region of hyperendemic vivax and falciparum malaria (Moulton 1941: 16; Thaeler et al. 1953: 991). The object of the study was to determine whether or not any G-6-PD deficiency was present in the population. Enzyme deficient males were detected by Motulsky's rapid screening test (Motulsky 1961: 166). This photometric test is based on an enzyme catalyzed reduction reaction, where brilliant cresyl blue indicator loses its color in the reduced state. The length of time required for decolorization is the significant feature for detecting the enzyme deficiency. Under standard conditions, time taken for decolorization is less than 100 minutes for those with normal G-6-PD activity, while enzyme-deficient samples require over 120 minutes. In this study the anti-mode of the bimodal distribution fell at 110 minutes. All samples requiring less than 100 minutes were considered normal, while those requiring 120 minutes or more were considered positive for the deficiency. Most enzyme-deficient samples decolorized in between two and four hours. Blood samples were obtained by finger puncture.

Of the 313 Creole schoolboys tested, 57, or 18.2%, were found to be enzyme deficient; of the 174 Mosquito boys tested, 33, or 18.9%, were enzyme deficient. Using these figures, the incidence of deficiency in the combined groups is calculated at 18.5%. The Creole samples were drawn for the most part from the town of Puerto Cabezas, while Indian samples were collected at two Mosquito villages and one refugee camp located within a 45-mile radius of the town. All samples were analyzed in the laboratory of the

Hospital Moravo in Puerto Cabezas. The results are summarized in Table I.

TABLE I

FREQUENCY OF G-6-PD DEFICIENCY IN CREOLE AND MOSQUITO POPULATIONS

	No. Males Tested	No. Males Deficient	Frequency (%)
Creole	313	57	18.2
Mosquito Indian	174	33	18.9
Total	487	90	18.5

It must be pointed out that this estimate may be slightly biased in that it was impossible to ascertain genealogical connections between subjects, many of whom were related, and it is quite possible that some families are represented by several members, some by only one. This would result in higher sample results than the actual population warrants if enzyme-deficient families were disproportionately represented or in lower sample results than is actually so if normal families were disproportionately represented.

Of the 313 Creole boys attending schools in Puerto Cabezas, 81 were not native to the immediate area but represented numerous villages situated along the coast from Bluefields, about 130 miles to the south, to Cabo Gracias a Dios, about 80 miles to the north. Of these 81, 16 were deficient. 135 samples were collected at a government refugee camp where Indians from the Rio Coco area, the boundary river between Nicaragua and Honduras, were stationed as a result of dislocation after the recent settlement of a long-standing boundary dispute between the two countries. 26 of these 135 samples were enzyme deficient. These figures indicate the widespread occurrence of the trait along most of the Nicaraguan east coast.

Testing with chi-square indicates no significant difference between Mosquito and Creole populations with regard to frequency of G-6-PD deficiency ($x^2 = .043$, not significant at .05 level with ldf.). Rather, the separation of the population into these two groups is based on geographical location and historical events. The bulk of the Creole population is in port towns along the coast, while the smaller Mosquito villages are scattered along the coastal lagoons and in the interior along the many rivers that drain the lowland coastal plain. The Creole population is largely the result of relatively recent Negro and mulatto immigration from the West Indies, particularly Jamaica. The Mosquito are representative of the indigenous population but also have a long history of admixture with Negro blood. Early writers repeatedly mention a large slave ship en route from the Guinea Coast to Brazil which wrecked near Cabo Gracias sometime between 1640 and 1650 (Conzemius 1932: 17-18; Bard 1855: 234). The escaped Negroes were captured by the Indians and intermarried with them, their children being accepted as full tribal members. Thus contemporary writers noted a considerable admixture of African and Indian blood all along the coast as early as the end of the 17th century (Bard 1855: 338). Negro-Mosquito contacts continued throughout the 18th and 19th centuries; Thaeler estimates that 50% of the present Mosquito population have Negro ancestors (Thaeler et al. 1953: 991).

An explanation for the occurrence of the enzyme deficiency among the Creole, a negroid population, can be found in the fact that the bulk of this population are the descendants of slaves brought to Caribbean plantations from West Africa and the Congo. Studies cited by Motulsky of West African and Congolese tribes show a considerable percentage of enzyme deficiency among some of these populations today (Motulsky 1961: 169). It does not seem unreasonable to suggest that tribal ancestors also were enzyme deficient and that those who were transported to the New World as slaves carried the trait with them. The widespread occurrence of vivax and falciparum malaria in the West Indies and along the Atlantic Coast of Central America would provide a selective agent for the maintenance of the G-6-PD deficiency gene.

The occurrence of the deficiency among the Mosquito Indians raises the possibility that the trait was introduced by population admixture and/or maintenance of an indigenous characteristic. To date population studies among indigenous New World groups have revealed no deficiency in Alaskan Eskimos, American Indians, or highland Peruvian Indians. On the other hand, 2% incidence of the deficiency has been reported among Caribs and 18% among Oyana Indians in Surinam (Motulsky 1961: 168). Given this information, it seems possible that at least some of the deficiency found among Mosquito Indians could be attributed to an indigenous characteristic. On the other hand, it is also possible that the trait was introduced through Negro admixture. It may be that both admixture and maintenance of an indigenous trait are involved.

In view of the amount of Negro admixture it is interesting to note that there have been no known cases of drug-induced hemolysis among Eastern Nicaraguans such as were found among some American Negro soldiers with G-6-PD deficiency (Thaeler et al. 1953; Wallace 1962). Nor is there record of any case of hemolytic anemia induced by virus or bacterial infection (Wallace 1962). Information is lacking with respect to the occurrence of hemolytic disease of the newborn—the third disadvantage associated with the G-6-PD polymorphism.

Acknowledgments

This project was financed by a National Science Foundation Summer Fellowship. The author wishes to express her thanks to the National Science Foundation and to Dr. E. A. Wallace and the staff of the Hospital Moravo, Puerto Cabezas, Nicaragua for their support and cooperation. The survey was conducted during the summer of 1962.

LITERATURE CITED

Alvarado, C. A. 1956. Status of the Antimalaria Campaign in the Americas, V Report. Washington, D. C.: Pan American Sanitary Bur.

Bard, S. A. 1855. Waikna: Adventures on the Mosquito Shore. New York: Harper & Bros.

Conzemius, Eduard. 1932. Ethnographic Survey of the Miskito and Sumu Indians of Honduras and Nicaragua. Smithsonian Inst., Bur. Am. Ethnol. Bull. 106.

Motulsky, A. G. 1960. Metabolic Polymorphisms and the Role of Infectious Diseases in Human Evolution *in* Lasker (ed.) The Processes of Ongoing Human Evolution. Detroit: Wayne State Univ. Press.

—— and J. M. Campbell-Kraut. 1961. Population Genetics of Glucose-6-Phosphate Dehydrogenase Deficiency of the Red Cell *in* B. S. Blumberg (ed.) Proc. Conf. Genetic Polymorphisms and Geographic Variations in Disease. New York: Grune and Stratton.

Moulton, F. R. (ed.). 1941. A Symposium on Human Malaria with Special Reference to North America and the Caribbean Region. Amer. Assn. Advancement Sci. Publ. 15.

Thaeler, A. D., Jr., J. Arnold, and A. S. Alving. 1953. A Clinical Study of Primaquine in the Treatment of Malaria Among the Miskito Indians of Nicaragua. Am. Jour. Trop. Med. Hyg. 2: 989.

U. S. Bureau of the Census. 1945. Nicaragua Summary of Biostatistics. Wash., D. C.

Wallace, E. A. 1962. Personal communication.

Papers of the Michigan Academy of Science, Arts, and Letters
Vol. XLIX, 1964 (1963 Meeting)

ASIAN STUDIES. A. SOUTHEAST ASIA
SILENCE AT LIFE'S NOON *

ANDRIES TEEUW

The University of Michigan
The University of Leiden

The title of the story with which I am going to deal refers to the critical stage which the life of the main character, a young author, has reached. At the beginning of the story we witness his departure from a friend after a rather vehement discussion; and the friend gives him a book by Gorki as a mirror for his own life. Then during a quiet night his past and present life passes before his imagination, and he knows that he has to face the fact that it has been a failure in all respects—that he is lonely, threatened by everything and everybody. " In this silence at the noontide of his life, even the most impossible things come crowding in upon him; everything wants to dominate him, at least part of him, if only the toes of his feet." There seems hardly a possibility of survival left: physically, humanly, artistically he is an outcast, whose fate it is to be destroyed. This is the inescapable doom of which he becomes aware at this moment of silence at life's noon.

The *poète maudit*, the doomed poet, is no new theme in literature but it seems worthwhile to analyze in more detail the Indonesian version of this doomed artist. In this particular story, a number of components can be distinguished which together make up the total picture. First of all there is the doom to be lonely. Society denies him the possibility of taking care of a family, because being a writer one cannot earn a living. He has tried it—married a girl and started a life together with normal human expectations. It soon turned out to be a mess, as she wanted him to take care of her family first of all, whereas he knew this would be the end of his career as an author. And now there is another woman—apparently understanding, not demanding, not wanting anything but to take care of him.

* The short story with which this paper is concerned, *Sunji-senjap disiang hidup*, was published in the magazine *Indonesia* VIII (6), June 1956. A more general survey of the oeuvre of Pramoedya Ananta Toer, including biographical and bibliographical information, will be given in a book on modern Indonesian literature now being prepared by the author.

He knows that he deeply longs for this safety, this protection, but he is also deeply suspicious of her, of himself, of the situation; is not it impossible to enter into such a relationship without demanding, wanting, without losing one's freedom? "A writer had better stay alone, because by doing so he can more rapidly achieve the final result of his existence, which is destruction."

The author not only is doomed to be single—he is forced by society to be solitary. No one ever really accepted him—his father hated him, his friends thought him arrogant. Since boyhood no hand was ever extended to lead him, and when he had made his own life, people had resisted him or they had laughed at him, just because he had wanted to work to fulfill his vocation. He also compares himself with his friend Bhisma, who had no ideals to follow, no responsibilities to fulfill, who had never troubled himself with having a wife and children to care for. He had only taken care of himself—and for that very reason everybody had accepted him; his life was a success, just because he had abstained from doing anything which really mattered. Or he thinks of his publisher, who carelessly drops in before he goes on another trip to the U. S. A. to see whether the author has completed his manuscript—but he does not even think of giving him an advance payment on his royalties. He expects the doomed writer to fulfill his doom and write—but whether he goes on living is of no interest to him. Virtually anybody but the writer is able to subsist in this society—the girl whose embroidery earns her a living as well as the man who grows vegetables on his small piece of land in the midst of the town of Djakarta. Only to the author existence is denied by society.

Still, there is more to the doom of the author than just the outward circumstances, however hurting and destructive. He faces a deeper crisis within himself, in this crucial moment. He knows that he has failed within himself to live up to his vocation. As an author he was basically more concerned with himself, his feelings and thoughts, than with anything else: while the misery and suffering of so many thousands of the population of Djakarta were all around him, he had shut off himself, he had been really arrogant. Because no one ever cared to help him in his troubles he had refused to be really involved in the misery of others. And it may be for this very reason, that as an author he by now feels so impotent. His soul is full of ideas, plans, emotions—and in this still night he endeavors again to bring out some of them—but the cry of a baby next door is enough to stop him; his confusion is too strong. He knows that he is in a battlefield, that the conflicts are within himself. He strongly becomes aware of the nature of these conflicts when he opens a cupboard: there lies a kris, wrapped in a white cloth—the only heirloom which he ever received from his parents. He always becomes confused when confronted with this kris because he does not know when it is the month Sura to give it the ceremonial bath, nor who can take care of this lustration, in a town like Djakarta, which has so completely lost all traditional values. "A pinch of superstition together with a life of materialism, all these years, have caused his soul to split into two parts, dangling apart, each heading in a different direction, sometimes having no connection whatsoever."

There is a deep longing in his heart for the home, the village, and the countryside. In all his misery there is one small happiness left. His hut, small and leaky as it may be, gives him a view of one of the few fields left in the midst of the town of Djakarta. This view often sets him free when he is depressed by people and things around him; it makes him feel part of a natural unity—so small a part of so grand a nature that his ambitions and sufferings really no longer matter.

But he also knows that this last piece of happiness is not going to last any longer than this field. This field, so green because of its abundantly growing vegetables, will have disappeared and become a white building, a factory, of which later on the newspapers will boast that it is the largest drug factory in Southeast Asia. By then the width and glory and greatness of that little piece of nature will have been robbed from his view, and again he will live like the others, a man of the city who lives unnoticed among the stone buildings.

And there is no escape—the fact of being an author, and having to be one is all too clear; the alternative to be a farmer is no longer left. As a writer he is the mirror of his society, and that society has entered upon a road of no return; the confusion in the soul of the author is actually the foreboding of the things which are irrevocably going to happen in his society. It is "the first stage of the explosion which is the result of the mixture of East and West, where the level of the West is demanded, but the eastern heritage still does not accept it. The East, which has to make concessions continuously, while the West keeps demanding a place within the self, within the soul, even though this soul does not really know the West. The only thing I know definitely is that this self has become nothing but the place which can act as a battlefield, where the struggle will go on, endlessly, explosion after explosion." And for the author doomed to be this battlefield, there is no alternative left but to register the shocks and explosions until some day he will succumb to them.

The analysis of this story will have shown, I hope, that it is a revealing piece of literature, only very inadequately characterized as fiction. Its relevancy to the reality of the author's personal life and to the society in which he lives is obvious. Pramoedya Ananta Toer published this story in the middle of 1956. Before, he had published a large number of novels and short stories, the main theme of which was the revolution, the struggle for freedom of the Indonesian people. More specifically (for his works were by no means political novels) , his concern was the role and responsibility of the individual in that struggle. Especially his early work, for the larger part written while he was kept in jail by the Dutch, is one loud cry for human dignity amidst the madnesses and horrors of this war, for integrity in the subsequent period of confusion and corruption which so clearly went against all ideals of real freedom.

This short story then, seems to be a personal account of ten years of writing in a society which makes demands on a writer, but does not seem to have a place for him. The disappointment, the frustration is certainly not fictitious, but very real to this author, even though he writes this story in the third person. If one happens to know more about the life of the writer, it soon becomes clear that, like many of his earlier writings, this too bears a strongly autobiographical character: the hatred of the father, the failure of the marriage, the accusation of arrogance by his friends, even the hut in which he lives, these all are in no sense fictitious but most real experiences in his life. And even the silence at life's noon seems to be more real than perhaps the writer himself was aware of when writing this story. The story was written at a crucial moment in Pramoedya's life: a few months later he was invited to attend the commemoration of the 20th anniversary of the death of Lu Sin, the Chinese Gorki, in Peking, and from this visit he returned as a man convinced of the validity, in the Indonesian situation, of a number of the communistic ideals about society and art. He became active in various kinds of social activities, and it seems as if his creative writing at least came to a temporary stop—as if he for once decided to replace the dream of the writer by the deed of the revolutionary—and by doing so tried to escape the doom of the poet. Even if this escape is only temporary, the story certainly marks a deep break in the oeuvre of this novelist—perhaps one should say marks the end of his youth.

The story is not only relevant to this particular individual writer, but it is certainly as relevant to Indonesian society as a whole. The country had gone through a Dutch colonial period, then the Japanese occupation, the four-year struggle for independence against the Dutch, and the first chaotic years of freedom, dominated by conflicts, corruption, and confusion. The very problems to which sociologists would refer in abstract terms as adjustment, acculturation, urbanization are very concretely evoked by this author in a literary form: the kris, symbol of the traditional beliefs, no longer maintained or understood, but still vaguely present; that little piece of land in the midst of the huge, crushing town, doomed to disappear in the name of progress and civilization. It is striking indeed how often this theme of the conflict between town and village occurs in modern Indonesian literature. To quote one other example, a poem by a young poet who has lived in Djakarta for some time, then he returns for a visit to his family in a village somewhere in the countryside: everybody at home reacts in a different way to his visit—as he says.

> Grandmother laughingly asks: I had expected you during the
> fasting month
> Mother's emotion is only visible in an involuntary smile
> Father is just silent

And then he tries to play with his younger brother:

> "Little brother, come on. Let's try to build a bridge,
> I am so tired with Djakarta, Djakarta."

But at the same time the artificiality is evident—there is no bridge with the past, the loneliness is inescapable and irrevocable.

Modern Indonesian literature is young, young as its nation, young as the language, the Bahasa Indonesia in which it is written. By its mere existence it already shows the break with the past: its forms, short stories, novels, free verses, as well as its contents, problems, and issues of the individual human being are as far from traditional Malay and Javanese literature as modern Djakarta is from the old Indonesian society. A story like Pramoedya's does not intentionally try to depict the reality of this development, but the very fact that as a piece of literature it transcends reality makes it more revealing, more relevant to the real human problems involved in the coming into being of this new nation than many a long academic treatise would ever be able to do.

PAPERS OF THE MICHIGAN ACADEMY OF SCIENCE, ARTS, AND LETTERS
VOL. XLIX, 1964 (1963 MEETING)

SOME ASPECTS OF INDIAN OCCUPATIONAL MOBILITY IN MALAYA

EDWARD J. VANDER VELDE
The University of Michigan

RECENT studies of the Indian community in Malaya have been concerned primarily with the role of Indian immigrant labor in the development of commercial agriculture, notably rubber production.[1] These works have tended to obscure the fact that a substantial portion of Malaya's Indian population has been active in commercial and servicing functions and has made a significant contribution to the expansion of the nonprimary sector of the Malayan economy.[2] Today, over 40% of the economically active Indian population is found in secondary and tertiary associated occupations; Indian participation in the latter occupations, in particular, represents a considerable increase since 1947. This development seems to indicate a modification of the traditional pattern of Indian society in Malaya. It is, therefore, essential that an alternative examination of the occupational structure of the Indian community as a critical element in a complex socioeconomic organism be made. This paper is an attempt at such a study. Hopefully, a foundation for a more complete understanding of the role of the Indian within Malaya's plural society will thereby be laid.

Indians presently comprise 11.7% of the population of Malaya, which is also composed of 49.8% Malaysians and 37.2% Chinese.[3]

The Indian community, in contrast to the Chinese and Malaysian communities, reflects a slightly greater degree of economic activity

[1] See J. Norman Parmer, *Colonial Labor Policy and Administration* (Locust Valley, N. Y., 1960); R. N. Jackson, *Immigrant Labour and the Development of Malaya, 1786-1920* (Kuala Lumpur, 1961); Usha Mahajani, *The Role of Indian Minorities in Burma and Malaya* (Bombay, 1960); C. Kondapi, *Indians Overseas, 1838-1949* (New Delhi, 1951). The last work is still an indispensable source of information on overseas Indian communities.

[2] E. g., it is estimated that about 20% of the total private investment fund in Malaya is supplied by the South Indian Chettiar community.

[3] The 1957 census enumerated 731,724 Indians, 3,125,474 Malaysians, and 2,333,756 Chinese in a total population of 6,278,758. Malaya, Department of Statistics, *1957 Population Census of the Federation of Malaya* by H. Fell, Report No. 14 (Kuala Lumpur: n. d.), pp. 56-57.

proportional to its numbers in the total population, forming 15% of the economically active population, while the latter represent 35.7 and 47.3% respectively. In general, however, the economic pattern of the Indians more closely parallels that of the Chinese than the Malaysians, a phenomenon most apparent in tertiary occupations.

Immigration and the Evolution of the Indian Occupational Pattern

Indirectly, the Indian occupational pattern is a result of unsuccessful attempts to secure a large and stable labor force, a necessary concomitant of the tremendous expansion of commercial rubber production in Malaya in the early 1900's, from the indigenous Malay and alien Chinese communities. This situation forced a turn to an alternative source of cheap labor, and Indian labor immigration was actively fostered to fill the needs of the estates and various government departments. Consequently, the livelihood base of Malayan Indian society has been more closely associated with economic activities of an international or national character, rather than those of a local nature, involving, primarily, specialized labor skills and, secondarily, clerical and mercantile talents.

Indian labor immigration comes under three basic headings: "indentured immigration," which was the mainstay of early commercial agricultural production but was abolished in 1910; "*kangany* recruitment," which accounted for 70% of all South Indian laborers recruited prior to 1938, when it was banned by the Government of India; and "assisted voluntary labor immigration," which became increasingly important after the mid-1920's and involved the use of monies collected under the Indian Immigration Fund to subsidize the emigration of independent laborers, but was also banned in 1938. Substantial numbers of Indian laborers emigrated independently on their own resources, and between 1946 and 1953 this form of immigration was particularly important in refilling the ranks of the war-decimated Indian labor force.[4]

In only a few instances was the immigration of commercially or professionally oriented Indians actively encouraged.[5] Although such

[4] See Kondapi, *op. cit.*, pp. 8-52 and Parmer, *op. cit.*, pp. 19-27, 38-78, for detailed discussions of the systems of Indian labor immigration and their manner of operation.

[5] An acute shortage of trained administrative and clerical staff in the early

Indian immigrants were never as numerous as labor immigrants, they became increasingly prominent in the post-Depression and post-World War II periods. They have long formed the most influential segment of Malayan Indian society.

Thus, Indian immigration to Malaya since the late-19th century was essentially a movement of laborers, predominantly South Indian in origin, from India to fill the labor requirements of a rapidly expanding economy based principally upon commercial rubber production. During this period, over 4 million Indians immigrated to Malaya, but the ephemeral nature of the migration is clearly visible in the fact that over 2.8 million returned to India.

Throughout the period of Indian immigration to Malaya, there has been an apparent correlation between occupation and the ethnic group, distinguished on the basis of language and/or religion, to which the immigrant belonged; the latter condition was often related to the mode of immigration adopted by the migrant. In practice, this meant that nearly all labor-oriented immigration was assisted in some manner and originated almost exclusively from various districts of the old Madras Presidency.[6] Conversely, nonlabor immigration included persons from both northern and southern India who commonly immigrated on their own resources as independent immigrants. By the 1920's, the general economic pattern of the Indian community was clearly identifiable on this basis. South Indians—Tamils, Telugus, and Malayalis—were mainly laborers on estates and in such government services as public works, railways, and sanitary boards. Other South Indians were prominent in such nonlaboring activities as retail trade (Tamils and Muslim Marakkyars), estate clerks (Indian Tamils), government departments (Indian and Ceylon Tamils), and banking and finance (Chettiars). Sikhs, Punjabis, and Pathans were commonly found in quasi-military occupations and small-scale moneylending activities. Sindhis, Gu-

1900's prompted an appeal by the Government of Malaya to the governments of India and Ceylon to send personnel for service in Malaya. Consequently Tamils from both countries found positions as skilled employees in a variety of government departments, notably the Federated Malay States Railways. Similarly, Sikhs and Pathans were encouraged to emigrate on a semisponsored basis to fill positions in the police forces and other quasi-military occupations. See Kernial S. Sandhu, "Indian Immigration to Malaya, 1786-1957," in K. G. Tregonning (ed.), *Papers on Malayan History* (Singapore, 1962), pp. 52-53.

[6] The Government of India had restricted all recruitment of Malaya-bound labor to Madras Presidency.

jeratis, Marwaris, and other North Indians had entered into a variety of retail trading occupations, while Bengalis preferred clerical occupations.

The general occupational pattern of the Indian community between 1931 and 1957 is revealed in Tables I and II.

TABLE I [7]

DISTRIBUTION OF INDIANS BY MAIN INDUSTRIAL GROUP, 1931-57

	Persons ('000)			Percent Distribution by Main Industrial Group		
Year	Primary	Secondary	Tertiary	Primary	Secondary	Tertiary
1931	221.0	16.4	142.8	58.1	4.3	37.6
1947	189.2	23.7	88.0	62.8	7.9	29.2
1957	181.3	22.4	101.1	59.5	7.3	33.2

The 1947 figures in Table I show the greatest degree of variation, but this was the result of two external events. The Depression caused a net outflow of more than 190,000 Indians, the majority of whom were forced off the estates by the world-wide decline in commodity prices. As a result, despite the resumption of Indian immigration in 1934, the Indian estate population never again reached pre-Depression levels. Moreover, the 1947 census was conducted shortly after the termination of World War II, at which time the Indian community had not recovered from the disastrous effects of the war, during which there was an actual net decline in the Indian population. Taking these extraneous factors into consideration, it is observed that relatively little movement between the main industrial groups has taken place.

From Table II, it is evident that since 1931 there has been a considerable degree of stability in Indian participation in primary and secondary industries. Among tertiary industries, however, a definite shift toward commerce and finance and professional services is observable by 1957. This development is indicative of the changes presently taking place in the traditional structure of Malayan Indian

[7] Malaya, *A Report on the 1931 Census and on Certain Problems of Vital Statistics* by C. A. Vlieland (London, 1932), pp. 273-276, 298-301, 322-324; *A Report on the 1947 Census of Population* by M. V. Del Tufo (London, 1949), pp. 442-445; 1957 Census, *op. cit.*, p. 126.

TABLE II [8]

DISTRIBUTION OF INDIANS IN SELECTED PRIMARY, SECONDARY, AND TERTIARY ACTIVITIES, 1931-57

Category	Percent Distribution within Main Industrial Group Year 1931 [a]	1947	1957
PRIMARY INDUSTRIES			
Rubber	85.6	83.4	83.1
Coconut	4.3	5.7	5.1
Oil palm	1.9	3.6	4.1
Tea	–	–	1.5
Rice	1.0	0.6	0.3
Mining and quarrying	2.9	3.2	3.8
SECONDARY INDUSTRIES			
Mfg. of clothing	7.6	5.8	5.9
Mfg. food, drink and tobacco	13.2	14.9	13.3
General mfg. (metals, machinery and vehicles)	22.8	14.7	16.1
Mfg. and repair of railway equipment	–	7.0	5.7
Gold and silver smithing	7.3	3.4	4.9
Building and construction	–	54.9[b]	54.9
TERTIARY INDUSTRIES			
Transport and communication	15.2	16.0	15.9
Railways	5.8	6.2	5.2
Water transport	2.3	3.3	3.7
Road transport	5.8	4.8	3.9
Communications	0.4	2.0	2.4
Commerce and finance	16.5	25.7	32.4
Retail trade	15.2	22.8	27.8
Banking, moneylending, etc.	1.3	1.3	2.1
Personal service	15.2	19.3	18.4
Domestic	10.2	8.6	4.5
Barbering	1.7	2.3	3.4
Laundering	1.2	0.9	1.2
Restaurants, coffee houses, etc.	–	8.4	8.1
Public administration	2.7	20.0	14.6
Professional services	2.6	3.8	14.5
General clerical [c]	3.6	5.8	12.1
Laborers (nonag.)	35.8	25.8	32.2
Foreman and overseers	2.1	1.7	8.2
Watchmen (*Chowkidar*)	3.2	4.9	3.5

[8] 1931 Census, *op. cit.*, pp. 273-276, 298-301, 322-324; 1947 Census, *op. cit.*, pp. 442-445, 477-482; 1957 Census, *op. cit.*, pp. 102 110, 128 137. The reader is cautioned in placing undue weight upon these percentages. In many instances the number of persons in a particular category may have increased slightly, but this may have resulted in a percentage decline due to a large population increase in the main industrial group. The primary purpose of the table is to indicate the general trend which has taken place over the tri-censual period within the occupational structure of the Indian community.

N.B.: [a] The percentages for 1931 were derived from the classification of Indians by "occupation" and those for 1947 and 1957 from the classification by "industry." Consequently, these figures are not completely analogous.

[b] The 1957 census classed all Public Works Department under Building and Construction (*i.e.*, a secondary industry). The same procedure was followed in this table for 1947.

[c] The percentages for the following categories are all based on classification by "occupation," and should not be confused with the divisions under tertiary industries above. They are, however, percentages for certain occupations primarily in secondary and tertiary industries.

society, and it may be best understood through an examination of that structure.

TRADITIONAL INDIAN SOCIETY IN MALAYA

Prior to World War II, 75% of all Indian arrivals in Malaya were from South India, and 98% of the labor migration came from this group.[9] Thus the Indian community has consisted mainly of peasant immigrants from the rural agricultural districts of South India, equipped with traditional peasant values, possessing a limited economic horizon, largely illiterate and representative of the lower castes of India's stratified society. The great majority of them were short-term migrants, who commonly left their families in India and intended to return to them after two or three years.[10]

The isolation of the estate as a self-sufficient economic unit was ideally suited to the reestablishment of the traditional sociocultural complex of the Indian village.[11] The re-creation of its system of interlocking personal ties and relationships assured the Indian immigrant of a tolerable stay in Malaya and helped to surmount his initial reluctance to emigrate. He knew he would be with his fellow caste members, and all would occupy the same economic status level.[12] Nor was life in the coolie lines without other compensations; housing, fuel, water, medical services, and often entertainment and a food allowance in the form of a rice subsidy were all provided. These elements of social and economic security combined with the maintenance of equal wage rates for agricultural and nonagricultural labor discouraged the growth of a competitive Indian labor market. The manipulations of the *kangany*, who oper-

[9] Sandhu, *op. cit.*, p. 71.

[10] "The Indian peasant was not naturally inclined to emigrate. In fact the migratory instinct was practically nonexistent in him. Even under the most desperate circumstances he always left his land with the idea of returning to it." *Ibid.*, p. 48.

[11] Thus, the temple remained the center of the South Indian laborer's social life in Malaya, and the *panchayat*, the traditional organ of village elders charged with maintaining order within caste and mediating community and individual disputes, was reconstituted as an integral part of estate society.

[12] Norton Ginsburg and Chester Roberts, *Malaya* (Seattle, 1958), pp. 343-344.

ated ". . . through a compound of family ties, personality, friendship and admiration for (his) worldly wisdom," also helped to suppress any expression of higher economic aspirations which the Indian laborer might have possessed.[13]

In marked contrast to the Indian laborer, the nonlabor, commercially or professionally oriented immigrant usually had some previous experience in his chosen trade or occupation, was representative of a higher caste group, and often possessed a little capital which enabled him to become self-employed. These immigrants were frequently accompanied by their families and generally remained longer in Malaya, though the high return rate among this group suggests that they, too, were unwilling to sever all bonds with India. Situated mainly in urban areas, this element of the Indian community was not isolated in the same physical sense as the estate dwellers, but they were isolated insofar as they tended to concentrate in definite sections of the city or town, establishing social and business relationships within their own groups and raising their children within the framework of Indian customs and traditions.[14]

The main source of persons entering into commercial and servicing occupations has been the children of those already established in these fields, and, until 1953 when Malaya restricted alien immigration, continued immigration from India. This development necessitated the transference of certain sociocultural and economic values from parents to children, its success being predicated upon the tangible element of higher status which is accorded to " white collar " and commercial occupations within Malayan Indian society and the greater capability of the nonlaboring community for providing its younger members with an adequate education. Thus, " in the Malayan Indian community it is . . . considered proper that the father should choose his son's occupation. It must . . . be remembered that the son's range of choice is in any case restricted for all but the minority. If he is from the working class, the educational

[13] Parmer, *op. cit.*, pp. 68, 258-259. Kondapi notes that ". . . the common unwillingness of the labourer to sever the ties and uproot himself from particular areas or estates dissuaded the labourer . . . from seeking employment elsewhere." Kondapi, *op. cit.*, pp. 97-98.

[14] Ginsburg and Roberts, *op. cit.*, p. 343. B. W. Hodder (" Racial Groupings in Singapore," *Malayan Journal of Tropical Geography*, 1 (1953), 33) distinguishes four separate Indian groups in Singapore, all reflecting the trading character of the urban Indian community.

opportunities open to sons of Tamil-speaking labourers are not very wide. Furthermore, the small family business is still of great importance among Indians of Singapore and Malaya, and there is a strong pull on young people to follow their father's trade, or go to the shop of some relative or neighbor." [15]

The Increasing Occupational Mobility of the Indian Community

Prior to World War II, the factors noted above worked to create a surprising degree of occupational stability among the Malayan Indian community. But the greater degree of permanent settlement observable among the Indian population between 1921 and 1957 and the corresponding decline in the monosexual character of the community predicted that occupational mobility would become an important phenomenon.[16] These developments were concomitant with an increased degree of stability in the demands of the estate industry for Indian labor prompted by the modernization of many aspects of rubber production. The general trend was also reflected in the growing proportion of the Indian population found in urbanized areas.[17]

While the lack of current information on this important trend indicates the need for additional fieldwork on the Indian community, a preliminary survey of social conditions in Singapore done in 1947 lends support to the thesis that occupational mobility within Malayan Indian society is increasing, and in the direction of greater participation in tertiary activities. The native-born Indian, especi-

[15] A. F. Wells, " Patterns of Authority in the Family in Malaya," *Transactions of the Third World Congress of Sociology*, 4 (Amsterdam, 1956), 244. An earlier social survey of Indian shopworkers in Singapore disclosed a similar tendency among South Indian shopowners to recruit their assistants locally. North Indian merchants were observed to have acquired their employees directly from India on a contractual basis. Ginsburg and Roberts, *op. cit.*, pp. 340-341.

[16] The percentage of Malayan-born in the total Indian population increased from 12.1% (1921) to 21.4% (1931) to 51.6% (1947) to 64.7% (1957). For the same censual periods, the number of females per 1000 males increased from 424 to 514 to 687 to 746. 1947 Census, *op. cit.*, pp. 58, 85; 1957 Census, *op. cit.*, pp. 9, 16.

[17] The percentage of the Indian population found in centers with populations greater than 1000 increased from 25.9% in 1931 to 41.1% in 1957; for centers greater than 10,000, it increased from 25.8% in 1947 to 30.6% in 1957. 1957 Census, *op. cit.*, p. 11.

ally of the second generation, has exhibited a pronounced shift away from unskilled labor toward an increased participation in commercial and servicing occupations; 35% were found in clerical occupations, whereas 33% and 11% respectively were semiskilled and unskilled workers. Among the immigrant Indian wage earners, 15% were found to be shopkeepers or assistants, 30% were semiskilled workers, and 32% were unskilled laborers.[18] Second generation estate children have also become a particularly important source of persons entering the commercial and servicing occupations.

The rising level of Indian literacy in English, a prerequisite for many positions in government and public service, as well as commerce and finance, may also be interpreted as an indicator of an increased degree of occupational mobility. Between 1947 and 1957, the Indian community experienced the greatest percentage increase in English-literate males in the 20- to 40-year age group, *i. e.* those most likely to have demonstrated a break with the community's traditional occupational pattern.[19]

Conclusions

It is apparent from the foregoing that the economic pattern of the Indian population was a reflection of the interaction of both internal and external factors within the complex structure of Malayan Indian society. The Indian immigrant's tendency to recreate in a Malayan setting many of the elements of traditional Indian village society—particularly the relatively rigid definitions of the roles of members of the family and the social order at large, combined with pressure for conformance with traditional patterns of action reinforced by the economic implications of caste stratification—inhibited the movement of Indians into commercial, industrial, professional, and servicing occupations. Moreover, the lack of educational opportunities and the remittance of earnings to the family in India, a manifestation of the ties to the home village retained by the Indian immigrant, blocked the acquisition of the skills and capital essential for occupational mobility. External forces also

[18] Cited in Ginsburg and Roberts, *op. cit.*, pp. 339-340 from *A Social Survey of Singapore: A Preliminary Study of Some Aspects of Social Conditions in the Municipal Area of Singapore, December, 1947* by the Department of Social Welfare, Colony of Singapore.

[19] In this category, Indian literacy in English increased 8.3% to 21.3%; the Chinese increase was 5.8% to 15.8%, while the increase in the Malaysian community was only 1.5% to 6.5%. 1957 Census, *op. cit.*, pp. 93-95.

prevented expression of higher economic aspirations; Indian immigration was necessary to fill the needs of a rapidly expanding commercial agriculture economy, and the development of a large commercially oriented Indian community which might compete with European business was not desired. To this end, a variety of pressures was used by employers of Indian labor to maintain the status quo.

The rapid expansion and diversification of the Malayan economy in association with a high degree of urbanization appears to have profoundly influenced the traditional structure of Malay's Indian community. Increased educational opportunities and the rising prestige attached to commercial and professional occupations has made it possible for the native-born Indian in particular to become a member of an emerging " Malayan " society. The greater Indian participation indicated by the 1957 census in professional services, commercial and financial activities, and general clerical occupations is partial evidence that such a development is taking place. But, on the whole, much more needs to be known about the organization and operation of traditional Malayan Indian society and its present position vis-à-vis the Malayan-born component of the community, before it may be decisively concluded that a changing occupational pattern reflects the permanent disruption of the Indian socioeconomic organism.

PAPERS OF THE MICHIGAN ACADEMY OF SCIENCE, ARTS, AND LETTERS
VOL. XLIX, 1964 (1963 MEETING)

B. JAPAN

KINOSHITA JUNJI'S USE OF TRADITION IN MODERN JAPANESE DRAMA

KAREN WOODARD BRAZELL
The University of Michigan

THE revolution in Japanese culture that followed the Meiji Restoration shook the very foundations of Japanese literature. The bonds of tradition were discarded; Western works were eagerly translated and slavishly imitated. Some of the earliest efforts at a new literature now seem amusing. In the field of prose fiction, however, the first two decades of the 20th century saw the emergence of such literary giants as Natsume Sōseki, Mori Ōgai, Nagai Kafu, Akutagawa Ryūnosuke, and Tanizaki Junichirō, men who freed modern Japanese fiction from the fetters of imitation and endowed it with an independent spirit of its own. Through the efforts of these men, modern Japanese fiction has earned its place in world literature. The same is not true of modern Japanese drama, however. It too responded to the flood of western ideas, threw off tradition, and went through the same stages of translation and imitation of Western works. But in prewar Japan few really successful playwrights emerged, and no new and independent tradition developed.

The nature of drama itself and the nature of the Japanese dramatic tradition help to explain why a new tradition has been slow in developing. Drama includes not only the art of playwriting but also the arts of acting and stagecraft. These arts, which were highly developed in pre-Meiji Japan, differ widely from their Western counterparts. Japanese dramatic tradition emphasizes stylized and formalized elements; it is presentational rather than representational; and the actor is more important that the playwright. It is not a realistic theater. Therefore, at the turn of the century when Tsubouchi Shōyō called for realism in all forms of literature, he struck at the very heart of traditional drama. To meet this new requirement Japanese drama had to turn to new methods of playwriting, acting, and producing. It had to permit women on the stage, develop new techniques of acting, build new stages, and introduce many new dramatic conventions and techniques. Before a modern Japanese drama could be developed, the myriad techniques of modern drama had to be learned. This required time. Today, the many fine Japanese motion pictures offer convincing proof that they have assimilated—and even improved upon—these dramatic techniques.

What happened in the Japanese theater between 1868 and World War II? How were the techniques of modern drama learned? The process was one of broad experimentation. *Shimpa,* a type of theater halfway between Kabuki and *shingeki* (the Japanese term for modern drama), prospered briefly. Tsubouchi not only advocated realism, but he also took an active part in developing a realistic theater with his own playwriting, his translations (notably of Shakespeare), and his work in acting and producing. Osanai Kaoru, another leader in the development of a modern theater, was especially influenced by such modern realists as Chekhov, Ibsen, and Hauptmann. He was an active leader of the *Jiyū Gekijō* (Free Theater) and the *Tsukiji Shōgekijō* (Tsukiji Little Theater), two experimental theaters. Proletarian drama was tried by leftist writers, while other playwrights, led by Kishida Kunio, experimented with psychological realism. Kikuchi Kan won fame with plays that often dealt with social problems. *Chichi Kaeru* (Father Returns), his most popular play, presents the problems of the family in modern society. Thus, in prewar days, almost everything was tried and much was learned, but no lasting Japanese tradition was born.

After World War II, however, new life began to stir within the Japanese theatrical world. Today, conscious attempts are being made to develop a modern theater that is also Japanese. Dramatists who are well versed in Western drama are now looking back to the Japanese traditions that were largely cast off with the discarding of feudal institutions.[1] Two of the most successful of these attempts to combine Japanese and Western traditions are those of Mishima Yukio and Kinoshita Junji. Mishima has written modern versions of Nō plays.[2] His plays, which are extremely modern in content

[1] Yoshida Seiichi in *Gendai Bungaku to Koten* (Contemporary Literature and the Classics) (Tokyo, 1961), esp. p. 66, argues that this conscious turning to tradition is characteristic of postwar literature in general.

[2] These modern Nō plays are, of course, only a small part of Mishima's work, which includes fiction as well as drama. For translations of the first five of Mishima's Nō plays see *Five Modern Nō Plays*, trans. Donald Keene (New York,

and technique, retain many structural and thematic features of Nō. In his modern version of *Kantan*, for example, instead of a traveler dreaming of becoming the emperor of China, a young man dreams of becoming a financial tycoon and a dictator. Kinoshita's efforts have been in a different direction.[3] He has used the structure of Western drama but has drawn on folktales to give his plays Japanese content. These two dramatists have approached the problem from different angles, but they have both created a type of modern Japanese drama. In this paper I will concentrate on the method used by Kinoshita Junji. This method can best be revealed by an examination of his one-act play, *Yūzuru* (Twilight Crane), probably the most popular play in postwar Japan.[4]

Before turning to this analysis it is necessary to say a word about the subject matter of drama in general. The subject matter of Japanese traditional drama is seldom entirely the invention of the playwright. The dramatist usually drew upon existing myth, legend, literature, or history for his material. Zeami, the great perfector of the Nō, says that the first step in writing a Nō play is the choice of a "seed," a seed being the story out of which the action of the play grows.[5] Chikamatsu, as well as lesser Tokugawa playwrights, continued to use historical and literary materials for the content of their plays. But they also drew on events so timely that the theater became an excellent place to learn the latest gossip.[6] Thus Kinoshita's and Mishima's use of existing materials for their subject

1957). Keene's introduction discusses briefly the way Mishima has used the original material in his plays.

[3] This paper deals only with the folktale plays of Kinoshita, although he has also written other types of modern drama. He has written about ten folktale plays, not counting a dozen or so radio plays, many of which are also based on folktales. His folktale plays are collected in *Yūzuru* (Twilight Crane) and *Onnyoro Seisuiki* (The Rise and Fall of Onnyoro), Vols. I and II of *Kinoshita Junji Sakuhinshū* (Collected Works of Kinoshita Junji) (Tokyo, 1961). These volumes will hereafter be referred to as *Works I* and *Works II* respectively.

[4] Statistics given in Kinoshita Junji, *Yūzuru Sōgōhan* (Twilight Crane, A Complete Edition) (Tokyo, 1953), pp. 443 and 451, indicate that from October 1949 to April 1952 alone the Budō no Kai Troupe presented this play 145 times to 137,100 people throughout Japan. About 2000 amateur groups are thought to have produced it from 1949, the year of its first publication, to 1953. To date it has been published 13 times, translated into English, Chinese, and Russian, and used in nine different language textbooks for Japanese high-school students.

[5] Arthur Waley, *The Nō Plays of Japan* (London, 1954), p. 35.

[6] Earle Ernst, *The Kabuki Theater* (New York, 1956), pp. 208-209.

matter is no departure from Japanese tradition. It is, in fact, a practice common to drama throughout the world. Originality of subject matter is not an essential element of any drama.[7] Greek drama presupposes an acquaintance with the mythology and legends on which it is based. Most of Shakespeare's plays were drawn from literary or historical sources. Classical French, German, Spanish, Chinese, and Javanese drama all make use of stories or fables already current in the culture. Even today, when originality is so highly esteemed, drama continues to draw on older sources. Movies like *Phaedra* and *Orfeo Negro,* plays like *The Cocktail Party* and *Mourning Becomes Electra,* and even the musical *West Side Story* are but a few examples. The originality of subject matter is thus less important to drama than the nature of the materials chosen and the way the dramatist uses them.

Kinoshita has a reason for basing plays on folktales. He sees in Western culture a common heritage of Greek and Christian mythology from which authors can draw materials and motifs. The counterpart of this in Japan seems to Kinoshita to be the mythology of the *Kojiki* and *Nihonshoki* and the large store of Japanese folklore. By drawing on this common heritage and adding meanings pertinent to modern life, Kinoshita feels that modern, living folktales can be created. This is what he attempts to do in his folktale plays.[8]

The folktale Kinoshita uses in *Yūzuru* is not only a tale found in many parts of Japan, but, as shall be demonstrated later, its themes are common in Japanese folklore. The tale has many different variations, but the Sado Island version which first inspired Kinoshita goes as follows: A poor young man removes an arrow from the body of a wounded crane. A few days later a young maiden comes to his house and offers to become his wife. She then weaves

[7] Wolfgang Kayser has a very good discussion of this point in *Das Sprachliche Kunstwerk: Eine Einführung in die Literaturwissenschaft* (Bern, 1959), pp. 55-59. He goes as far as to say, "Gerade beim Drama wird man feststellen, dass die Erfindung des Stoffes eine Ausnahme darstellt" (p. 55).

[8] A good summary of these ideas of Kinoshita is in *Engeki no Dentō to Minwa* (Folktales and the Tradition of Drama), Vol. I, *Kinoshita Junji Hyōronshū* (Collected Criticism of Kinoshita Junji) (Tokyo, 1961), pp. 194-197. Kinoshita is in fact a scholar of Western literature. He did his college work in English literature, has traveled in the West, and has done some translation. For a chronology of his life and works see: Vol. 50, *Gendai Nihon Bungaku Zenshū* (Anthology of Modern Japanese Literature) (Tokyo, Chikuma Shobō, 1956), pp. 429-431.

him a beautiful piece of cloth and tells him to sell it. Later on the young man wants more cloth, so the wife, telling him as before not to peek at her while she is weaving, begins to make the cloth. Breaking his promise he looks into the weaving room and sees a crane at the loom. His wife, the crane, sees that he has looked and flies away.[9]

The story given in *Yūzuru* is an expanded version of this tale. The play opens with a scene in which Yohyō and his wife Tsū are happily playing games with a group of children. When they go off two villagers, Unzu and Sōdo, come on. Their conversation reveals how Yohyō once helped a crane and how Tsū (the Japanese word for crane is *tsuru*) became his wife and wove him some beautiful cloth. The villagers want to get hold of some of this cloth and sell it to their own advantage. Yohyō is hesitant about asking for more cloth because each time Tsū weaves she grows thinner. Urged on by Unzu and Sōdo and by his own desires to make a lot of money and to see the capital, Yohyō finally asks Tsū for the cloth in a very tender supper scene. Tsū, who feels it is the only way she can keep her husband, agrees to make one more piece. As usual she makes Yohyō promise not to look while she is weaving. Unzu and Sōdo reenter and, ignoring Yohyō's attempts to stop them, peek into the weaving room. They see a crane weaving with feathers plucked from its own body. Satisfied that the cloth is really the crane feather cloth that they had hoped it was, they go home to await its completion. Yohyō, unable to restrain his curiosity and anxiety, peeks into the weaving room. All he sees is a crane, so he goes off in frantic search of Tsū. The next evening Tsū, emaciated and pale, comes out of the room carrying two pieces of cloth. She gives them to Yohyō telling him to treasure one of them always in memory of her. She says that as he has looked at her she must leave him, and before he realizes what is happening, she is gone. Unzu, Sōdo, and the children come on. A child points out a crane barely managing to fly away, and they all stare at the disappearing bird.

To see how Kinoshita has used this folktale we must look at the ways his play differs from the tale. First, Kinoshita's ordering of the action of the play is quite different from the sequence of events in the folktale. The play starts with the end of the tale and deals directly only with the final request for cloth and the tragic consequences of that request. All the other events—the wounded crane.

[9] This is a summary of the story quoted by Kinoshita in *Yūzuru Sōgōhan*, pp. 61-62.

the arrival of Tsū, and the weaving of the first pieces of cloth—are brought out indirectly and gradually. They are not actually enacted. This concentration on a single incident and the subsequent unity of plot is what gives the material dramatic structure. Kinoshita handles his materials in the manner of Sophocles who began *Oedipus Rex* at the end of the myth and permitted what had gone before to be revealed only gradually.[10] Kinoshita's method is indeed much more along the lines of Greek than traditional Japanese dramaturgy. Plot is meager in Nō; and in Kabuki and the puppet theater the structure of the story is so loose that whole acts can easily be left out.

Another difference between the tale and the play is the creation of new characters in the latter. Kinoshita has added the villagers, Unzu and Sōdo, and the children. These characters function in the structuring of the plot as well as in the expression of the themes. The villagers make it possible to reveal to the audience what has happened before the play opens. They carry the burden of the exposition in the first part of the play. The children function as a kind of chorus, a feature common to both traditional Japanese and to Greek drama. However, the children are more like a Greek chorus than a traditional Japanese one. The chorus in traditional Japanese drama takes no part in the action of the play; rather it serves as an impersonal narrator, or it chants lines for the actors while they act or dance.[11] In Greek drama, according to Aristotle,[12] the best chorus was the one that functioned as an actor. Kinoshita's chorus does precisely this. The children who open the play set an atmosphere of cheerfulness and innocence, but the song they sing also foreshadows one of the central actions, the weaving of the cloth. Their song goes: [13]

> Grandpapa, we will weave you a quilt, hey!
> Grandmama, we will weave you a quilt, hey!
> Chin kara, kan kara, ton, ton, ton.

The last line represents the sound of the loom, which is an im-

[10] For a discussion of Sophocles' use of this myth see Francis Ferguson, *The Idea of a Theater* (Princeton, 1949), pp. 14-19.

[11] Richard N. McKinnon, "The Nō and Zeami," *Far Eastern Quarterly*, 11 (1952), 355-356.

[12] *Criticism: The Major Texts*, ed. Walter Jackson Bate (New York, 1952), p. 31.

[13] Kinoshita found this song in a book of Sado Island folktales. See *Works I*, p. 228.

portant sound effect later in the play. After this cheerful introduction Unzu and Sōdo express their desire to get hold of a piece of Tsū's cloth. Halfway through the play the children enter again, this time trying to play games with Tsū. The children are still gay, but Tsū's suspicion of the villagers dampens the mood. The song they sing to accompany their game this time foreshadows the end of the play. The song begins:

Squat down, blind your eyes.
There in the cage is a little bird.
When oh when will it fly out?
When dawn breaks or at twilight?
Slip and slip and slide, slide, slide.

The relation of the song to the play is heightened by the fact that as they play this game Tsū is "it" and is caged in by the circling children. As Tsū is saying farewell at the end of the play, the first song is again heard. The children themselves come running on, see the crane, and run off following it. The song is heard again as Yohyō, Unzu, and Sōdo stand gazing at the disappearing crane and the curtain falls.

So far I have shown how the changes Kinoshita made in the folktale have resulted in giving the material a dramatic and highly unified structure. Kinoshita's changes have also added to the meaning of the play. The folktale itself contains two motifs common in Japanese mythology and folklore. The first is called *ongaeshi*, the return of a kindness. In Japanese folklore this often takes the form, as it does here, of an animal returning the kindness of a man. An old and famous example is the story of Urashima Tarō, a kind-hearted lad who rescued a turtle and who was then escorted to the Kingdom of the Sea in return.[14] The transformation of animals into people for both good and harmful purposes is also a common feature of Japanese folklore. The "Pandora's box" motif is found in folklore throughout the world. In the Urashima Tarō story, for example, the hero receives a box which he is forbidden to open. He does open it, of course, and immediately turns into an old man and dies. Another example even closer to the *Yūzuru* incident is found in a story in the *Kojiki* where the daughter of the God of the Sea tells a prince not to look at her in childbirth.[15] He looks and sees

[14] Trans. in Yei Theodora Ozaki, *Japanese Fairy Tales* (New York, 1903), pp. 25-42.

[15] Trans. in *Anthology of Japanese Literature*, ed. Donald Keene (Rutland, Vt., 1955), pp. 54-58.

that she has turned into a crocodile. Because he has seen her in her native shape, the princess leaves him and returns to the sea. Many other examples could be readily found to illustrate the fact that these motifs are common ones in Japanese folklore.

Kinoshita has left both of these motifs in his play. Tsū makes it clear in a soliloquy that she came to Yohyō to repay his kindness. In the play as in the tale, Tsū's leaving is a direct result of Yohyō's peeking after he had been forbidden to do so. But in the play Yohyō's motivation is much clearer than in the tale. He is tempted to look when Unzu and Sōdo look first and say that there is a crane weaving. Kinoshita, in developing both of the motifs found in the original tale, also adds new meaning. In the play Tsū cannot understand the greed of the villagers, the greed that is growing in Yohyō. In the tale the wife tells her husband to take the cloth and sell it, but in the play "money" is something Tsū cannot comprehend.

> What is this? . . . . Everything is for the sake of this . . . this . . . . money. I just thought that you would like to see some beautiful cloth . . . . you were so thrilled when you saw it that I too was very happy. Only for that did I waste away my body in weaving. But now I have no other way to keep you.

This is clearly what Aristotle would call reversal of intention: Tsū weaves the cloth to make Yohyō happy, but the intervention of greed thwarts her aim, and the effect of the action is unhappiness. The lack of understanding, or as one Japanese critic put it, the "dis-communication," between Tsū and the materialistic villagers is heightened by the language they use.[16] The villagers speak a kind of general dialect invented by Kinoshita to enhance the universality of this play. Tsū's language is much closer to standard Japanese. The following two examples illustrate this:

> Unzu: Sō da. Tada na, ano nyōbō, keshita ottoru tokoro michanaran chūte hataya ni hairu ge na.
>
> Tsū: Anata wa dandan ni kawatte iku. Nan da ka wakaranai keredo, atashi to wa betsu no sekai no hito ni natte itte shimau.

---

[16] *Yūzuru Sōgōhan*, p. 58.

It is interesting to note that there is no conversation between Tsū and the villagers. They try to talk to her once, but she does not answer.

In the play Tsū is indeed repaying a debt, but her feelings are more than those of gratitude. She loves Yohyō and hopes to live with him in the innocent and cheerful world represented by the children. Her attempts to make him happy go to the extent of weaving cloth for him from the feathers of her own body. When she realizes that he has spied on her and that she must leave him, she risks death to weave him two pieces of cloth, one to sell and one to treasure as a keepsake. The love portrayed in Yūzuru is a much deeper emotion than the *ongaeshi* dealt with in the original tale. Yet Tsū remains an outsider. Her pure love and the idyllic life she dreams of cannot be reconciled with the materialistic society she finds herself in.

The most tragic character in the play is Yohyō. He is tragic because he is human and ignorant. He was once a very hard-working and kindhearted young man. Although the cloth Tsū wove made him happy, he was too much attracted by the materialistic world to resist selling it for money. A small amount of money led to the desire for more, and the urging of the villagers gradually drags him into their world and away from the world of Tsū and the children.[17] His love for Tsū is apparent, but so is his ignorance. He does not realize that his love and his greed are incompatible. He is sincerely concerned about Tsū's growing weakness and her unhappiness, but he is excited by prospects of a fortune and a trip to the capital. When in the last scene he stands dazed watching the crane fly away and clutching the cloth Tsū wove, he still does not seem to be quite aware of what he has had, what he has lost, and what he has traded it for.

Kinoshita Junji has created a masterful play from a simple Japanese folktale. He has drawn on Japanese folklore for the content of the play but has turned to Western tradition for a method of structuring this material. For modern Japanese drama, however, the importance of his attempt lies in the fact that he has consciously drawn on Japanese tradition, combined it with elements of Western tradition, and has come up with a successful modern Japanese play.

[17] The theme of money can be seen on a social and economic level with money and merchants upsetting village structure and life. For a discussion of this see *Yūzuru Sōgōhan*, p. 373.

PAPERS OF THE MICHIGAN ACADEMY OF SCIENCE, ARTS, AND LETTERS
VOL. XLIX, 1964 (1963 MEETING)

REGIONAL VARIATIONS IN JAPANESE INTERNAL MIGRATION

JOHN D. EYRE
University of North Carolina

BETWEEN 1955 and 1960, when Japan's population increased 4.6%, rising 4,142,972 to a total of 93,418,501, 26 of the 46 Japanese prefectures registered population losses.[1] As Fig. 1 shows, 15 prefectures had losses in excess of 2%, with maximum losses of more than 4% in Kagoshima and Shimane. Prefectures which gained were fewer, but their rate of increase in some cases was spectacular: more than 20% for Tokyo, and 10-20% for Kanagawa, Aichi, and Osaka. In number, losses ranged from 62 (Fukui) to 81,008 (Kagoshima) and gains from 4197 (Nara) to 1,644,588 (Tokyo). Data for other prefectures is given in Table I. While prefectures have exhibited sharp variations in rate of growth ever since the first formal national census of 1920, the trend toward net losses is relatively new and apparently is growing stronger. Only seven prefectures declined in numbers during 1950-55, and one of them, Yamagata, had a net gain in the following five-year period.[2]

A first reaction is to relate these losses to the drastic changes in the Japanese rate of population increase that have taken place since the late 1940's. Long described as having an "Oriental birth rate and an Occidental death rate," Japan has had its death rate decline further and its birthrate take an even sharper drop. As a consequence, it now has one of the lowest rate of natural increase in the world. However, even with this "demographic revolution," none of the prefectures has a rate of increase below zero. It can be assumed, therefore, that the population losses and part of the gains in prefectures during 1955-60 were the result of large-scale internal migration. It is with the volume and direction of these movements and their bearing on the 1955-60 population changes that this paper

[1] Bureau of Statistics, Office of the Prime Minister, *1960 Population Census of Japan, Preliminary Count of Households and Population* (Tokyo, 1960) and *Final Count of Population* (Tokyo, 1960), pp. 2-3.

[2] Bureau of Statistics, Office of the Prime Minister, *Population of Japan, Summary of the Results of the 1955 Population Census of Japan* (Tokyo, 1960), p. 79.

TABLE I

JAPANESE POPULATION CHANGES BY PREFECTURE, 1955-60 *

Prefecture	1955 Population	1960 Population	Change (* indicates loss)
Hokkaido	4,773,087	5,039,206	266,119
Aomori	1,385,523	1,426,606	41,083
Iwate	1,427,097	1,448,517	21,420
Miyagi	1,727,065	1,743,195	16,130
Akita	1,348,871	1,335,580	13,290 *
Yamagata	1,353,649	1,320,664	32,985 *
Fukushima	2,095,237	2,051,137	44,100 *
Ibaraki	2,063,944	2,047,024	16,920 *
Tochigi	1,544,987	1,513,624	31,363 *
Gumma	1,616,142	1,578,476	37,666 *
Saitama	2,260,493	2,430,871	170,378
Chiba	2,205,153	2,306,010	100,857
Tokyo	8,039,214	9,683,802	1,644,588
Kanagawa	2,919,497	3,443,176	523,679
Niigata	2,473,492	2,442,037	31,455 *
Toyama	1,021,121	1,032,614	11,493
Ishikawa	966,187	973,418	7,231
Fukui	752,758	752,696	62 *
Yamanashi	807,044	782,062	24,982 *
Nagano	2,019,860	1,981,433	38,427 *
Gifu	1,586,334	1,638,399	52,065
Shizuoka	2,650,435	2,756,271	105,836
Aichi	3,769,209	4,206,313	437,104
Mie	1,485,582	1,485,054	528 *
Shiga	853,734	842,695	11,039 *
Kyoto	1,933,886	1,993,403	59,517
Osaka	4,619,583	5,504,746	885,163
Hyogo	3,620,947	3,906,487	285,540
Nara	776,861	781,058	4,197
Wakayama	1,006,819	1,002,191	4,628 *
Tottori	614,259	599,135	15,124 *
Shimane	929,066	888,886	40,180 *
Okayama	1,689,800	1,670,454	19,346 *
Hiroshima	2,149,044	2,184,043	34,999
Yamaguchi	1,609,839	1,602,207	7,632 *
Tokushima	878,109	847,274	30,835 *
Kagawa	943,823	918,867	24,956 *
Ehime	1,540,628	1,500,687	39,941 *
Kochi	882,683	854,595	28,088 *
Fukuoka	3,859,764	4,006,679	146,915
Saga	973,749	942,874	30,875 *
Nagasaki	1,747,596	1,760,421	12,825
Kumamoto	1,895,663	1,856,192	39,471 *
Oita	1,277,199	1,239,655	37,544 *
Miyazaki	1,139,384	1,134,590	4,794 *
Kagoshima	2,044,112	1,963,104	81,008 *

* Source: *1960 Population Census of Japan, Preliminary Count*, pp. 2-3 and *Final Count*, pp. 2-3.

is concerned. Discussion is restricted to the interprefectural movements of persons who were making a permanent change in residence.

INTERNAL MIGRATION DATA

Besides the computations that can be made from census materials, Japanese internal migration can be traced in detail and with considerable dependability with two bodies of data. One consists of statistics collected in connection with the issuing of rice ration books under regulations of the Foodstuffs Administration Law. In operation since 1948, the law was put into effect to equalize the distribution of the main Japanese food. The issue and cancellation of ration books provides valuable information about the size of local rice demands and, as a by-product, a tally of registrant movements. Statistical compilations are issued annually by the national government for the period November-October.

A second set of data is obtained through the operations of the Resident Registration Law, which requires all persons who establish permanent residence to register with the local unit of government. Upon moving, the registrant must cancel his registration and re-register at his new address. Although dating back to 1914 as formal legislation, the law was put in its present form in 1951. Its main purpose is to yield population information that will help local and national administrative agencies to provide for and to anticipate the need for critical public services. Annual compilations of statistics covering the origin (prefecture) of new registrants, the destination of registrants who have moved, the month of the transfer of residence, and the sex of the migrant have been issued since 1954. The 1960 resident registration report [3] is the primary source in tracing internal migration in this paper.

The 1960 report revealed that 5,652,659 persons changed residence during the twelve-month period. In round figures, this was a 300,000 increase over 1959 movements. Table II shows that 6.1% of the total national population were involved, higher than any year of record except 1954. One unknown detail that could have an important bearing on the exactness of these figures was the number of multiple moves made by individuals during the year. In sex, 52.9% of the migrants were men and 47.1% women, or

[3] Bureau of Statistics, Office of the Prime Minister, *Jūmin tōroku jinkō idō hōkoku nempō* (Annual Report on Movements of Resident Registration Population), 1960 (Tokyo, 1962).

roughly 112 men to every 100 women. One additional statistic pertinent to this paper was the volume of intraprefectural as opposed to interprefectural migrants. As Table III shows, the former were

TABLE II

NUMBER OF MIGRANTS SHOWN BY RESIDENT REGISTRATION, 1954-60 *

Year	Number of Migrants	Percent of National Population
1954	5,500,000	6.2
1955	5,140,000	5.8
1956	4,860,000	5.4
1957	5,270,000	5.8
1958	5,290,000	5.8
1959	5,360,000	5.8
1960	5,650,000	6.1

* Source: *Jūmin tōroku jinkō idō hōkoku nempō* (1960), p. 6.

TABLE III

INTRA-PREFECTURAL AND INTER-PREFECTURAL MOVEMENTS SHOWN BY RESIDENT REGISTRATION, 1955-60 *

Year	Intra-Prefectural Movements	Percent of Total	Inter-Prefectural Movements	Percent of Total
1954	3,150,000	57.2	2,350,000	42.8
1955	2,910,000	56.7	2,230,000	43.3
1956	2,740,000	56.3	2,120,000	43.7
1957	2,890,000	54.8	2,380,000	45.2
1958	2,910,000	55.0	2,380,000	45.0
1959	2,920,000	54.4	2,440,000	45.6
1960	2,970,000	52.6	2,680,000	47.4

* Source: *Jūmin tōroku jinkō idō hōkoku nempō* (1960), p. 7.

52.6% and the latter 47.4% of the total movement. However, over the seven-year span since 1954, the interprefectural movements have increased steadily in absolute and relative strength.

Analysis of 1960 Migration

Although the reasons behind internal migration are admittedly complex and involve a variety of "push" and "pull" factors, they tend to be heavily economic. In Japanese rural areas, birthrates remain relatively high and a dense population is already pressing hard on the cultivated land resource. These conditions result in a steady stream of young people moving toward the cities in search of greater economic opportunity. At the same time, the demand for specialized skills in commerce and manufacturing and the changing economic conditions in urban areas stimulate a heavy and continuous flow of migrants between the many large Japanese cities. The zone of maximum urban opportunity extends westward from Tokyo and the Kanto Plain through the Inland Sea borderlands to northern Kyushu. Turning again to Fig. 1, we see that this zone contained 14 of the prefectures that registered population gains during 1955-60, including all the prefectures with more than 6% increase.

The magnetic attraction which this urban-industrial zone had for migrants in 1960 is shown in Fig. 2. In that year, only seven prefectures had an excess of in-migrants over out-migrants. These included the four Kanto Plain prefectures clustered around the Tokyo-Yokohama urban complex; Aichi, with its great city of Nagoya; and Osaka and Hyogo, which contain the contiguous urban areas of Osaka and Kobe, at the eastern end of the Inland Sea. A number of prefectures adjacent to or near these seven suffered only small losses through excessive out-migration. The most striking negative net migration, where out-migrants topped in-migrants by 10% or more, was in the heavily agricultural prefectures of northern and central Honshu, Shikoku, and Kyushu. Two interesting cases are Fukuoka, in northern Kyushu, and the extreme northern prefecture of Honshu, Aomori. The former is generally considered the western anchor of the urban-industrial zone and had substantial population growth during 1955-60. The negative net migration in 1960 is accounted for in part by serious depression in the important coal-mining industry and the problems of industrial expansion in the face of serious locational handicaps. In Aomori, the population gains during 1955-60 in spite of relatively heavy out-migration were caused mainly by the high birthrate, one of the highest in Japan. As this case indicates, birthrates merit equal attention with the level

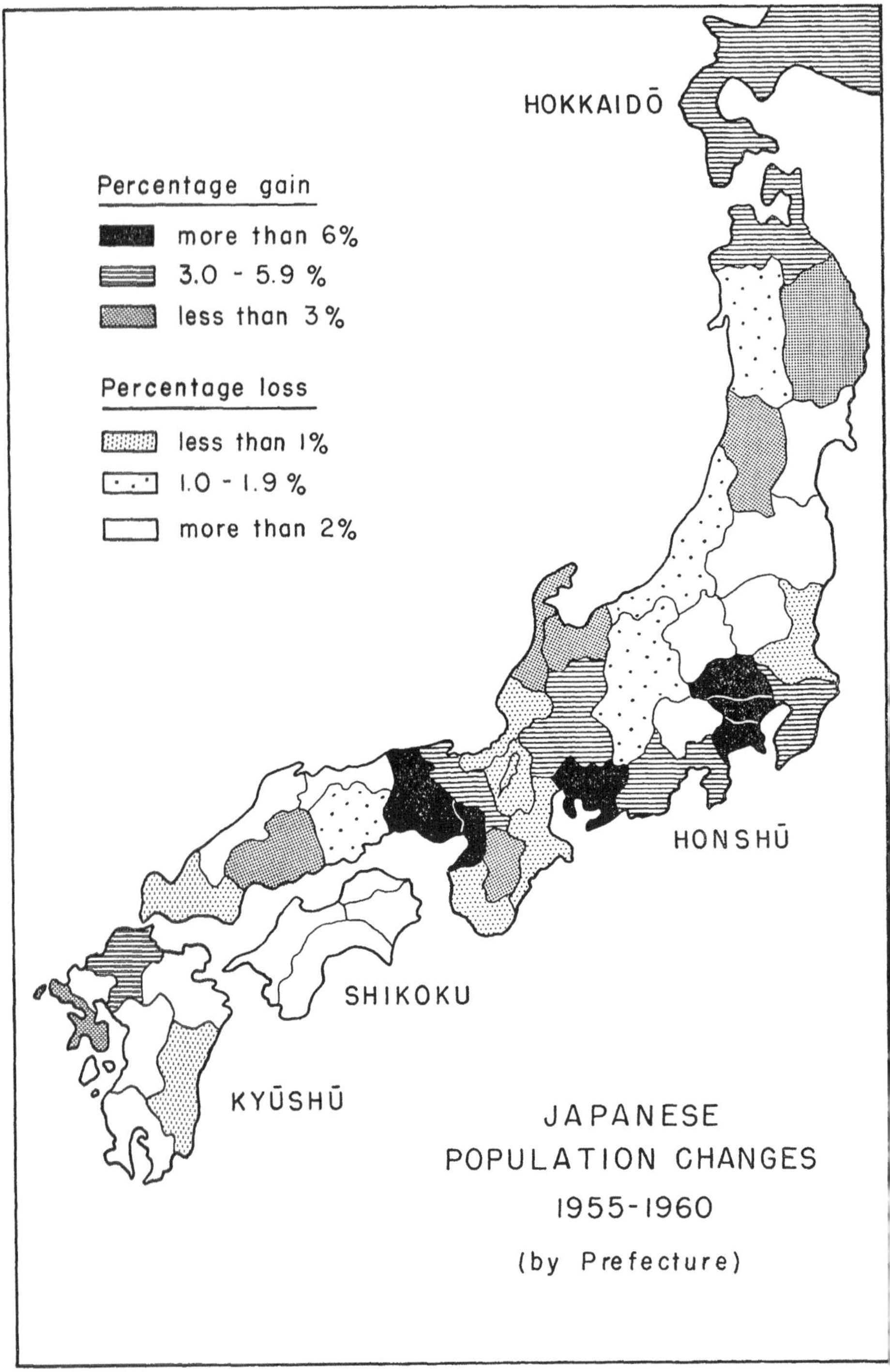
HOKKAIDŌ
Percentage gain
more than 6%
3.0 - 5.9 %
less than 3%
Percentage loss
less than 1%
1.0 - 1.9 %
more than 2%
HONSHŪ
SHIKOKU
KYŪSHŪ
JAPANESE
POPULATION CHANGES
1955-1960
(by Prefecture)

Fig. 1.

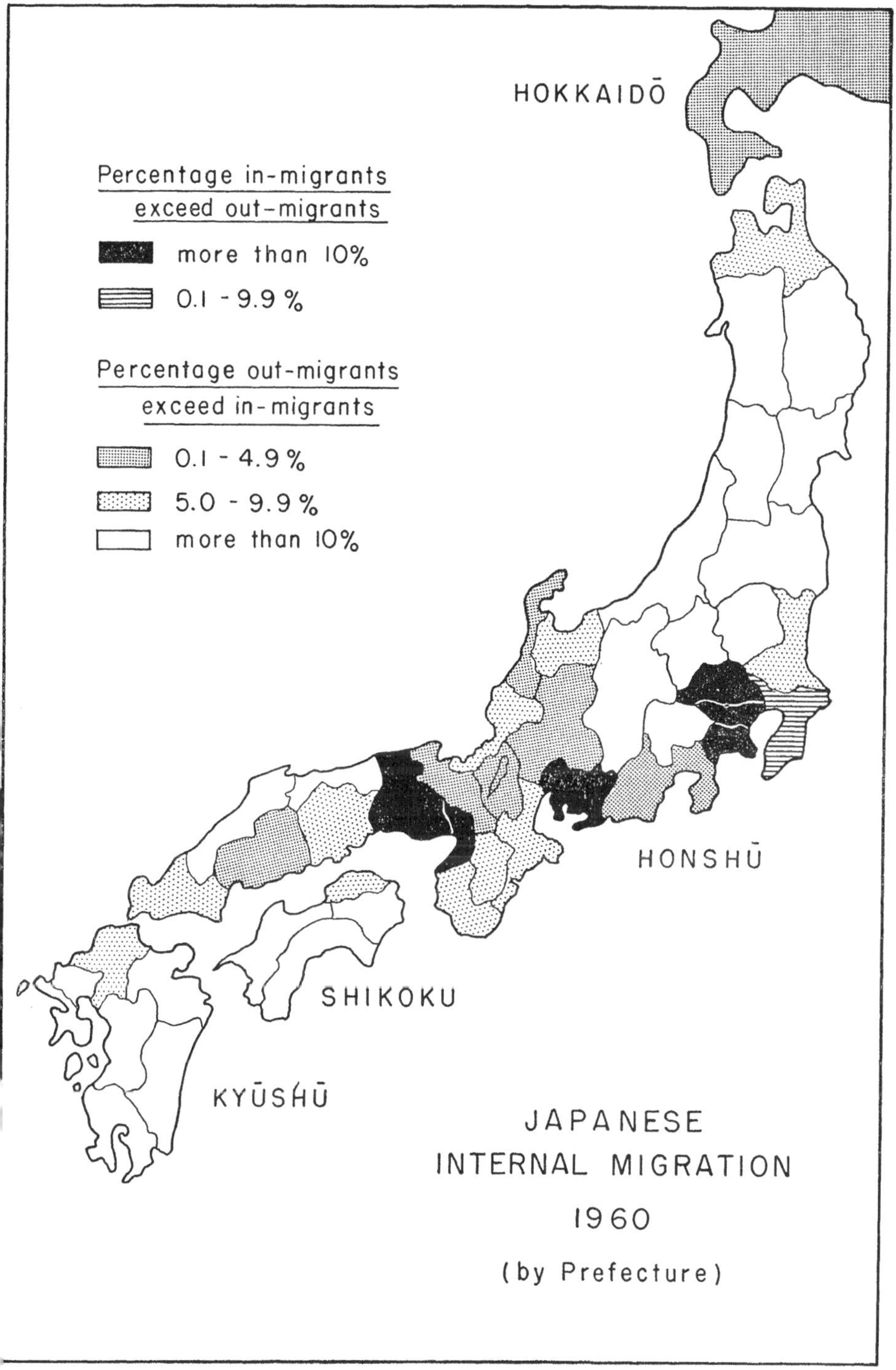
HOKKAIDŌ
Percentage in-migrants exceed out-migrants
more than 10%
0.1 - 9.9 %
Percentage out-migrants exceed in-migrants
0.1 - 4.9 %
5.0 - 9.9 %
more than 10%
HONSHŪ
SHIKOKU
KYŪSHŪ
JAPANESE
INTERNAL MIGRATION
1960
(by Prefecture)

FIG. 2.

TABLE IV

JAPANESE INTERNAL MIGRATION BY PREFECTURE, 1960 *

Prefecture	Number of In-Migrants	Number of Out-Migrants	Net Balance (* indicates loss)
Hokkaido	53,778	68,087	14,309 *
Aomori	18,684	31,364	12,680 *
Iwate	18,251	32,966	14,715 *
Miyagi	30,026	49,276	19,250 *
Akita	14,981	35,245	20,264 *
Yamagata	19,546	36,747	17,201 *
Fukushima	28,388	64,879	36,491 *
Ibaraki	37,133	54,145	17,012 *
Tochigi	23,338	41,461	18,123 *
Gumma	23,981	40,542	16,561 *
Saitama	101,007	66,962	34,045
Chiba	86,883	69,710	17,173
Tokyo	599,477	388,764	210,713
Kanagawa	197,998	104,663	93,335
Niigata	30,882	63,605	32,723 *
Toyama	12,834	20,503	7,669 *
Ishikawa	14,459	19,297	4,838 *
Fukui	10,485	16,497	6,012 *
Yamanashi	13,325	24,023	10,698 *
Nagano	27,892	50,185	22,293 *
Gifu	40,506	42,466	1,960 *
Shizuoka	58,104	62,215	4,111 *
Aichi	157,965	89,358	68,607
Mie	29,702	37,374	7,672 *
Shiga	21,755	23,810	2,055 *
Kyoto	51,511	57,286	5,775 *
Osaka	299,785	148,790	150,995
Hyogo	140,988	106,629	34,359
Nara	18,723	24,231	5,508 *
Wakayama	19,320	24,832	5,512 *
Tottori	10,720	18,626	7,906 *
Shimane	14,328	27,776	13,448 *
Okayama	29,853	42,171	12,318 *
Hiroshima	46,719	53,034	6,315 *
Yamaguchi	34,564	50,521	15,957 *
Tokushima	11,774	25,551	13,777 *
Kagawa	17,212	29,337	12,065 *
Ehime	24,788	47,789	23,001 *
Kochi	12,244	25,435	13,191 *
Fukuoka	100,756	131,472	30,716 *
Saga	22,552	43,396	20,844 *
Nagasaki	34,774	65,783	31,009 *
Kumamoto	32,429	61,994	29,565 *
Oita	23,858	41,454	17,596 *
Miyazaki	25,534	40,873	16,339 *
Kagoshima	36,847	78,595	41,748 *

* Source: Compiled from *Jūmin tōroku jinkō idō hōkoku nempō* (1960), pp. 52-97.

of economic activity for the analysis of the migration characteristics of each prefecture.

Two features of the 1960 internal migration pattern not revealed by Fig. 2 can be traced in Table IV. One is that even for the most agricultural and geographically remote prefectures which have a heavy out-migration there was also a heavy in-migration. This suggests unusually strong social and economic ties between migrants and their home prefecture. Involved are the return of young men and women who have gone to the cities or elsewhere for seasonal or short-term work, movement inspired by marriage and family obligations, and the seasonal movement of students to and from the urban universities and other schools. A second and related feature is that a number of prefectures, including Hiroshima, Ishikawa, Gifu, Shizuoka, and Shiga, had a large stream of migrants in and out even though their net loss was relatively small.

TABLE V

POSITION OF THE SIX LARGE CITIES IN INTERNAL MIGRATION, 1960 *

City	Number of In-Migrants	Number of Out-Migrants	Net Gain	Percentage of Total Gain
Tokyo	585,468	410,852	174,616	50
Osaka	224,140	157,302	.66,838	19
Nagoya	98,875	57,925	40,950	12
Yokohama	96,897	53,177	43,702	12
Kyoto	46,751	41,426	5,325	2
Kobe	64,433	43,777	19,656	6
Totals	1,116,546	764,459	351,087	100

* Source: Compiled from *Jūmin tōroku jinkō idō hōkoku nempō* (1960), pp. 98-104.

DRAWING POWER OF THE LARGE CITY

The six Japanese cities (Tokyo, Osaka, Nagoya, Yokohama, Kyoto, and Kobe) with more than 1 million population are the focal points in the internal migration pattern. In 1960, they were on the receiving end of 23% of all migrants, totaling 1,116,546 persons. More impressively, they were the target of 54% of all inter-prefectural migrants. The reverse flow away from the six cities was 764,459 persons, 16% of all migrants or 33% of the inter-prefectural totals. The share of each city is shown in Table V.

Tokyo, or actually the 23 wards in the eastern part of the prefecture-level Metropolis of Tokyo, is far and away the dominant urban destination and source of migrants. In 1960, it realized as much growth through net in-migration as the other five large cities combined. Fig. 3 shows that the bulk of the in-migrants came from prefectures in the Kanto Plain, northern Honshu, and Hokkaido. Niigata, Yamanashi, and Nagano in central Honshu were also well represented. Some idea of the drawing power of Tokyo can be derived from the percentage of out-migrants from various prefectures that it attracted: more than 60%, Chiba; 50-60%, Ibaraki, Tochigi, and Saitama; 40-50%, Yamagata, Fukushima, Gumma, other parts of the Metropolis of Tokyo, Kanagawa, Niigata, and Yamanashi. Tokyo drew at least small numbers of migrants from every prefecture, with appreciable numbers from the more urban-industrial prefectures of the Inland Sea borders and the agricultural prefectures of central and southern Kyushu. In general, the smallest numbers came from Nara and other prefectures in western Honshu and Shikoku.

Osaka, Japan's second largest city and the acknowledged economic leader of western Japan, was the center of the migration patterns outlined in Fig. 4. Most in-migrants came from an almost solid block of prefectures west of Nara and Kyoto, including those on Shikoku and Kyushu. The strong showing of Kagoshima may come as a mild surprise, but it has a normally heavy outward flow of migrants and had a large share of them move toward Osaka. Osaka-oriented movements thin gradually eastward into central Honshu with only a trickle coming from the prefectures of the Kanto Plain and northern Honshu in Tokyo's main zone of influence. One noteworthy point is that Osaka did not attract as heavy a percentage of the out-migrants as did Tokyo from the prefectures in their respective zones of influence. Only other parts of Osaka Prefecture and neighboring Nara had as much as 30% of their out-migrants bound for Osaka, while the remainder headed toward other cities to the east and west. As with Tokyo, out-migration patterns mirror the in-migration pattern on a reduced scale.

Nagoya's ability to carve out an economic domain, the so-called Chukyo Economic Sphere, in central Honshu between the spheres of Tokyo and Osaka is reflected in the migration patterns shown in Fig. 5. In-migrants in the largest number came from the three prefectures of Aichi, Gifu, and Mie. Small flows came from western

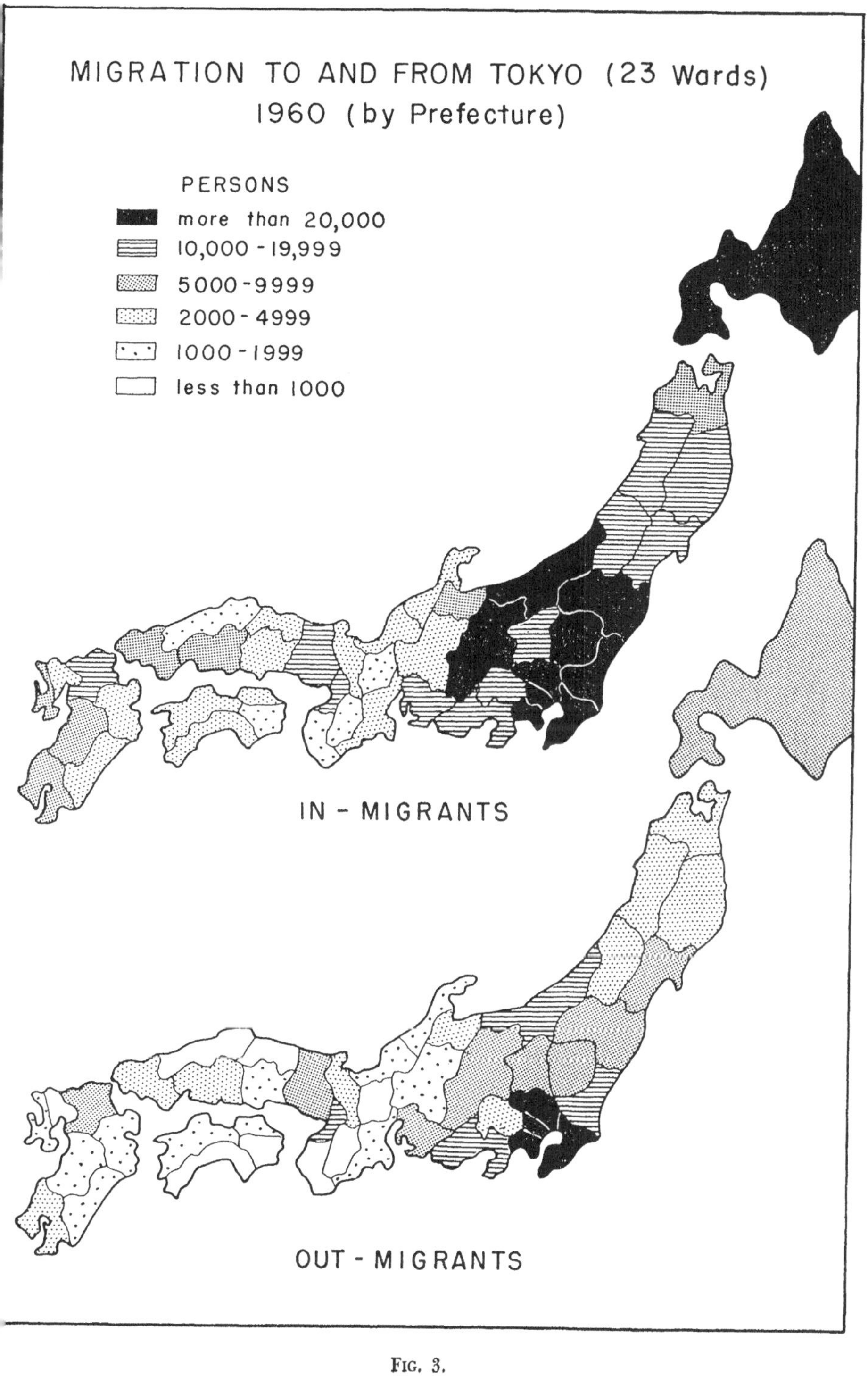
MIGRATION TO AND FROM TOKYO (23 Wards)
1960 (by Prefecture)
PERSONS
more than 20,000
10,000 - 19,999
5000 - 9999
2000 - 4999
1000 - 1999
less than 1000
IN - MIGRANTS
OUT - MIGRANTS

FIG. 3.

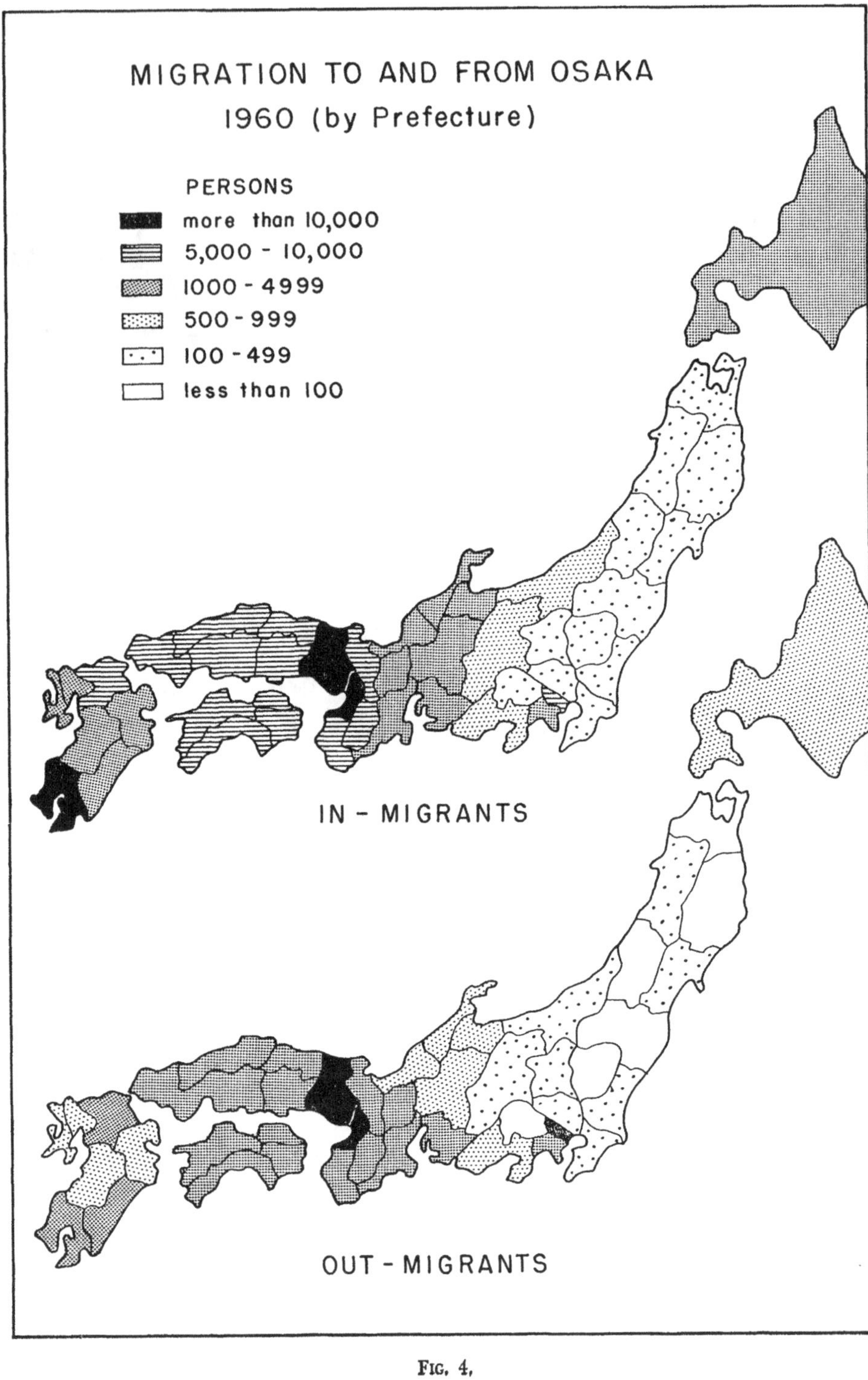

MIGRATION TO AND FROM OSAKA
1960 (by Prefecture)
PERSONS
more than 10,000
5,000 - 10,000
1000 - 4999
500 - 999
100 - 499
less than 100
IN - MIGRANTS
OUT - MIGRANTS

FIG. 4.

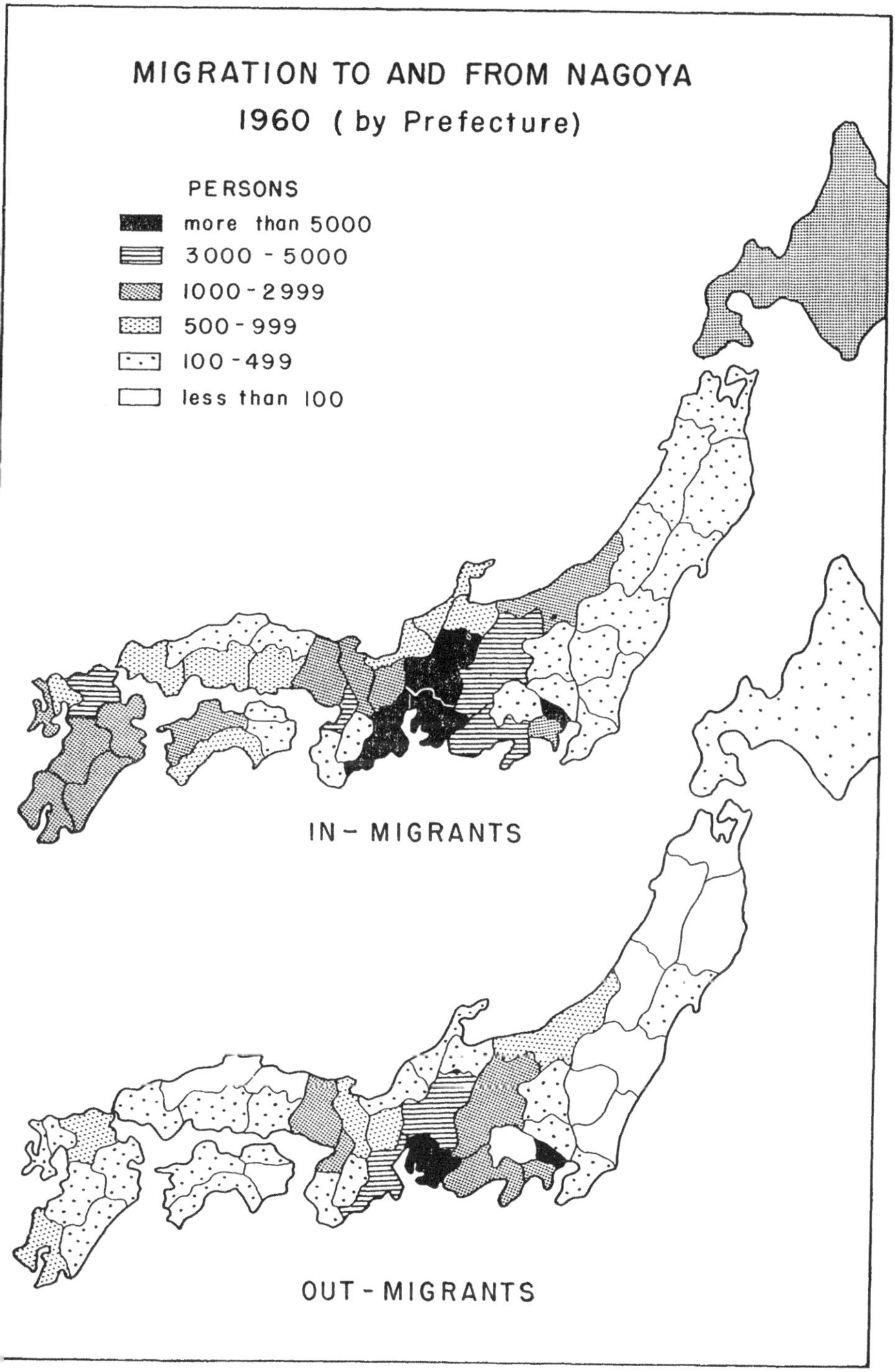
MIGRATION TO AND FROM NAGOYA
1960 (by Prefecture)
PERSONS
more than 5000
3000 - 5000
1000 - 2999
500 - 999
100 - 499
less than 100
IN - MIGRANTS
OUT - MIGRANTS

Fig. 5.

Honshu, where Osaka is the primary target, but Nagoya attracted a considerably larger movement from Kyushu. In keeping with the Tokyo and Osaka cases, population exchanges between the large cities were heavy. The out-migration from Nagoya emphasizes the importance of its relationships with central Honshu prefectures.

Of the three remaining large cities, Yokohama had a migration pattern like Tokyo's and Kobe like Osaka's, although on a smaller scale. Kyoto was rather unusual in its very small movements and the narrow margin of in-migrants over out-migrants. A partial explanation is the severe competition from nearby Osaka and Kobe. Its main migration interaction was with a small block of adjoining prefectures and a few others in the Inland Sea area.

Summary

The internal migration patterns revealed by resident registration data for 1960 are for a thin slice of time and may well differ in detail in subsequent years. However, they do show dramatically the role of the prefectures and large cities in the urban-industrial zone between the Kanto Plain and northern Kyushu in attracting and dispatching flows of migrants. Main movements presumably came as the result of geographical inequities in economic opportunity which will carry over into the future. The continuation and intensification of 1960 trends may well result in even greater inequities in the rate of prefectural population growth than were recorded during 1955-60.

PAPERS OF THE MICHIGAN ACADEMY OF SCIENCE, ARTS, AND LETTERS
VOL. XLIX, 1964 (1963 MEETING)

C. SOUTHERN AND EASTERN ASIA

IMPACT OF INTERNATIONAL RELATIONS ON THE COMMUNIST PARTIES OF THE MIDDLE EAST

RAIS A. KHAN
The University of Michigan

A widely held belief in the study of Communist expansion is founded on the conception that underdeveloped countries that have been exposed to outside influence offer a favorable breeding ground for Communism. Slow and lopsided progress of industrialization and economic development of these countries contrary to their desire for speedy and quick results gives rise to the competition between Communism, which promises to work wonders, and the West, which promises slow progress or nothing at all. The development of Soviet Central Republics acts as the attraction of Communism which with its system of planned and centrally controlled economy has made such wide progress in so short a time. This fact could also be regarded as a possible explanation of the growth of Communist influence in the Middle East.

The implication thereof would be that Communist propaganda in the Middle East has successfully won popular support for its political and economic doctrines with the party as the fountainhead of these ideas. The theme of this paper, however, is that Communist parties in the Middle East have played an unsuccessful role in the expansion of Communist influence in the area. The primary element has been the entire trend of internal and external political developments over the past decade which, as a reaction, have brought Communism, but not Communist parties, in the limelight of Middle Eastern politics. In other words, international relations of the Middle East have reflected on the Communist parties of the area rather than the parties having any significant influence on the conduct of these affairs.

The general Communist (Soviet) attitude toward the Middle East has been more or less guided by the mythical will of Peter the Great of Russia, in which he bade his successors:

To approach as near as possible to Constantinople and India. Whosoever governs these will be the sovereign of the world. Consequently, excite continual wars, not in Turkey, but in Persia. Establish dockyards on the Black Sea, seize upon little pieces on this sea as well as on the Baltic Sea, which is doubly necessary for the attainment of our project. And in the decadence of Persia, penetrate as far as the Persian Gulf, re-establish India, which is the depot of the world. Arrived at this point, we shall have no longer need of England's gold.

Sweden being dismembered, Persia subjugated, Turkey conquered, our army reunited, the Black Sea and the Baltic guided by our ships, we must then propose separately, and very secretly, first to the Court of Versailles, then to that of Vienna, to share with them the empire of the universe. If one of the two accepts, which is nearly certain by flattering her ambition and national vanity, to make use of her for crushing the other. Finally, annihilate in her turn the one which remains by commencing a struggle—which cannot be perilous, Russia possessing already all the Eastern or Greater portion of Europe.[1]

In writing this will, Peter the Great was laying down a general line of policy for his successors for the eastward expansion of Russia. The Soviets, since coming into power, have followed very much the same line of policy with, of course, changes here and there as the time and situations demanded. Communist influence in the Middle East today might be regarded as the result of this role of the Soviet Union in the international relations of the Middle East, and the present status of the Communist parties in the area the direct outcome of these trends of international relations.

Simultaneously with the coming into power of the Bolsheviks in Russia, the Middle East was witnessing the rising tide of Arab unity. Communism's first competitor in the Middle East, therefore, was the idea of Arab unity. And it is in company with this rising tide of Arab unity that Communism in the late 50's has gained its present influence in the area. It is, therefore, relevant to note two basic impulses in the movement for Arab unity which offered to the Communists a favorable ground for making common cause with the nationalists. (Yet, it is also worthy of note that Arab unity, generally regarded by the Middle Easterners as synonymous with Arab nationalism, drew its first inspirations from the liberal political philosophy of the West; and the two basic impulses, in effect, were a means of protecting against the Western domination of the area.) First, the rejection impulse, under the influence of which the idea of Arab unity was conceived as an urge to preserve theretofore existing unity and as one manifestation of a total rebellion against

[1] Sir Percy Sykes, *A History of Persia* (London: Macmillan, 1930), II, pp. 336-337.

portents of a slowly unfolding Allied betrayal at the Versailles Peace Conference. This imparted to the then embryonic idea the streaks of retroactiveness which characterized the early turbulent years of the Arab body politic. The second, the restorative impulse, which was mother to the idea of Arab unity, oriented this idea and turned its gaze toward the immediate past. For, despite the abhorrent nature of many of the features of that past, it was nonetheless considered superior to the situation created by the postwar settlement, in that it had at least preserved Arab unity. Hence, the nostalgic longing for the recent past and the demand that the pre 1918 unity be restored.[2] Later on this embryonic idea took the shape of a demand for a centralized and united territory of the Arab part of the Middle East.

The end of World War I witnessed the creation of a series of whole new states with their boundaries determined by the Western powers. The dividing lines in some cases were official and did not correspond to the natural factors. No clear national consciousness has yet developed in the Arab countries, nor do the people identify themselves completely with the territory within which they live. Local and regional nationalisms compete with the Pan-Arab ideal. The mosaic of tribes and ethnic and religious groups has not yet been consolidated into an integral whole, or into clear and distinct parts. This entire process of development into nations and states is still under way. Of the three constituent elements of a modern state—territory, nation, and constitution—the territory has been circumscribed by an alien will, and the laws, with the exception of the law of persons, have been taken over almost completely from European countries, while the people struggle to achieve the future form of a nation.[3]

The Communists have exploited the Pan-Arab agitation and they have sought, whenever possible, an alliance with the nationalists. Communist parties alone cannot hope, at the present time, to grasp power in the Arab countries. They have not succeeded in influencing the trend of affairs in the area. More often than not, it has been the trend of affairs, especially of an international nature, that has determined the course of action of the Communist parties.

[2] Fayez A. Sayegh, *Arab Unity, Hope and Fulfillment* (Devin-Adair Co., 1958), p. 64.

[3] Hans E. Tutsch, "Arab Unity and Arab Dissensions" *in* Walter Z. Lacqueur, ed., *The Middle East in Transition* (New York: Praeger, 1958), p. 13.

Instead of engendering causes for excitement, it uses the wave of national excitement to steer the masses toward its special goal.

> The Communists have shed the heavy overcoat of the agitator, which is the general philosophy of their conduct in the Arab countries and changed it for the emperor's purple; instead of sneaking in through the backdoor, they march in full daylight through the main portal accompanied by a host of nationalist train-bearers. They cleverly use the fascination of power which radiates from their strong leadership; they boost the Arab's self-confidence by courting their favour and, since the West is preoccupied with the growing Communist influence, confer on them a greater possibility to maneouver in the field of foreign politics.[4]

Communism, as it reaches the Arab nationalists, is not a creed of social revolution aiming at justice for the underdog and based mainly on the aspirations of the peasants and workers. It becomes the ideology and the technique of the ruling class or aspirants to power, or of soldiers and bureaucrats. These ideas and techniques, far from leading toward the establishment of a new Communistic form or even of a relatively more advanced form of a society, "serve to provide the military-bureaucratic classes with instruments and rationalizations for re-establishing the traditional Middle Eastern despotism where political power is based on military force."[5] This was demonstrated by Gamal Abdel Nasser adopting a pro-Soviet policy vis-a-vis the West, and secondly, to placate Communist influence in Egypt and keep it from creating new troubles for him. At the same time, known Egyptian Communist leaders and the Party itself were very strongly dealt with. Abdel Karim Qassem of Iraq utilized the Iraqi Communist Party and its techniques to strengthen and stabilize his new regime in Iraq. When he realized that the rope allowed to the Communists was getting too long and might be used for hanging the new Iraqi regime, and as soon as his own need of the Communist Party was over, he came down upon them with a hard hand. "What probably appeals to the Arab intelligentsia is not straight Leninist ideology but the local adaptations, the potent mixtures of populist, nationalist, and Communist elements that permeated the writings, speeches, and actions of Egyptian, Syrian, and Iraqi leaders."[6]

[4] Tutsch, p. 23.

[5] Walter Z. Lacqueur, "Nationalism and Communism in the Arab World: A Re-Appraisal" *in* Walter Z. Lacqueur, ed., *The Middle East in Transition* (New York: Praeger, 1958), p. 444.

[6] Lacqueur, *Nationalism and Communism in the Middle East* (New York: Praeger, 1956), p. 291.

So long as the Middle East was organized in empires and dynasties within which religious communities carried on an autonomous existence, the problem of political stability was not a serious one. But the eventual confrontation between secular nationalism on the one hand, and Islamic fundamental and communal particularism on the other, produced a major crisis which is as yet unsolved. The concept of nation-state introduced into the Arab Islamic area by preachers of the gospel of secular nationalism was a decided novelty which so far does not seem to be the final political form agreed upon in the Middle East. In this confused and fluid state of political transition, Middle Eastern leaders utilize Communism and Communists for stabilizing their own position and at the same time seeing to it that the reins of political power do not pass under Communist control. Except perhaps for those who are regular members of secret Communist parties (all Middle Eastern countries have either refused to permit, or have suppressed, Communist organizations), most of their followers may be classed as sympathizers who probably have certain mental reservations about the doctrine. Communism's opposition to religion and its disdain for other mores, to say nothing of its affiliations with a foreign power, are some of the factors which still make many hesitate to accept in toto the Communist creed. One can agree with Majid Khadduri that, despite these reservations, "If these sympathizers were forced to choose between the existing regimes and Communism, it is not unlikely that their choice would fall on Communism." [7]

Early Communist endeavors to gain a foothold in the Middle East were haphazard and uncoordinated. Support in money and arms was given to Mustafa Kemal and the Turkish nationalist movement. The new Soviet Government signed treaties with Persia and Afghanistan in 1921 and almost simultaneously an attempt was made to establish a Soviet republic in Gilan. A little later, an elaborate network of propaganda, espionage, trade representations, and trade subversion was laid down in Persia and on a smaller scale in Afghanistan. Inducements were offered to Azari, Turkeman, Uzbek, Tajik, and other tribal minorities of Persia and Afghanistan to migrate to the Soviet Union. The Soviet Government greatly overestimated the attraction of the Revolution, and most of these early plans were ill-conceived and badly executed. The Turks were

[7] Majid Khadduri, "The Role of the Military in Middle Eastern Politics," in *The American Political Science Review*, 47 (3) (June 1953), 516.

ready to accept material support in order to reestablish themselves, but once they had done this, they wished to have no more to do with the Russians, whom they had always feared and distrusted and whose agents and representatives now seemed mostly to belong to races which were the object of their traditional dislike. The Persians were at first mildly impressed by Soviet protestations of goodwill and of their intention to cancel all concessions granted to the Tzarist Government. But they were alarmed at the landing of Soviet forces in Gilan and by the number of Soviet spies and trade agents who swarmed into the cities of north Persia. It was particularly unfortunate for the Soviet Government that their designs in the Middle East happened to coincide with the appearance of two of the strongest and most determined rulers Turkey and Persia had ever known, Mustafa Kemal and Reza Shah. The new states in the Middle East that had arisen in the wake of the disintegrated Ottoman Empire were mostly under British and French mandatory control and military occupation and, thus, denied to Soviet penetration. The year 1920 really marks the end of the first phase of Communist expansionist endeavors in the Middle East. " It had failed partly because of the rise of the Middle Eastern nationalism, partly because of Western opposition, but most of all, in my opinion, because of Soviet miscalculations and precipitancy." [8]

The young Soviet Republic had to face the Eastern " pan-movements " in both domestic and foreign policy in the Middle East. Soviet approach to the movements was not only ill-conceived and badly executed but was also not entirely consistent. Both at the Baku Congress in 1920 and at the Second Comintern Congress, some weeks earlier, Soviet opposition to the Pan-Islamic movements was clearly expressed.[9] Hans Kohn argues that the Baku Congress was a failure in its attempts to make any real progress in the Middle East, especially in its efforts to organize.[10] This was evident from the fact that at the Congress of the Third International in Moscow, in the summer of 1921, the Middle East, with the exception of Turkey and Persia, remained unrepresented. Even within the Communist parties in these two countries, there were internal conflicts

[8] G. E. Wheeler, " Russia and the Middle East " in *International Affairs,* 35 (3) (July 1959), 297.

[9] Lacqueur, *Soviet Union and the Middle East* (New York: Praeger, 1959), pp. 35-36.

[10] Hans Kohn, *Nationalism and Imperialism in the Hither East* (New York: Harcourt, Brace, and Co., 1932).

and rivalries.[11] From the Soviet policy point of view, Zinoviev clearly repudiated anything in the nature of pure nationalism, however it might be disguised.[12]

Added to this none-too-friendly attitude of the Communists toward the pan-movement was their conflict of interest with Arab nationalism in the early period, which was modified only in the late 1950's, as evidenced by the general Communist attitude toward the Wafd Party. The Communists are probably right in thinking that antiimperialist slogans, leaning toward a planned economy, and partial nationalization, are not equivalent in the long run to a Communist monopoly of political power. The recent phase of bourgeoisie Arab nationalism may be progressive, but at the same time it presents obstacles to the complete transformation of Arab society, a consummation the Communists may be willing to see delayed, but not indefinitely. The Communists in the Arab world, in other words, want power, but the " nationalists " have no intention of relinquishing it. This, then, is the cause of an apparently irreconcilable conflict, and an open break between Communism and Pan-Arab nationalism would seem to be inevitable.

In relation to the Wafd Party in Egypt, which, at one time, was the guiding spirit of Arab " nationalism," Communist attitude was quite cool if not directly hostile. There have been a great many fluctuations in the attitude toward the Wafd. Since World War II the Wafd has been described in Soviet historical writings as that part of the Egyptian bourgeoisie that wanted to compromise with British imperialism and was only occasionally prevented from doing so by pressure from below. In 1955, with the general modification of the line on the national movement in the East, the Wafd might have expected to benefit retrospectively. But since, by 1955, the Wafd was no longer in existence, Soviet historians saw no compelling reason to make the kind of concessions that had to be made, for instance, to the Indian Congress Party; the appraisal of the Wafd has not, therefore, improved appreciably since 1955.

Even with regard to Arab nationalism, Communist attitude was modified not in an attempt to fit political developments in the Communist program but to cut the Communist coat to fit political

[11] Nasrollah Fatemi, *Diplomatic History of Persia* (New York: Russel F. Moore Company, 1952), p. 177.

[12] Lacqueur, " The National Bourgeoisie " in *International Affairs*, 35 (3) (July 1959), 330.

developments. In the early postwar period, Soviet Eastern experts did not favor the movement for Arab unity as was embodied at the time by the Arab League. But there was also no outright rejection. The general consensus in 1946 was that the Soviet attitude would depend on whether the Arab League followed a " reactionary course of action or a progressive one." In 1947-48, with the intensification of the Cold War, the Soviet approach became much more critical. The Arab League was defined as a British agency, a " reactionary bloc," an " instrument in the struggle against national liberation movements in the Middle East." It was said to have played an " ignominious role " in the unleashing of the Palestine War and in persecuting the progressive elements in the Arab world. After 1950, considerable attention was devoted to the struggle for national liberation by the people of the Near East, but the emphasis was always on the activities of the " Partisans of Peace " and other front organizations; there was as yet no sympathetic approval of the policies of Arab governments or the movement for greater Arab unity. The Arab League, originally described as a tool in the hands of British imperialism, was by 1950 generally characterized as a toy in the hands of the American imperialists.[13]

As far as Communist following in the Middle East is concerned, they did not succeed in finding many Arab adherents in the beginning. Jews generally played a prominent part in the Communist movement in the Middle East. During the last ten years, however, the situation has considerably changed as the result of the emigration of many Jews to Israel and because it became difficult for the Jews to be active in Arab countries.[14] Consequently the parties had some Arab followers but mostly collaborators. Among the collaborators there were great differences. On the one hand, there were zealots, reactionary groups, the Moslem Brotherhood, and the Qijanis, who refuse to take cognizance of anything new and who wish to escape to the past; and, on the other hand, the " Herodians " like Ahmad Husain and Dawalibi, who wish to combat the West by imitating it. Political leaders of this type appear now to have been merely precursors of a general trend which has subsequently assumed far larger proportions and political importance: the alliance between Communism and extreme Arab nationalism. And there are many in between these groups.[15]

[13] Lacqueur, *Soviet Russia and the Middle East*, p. 150.
[14] Lacquer, *Nationalism and Communism in the Middle East*, p. 234.
[15] *Ibid.*, p. 259.

Most of this class of collaborators and halfhearted followers came from the intelligentsia, for whom an acceptance of Communism meets with many reservations. Communist following among the peasants and workers, the real base of a traditional Communist movement, is neither very effective nor significant. In his report to the National Conference of the Communist Party in Syria and Lebanon, which was held on December 31, 1943 and on January 1-2, 1944 the Syrian Communist leader, Khaled Bakdesh, while admitting the weakness of the Communist movement in these countries, also explained some of the reasons for the skeptical attitude of the people toward the Party:

> The new position of the Communist Party which has resulted from the decision concerning the dissolution of International Communism, undoubtedly lent new vigor to the national and nationalist movement in both Syria and Lebanon. In the first place, this new position makes it easier for the party to pursue its great national task of combatting Fascism and Nazism, which constitute the greatest threat to the national and independence movement. In the second place, it puts an end to the excuses of those who have been making hints about the relationship of the Communist Party with International Communism, and using it as an excuse for keeping many of the national forces and youthful talent away from the national life and refusing to cooperate with them. On the other hand, the dissolution of International Communism and the complete independence, both operational and official, of the Communist Party within its national boundaries will no doubt remove any of the cautiousness displayed by many of the educated national elements and of our brethren, the members of the Syrian Nationalist Party, who have always been in accord with the scientific theories of the Party concerning national liberation, and who admired the Party's activities and the righteousness of its course. But the fact that they viewed the Party as a branch of International Communism made them hesitate to rally around the Communist Party and support its national efforts and struggle. The Communist Party shall endeavour more than ever to fill the ranks of national forces for the sake of national unity so that our country can meet the forthcoming events with unified national forces.

Although this statement was made in particular reference to the situation existing in Syria and Lebanon, it was by and large true of the whole Arab world. By 1945, Soviet legations had been opened in Cairo, Beirut, Damascus, and Baghdad and a definite foothold in the Arab world thus established. By the end of 1947, it must have become clear to Russia that no progress could be made with her immediate neighbors, Turkey and Persia, in the Middle East, who still constituted the lines of approach to her main objectives of the Eastern Mediterranean and the Persian Gulf. It was probably at this point that Russia decided to by-pass Turkey and Persia for the time being and to concentrate on the Arab countries. There, the rising tide of nationalism, hostility to the West, and, particularly, to the Western defense plan and Arab-Zionist conflict seemed to offer much better prospects for exploitation. It persisted in regarding them as "bourgeois" and reactionary and as "the lackeys of imperialism." But the only means which it now cared to use were diplomatic pressure on governments and radio propaganda directed toward the masses. Local Communist parties working among the intelligentsia, proletariat, and peasants were useless, and effective power was in the hands of the "national bourgeoisie" who controlled and very often composed the armed forces and who were hostile to the parties. In general, Soviet policy and Communist parties remained sterile. In 1951, for example, threatening Soviet notes to Egypt, Syria, Iraq, Lebanon, and Israel in connection with their possible adherence to the proposed Middle East Command were ignored by the governments concerned and were condemned by them as interference in these countries' internal affairs. In effect, up to 1951, Communists still had next to nothing in the Middle East.

A change of attitude toward the Middle East in the Soviet Union was discernible by the beginning of 1952. But the existence of certain very obvious contradictions in the official Soviet attitude, writings of Soviet intellectuals, and the policies of local Communist parties in the Middle East prevented Communist influence in the Middle East from making much headway. The Soviet Union had voted for the Palestine partition resolution, had sent arms to Israel through Czechoslovakia, and was among the first few countries to grant diplomatic recognition to Israel. The Communist Party of Palestine had, however, always been opposed to the Zionist movement, and the attitude was not substantially modified even after the establishment of Israel. Among the Soviet intellectuals, Klimovich, who has probably written more books and articles against Islam than any man since the Middle Ages, went on record in October 1956 as reiterating the traditional definition of Islam: "an antiscientific, reactionary world concept, alien and inimical to Marxism-Leninism. Islam is in opposition to the optimistic and life-affirming materialist teaching; it is incompatible with the fundamental interests of the Soviet peoples."[16] During the summer months of 1956, anti-Islamic propaganda was stepped up; there

[16] Quoted by Walter Z. Lacqueur, *Soviet Union and the Middle East*, p. 177.

were frequent appeals to " raise the level of scientific, atheistic, propaganda "; meetings were convened in which the harmfulness of Islam in general and such specific obligations as the fast, in particular, were denounced. Occasionally, some of the international ramifications were also mentioned such as the " reactionary role " of Pan-Islamism, Pan-Turkism, and even Pan-Arabism. This anti-Islamic propaganda reflected adversely on the position and influence of the local Communist parties by antagonizing the orthodox conservative classes, religious organizations, and the illiterate masses whose main center of allegiance and loyalty is religion.

As a result Communist parties in the Middle East had not made many gains in their influence, prestige, and popularity up to 1956. One testimony to this is the Report of the Central Command of the Communist Party in Syria and Lebanon in January 1951.

> Does our activity lead with the required speed toward our party's becoming a strong party of the masses among the workers and the peasants, enjoying their confidence and adherence and able to move and muster them for the struggle, for our slogans?
>
> a. *Errors in the Present General Organization of the Activity of Our Organizations and Some of the Historical Causes thereof:* Our progress among the working class and the peasantry has been very slow and if it continues at this rate it cannot produce the necessary results in the required speed and in conformity with the requirements of the development of the international and home situation. Therefore, in our political and organizational activity and our work among the masses, we must make a fundamental change; it marks a decisive shift which some comrades call a " revolution." In the past the activity of our party has been distinguished by two things:
>
> (1) Nearly 75% of its activity, whether in propaganda, organizing on daily work, has been confined to petty bourgeois elements in cities and villages on the one hand and selected workers on the other. But the broad masses of workers and fellahin [peasants] have received hardly any attention and have been the object of little of our political and organizing activity. There are town quarters (consisting mostly of petty bourgeoisie) and villages (likewise consisting principally of petty bourgeoisie, since most of them are medium and small landowners) where the Party works hard and continuously while there are other quarters and centers where workers congregate, as well as villages inhabited by poor fellahin, where there is either no party activity or only occasional activity.
>
> (2) The main attention of our organizations is directed towards the creation of a lot of sound and fury around the Party and its slogans rather than toward the building up of bases and foundations among the workers and masses of fellahin, especially the poor among them. It is very well to make a big noise but noise alone has no lasting effect. The best example of this is the peace movement. Despite our great participation, with all of the

> supporters of peace, and in spite of our making a big noise along the slogans of the Partisans of Peace movement, we have not yet succeeded in carrying out our duty to center the peace movement on the firm base of the masses; for in order for the bases to be firm, strong, real and effective they must rest specially on the masses of workers and fellahin. . . .
>
> Therefore, we can say that (1) the general milieu and atmosphere in which our party works are not yet proletarian, but are still generally petty bourgeois in character; (2) our effort is not being expended in the circles where it should be expended; (3) and the tendency to make as much noise as possible, without using this noise to build up organized mass foundations for the party among workers and peasants, specially the poor, still prevails in the Party.[17]

There could be no better testimony to the fact that Party propaganda and "noise-making" had failed to gain its purpose, that it had not been successful as a means of penetrating the workers and *fellahin* and strengthening and broadening the base of the Party among them.

It was not until 1955 that it became clear that a new policy was on the way from the Soviet Union. The tour of Bulganin and Khrushchev to India and Burma and Afghanistan and Shepilov's tour to the Middle East were the first signs that a vital and epoch-making decision had been taken in the Kremlin, "namely that at any rate for the present the Soviet Government so far from continuing its denigration of 'bourgeois-nationalist' Governments would support these Governments wherever and whenever it thought necessary."[18] Luckily for the Soviet Union, the two years preceding and the three years following were somewhat crucial and troubled years in Middle Eastern international relations and offered the Soviet Union enough troubled waters to fish in. It was this change of heart and mind in the Kremlin and its entry into the domain of the Middle Eastern international politics that achieved for Communism what the activities of Communist parties of all the Middle Eastern countries put together had failed to achieve in the past decades, namely the establishment and expansion of Communist influence in the area. This new influence, however, remained confined to the domain of international politics. Its introduction did not in any way internally improve the position of the Communist parties. If anything, it put them in a worse situation inasmuch as Middle Eastern leaders in no way desired to share power with them or to allow them to become a threat to the existing order of political

[17] *The Middle East Journal,* 2 (Spring 1953), 208-209.

[18] Wheeler, *op. cit.,* p. 300.

institutions. While, therefore, they walked arm in arm with the Kremlin internationally, at home they tightened their control of internal Communism more strongly than before. This is evident from the strong measures of President Nasser in Egypt and, in later years, of Premier Abdel Qassem of Iraq.

The new policy came formally into being with the Twentieth Party Congress, held in February 1956. This important statement of policy did not receive much attention in the West. The new policy meant the adoption of new methods far more difficult to counter than any used previously. In effect, the Soviet leaders seemed to have decided that the countries of the Middle East could most conveniently be brought under Soviet influence not by internal subversion against the real rulers of these countries. At the same time local Communist parties, so long the outposts of Soviet influence, were now to play a different role. They would no longer act as bullies to the "national bourgeoisie" with threats to its vested interests but were to join with all other parties supposed to be fighting for nationalism against Western imperialism.[19]

The new Soviet Government that replaced Stalin expressed a desire for friendly cooperation, for more trade, for the settlement of troublesome questions still in dispute. Turkish and Iranian representatives were informed that in some of its policies toward their countries, Stalin had been wrong.[20] The territorial claims on Turkey put forward in 1945 were withdrawn. Negotiations for the demarcation of unsettled portions of the Soviet-Iranian border and for the return of the Iranian gold long held in Moscow were started with the Iranian Government. Subsequent to the withdrawal of Britain and France from the Middle East, the European Powers had not been able to bring about stability of relationship or to promote settlement of their disputes with the Middle Eastern countries by the time the Soviet Union appeared on the scene as the great friend of the Arabs. And it appeared, with the added advantage, that during the past few years, while the Western Powers were trying to organize the Arabs into pacts and blocs, the Soviet Union had not apparently been trying to organize anything or to coerce anybody.

Up to 1955, Soviet propaganda had been calling Abdel Nasser

[19] *Ibid.*

[20] John C. Campbell, *Defense of the Middle East* (New York: Praeger, 1960), p. 208.

a fascist dictator who persecuted "democratic" elements such as the Communists. Overnight, he became a patriot, a national leader fighting for his people's right against the imperialists. The old question, much debated in the days of the Comintern, whether Communists should work together with the bourgeois nationalists, seemed to be resolved in the affirmative, at least for the current phase of Soviet strategy. Having made their decision to use Arab nationalism as the path to penetration of the Middle East, the Soviet leaders followed the line of supporting all elements, even the most revolutionary, whose weight could be thrown in the scales against the West.

The Soviet Union refused to discuss the Egyptian arms deal and called it a "purely commercial transaction." Then they made offers to Western Powers to negotiate on the whole Middle Eastern question. On the occasion of the Khrushchev-Bulganin visit to England in April 1956, the Soviets issued a statement of willingness to cooperate for peace and stability in the Middle East. At London, Khrushchev is reported to have broached the possibility of an agreement to restrict the delivery of arms to the Middle East. During the Suez Crisis, in November 1956, Moscow proposed joint Soviet-American intervention to "stop the aggression"—a suggestion evidently based on the concept that those two powers could and should keep order in the Middle East and settle its problems. Then in February 1957, came a return to the idea of a four-power conference and an overall settlement in the Middle East, followed by a number of proposals to the effect that the four great powers agree to renounce the use of force in the area.

Soviet penetration in the field of political developments in the Middle East has been matched by the development of closer cooperation in the field of economic activity through expanded trade, grants or loans, and offers of economic assistance, both fulfilled and unfulfilled. Despite some serious disappointments and disillusionment of the Middle Easterners in this field of relationship with the Soviet Union, economic activities of the Communist countries have had a substantial impact on the public mind and the commercial community of the Middle East. The economic impact has resulted in the commitment of large quantities of Middle Eastern exports to Communist markets for many years to come, thus giving to the countries of the Middle East a feeling of freedom from dependence on Western powers for the export of Middle Eastern products.

There has also been the psychological impact upon Middle Eastern public officials and business leaders who have been impressed by the fact that the Soviet Union and other Communist countries can even consider the program which they have initiated, just as they have been impressed by the postwar rate of economic growth of Communist countries. The fact that the Soviet Union, which during the 1920's and 1930's leaned heavily on American, British, and German technical assistance, which after the war dismantled no small part of East European industry with which to reconstruct its own economy, and which sought American credit and received substantial UNRRA funds following the war, can now offer technical assistance and provide credit for capital goods, strikes the Middle Easterners as a phenomenal development, worthy of admiration. Accompanying this new posture of the Soviet Union has been an almost respectful and friendly attitude on the part of the Soviet and other Communist leaders and an often stated willingness to assist underdeveloped countries in their process of development.[21]

It would not be absolutely correct to say that Soviet strategy in the Middle East proceeds according to a specific or detailed master plan. "Many Soviet moves in the Middle East have been reactions to those of other powers or the product of circumstances, favorable or unfavorable, as they arose. At this particular juncture in history, from 1955 to 1960, the circumstances were specially favorable. Khrushchev and his colleagues have found themselves at a point much further advanced toward domination of the Middle East than Stalin or the Czars had ever reached." [22]

Behind their apparent success in the Middle East are certain formidable factors in the Soviets' favor. First, the West is morally discredited in the eyes of the Arabs. It is regarded as predatory and imperialistic, ever ready to pounce upon defenseless Asians and Africans, and prevented from doing so only by fear of Russian reprisals. It is pictured as a rabid warmonger, eager to drop atomic bombs and again unleash the ultimate horror on a peace-loving world. Second, there is the Soviet claim to have found a "scientific solution" for all Eastern ills, including colonialism, and the new Soviet reputation for technological skill with its impressive shop

[21] Robert L. Allen, *Middle Eastern Economic Relations with the Soviet Union, Eastern Europe, and the Mainland China* (Charlottesville: Woodrow Wilson Department of Foreign Affairs, Virginia, 1958), p. 71.

[22] Campbell, *op. cit.*, p. 214.

window of the six Muslim Republics abutting on the Middle East. Finally, but equally important, is the new Soviet ability to draw on its now vast academic resources for organized research into every kind of Middle Eastern problem and for the training of expert personnel for employment in Middle Eastern countries.

An additional factor thought to favor the alliance between nationalism and Communism is the existence of a common enemy, namely the West. Soviet observers realize both that this is a prerequisite for an alliance with nationalism in the Arab world but that it is steadily becoming unreal. Arab writers have on occasion indicated the increasingly nonexistent nature of the "western danger" and have, in the process, expressed doubt about the future role of the Communist parties. In the words of Hussein Heykal, editor of *Al Ahram*: "In the past Communist organizations have stood by the nationalist forces in the bitter struggle against imperialism, pacts, imperialist collaborators, and feudalism. The struggle has ended or is about to end. What will be the attitude of the Communists in the future? Will they be deflected from the clear nationalist line in order to raise red flags in the Middle East? Or will they keep their mouths shut?" [23]

The latter half of the 1950's witnessed the awakening of the Middle East to the new social and political forces that were replacing the "old order" as a part of the phenomenon of the "Rise of New Asia." The rise of these new forces led to the emergence of a revolutionary situation in the Middle East. Communist parties in the area had made little or no direct contribution to the development of the new forces or the creation of this revolutionary situation. The entire political scene was further exacerbated by the deepening clash between Arab "nationalism" and Western political and economic interests. The change in Soviet tactics after 1955 coincided with the emergence of the fast-growing pattern of the Middle East's domestic and international politics. It was this coincidence and the short-lived honeymoon between Soviet policy and Arab "nationalism" which is largely responsible for the expansion of Communist influence in the area.

[23] *Al-Ahram* (Cairo, Jan. 4, 1959).

PAPERS OF THE MICHIGAN ACADEMY OF SCIENCE, ARTS, AND LETTERS
VOL. XLIX, 1964 (1963 MEETING)

SOUTH KOREAN CONSTITUTIONAL DEVELOPMENT: THE MEANING OF THE THIRD REPUBLICAN CONSTITUTION *

C. I. EUGENE KIM
Western Michigan University

No political system can properly be understood without a more schematic and comprehensive, conceptual analysis of the working of that system than a merely categorical constitutional analysis. Yet, a constitution is the operational framework of a political system and a close examination of the constitution and its history tells a great deal about that political system. This is particularly illuminating in the case of Korea, which in a short span of roughly fifteen years has experienced three republics and three constitutions. Syngman Rhee had his constitution. The Second Republic headed by John M. Chang had another. The military junta in power since its successful coup of May 16, 1961 has a constitution that was adopted by a referendum on December 17, 1962 and is to come into effect "on the date of the First Convocation of the National Assembly . . ." (Supplementary Rules, Art. 1).

The principal interest of this paper is in the third constitution, particularly that aspect of the constitution which is indicative of both the power alignment within the military junta in the framing of the constitution and this military junta's intention to legitimize its position of power under the constitution. The other two constitutions are discussed only for historical interest, as each constitution is progressively based on the lessons learned from the working of the previous one.

THE FIRST CONSTITUTION

The First Republican Constitution was drafted by the Constitution Drafting Committee appointed by Syngman Rhee in his capac-

* For this paper, the author is indebted to his students in Comparative Government at Western Michigan University who took their assignment on the third Korean constitution seriously and pointed out the various aspects of that constitution which the author would have otherwise missed. Also thanks are due to Milton Greenberg and Frank L. VanVoorhees, both of the Department of Political Science, Western Michigan University, who read the whole manuscript and offered suggestions.

ity as temporary chairman of the First National Assembly, which was elected on May 10 and met on May 31, 1948. The constitution drafted by this committee was adopted on July 12 and came into effect on July 17. On August 15 the Republic of Korea was proclaimed. Syngman Rhee was elected president by the National Assembly by a vote of 180; Yi Si-yong, an 80-year-old member of the old Korean nobility, was elected vice-president by a vote of 133; each out of a total of 196 votes. (The former was elected on the first ballot and the latter on the second).

This constitution was made up of 103 articles in 10 chapters. The preamble to this constitution stated:

> We, the people of Korea, possessing a glorious tradition and history from time immemorial, follow the great spirit of independence as manifested in the establishment of the Republic of Korea and by proclamation thereof to the whole world by the March 1st Movement in the year of Kimi (T. N. A. D. 1919).
>
> Now at this time we are engaged in the re-establishment of a democratic and independent State and are determined:
>
> to consolidate national unity through justice, humanity and fraternity,
>
> to establish a democratic system of government eliminating evil social customs of all kinds,
>
> to afford equal opportunities to every person and to provide for the fullest development of the capacity of each individual in all the fields of political, economic, social and cultural life,
>
> to require each person to discharge his duties and responsibilities,
>
> to promote the welfare of the people at home and to strive to maintain permanent international peace and thereby to ensure the security, liberty and happiness of ourselves and our posterity eternally. . . .

To accomplish these purposes a government was instituted that in spirit reflected the principles of Western democratic republics—popular sovereignty, guarantee of individual rights, the rule of law, a limited government, the separation of powers with checks and balances, judicial review, civilian control of the military, etc. There was to be local autonomy, but local autonomous bodies exercised only those powers delegated to them " by the State."

This constitutional embroidery was a delicate piece of work, but the Korean tradition of government of men rather than law was well reflected in the way that this fundamental law of the land was framed and approved. A handful of legal experts (Yu Chin-ho, a Japanese-trained professor of law, headed the list) was, as previously mentioned, appointed by Syngman Rhee, and within a month's time it was written and presented to the National Assembly for approval. The National Assembly quickly approved it after only perfunctory debate. This constitution was never seriously discussed within or without the assembly. It was a paper-work brainchild of a few legal experts who worked under the direction of Syngman Rhee. Within it there was an interesting intermeshing of Syngman Rhee and his personality on the one hand, and of Korean legal experts—then mostly trained in the European and Japanese tradition—on the other.

In the final analysis the first constitution was parliamentary in its orientation, with a strong executive office superimposed upon it. Later, the constitutional amendments of 1952 that were forced through the National Assembly (then meeting at the temporary capitol of Pusan in the course of the Korean War) were to make the executive more independent of legislative control and to weaken the power of the National Assembly. Thus, in the 1952 amendments, the president was to be elected popularly and not by the National Assembly as before. A second legislative chamber, the House of Councilors, was provided, making the legislative body bicameral. The former National Assembly was to become the House of Representatives. In 1954, amendments exempted Syngman Rhee from the constitutional limitation of no more than two consecutive terms for president.

As it stood at the time of Rhee's downfall in 1960, the first Republican constitution set up a structural framework of government of three separate branches with mutual checks and balances. The president and the vice-president were elected through popular, direct, universal, and secret suffrage, but each on a separate ticket. The relationship between the National and other levels of government was unitary.

It was President Syngman Rhee, however, who took initiative in national policy matters. He, furthermore, enjoyed a broad "crisis power" stipulated in Article 59 of the constitution. It read:

> When, in time of civil war, in a dangerous situation arising from foreign relations, in case of national calamity or on account of a grave economic or financial crisis, it is necessary to take urgent measures for the maintenance of public order and security, the president shall have power to issue orders having the effect of law or to take necessary financial disposition, provided, however, that the President shall exclusively exercise such power only if time is lacking for convening of the National Assembly.
>
> Such orders or disposition shall be reported without delay to the National

Assembly for confirmation. If confirmation of the National Assembly is not obtained, such orders or dispositions shall lose their effect thereafter and the President shall promulgate such nonconfirmation without delay.

Assisting him in the exercise of his broad executive power was a collegiate body called the State Council. It was a form of cabinet (not all the members were heads of the executive departments) with more power than the cabinet of an American president, but less than that of a British or a fourth Republican French cabinet. Among the members of the first State Council were included all of the cabinet ministers—Home Affairs, Foreign Affairs, National Defense, Finance, Commerce and Industry, Social Affairs and Health, Transportation, Communications, and Reconstruction.[1] In addition to these nine ministries, assisting the president were the office of Public Information, the office of Legislation, and the General Affairs Bureau of the State Council. Included in this latter agency was the Civil Service section that had replaced the Civil Service Committee established under the original constitution.

The Second Constitution

After the demise of Syngman Rhee's First Republic of Korea in 1960 (as a result of the student uprisings against the election riggings of March 10, 1960) the Second Republic of Korea came into being after a brief period of interim caretaker government. A new constitution was drawn up by June 15 of that year and all precautions were taken against another autocratic rule of the Syngman Rhee type. The former presidential system was replaced by a cabinet system, and the former presidential powers were to be shared now by the president and the prime minister. Furthermore, the executive powers of the president and the prime minister as prescribed in the new second constitution were made less extensive and more dispersed than they were under Syngman Rhee, since other agencies were set up to handle certain of the functions formerly exercised by the president. A liberal tendency in the new constitution may also be seen in more specific guarantees of civil liberties.

More specifically, Chapters III and IV of the second constitution based the framework of the government on a cabinet as against

[1] The Ministries of Social Affairs and Health were merged in 1954 into a single ministry of Social Affairs and Health. The Ministry of Reconstruction was created in 1954.

a president. The president was to be elected by a joint session of both Houses, his functions were to be titular and ceremonial, and the executive powers were to be exercised by the State Council. This Council was to be headed by the prime minister, who was to be nominated by the president with the consent of the lower chamber (the House of Representatives).

With regard to the controversial emergency power of the chief executive in the first republican constitution, the Second Republican Constitution made the following modifications:

> The President shall, upon the resolution of the State Council, proclaim a state of siege.
>
> However, the President may refuse to proclaim the state of siege despite the resolution of the State Council, if he deems it improper.
>
> When the state of siege has been proclaimed, special measures may be taken on the rights of people and the powers of administrative authorities or courts in accordance with the provisions of law (Art. 64).

In the area of civil liberties, Article 10 of the first republican constitution read: "All Citizens shall be free from restrictions, except as specified by law, on domicile or the change thereof, and from trespasses on and searches of private premises"; and Article 11 stated: "The privacy of correspondence of all citizens shall not be infringed upon." The new second republican constitution deleted the qualifying clauses from both articles and made them more absolute. In Article 28, however, it added clause 2, which read: "Laws imposing restrictions upon the freedoms and rights of free citizens shall be enacted only when necessary for the maintenance of public order and welfare." "Such restrictions," it continued, "shall not infringe upon the essential substances of the freedoms and rights and nothing may be provided by law with regard to license or censorship on speech and press, nor for the permission for assembly and association."

Other provisions in the area of civil liberties were contained in this new constitution: "The political parties shall enjoy the protection of the State in accordance with the provisions of law. However, if the purposes or activities of a political party are contrary to the basic democratic rules of the Constitution, the Executive shall accuse it upon the approval of the President, and the Constitutional Court shall order the political party to dissolve by adjudication" (Art. 13, Cl. 2). "The political neutrality and personal status of the public officials shall be guaranteed in accordance with the provision of law" (Art. 27, Cl. 2). "The law referred to in our preceding paragraph shall include provision concerning organization necessary for securing the impartiality of the police" (Art. 75, Cl. 2).

In order to uphold the principle of an independent judiciary, Article 78 of the first republican constitution, which read, "The Chief Justice of the Supreme Court shall be appointed by the President and with the consent of the National Assembly," was changed to read (in Article 77 of the second republican constitution): "The Chief Justice and Justices of the Supreme Court shall be elected by the electoral college, composed of persons qualified as judges, and shall be confirmed by the President." Also, the third clause of the same Article stated: "The judges other than those prescribed in the Paragraph 1, shall be, upon decision of the Justices Council, appointed by the Chief Justice of the Supreme Court." The second republican constitution also provided a constitutional court as a standing body, whereas the previous constitution made it an *ad hoc* body. Also provided in the Second Constitution was a central election committee "for the purpose of a fair election" (Art. 75-11). With regard to local autonomy, the second constitution stipulates specifically that "heads of Eup and Myun shall be elected by the direct suffrage of the inhabitants," whereas formerly these local government officials were appointed by the chief executive.

The Third Constitution

This second republic, though it took seriously the lessons of the failure of Rhee's autocratic rule, did not last long. On May 16, 1961 a small group of South Korea's powerful armed forces seized power in Korea, thus uprooting Premier John M. Chang's government of the Second Republic, a government which was a little less than a year old.

The coup took place on the 16th of May. The next day, the military junta of 30 men, made up mostly of generals and colonels, took charge of national affairs, permanently dissolving the National Assembly and prohibiting all political activities, organized and unorganized alike. On the same day this revolutionary committee announced their public pledges for revolution, stating the six basic points of their policy—a stronger anticommunistic stand, a stronger tie with the U. S., a strong effort to eradicate all corruptions and social evils and further create a fresh morale with a view to rejuvenating the demoralized national spirit, and a concentrated effort to establish a self-supporting economy. The junta furthermore pledged that "At any time upon completion of the afore-stated missions, we will turn over the control of the government to clean conscientious civilians and will return to our proper duties."

On the 19th this committee became the Supreme Council for National Reconstruction. The military cabinet, headed by Lt. General Chang To-yong as premier, was announced on the 20th, eventually to be replaced by the real movers of the coup, then Brigadier General Park Chung-hee and Colonel Kim Chong-pil. President Yun Bo-sun of the Second Republic was momentarily forced to remain in his office as a symbol of legitimacy for the new regime.

In this six-day period, 2014 persons were arrested for political and other reasons. Also arrested were 13,387 hoodlums and street toughs. In the same spirit, 49 out of 64 dailies published in Seoul were ordered to close. Altogether 1170 out of 1573 dailies, news services, weeklies, monthlies, and other public information agencies in operation in South Korea at the time of the coup were deprived of their registration under Supreme Council Proclamation No. 11 and Ministry of Public Information Ordinance No. 1. According to the Supreme Council, "the measure was intended to clean out those phony journalists and press, to establish a fresh order in journalism and develop a truly democratic press."

It is this military junta, then, that has sponsored the third republican constitution, promulgated after about one and a half years in power.

The third constitution expresses many of the same goals as the two earlier ones. In the formulation of the basic structural framework within which these goals are to be achieved, the framers of the new constitution have, no doubt, benefited by these earlier experiences. The preamble to the new constitution states: "Do hereby amend, through national referendum, the Constitution, ordained and established on the Twelfth Day of July in the year of Nineteen Hundred and Forty-eight, A. D." What is missing in the Preamble, however, is the great spirit of independence "as manifested in the establishment of the Republic of Korea" following the March 1 Movement in 1919. The third constitution reflects the revolutionary nature of the military junta of young generals and colonels.

It inherits, at least on the surface, all the democratic constitutional principles of the West, particularly those of the U.S. as found in the earlier constitutions. Furthermore, it conscientiously tries to avoid the abuse and corruptions of the Rhee government and at the same time strives toward a stronger government than the ineffective cabinet system of government of John M. Chang. One finds intermeshed in these efforts the junta's distrust of "politicians," their messianic assumption of responsibility of guidance for a better Korea, and their subtle power manipulations to hold on to the reign of government that they seized through the military coup.

In this third constitution the basic framework of government is now to be presidential again, with the separation of powers between the three different branches of government mutually checking and balancing. But this presidential system, strange enough to say the least, is monarchical in its setup and the president is weaker than was Rhee.

A new president of the third republic would not have all the powers of Rhee, even though he has the controversial emergency power of a president as stipulated in Article 73. The Judiciary, as in the second republic, is given greater independence, and the presidential power of appointments is limited only to the chief justice of the Supreme Court. Even here, the president must appoint a person recommended by the Judge Recommendation Council (Art. 99).

The weaker role of the president of the third republic as compared with that of the presidential system of Rhee is more clearly manifested in Chapter III, Section II of the new constitution, which deals with the Executive. It is not in the president that the executive power is vested, as in the U.S., but "in an Executive headed by a president" (Art. 63 (1)), and the prime minister is "to supervise under order of the President the executive ministries of their administration" (Art. 89). The prime minister, along with the president, has power to issue ordinances (Art. 90). Furthermore, the constitution says explicitly: "The Acts of the President in accordance with law shall be executed by written documents, and all such documents shall be counter-signed by the Prime Minister and by the members of the state council concerned. This applies also to military affairs" (Art. 40).[2] Also, "The President shall

[2] The significance of this article in the power relationship between the president and the prime minister and the others in the State Council will be better

not hold concurrently the offices of the Prime Minister, the State Council, the head of any executive ministry, or other official and private positions as determined by law nor engage in private business" (Art. 81).

As to the nature of the State Council, Articles 83 and 84 stipulate: "The State Council shall deliberate on important policies that fall within the powers of the Executive," and it is to be made up of "the President, the Prime Minister, and the members of the State Council whose number shall be no more than twenty and no less than ten" (Art. 83, Sec. 1 and Sec. 2). Here, the president alone appoints the prime minister, but he appoints the other members of the State Council "upon the proposal of the Prime Minister." The president is the chairman of the State Council, but in the final analysis he has one vote in the deliberations, the same as each of the State Council members.

The junta's fear of irresponsible, individual, political manipulation of "politicians" is reflected in the following provisions. "Any person desiring to become a candidate for the National Assembly shall be recommended by the political party to which he belongs" (Art. 36, Sec. 3). "A plural party system" is constitutionally guaranteed (Art. 7, Sec. 1), and political parties are to be protected by the state. However, "If the purposes or activities of a political party are contrary to the basic democratic order of the Constitution,

---

understood when we compare it with the articles to the same effect in the constitutions of the Fifth French Republic and of the West German Republic. In the Fifth French Republican Constitution, we find: "The President of the Republic signs such ordinances and decrees as have been considered by the Council of Ministers" (Art. 13); "The acts of the President of Republic other than those provided for in Articles 8 (para. 1), 11, 12, 16, 18, 54, 56, and 61 are countersigned by the Prime Minister and, where necessary, by the Ministers responsible for their execution" (Art. 22). In the West German Republican Constitution, it is stated: "Orders and decrees of the Federal President require for this validity the countersignature of the Federal Chancellor or the appropriate Federal Minister. This does not apply to the appointment and dismissal of the Federal Chancellor, the dissolution of the Bundestag under Article 63 and the request under Article 69, paragraph 3" (Art. 58). In a nutshell, then, the acts of the president of a third republic of Korea, without qualifications, are to be countersigned by the prime minister and by the members of the state council concerned and to that extent, the prime minister, though he is to be appointed by the president who is popularly elected, is in the position of playing as powerful a role as the president himself. The power of his office can be equal to, or to some extent greater than, that of the federal chancellor of the West German Republic.

the Government shall bring an action against it in the Supreme Court for its dissolution and the political party shall be dissolved in accordance with the decision of the Supreme Court" (Art. 7, Sec. 3). "A person shall lose his membership in the National Assembly during his tenure when he leaves or changes his party, or when his party is dissolved" (Art. 38).

Of great damage to the legislative role of the National Assembly of "politicians" is the constitutional curtailment of the legislative power of holding the purse strings: i. e., "The Executive shall formulate the budget for each fiscal year and submit it to the National Assembly..." (Art. 50, Sec. 2), and "The National Assembly shall, without the consent of the Executive, neither increase the sum of any item of expenditure nor create new items of expenditure in the budget submitted by the Executive" (Art. 53). Also, the executive, along with the members of the National Assembly, can introduce any bill (Art. 48). In these various ways this one-house legislative body is to be deprived of all its legislative weapons and relegated to a consultative and supervisory position only (Art. 57), even if it enjoys its fixed term of four years, since the president is not given the power to dissolve it (Art. 37).

There are some other novel features in the third constitution. A constitutional amendment is made more difficult, requiring for ratification "the affirmative votes of more than one half of votes of more than one half of all voters eligible to vote for the election of the members of the National Assembly" (Art. 121). The section dealing with the election management is expanded (Arts. 107, 108). A new board of inspection is established under the president "to inspect the closing accounts of revenue and expenditures, the accounts of the State and other organizations as determined by functions of the executive agencies and public officials" (Arts. 92-95). A tighter control over the mass media is to be exercised (Art. 18), despite the constitutional guarantee of freedom of speech and press.

Far-reaching implications are in the term "the State," meaning in this new constitutional framework the *independent* executive body. It is "the State" that is "to guarantee fundamental rights of the people to the utmost" (Art. 8). It is "the State" that shall assign "a counsel to the use of the defendant as provided by law" (Art. 10, Sec. 4). It is "the State" that is to guarantee "independence and political impartiality of education" (Art. 27, Sec. 4). It is also "the State" that is to regulate and coordinate and promote economic activities (Art. 111, Sec. 2).

In the final analysis, then, the third constitution is a paper work of the military junta that in essence is designed to perpetuate its own authoritative decision-making role. To that extent, it is an élitist government that is to be instituted under constitutional auspices, strongly reflecting distrust of and contempt for the masses, " politics " and " politicians "; strongly reflecting impatience in the parliamentary decision-making process.

In this national socialistic regimentation, the president, elected by a universal suffrage, is not alone in the predominant position of power. He is to share it with the Council of Ministers which, under a weak presidential leadership, may prove to be an independent force of its own within the government. This potential power struggle within the executive branch is especially likely should the president feel it necessary to appoint as prime minister a person not of his own choice but the head of a majority party in the Assembly.

How a government established under this third constitution will operate is yet to be seen.[3] The way it stands now, however, there are many loopholes in it and, despite all it intended to remedy, these loopholes seem to reflect a divided house of the junta. If this constitution is to be successful in its operation, there is need of a strong personality in the presidency, but in the junta power composition, General Park, head of the current junta government, can hardly be the Korean de Gaulle in his charismatic quality. Despite all the propagandizing efforts to create a hero and messianic image of himself, General Park has failed to produce among the people a collective attitude of " leave everything to big Charlie." Despite the modernizing role that the military oligarchy in Korea has assumed since his successful coup, the military oligarchy has been constantly haunted by that Korean Confucian tradition of civilian supremacy as shown in a distasteful connotation of *kunin chongch'i* (military politics) in the Korean political vocabularies. Korea in transition has yet to find a leader, charismatic in quality, truly dedicated to pulling her away from a threat of disintegration.

In summary, certain theoretical implications in this brief history of South Korean Constitutional Development are in order. Korea is a political laboratory. Since its republican inception, it has been

[3] Cf. Clinton L. Rossiter, *Constitutional Dictatorship, Crisis Government in the Modern Democracies* (Princeton University Press, 1948). Mr. Rossiter's discussion of Article 48 of the Weimar Constitution is of particular interest.

going through the traumatic experiences of a new underdeveloped country in the 20th century. There is a problem of overpopulation. There is a problem of underdeveloped economy. There is a problem of political immaturity. The Korean War was no help. Out of these, Korea has experienced three republics and three constitutions.

Each republic has had its own constitution. This itself is a significant phenomenon. Each republic, despite its frequent usurpation of power, particularly in the Syngman Rhee era, has sought constitutional sanction. Secondly, each constitution reflects the Western democratic political principles and in it is also found certain particularistic elements that tailor it to serve one particular regime. Each new regime therefore feels that it must have its own constitution. There is another important theoretical implication. Inasmuch as each constitution has been so identified with that regime that authored it, the constitution becomes as much a scapegoat of the wrong of that regime as those who have worked under it. Syngman Rhee was wrong and autocratic and so was his constitution. John M. Chang was wrong and this time ineffective in his leadership and so was his constitution, et cetera. In other words, a constitution in Korea appears to be regarded more as a tool of political manipulations than as a permanent, fundamental document of a country. And, when the author of a constitution goes, the constitution itself goes.

The third constitution is no exception to the above theoretical implications. To the extent that this is true, a regime other than the one that authored this third constitution in an effort to transform their role from " caretaker " to " de jure " may find it difficult to operate.[4]

[4] See Edward Shils, " The Military in the Political Development of the New States," *in* John J. Johnson, ed., *The Role of the Military in Underdeveloped Countries* (Princeton University Press, 1962), especially p. 58.

Papers of the Michigan Academy of Science, Arts, and Letters
Vol. XLIX, 1964 (1963 Meeting)

ECONOMICS

KENNEDY'S TAX REFORM

ROBIN BARLOW
The University of Michigan

In 1954 Representative Reed of New York, one of the sponsors of the tax reform passed by Congress in that year, hailed the measure as " the most monumental legislative undertaking in the history of the Congress." It is almost inevitable that a major effort to overhaul the federal tax structure will involve a highly complex piece of legislation, since that tax structure is itself highly complex. This is not a problem that can be solved by a stroke of the pen. Nine years after the Eisenhower reform, the Kennedy Administration has presented a new set of proposals to Congress, and one can be sure that the input of labor which has gone into the Kennedy bill is not less than the half-million man-hours claimed by Representative Reed for the 1954 measure.

In evaluating this complex proposal, I shall first explain why there is a need for tax reform. I shall then outline the main features of the proposed reform and show how it falls short of the ideal reform which has been contemplated in the academic literature for many years. Finally, I shall offer some thoughts on the manner in which the bill is being received by Congress and the public.

That the federal tax structure stands in urgent need of reform is widely admitted. There is general agreement among economists that the present structure of federal taxation and the present structure of federal expenditures jointly act to prevent the economy from attaining full employment of resources. Either taxes should be reduced or expenditures should be raised. I shall not discuss here the relative merits of these two alternative approaches, except to point out that the choice between them rests basically on a judgment about whether society needs more private goods or more public goods. The Kennedy Administration has chosen the solution of reducing taxes and restraining expenditures.

Hence a general reduction in tax rates aimed at solving the unemployment problem is one defensible type of tax reform. There remains the question of redistributing the total tax burden between individuals so as to promote other ends such as equity and long-run growth. In this area there is much to be done. At present federal income taxation consists of applying a set of very high marginal tax rates to a tax base which has shrunk through erosion to less than one half of personal income broadly defined. The severity of the rate structure and the multiplicity of deductions, exemptions, and exclusions have in combination produced many unfortunate results. The high marginal rates have perhaps discouraged work-effort and risk-bearing, although it must be admitted that the empirical evidence is not conclusive on this point. The existence of the loopholes has meant unjust discrimination in favor of those taxpayers who have been in a position to take advantage of them. The effort to take advantage of the loopholes has resulted in a considerable misallocation of resources. It has resulted in resources being diverted toward the favored sectors, such as the oil and gas industry, real estate, and the breeding of livestock. It has immobilized assets: for example, the assets of elderly investors who know that they will escape capital gains taxation if they keep their property in its existing form until death. It has absorbed an inordinate amount of the time and talent of the legal profession and the tax-paying public. No rational person can deny that the process of erosion has got out of hand and that there is an urgent need to begin the reverse process of reclamation.

The Kennedy Administration proposes to reduce the general level of taxation, to allow some more erosion of the tax base, and to undertake some reclamation which will result in a net increase in the size of the tax base. In dollar terms, the reduction in personal and corporate income tax rates would *per se* cause total tax collectors to fall by $14.0 billions; the additional erosion (or, to use the Administration's terms, measures for " the relief of hardship and encouragement of growth ") would cause a further decline of $0.8 billions; and the reclamation (" base broadening and equity ") would recapture $3.9 billions. Hence there would be a net tax reduction of $10.9 billions before feedback.[1] This decrease is the result

[1] *Hearings before the Committee on Ways and Means on the Tax Recommendations of the President,* Part I (Washington, D. C.: U. S. Government Printing Office, 1963), p. 63. The estimate of a $10.9 billion cut ignores the revenue gains from the unlocking of appreciated assets due to revisions in capital gains taxation ($0.6 billions) and from the acceleration of corporate tax payments

of many individual proposals, and the most significant of them are summarized in the following paragraphs.

In the field of corporate income taxation there are several proposals. The 1962 session of Congress enacted two important changes in the corporate income tax, namely the liberalization of depreciation schedules and the investment tax credit, both of which are expected to afford a considerable stimulus to business investment expenditures. In 1963 the Administration proposes further tax concessions aimed at stimulating corporate growth. It recommends that the total cost of machinery and equipment used for research and experimental purposes be claimed as a current expense instead of being amortized over the useful life of the assets as at present. The top rate of the corporate tax would be lowered in stages from the present 52% to 47%. The rate on the first $25,000 of taxable income would be lowered from the present 30% to 22%. This latter proposal appears to be a relatively inexpensive way of attempting to win the allegiance of small business. The tax rate applied to capital gains received by corporations would be reduced from 25% to 22%.

The reduction in the corporate tax rates provides one of those occasions when the business community and academic economists can unite in applause. At least this is true for those academic economists who have believed that the corporate income tax is little better than a rather erratic form of selective sales tax.

The concessions to the corporate sector are partly offset by certain other proposals. Large corporations would be required to pay taxes on a current basis instead of on a delayed basis. The acceleration of payments is tantamount to a temporary increase in corporate taxation. Groups of corporate affiliates would be prohibited from claiming more than one $25,000 surtax exemption. At present a corporation can materially reduce its tax liability by organizing itself as a chain of affiliates, each one of which claims a separate surtax exemption.

The oil industry and other industries enjoying percentage depletion allowances have with good reason expected an assault on their tax advantages, and the Administration has not disappointed them. But the assault has come from an unexpected direction. The percentages of gross income which can be claimed as a deduction from taxable income—$27\frac{1}{2}\%$ in the case of oil and gas, and so on—would

($1.5 billions in 1965) and also ignores the revenue effects of the induced economic expansion.

be left unchanged, and the Administration would exploit another provision in the existing law, namely that which limits the depletion allowances to 50% of net income. At present many oil companies deduct the drilling expenses incurred in developing a property from other sources of income during the interval when the property is not yet yielding income and are then able to claim the full depletion allowances on the property when it becomes fully productive. The total allowances on a property often exceed half of the net income yielded during its life, if net income is adjusted to take account of initial development costs, as seems reasonable. The Administration would oblige the company to deduct drilling and development expenses only from the income of the property for which the expenses were incurred. Thus the net income from the property in its productive years would be reduced, and depletion allowances would more frequently be limited by the 50% ceiling. In justifying this proposal the Administration says that it is merely seeking to preserve the intention of the original legislation establishing the depletion allowances. Other proposals regarding the taxation of mineral resources would introduce restrictions on the transfer of unused foreign tax credits. All of these recommendations should be welcomed as a worthy attempt to deal with one of the major scandals of the American tax structure.

Some corporations, such as those taking advantage of real estate tax shelters, would be affected by certain other parts of the reform which are discussed below in connection with the individual income tax.

For the individual income tax the Administration proposes several major revisions which can be grouped into the three categories of rate reduction, base erosion, and base reclamation. The adjustment in the rates applied to ordinary income would replace the present range of 20-91% with a range of 14-65%. Thus each rate of tax applied to the successive brackets of taxable income would be reduced by about 30%, and in one sense the degree of nominal progressivity in the system would be left unchanged. The rates applied to long-term capital gains income would change from the present range of 10-25% to a range of 4.2-19.5%. This would be achieved by applying the ordinary income rates to 30% of the capital gains instead of to 50% of the gains as at present. Further encouragement to risk-taking would come from the proposal to remove the time limit of five years on the carrying forward of capital losses. The generosity toward capital gains income would be somewhat modified by extending from six months to one year the period during which assets must be held in order for any appreciation to qualify for treatment as a long-term capital gain.

The additional erosion of the tax base proposed by the Administration is intended to alleviate the tax burden on individuals who are felt to be unreasonably treated under the present law. The groups in the tax-paying public who would benefit thereby are low-income individuals, elderly individuals (especially those earning wages), and individuals with high medical expenses, with highly fluctuating incomes, with moving expenses, and with child-care expenses. Low-income families would be the main beneficiaries of the minimum standard deduction which the Administration wishes to introduce. The standard deduction, equal to 10% of adjusted gross income or $1,000, whichever is smaller, would be left unchanged, but a minimum deduction of $300 plus $100 for each dependent would be permitted. The intention of the Administration here is to lighten the tax burden on families with very low incomes, and the device of the minimum deduction is seen as achieving the desired result at minimum cost to the Treasury. Other means to the same end, such as lowering the tax rate applied to the first bracket of taxable income or raising the personal exemption from its present level of $600, would entail much greater revenue losses. Following this proposal, a single taxpayer would begin to pay tax when his income exceeds $900, whereas now he begins to pay tax when it is as low as $667.

Most elderly individuals would find their tax burdens reduced under the Administration's plan, quite apart from the effect of the rate reduction. Taxpayers aged 65 and over who itemize deductions would be permitted to deduct all of their medical expenses (except for the matter of the floor on total itemized deductions, which is discussed below), whereas at present they must subtract 1% of adjusted gross income from any drug expenses which they claim. In addition, the Administration proposes to grant all elderly individuals a tax credit of $300, which would replace the retirement income credit and the extra personal exemption of $600. The effect of this change would be to lighten the tax burden on elderly taxpayers with low and middle incomes and to increase the burden somewhat on those with high incomes (before taking account of the rate reduction). The repeal of the retirement income credit would also simplify the computation of tax liability to a considerable degree and remove the anomaly whereby an elderly individual earning wages pays more tax than one receiving an equivalent income in the form of dividends or interest.

Among the concessions in the Kennedy reform bill, the minimum standard deduction and the revisions in the treatment of elderly taxpayers have the greatest revenue implications. The other concessions involve relatively small revenue losses. The upper limit on medical deductions, which is now $5,000 for a single taxpayer, would be removed entirely. The discrimination against fluctuating incomes, which is inherent in a progressive tax, would be mitigated by the adoption of an averaging device. Employees receiving reimbursement for moving expenses from a new employer would be allowed to exclude such amounts from their taxable income. The child-care deductions allowed to working women, widowers, and divorced men would be liberalized.

Not surprisingly, the tax concessions in the reform bill are encountering few objections. It is true that they imply some further erosion of the tax base, but they nevertheless seem to be justifiable. Of the two measures which have an appreciable revenue impact, the minimum standard deduction can be seen as a device for minimizing the revenue loss of reducing taxes on very low incomes, while the favorable treatment of elderly taxpayers can be seen as conforming to the principle of ability-to-pay. An individual who, like the typical elderly taxpayer, cannot look forward to higher income in the future, does have in some genuine sense an ability to pay taxes which is less than that of an individual earning an equivalent income that is expected to rise.

We turn now to those proposals in the Kennedy reform which are intended to reclaim part of the base of the individual income tax. The proposals would close loopholes or withdraw privileges which have been deliberately granted by Congress to particular groups in the past or which have been unearthed by astute tax lawyers. The withdrawal of privileges can be expected to arouse protest from the dispossessed, and this in fact has occurred, but it must be asked whether the privileges should ever have been extended in the first place.

From the viewpoint of revenue yield the most important of the reclamation proposals is that which would oblige the taxpayer to subtract 5% of adjusted gross income from the total of his itemized deductions. Among the structural reforms, it is this proposal which has attracted the most criticism, since it is seen as discriminating against the virtues of homeownership, charitableness, and fidelity in the payment of state and local taxes. In evaluating the 5% floor we should take note of the expressed intention of the Administration, which argues that in order to secure a marked reduction in the marginal rates of tax—an overriding consideration—it is necessary to expand the tax base. The Administration also places a strong emphasis on the simplification for both the taxpayer and the Treasury which would result from the imposition of the floor. Many taxpayers would be induced to take the standard deduction instead of itemizing.

If we accept the view that a significant reduction in marginal rates is of high priority, then we must admit that the 5% floor is perhaps the best way of securing the required expansion of the tax base. The alternative means of expansion are not promising. Capital gains could be taxed at the full rates, but this would conflict with the objectives of stimulating risk-taking and growth. Personal exemptions or the standard deduction could be lowered, but this would result in unreasonable hardships for low-income families. Certain of the deductions could be abolished altogether (and the deductions for interest and state and local taxes present themselves as leading candidates for abolition), but the political resistance would be overpowering. Forms of money income presently excluded from the tax base, such as interest on state and local government securities, could be reclaimed, but the revenue impact would be minor. In short, the 5% floor appears to be the only satisfactory means of securing a large-enough increase in the tax base.

Nevertheless one must examine the equity implications of the proposal. If the starting premise is that all itemized expenses should legitimately be excluded from the tax base, then obviously enough the introduction of the floor results in unjust discrimination in favor of those who do not have the expenses. To justify the proposal it would be necessary to argue that only "extraordinary" itemized expenses should in fairness be deductible. It could be claimed that the great majority of families have routine or ordinary expenses in the form of charitable contributions, payments of interest and state and local taxes, and so on, and that only amounts in excess of the ordinary (defined as 5% of adjusted gross income) should qualify the taxpayer for favorable treatment. Ordinary virtue is its own reward. The argument is not totally convincing, because already one deduction is limited to what is extraordinary (the deduction for medical expenses exceeding between 3 and 4% of adjusted gross income) and if the Administration has its way, another deduction would be added to that category (only those casualty losses exceeding 4% of adjusted gross income would be eligible for deduction). Furthermore, if the equity of the 5% floor is justified in this way, then the standard deduction is seen to be too generous. A large number of taxpayers would be receiving deductions much in excess of what their virtue qualifies them for. There would therefore seem to be a strong case for reducing the standard deduction from its present level of 10% of adjusted gross income.

The protests against the 5% floor which have been heard from charitable organizations, real estate brokers, and others can be regarded as protests against the concept of "extraordinary" deductions, and should be judged in that light. The Administration is probably correct in insisting that donations to charity and purchases of homes would be little affected by the plan. Most taxpayers now itemizing would find that additional contributions to charity would continue to be fully deductible, since "the floor is not a ceiling." And donations should be stimulated by the higher level of prosperity which the tax program as a whole is designed to achieve.

My own opinion on the deductions is that most of them represent inefficient and inequitable subsidies to particular activities, and that it would be better to abolish some of them entirely. Given the political realities, however, the device of the 5% floor is perhaps the best way of reclaiming part of the tax base, so long as the standard deduction is correspondingly adjusted.

The reclamation proposal which would result in the next largest revenue gain is the repeal of the $50 dividend exclusion and the 4% dividend credit. The exclusion and the credit were originally introduced to provide some relief from the alleged double taxation of dividends caused by the levying of both personal and corporate income taxes, but the Administration appears to be on firm ground in asking for repeal. To the extent that the corporation income tax is shifted to consumers in the form of higher prices, there is no double taxation and no cause for extending relief to recipients of dividend income. Even if there is some double taxation, the 4% credit (the more important of the two integration devices) is the wrong way of providing relief. The credit means that two stockholders receiving the same amount of dividends but with different total incomes get the same tax relief in terms of dollars, whereas the principle of ability-to-pay demands that the poorer taxpayer should get the greater relief from double taxation. The credit is therefore the wrong way of solving a problem that in any case may not exist. To pacify unconvinced stockholders, the Administration points out that the source of dividends, namely corporate profits, would be more lightly taxed under its program.

The rest of the reclamation proposals seem on the whole to be quite justifiable. The repeal of the sick pay exclusion would end the inequity whereby a person who insists on returning to work despite medical problems ends by paying more tax than does a similarly incapacitated person receiving the same income who decides to stay at home. The value of stock options would be taxed as ordinary income and not as capital gain, and thus it would be recognized that the value of the option is a form of compensation which should be treated like any other form of compensation, such as salaries, bonuses, and commissions. Similar arguments apply to the proposals to tax employees at ordinary rates on the value of employer contributions to group term life insurance protection beyond $5000 (which presently is not taxed at all) and on the value of lump-sum pension payments (which presently is taxed at capital gains rates).

The appreciation in assets held until death, which is now subject only to federal estate taxes, would be taxed at capital gains rates. Similar treatment would be given to the appreciation in assets given away during the lifetime of the donor. Thus two important avenues of tax avoidance would be closed, and in the case of the taxation of capital gains at death a harmful "locked-in" effect would be much reduced. Real estate tax shelters would be removed, thereby ending the inequity by which some persons escape ordinary income taxation by claiming depreciation on hotels and office buildings much in excess of the actual decline in economic value. Similar shelters in the areas of mineral properties, livestock breeding, citrus groves, and timberland would also be removed.

Finally, there are some relatively minor adjustments among the itemized deductions, in addition to the introduction of the 5% floor which has been discussed above. The floor on deductible medical expenses at present ranges from 3% of adjusted gross income (if the taxpayer has no drug expenses) to 4% (if his drug expenses exceed 1% of adjusted gross income). The reform would fix the floor at 4% in all cases. Casualty losses eligible for deduction would also be subject to a floor of 4% of adjusted gross income. The ceiling on deductible contributions to charity now reaches 30% of adjusted gross income if total contributions to certain specified institutions (hospitals, colleges, and churches) exceed 10% of adjusted gross income. The reform would delete the qualifying clause in the previous sentence. At the same time, the provision which removes the ceiling when the sum of charitable contributions and income tax liability exceeds 90% of taxable income would be repealed.

The Kennedy tax bill can be seen as a serious and comprehensive attempt to reform the federal tax structure. It is interesting to see how the final product would compare with the ideal tax structure which is envisaged in some parts of the academic literature.[2] It can be plausibly argued that in meeting the needs for the provision of public services, for the redistribution of income, and for the attainment of full employment with price stability, the federal government can rely almost exclusively on an individual income tax. The income subject to tax at standard rates would be broadly defined and would, subject to administrative considerations, encompass all forms of money income, including realized capital gains, and all forms of income in kind. Unrealized capital gains would be taxed at death. No exclusions or deductions would be permitted: if certain activities such as medical research or homeownership needed encouragement, they would be directly subsidized instead of being indirectly subsidized through the income tax. In such a system there would be no place for a separate corporation income tax, since all income derived by individuals from corporate profits (dividends and capital gains) would be fully taxed under the personal income tax. There would be no place for estate and gift taxes, since all gifts and bequests would be taxed as income of the beneficiaries. There would be no place for excise taxes, which merely distort the allocation of resources between the production of alternative commodities, except in cases where the excise taxes approximated a user charge (gasoline taxes earmarked for highway construction) or reflected the social costs of consumption (liquor taxes).

The federal tax structure even as reformed by the Kennedy bill would obviously fall far short of the grand simplicity and neu-

[2] See, for example, Richard A. Musgrave, *The Theory of Public Finance* (New York: McGraw-Hill, 1959), Chapter 8.

trality of such a system. Nothing is proposed for the estate and gift taxes or for the excise taxes, although it appears that the Administration intends to make recommendations in these areas after the income taxes are dealt with. The separate corporation income tax remains in existence, thus leading to an erratic form of sales taxation to the extent that it is shifted, and leading to double taxation of dividend income to the extent that it is not shifted. Income from capital gains continues to receive favorable treatment, equity being sacrificed on the altar of growth, perhaps mistakenly. The income-splitting privilege, which constitutes a rather uneven form of subsidy to marriage, is left untouched. Except for the matter of the 5% floor, the itemized deductions are hardly challenged, despite their doubtful logic. The percentage depletion allowance continues as an unwarranted subsidy to the extractive industries. Many forms of money and nonmoney income continue to be exempt from taxation: interest on state and local government securities, imputed net rental income of owner-occupiers, most types of fringe benefits.

No reformer who is even remotely aware of the political situation in Congress would, however, criticize the Administration for failing to press for a complete reform. In the 1962 session of Congress even the mild and obviously justifiable proposal which would have introduced withholding of taxes on interest and dividend income encountered massive resistance. So long as the economy remains at the mercy of legislators who are more responsive to special interests than to the general interest, a reasonably thorough reform of the tax structure will be impossible. In the light of the political situation, the moderate reform recommended by the Administration can be seen to be an honest and intelligent attempt to reach the important goals of equity and growth. But to judge from the resistance offered to the measure, even moderation is too extreme.

The tax bill has been much maligned and has few supporters. I will conclude by considering some of the reasons for the almost uniformly hostile reception of the bill. In the first place, many of the more vocal critics are simply misinformed about what has been proposed. Some of the complaints from charitable organizations reveal that the 5% floor is thought to apply to charitable contributions alone, whereas in fact it would apply to the sum of all eligible expenses. Other critics have charged that the proposed closing of loopholes is nothing other than a policy of "soaking the rich." In fact the average taxpayer at all income levels would find that his taxes would be reduced if the reform bill as a whole were passed. Only a tiny minority of upper-income individuals who have been particularly energetic in making use of tax-avoidance devices would find themselves paying more tax.

It must of course be recognized that in percentage terms the tax reduction would be greater for the poor than for the rich. The Treasury estimates that the taxes of families with incomes less than $3000 would be reduced by an average of 42%, whereas the taxes of those with incomes exceeding $50,000 would be reduced by an average of 15%.[3] It may well be that some equalization of the tax reduction would be in order, but it must be stressed that equalization should not be achieved by rejecting the reform and retaining the loopholes. It should be achieved by adjusting the structure of tax rates. The reforms would represent a series of modest victories for the concept of interpersonal equity. It would be wise to adopt them and use the rate structure alone to achieve the desired degree of progressivity. The efforts of those who wish to distribute the tax reduction more evenly ought to be directed toward persuading the Treasury to reduce the rates applied to the top brackets of income and not toward perpetuating inequities like the real estate tax shelters.

The reform bill is in the unhappy position of being misunderstood, and it is also heavily handicapped by the fact that in large part it consists of withdrawing privileges from particular groups. The withdrawal of a tax privilege hurts a few and benefits many: the victims protest loudly but there is no counterprotest from the beneficiaries—the general public—because the gains to them from the closing of any single loophole are spread so thinly. Hence the reform bill has many enemies and few friends.

It is interesting to note that the movement for tax reform has lacked its muckrakers. None of the reformers has succeeded in arousing the public conscience on the matter of the serious inequities and the misuse of scarce resources. The public is not outraged, and hence the pundits who are saying that tax reform is doomed are probably right. But no satisfaction should be felt if the reform is mutilated or abandoned.

[3] *Hearings before the Committee on Ways and Means*, p. 625. The estimates allow for an increase in dividend distributions following the reduction in the corporate tax rate but do not take account of the changes in capital gains taxation.

Papers of the Michigan Academy of Science, Arts, and Letters
Vol. XLIX, 1964 (1963 Meeting)

GEOGRAPHY. A. REGIONALISM—CONCEPTS, METHODS, AND EXAMPLES

THE REGION IN FRENCH GEOGRAPHIC THOUGHT, c. 1900-1930

GEOFFREY J. MARTIN
Eastern Michigan University

It was the spirit of France in the late 1800's that was so indispensable to the rethinking if not the remaking of a discipline. The Paris which Baron Haussman had started to rebuild was showing the world that to be an artist was a serious business. Dumas Fils was converting current problems into plays, the Jockey Club had become a social arbiter, the people had faith in Ferdinand de Lesseps and French engineering when he proposed the construction of an interoceanic Panama Canal. A destroyed optimism experienced rebirth with the international exhibitions of 1889 and 1900, at the first of which scientific exhibits filled many buildings and the resurgent spirit of a new Paris reared itself in the form of the Eiffel Tower, and the second of which boasted yet more science and only casually managed to commemorate the 19th centenary of the birth of Christ. The pomp and pageantry of Paris released the tension and fired the energies of scientists. There were no pantheons or schemas championed by the establishment which inhibited freedom of thought. This was the age of Gaugin, Toulouse Lautrec, Maupassant, Maurice Donnay. A delightfully democratic intellectual ferment had been encouraged by the creation of l'École Normale Supérieure where the bourgeoisie might be inspired to become pedagogues, by *Das Kapital* and *The Origin of Species*, and by the inventions and discoveries of a continuing industrial revolution which added rationalism to romanticism. Paris was not only big enough to accommodate men of science wishing to find new directions for their thoughts, but it encouraged this continual groping at a time when intensity and universalism were standard intellectual equipment. Concomitantly, erudite thinkers were encouraged as never before to express their creativity in a language and purpose necessarily intense but expansive in philosophy. The social thought and clime of time and place inevitably form the environment to which intellectuals are bound, nor were late 19th century Parisian intellectuals an exception.

In the last three decades of the century France had breathlessly swept into Africa in a colonial venture of very meaningful economic proportion. In 1870 the French Colonial Empire boasted 701,250 square miles and a population of 5 million. By 1920 the Empire included 6,400,625 square miles of land and a population of 52 million. Perhaps the movement was incited by the fact that Europe " lost a mistress and gained a master "—a result of the Franco-Prussian War; perhaps the idea was an extension of de Lesseps' successful Suez Canal; perhaps French colonialism was opportunism; or perhaps such expansion was planned economic development. Whatever the reason, the fact of colonialism was a powerful boost to geographical societies throughout France and indeed Europe.[1] Government used French geographers in the hectic scramble for Africa and the quieter conquest of lands elsewhere in the world. Geography had become useful. It now began to find a more readily accepted position in schools and universities throughout France and Europe. More scholars could give their legitimate attention to geography, which inevitably meant a revitalization of a subject matter all too frequently associated with gazetteer or travelogue literature.

Probably French geographic methodology was influenced by the aggression and ambition which characterized the work of Friedrich Ratzel and which found acceptance and support in Germany, and was later championed by non-German neo-Ratzelians, chief among whom was Miss Semple. To the French geographers with their heightened sensibility it seemed as though Ratzel bullied and bruised rather than observed and offered, he generalized and created a world view from an insufficiency of data. Man on the land did not always fit into the " cases " he established, neither did the French accept the idea that man's activities were determined by environment. The French shrank from Ratzel's dogma, yet appreciated his attempt to make of human geography a science, and to that extent were positively influenced by the geographer from Leipzig. French geography, a product of intellectual humility and essentially creative, was perhaps inspired by reaction to the more grandiose scientific arrangements of her eastern neighbor.

[1] D. V. McKay, " Colonization in the French Geographical Movement 1871-1881," *Geographical Review*, XXXIII (2) (April 1943), 214-232.

In an area of little more than 200,000 square miles, France manages to provide much diversity. The mountain lands of the South, the Pyrénées, Alps, and Massif Central have proximity with the Mediterranean, yet with elevation sensibly shun its influence. The basin of Aquitaine and the Rhône-Saône corridor forge lines of human transaction through the mountains and provide corridors for the carriage northward of the Mediterranean vision and ideal. Brittany, simple, austere, remote, and Gallic, retains a landed peasantry, distant, yet only a few miles from the cosmopolitanism of the Paris basin. The gates of Metz-Verdun, the Meuse-Sambre, and Ypres receive the traffic of humanity and commerce from lands east. It was not possible to study different *genres de vie* as transhumance of the pre-Alps, chestnut culture of the southern Massif Central, seaweed collection of Brittany's Armor, and grape culture of the Paris and Aquitaine basins, without a careful comprehension of the environment and its historic necessity. To know himself better, the man of France had to study the environment which had made him a Frenchman: the Frenchman was obliged to study the diversity which has been France. The selection and delimitation of a manageable area suitable for study inevitably obliged French geographers to consider the advantages of studying the natural region.

These were the circumstances which made possible the birth and development of the French concept of the natural region. Fortunately at this time there was in the universities of France a number of most talented scholars, who were able to jointly develop the structure of the discipline in part owing to the spirit of the times, and in part to their ability to collaborate and accept intellectual leadership from Paul Vidal de la Blache. Thus the work of these geographers was not hampered by the habit of perpetual recommencement; rather was it the flow of a school of thought. Members of the school included Baulig, Blanchard, Brunhes, Cholley, Demangeon, De Martonne, Gallois, Siegfried, Sion, Sorre, Vacher, Vallaux. Indeed, by 1922 all 16 departments of geography in the French universities were chaired by men who were one-time pupils of Vidal de la Blache, or who were pupils of pupils of Vidal de la Blache.[2] No hard carapaces of indifference, these minds were alive scrutinizing the disordered condition of geography, not embarrassed

[2] See R. J. Harrison Church, "The French School of Geography," Chapter 3 of Griffith Taylor, ed., *Geography in the Twentieth Century* (New York: The Philosophical Library, 1951).

by what they found; these regionalists were geographers issuing from a minority, leaving behind in gazetteerism and description a death which bequeathed no corpse. It is hard indeed to overestimate the role played by Vidal de la Blache in the development of geography in France—either in his role as pedagogue, teacher, philosopher, or co-worker and human being. As historian de la Blache had perceived, the subject matter of history was converted to discipline through the adoption of the framework provided by chronology. Quietly, but surely, de la Blache recognized that the region might offer a vital framework to the subject matter of geography, a subject matter which so urgently needed a unifying theme. It was a happy blend of warm personality, extreme literary ability, erudition and perception as a scholar that won profound respect for de la Blache among his pupils, a respect that bordered on apostolic fervor. In attempting to erect the regional schema, there was no room for oppositions of a personal nature, for dissensions based on whim or caprice, and so the mutual respect enjoyed among the collaborators for the idea of the region was vital. Lucien Gallois writes of this scholarly cohesion in his introduction to Volume I of *Géographie Universelle*:[3]

> When larger and larger horizons open to geography, no person in France will have more clearly shown the way which, resolutely, she must take. His influence in the domain of our studies has been decisive. Through his teaching, and by his writing, he was truly a master. Those who have responded to his call have, for the most part, been his students. All have come to the ideas which he never sought to impose, but which imposed of themselves. This unity of doctrine assured us of unity in this vast enterprise.

This display of belief and determination, be it noted, was written after Vidal de la Blache had recently died, at a time when the First World War had interrupted the already advanced labors of the discipline-building *Géographie Universelle*, and while the French nation had been harassed and despoiled by Germany. Only a few months earlier, Lucien Gallois had written to Mark Jefferson in Ypsilanti, reflecting the despair of France and the disrupted organization of the French geographers and their endeavors. This hitherto unpublished letter was sent to a Jefferson who was a personal friend of many of the regionalists (several of whom he had met on the occasion of the 1911 Liverpool-Rome Excursion and the 1912 U. S.

[3] P. Vidal de la Blache and L. Gallois, ed., *Géographie Universelle*, Vol. I, A. Demangeon, *Les Iles Britanniques* (Paris: Librairie Armand Colin, 1927).

Transcontinental Excursion), who had exchanged numerous ideas in correspondence with them across the Atlantic, and who was a confirmed disciple of the French regional method. This Gallois was a very different person from the prewar Gallois who had jokingly commented of Jefferson's earlier correspondence with him, always in the French language, " you make at least one grammatical mistake in each letter that you send me."

> I know that the Germans pretend now that their troops had been the target of some shooting. In the region where I was, I can assure you that no one shot outside of the regular troops. And with what could the peasants have shot? With hunting rifles and revolvers against men whose gun range is 1,000 meters or more? When you have seen, as I did, the unhappy people running ahead of the enemy with all they could carry, in cars, in carts, often in baby carriages you realize that the horde did not think about resisting, and still less of provoking retaliation.
>
> What is the strangest thing in this war is the German mentality. It looks like a crazy wind which has blown on this country. . . .
>
> I apologize for letting myself be so involved. I did not expect to write to you at such a length, but how to restrain one's heart from talking? What I also want to tell you is that we are all just as resolute as on the first day and that we all feel that the future of our country is at stake. We will not fail. Our troops behaved admirably; our young men are and still will be heroic and how many fell and walked to their death without regret. It is a destroyed generation. Of all the ones you knew the youngest are in the army, Herbette, Baulig and the others. Demangeon and De Martonne are with me at the Geographical Service, Vacher who before the war had a very serious surgical operation is not in shape to walk. Maybe you remember a young student called Brienne who was with us at Brive. He was killed in the Argonne and how many more with him among our best students. Denis has had one brother killed but fortunately nothing has happened to him yet. There is no family that does not have its victims. I have a son who was severely wounded and whose right hand is still paralyzed and he thinks only about returning to the front. It is a beautiful youth and everywhere the spirit of the public is excellent. There will not be any faltering and since the war is a war of harassment we will see who will hold the longest. Neither the English nor we will give in. The Russians can move back but their army is always solid and their reserves in men inexhaustible.[4]

The idea of the region itself was not entirely a product of the fecund mind of Vidal de la Blache. Geologists in the early years of the century had been studying "pays," while in Germany both Ritter and Humboldt had written extensively on regions, though Ritter regarded life as an expression of a divine purpose and Humboldt saw "humanity as one whole existing for the attainment of

[4] Undated letter from Lucien Gallois to Mark Jefferson; translation mine.

one purpose, the free development of inner powers." De la Blache now proposed by personal example the appreciation of natural regions as the culmination of the synthesis of geographic entities. His *Tableau de la Géographie de la France* inspired numerous regional monographs including: *La Flandre* (Blanchard) ; *Plaine Picarde* (Demangeon) ; *Régions Naturelles et Noms de Pays* (Gallois) ; *Le Poitou* (Passerat) ; *Les Paysans de la Normandie Orientale* (Sion) ; *Les Pyrénées Méditerranéennes* (Sorre) ; *Berry* (Vacher) ; *Basse-Bretagne* (Vallaux), together with the work of Brunhes, and Vidal de la Blache's son-in-law, Emmanuel De Martonne. The monographs form a distinct era in geographic literature; they championed the French regional cause and method; they were at once descriptive and explanatory. The studies were the product of a method which, if not endowed with the rigors of a mathematical precision, was nevertheless intense, empirical, and unambitious of imperial design. The work had about it the air of a scientific neutrality; it was uncommitted to any metaphysical or theological interpretation of its findings, and it was free from all forms of control that those authorities can impose in the attempt to reconcile scientific finding with orthodoxy.

The geographer had to walk the region, preferably grow up with the region, know its highways and byways, meadows, streams, its every wrinkle in the earth's crust, and come to know intimately the way of life of a humanity distributed upon this portion of the earth's surface. Coming to know a region was the first principle of this school. Knowing about the world suddenly became less important than the appreciation of more local relationships. The patient, unambitious accumulation of facts in a selected region was the way to understanding. Life on earth in the natural region was being examined. The written monograph was descriptive, yet necessarily and simultaneously analytic. Fortunately, this school of geographers were gifted writers whose fine literary style was as vital to the development of a successful method as thoroughness, the guideline to regional comprehension, a thoroughness known alone by the conscience and appreciated only by the learned. The undertaking was modest; it sought, not universals, but quested for reality in a set of integrated phenomena. Thus the region was revealed with all the wonder with which the thinker discovers a thing he has seen all his life and yet has never known before.

Only a reading of the regionalists' literature can reveal the fullness and scope of the school's method and system.[5] The regionalists, with their painstaking care for detail, recognized the complexity of causality and the irregularity of phenomena at the earth's crust. They recognized that causes are frequently unique to a region and that an understanding of causality may not be transferred from one region to another without careful study and deliberation. From this attempt to read the activity of man on the land there developed a philosophy of possibilism which denied the extravagance of determinism. The regionalists asserted that the environment offers man possibilities. They rejected the Ratzelian assertion that the soil, "always the same, and always situated at the same point in space, serves as a fixed support to the changing aspirations of men. This it is, which governs the destinies of peoples with blind brutality." After Vidal de la Blache, the disciples asserted that the earth exerts no influences, but merely extends opportunities which man might adopt or reject according to his capacity. Possibilism was a point of view quite definitely opposed to Ratzelian determinism and neo-Ratzelian influence. The appreciation was largely left implicit in the work of the French geographers, and remained a philosophy ill-defined until Lucien Febvre, French historian and profound admirer of Vidal de la Blache, wrote "The Definition of Possibility" in a chapter entitled "Possibilities and Different Ways of Life," in his book *A Geographical Introduction to History*.[6]

The French geographers realized that this geography essentially constituted a program of tremendous tasks ahead rather than a collection of past achievements. Within the system of the school, the varying talents of the workers were given full expression and encouragement in developing methodologies not merely separated by that increment of difference which is the measure of the human gulf between personalities and points of view but difference which is the genius of a scientist searching for a new tool and fashioning this in his own unique way from his person, to further comprehension. These philosophic developments were not divisive, almost surely they were cohesive, giving variety and strength to the system.

[5] Two concise statements written by members of this French school include: Foreword to *Géographie Universelle*, Vol. I, *op. cit.*, pp. v-viii. Raoul Blanchard and Millicent Todd, Preface to *Geography of France* (Chicago: Rand McNally & Co., 1919), pp. 9-12.

[6] Lucien Febvre, *A Geographical Introduction to History*, Part III, Chapter 1 (New York: Alfred A. Knopf, 1925).

A discipline was in the making. It is worth noting that Vidal de la Blache gave particular emphasis to *genre de vie* as a product of the historical process; Jean Brunhes delimited and studied the essential facts of a region and, working from these individual phenomena, came to an appreciation of the region through the sum of its parts; Albert Demangeon displayed an interest in statistical comparison of regional data and so became perhaps one of the forerunners of the quantification proceduralism popularized later in the 20th century by U. S. geographers.

The regionalists circulated their work in the *Annales de Géographie,* in the numerous monographs offered by them, in the 15-volume *Géographie Universelle,* in the numerous publications of the French geographical societies, and on the occasions of the quadrennial meetings of the International Geographical Union.

The concept of the natural region developed by these French thinkers helped geography become less a description of the earth and more nearly a science of man on the land. The contribution of the regionalists was not alone attributable to one man's teachings and the emergence of a school of very talented collaborators; the product of the French geographers was also, in part, the product of an expansive and exuberant age in France, a scientific milieu, the rejection of ill-founded scientific thought, and the reality of human and physiographic diversity in France.

PAPERS OF THE MICHIGAN ACADEMY OF SCIENCE, ARTS, AND LETTERS
VOL. XLIX, 1964 (1963 MEETING)

A METHOD FOR THE MEASUREMENT OF AREAL CORRESPONDENCE [1]

ROBERT F. MINNICK
Central Michigan University

THE basic procedure required for developing understanding of total earth space is the procedure called partitioning or compartmentalization, that dividing of total earth space into compartments geographers call areal differentiation. Useful differentiation of total earth space or of any earth area is accomplished by applying a specific criterion according to some partitioning procedure. The result of such differentiation is a system of regions. The system is composed of regions related to one another via the partitioning procedure and the applied criterion. A region is a compartment of earth space internally consistent with the applied criterion of the differentiating procedure surrounded by area not consistent with that applied criterion.

A second important procedure is that of aggregation, the collecting together of contiguous places by applying a specific criterion to places according to some aggregating procedure. The result of such aggregation is a region or system of regions. The regions would be related to one another via the aggregating procedure and the applied criterion. A region is an aggregate of contiguous places internally consistent with the applied criterion of the aggregating procedure surrounded by area not consistent with that applied criterion.

A next step toward understanding of earth space and its component places is the measuring of the locational correspondence of different regions or of different regional systems. Devices exist for measuring how phenomena vary from place to place and for correlating those variations. However, there has been no device for the accurate measuring of the degree to which two regions or two regional systems occupy the same segment of earth space. Nor has there been a device for the accurate measuring of the degree to which a region changes its position in space.

[1] Here the method is applied particularly to population density and manufacturing areas in Michigan, Ohio, Indiana, Illinois, and Wisconsin.

This paper offers for consideration devices for making those measurements, a cartostatistical method for determining areal correspondence that is simple, meaningful, and clearly interpretable. The technique is not one designed especially for measuring the association of phenomena within a spatial framework but rather a technique to measure the degree to which two regions, parts of two different regional systems, occupy the same segment of earth space.

NOTATION AND COEFFICIENTS

Fig. 1 illustrates the employed notation, which is an algebra of sets.

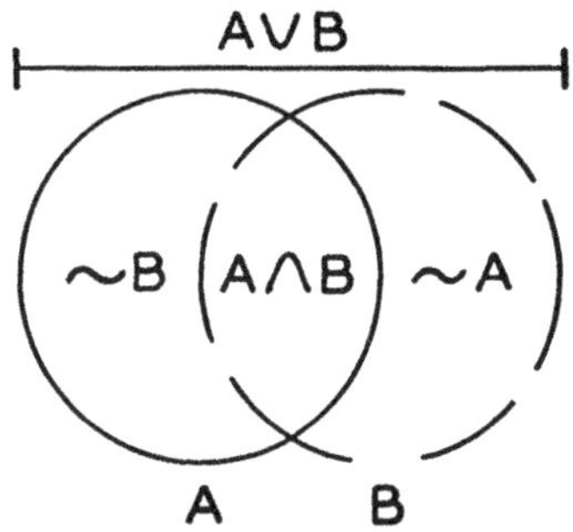

FIG. 1.

Region A = all area enclosed by the solid line whether intersecting region B or not.

Region B = all area enclosed by the dashed line whether intersecting A or not.

Region $A \cap B$ = the intersection of regions A and B; all area within both region A and region B; may be read "A cap B" or "A intersection B."

Region $\sim B$ = all of region A except where intersecting region B; $A - A \cap B$; may be read "A without B" or "A separate from B" or "complement of B."

Region $\sim A$ = all of region B except where intersecting region A; $B - A \cap B$; may be read "B without A" or "B separate from A" or "complement of A."

Region $A \cup B$ = the union of region A and region B; all area within either A or B or both A and B; the sum of the regions $\sim A$, $\sim B$, and $A \cap B$; may be read "A cup B" or "A union B."

It is possible to express the antecedent and consequent of set theory as a fraction $(A \cap B)/A$ in which the denominator may be read "if A" and the numerator may be read "then B" (literally "then both A and B"). Therefore, the expression $(A \cap B)/A$ would be read "if A, then B." Ordinary probability notation for this expression is usually read "the probability of B given A."

Substituting the concept region for the concept set, the notation $(A \cap B)/A$ would be read "where region A, there region B," with $A \cap B$ representing all area within region A whether or not it is in combination with region B. Similarly the notation $(A \cap B)/B$ would be read "where region B, there region A."

The most significant expression arising from this notation is the statement $(A \cap B)/(A \cup B)$ "where either region A or region B, there region $A \cap B$." Like the foregoing fractions this one assumes a range from zero to one. However, unlike the other fractions this one is an expression of the mutual correspondence of regions A and B, a coefficient of areal correspondence. The value of the coefficient is zero when there is no intersection of regions A and B; the value is one when the intersection is equal to the union of A and B, i. e., when there is no part of region A separate from region B and no B separate from A; the value is .5 when the intersection equals one half of the area of union.

The complement of the coefficient of areal correspondence is $1 - (A \cap B)/(A \cup B)$ or $(\sim B \cup \sim A)/(A \cup B)$. As such this complement is a coefficient of areal disjunction measuring the reliability of the hypothesis "wherever either region A or region B, there only one of the two." When it is applied to regional systems differentiated by use of identical criteria for different dates, the complement is a coefficient of relocation:

$$1 - (A_{1950} \cap A_{1960}) / (A_{1950} \cup A_{1960}) \text{ or } (\sim A_{1960} \cup \sim A_{1950})/(A_{1950} \cup A_{1960}) .$$

Maps and Coefficient Values

Maps are presented here primarily to demonstrate the coefficients. They were constructed by using standard isoplething techniques with county centers as control points and data obtained from the Census of Population.[2] Areas were measured by counting squares

[2] Bureau of the Census, *Census of Population, 1950* and *Census of Population, 1960.* U.S. Department of Commerce, Washington, D.C.

of a grid overlay; but in spite of the generalization and error in mapping and measuring, resulting differences among the maps ranged only from 0.6% to 3.1%.

1950 and 1960 population and manufacturing data for the East North Central U. S. were mapped, and the distributions were measured. For the five-state area in 1950 the population density was 124 per square mile, and the percent of the employed labor force engaged in manufacturing was 35. The figures for 1960 were 148 per square mile and 35%. Isopleths for these values indicate the limits for the distributions shown on the maps (Figs. 2-6).

Fig. 2 depicts little change in the distribution of above-average population density from 1950 until 1960. The coefficient of areal correspondence for the distributions of the two dates is .93; the coefficient of relocation is .07.

Fig. 3 depicts fairly substantial change in the distribution of above-average per cent of employment in manufacturing. The coefficient of areal correspondence for the distributions of the two dates is .56; the coefficient of relocation is .44.

Fig. 4 illustrates the relatedness for 1950 between the two distributions shown on the previous plates. The coefficient of areal correspondence is .51; and, of course, the coefficient of disjunction is .49.

Fig. 5 is the same as Fig. 4 except that it is for 1960. The coefficient of areal correspondence is .43; the coefficient of areal disjunction, .57.

Fig. 6 depicts the area of overlap shown on Fig. 4 and the area of overlap shown on Fig. 5. Consequently, the area of intersection is a region of continuing mutual occurrence of above-average population density and above-average percent of employment in manufacturing. The areas of disjunction are areas in which above-average condition is met in only one of the two years. The coefficients are .72 correspondence and .28 relocation.

Other types of statements may be made about the foregoing values. "Whenever either above-average per cent of employment in manufacturing or above-average population density occurs, both occur" is a statement with a reliability of 51% for 1950. The prediction that an area with population density above average in 1950 or in 1960 was above average in both years has a 93% chance of success. The probability that above-average percent of employment in manufacturing occurred in both 1950 and 1960, given that it occurred in one of those years, is .56.

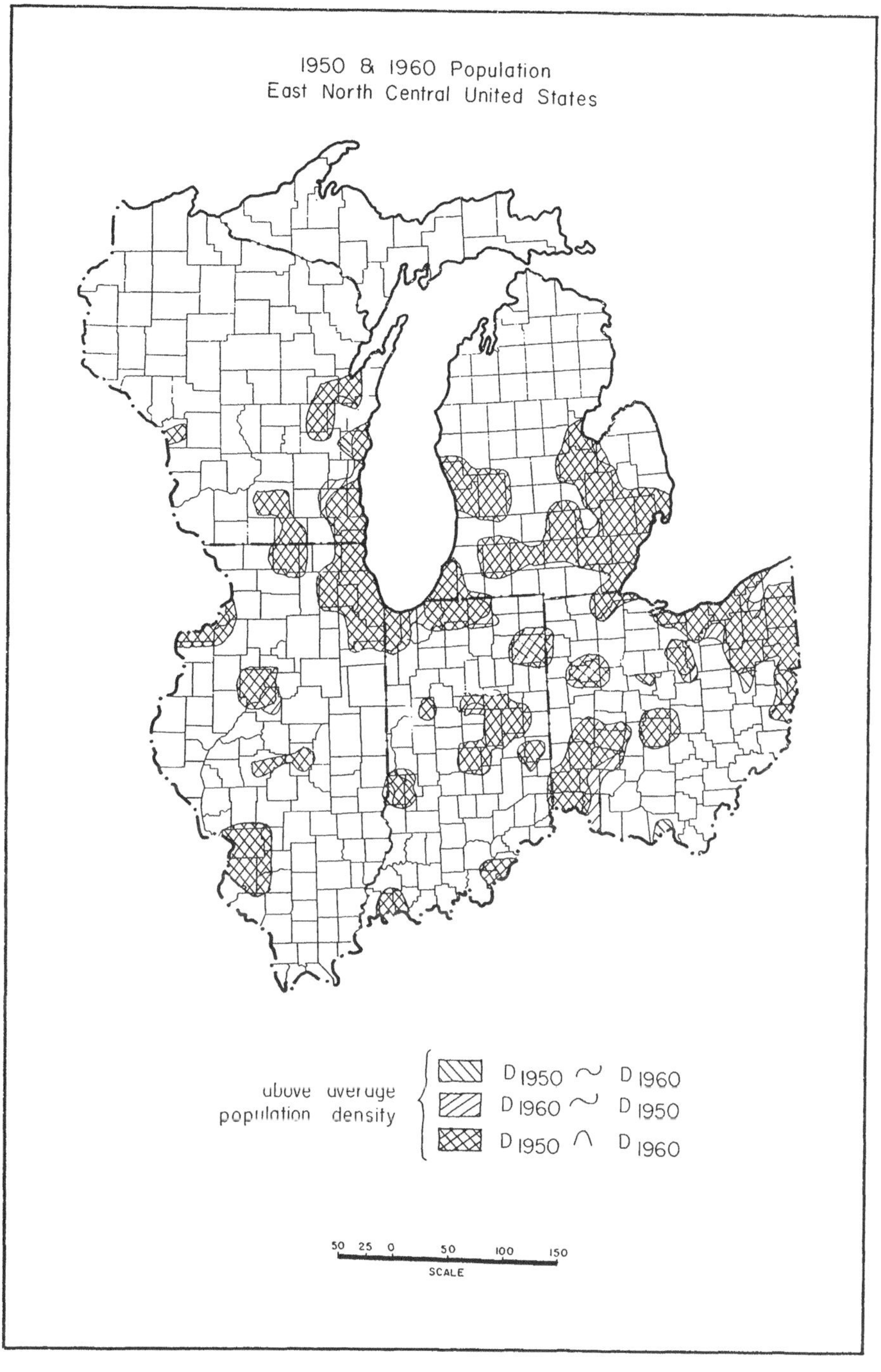
1950 & 1960 Population
East North Central United States
above average population density
D 1950 ~ D 1960
D 1960 ~ D 1950
D 1950 ∩ D 1960
50 25 0 50 100 150
SCALE

FIG. 2.

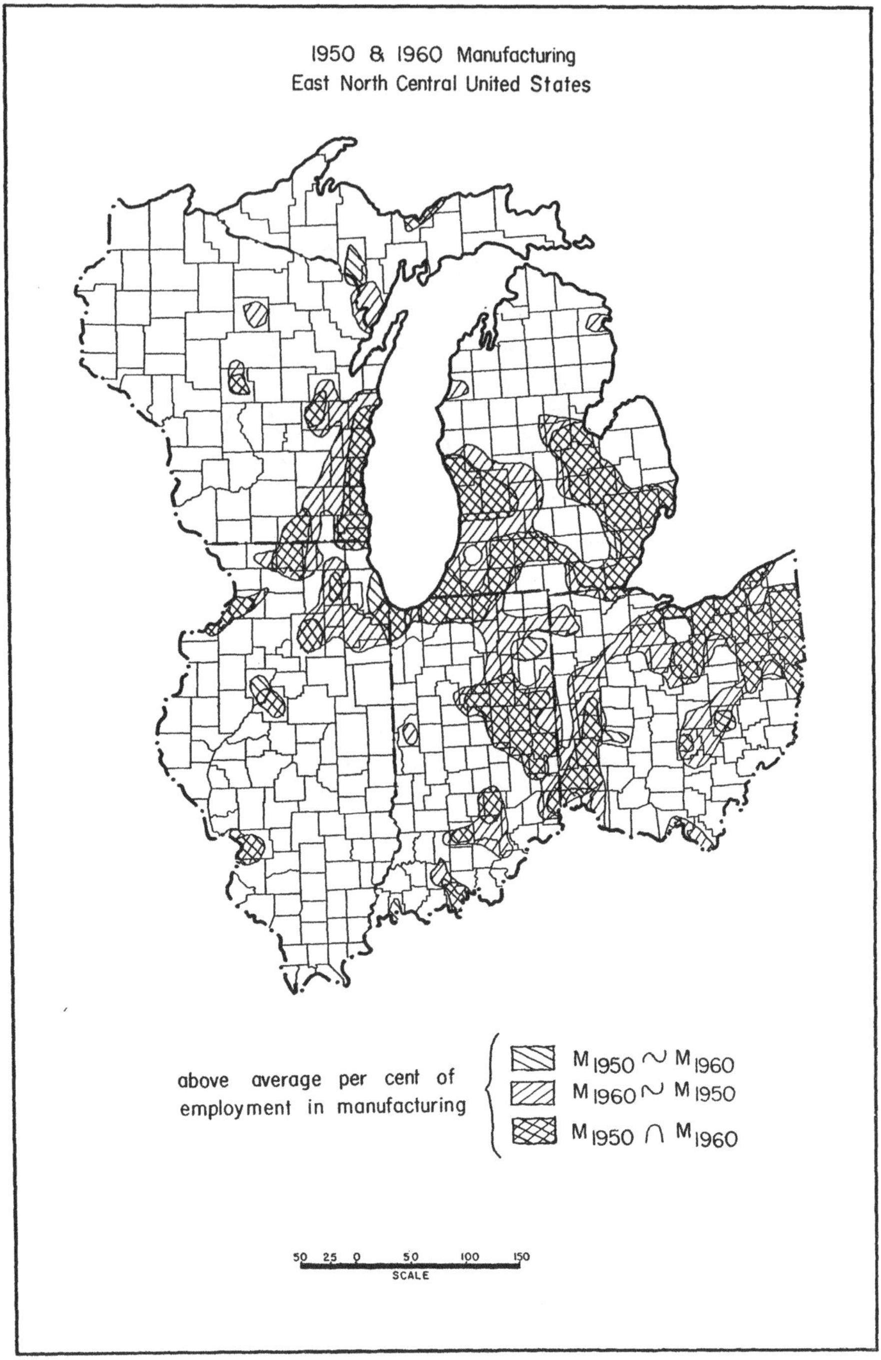
1950 & 1960 Manufacturing
East North Central United States
above average per cent of employment in manufacturing
M1950 ∼ M1960
M1960 ∼ M1950
M1950 ∩ M1960
50 25 0 50 100 150
SCALE

Fig. 3.

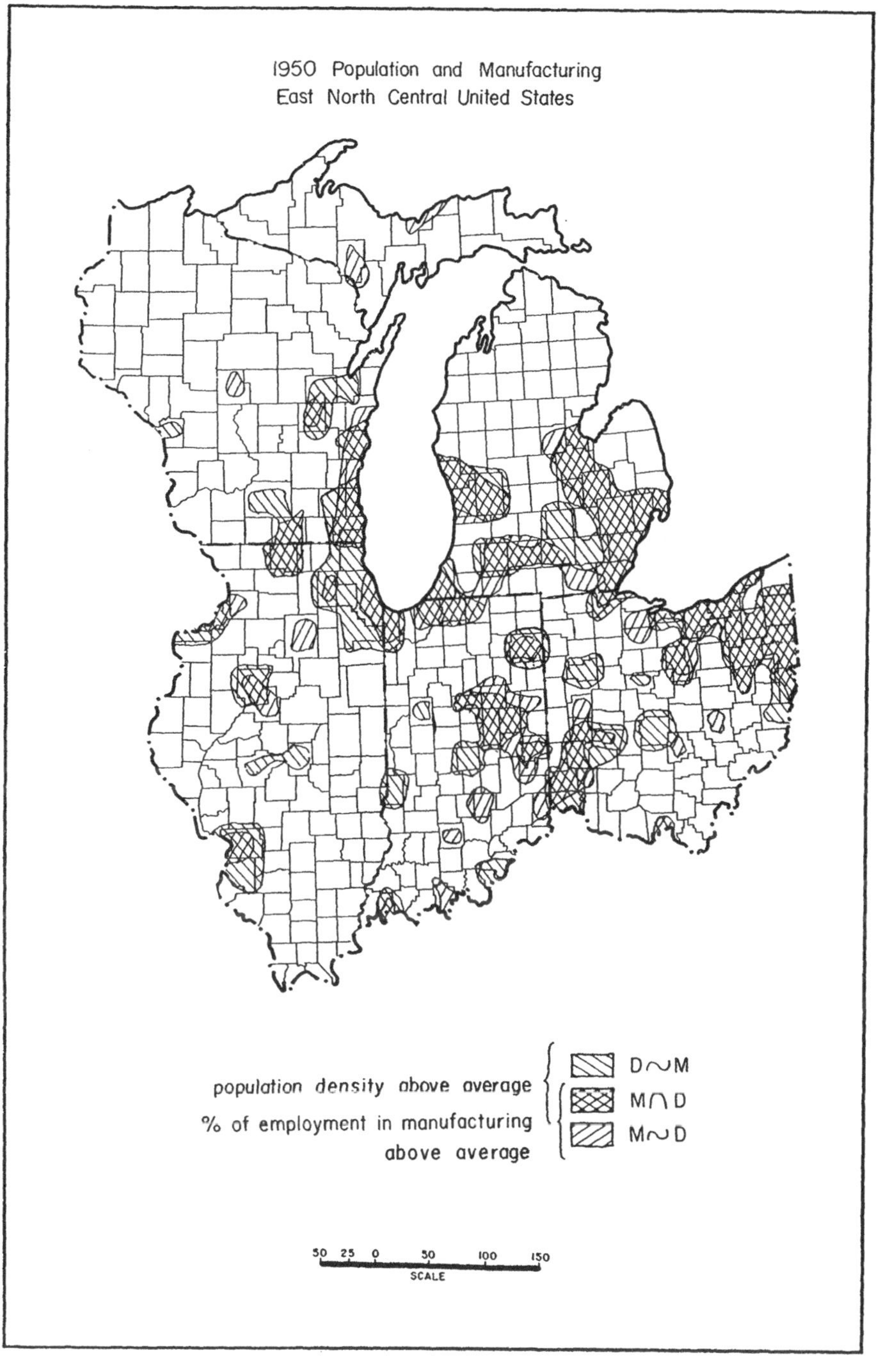
1950 Population and Manufacturing
East North Central United States
population density above average
% of employment in manufacturing above average
D∼M
M∩D
M∼D
50 25 0 50 100 150
SCALE

FIG. 4.

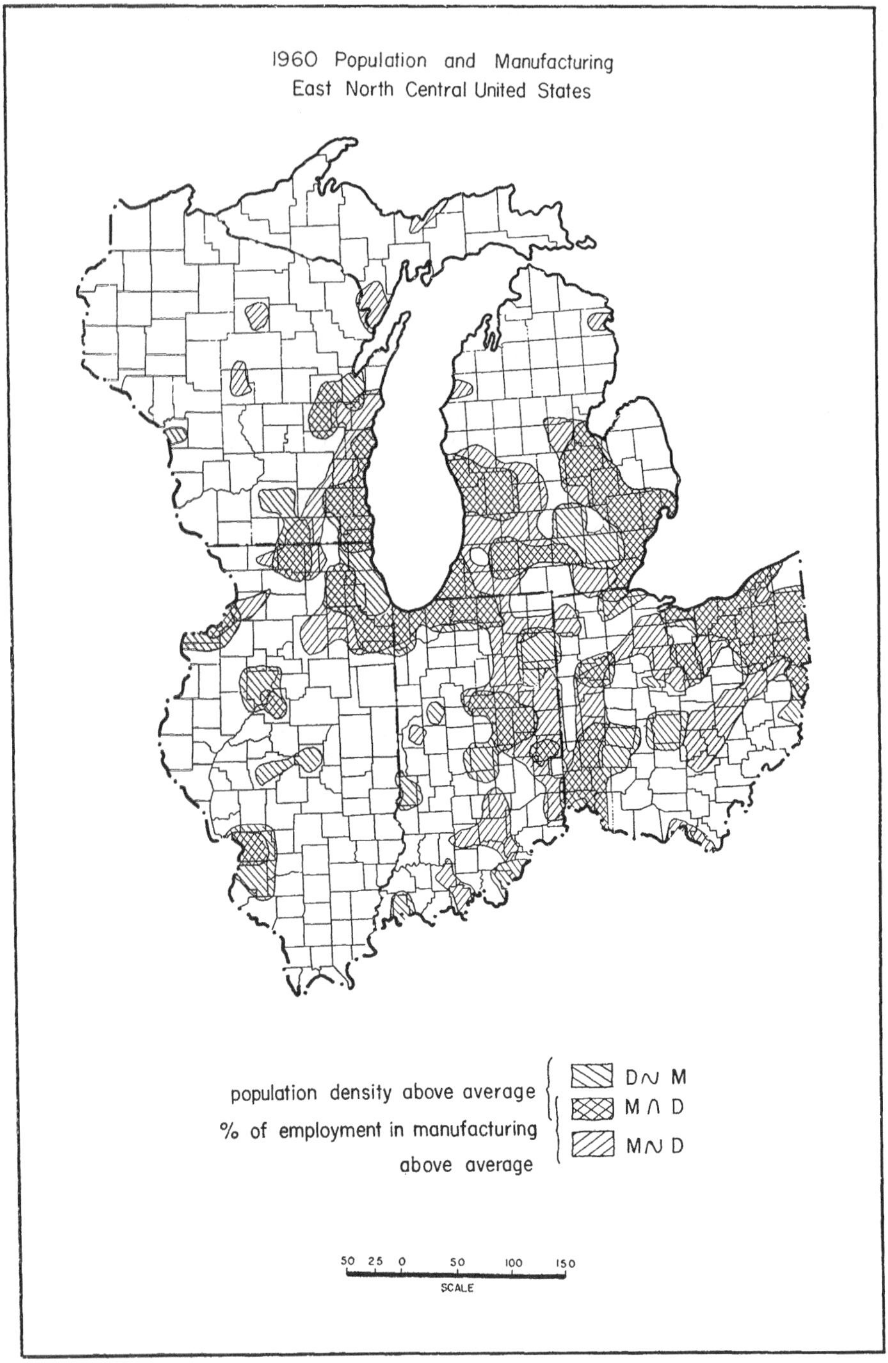
1960 Population and Manufacturing
East North Central United States
population density above average
% of employment in manufacturing above average
D ∩ ~M
M ∩ D
M ∩ ~D
50 25 0 50 100 150
SCALE

FIG. 5.

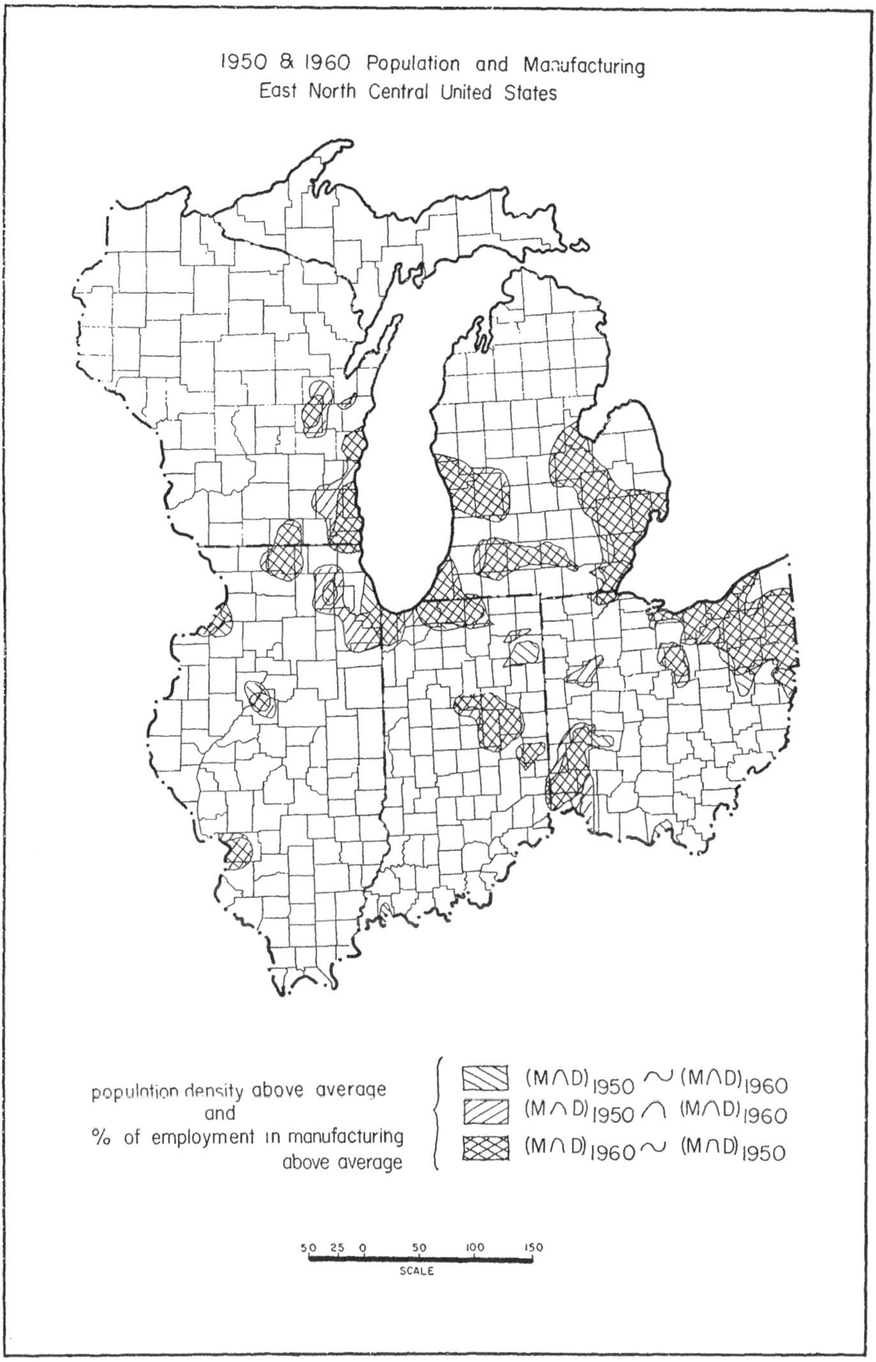
1950 & 1960 Population and Manufacturing
East North Central United States
population density above average
and
% of employment in manufacturing
above average
(M∩D)1950 ∼ (M∩D)1960
(M∩D)1950 ∩ (M∩D)1960
(M∩D)1960 ∼ (M∩D)1950
50 25 0 50 100 150
SCALE

FIG. 6.

Geographic Significance of the Coefficients and of the Notation

It has been my intent to present those two coefficients based on the intersection of two sets that would be of greatest significance to geography. However, there are 10 such coefficients ranging from zero to one that may be derived from the overlapping of just two regions. Moreover, when three regions overlap, over 400 meaningful coefficients may be derived. Any attempt to examine fully the relatedness of the distinctive segments of three such distributional patterns is futile if the effort is made using only word symbols rather than an orderly system of notation. I wish to encourage fellow geographers to become familiar with the mathematics of sets.[3] Its use as illustrated here depends only upon the ability to formulate meaningful hypotheses, to manipulate fractions, decimals, and percentages, and to become familiar with the system of notation. Incidentally, it is now being taught in some high schools.

It should be recognized that the expression $(M \cap D) / (M \cup D)$ is subject to change for every redefinition of either element. Holding M constant, how is the coefficient affected by redefining D; and conversely, holding D constant, how is the coefficient affected by redefining M? The range of values and the maximum values attained should be significant as analytical measures and also in regional integration. When "best fit" between M and D is determined, what other elements are present and at what levels? When manufacturing is mapped with urbanization and the coefficient is maximized, is the value of M the same as when mapped with population density or is the area of urbanization similar to the area of population density?

The coefficients should be particularly useful in historical geography and in study of the historical present to measure concurrences in the relocation of different elements. Studies of concurrent relocation using precise values should provide further insight into the processes by which areas become differentiated from one another, phenomena become regionally integrated, and regions become relocated.

[3] Committee on the Undergraduate Program, *Elementary Mathematics of Sets with Applications.* Mathematical Association of America, 1958.

Other Uses for the Coefficients

The coefficients presented here should be useful in any field of study considering intersecting sets and their elements (overlapping discontinuities and their characteristics). A few questions to which the measures might be applied are:

> To what degree are New York and Liverpool interdependent in international trade?
>
> To what degree is the international trade of New York and Liverpool mutually independent?
>
> To what degree has their interdependence changed?
>
> To what degree are Ohio and Illinois (or Cuyahoga County and Cook County) interdependent in migration?
>
> To what degree are boundaries (or other lines) coextensive?
>
> To what degree are the roles Catholic and Democratic roles for the same individuals or groups?
>
> To what degree are the traits brown-eyed and blonde traits for the same people?
>
> To what degree do distributions of grama and buffalo grasses correspond?
>
> To what degree are corn and oats elements of the same agricultural system?
>
> To what degree did the regionalization of wheat production change between 1930 and 1940?

Situations in which concepts can occur both separately and in combination are subject to the use of the coefficient of correspondence, whether that name is employed or some other such as coefficient of intersection or interdependence or interlinkage. Situations in which a concept changes through time are subject to the use of the coefficient of change or relocation.

Possibly as natural and social scientists we have not spent too much time and energy searching for and trying to explain relatedness of a unidirectional nature, but most assuredly we have spent far too little seeking out and examining interrelatedness.

Papers of the Michigan Academy of Science, Arts, and Letters
Vol. XLIX, 1964 (1963 Meeting)

B. GEOGRAPHY OF LATIN AMERICA

ECONOMIC DEVELOPMENT IN LATIN AMERICA—PUERTO RICO AS A CASE STUDY *

JOHN F. LOUNSBURY
Eastern Michigan University

In recent decades, one of the most serious and widespread problems confronting many areas of Latin America has been the rapid increase of population coupled with a relatively slow growth of economic development. In recent years, the percent of population increase in Latin America has been greater than that of any other part of the world. Several countries have experienced an annual increase of 3 to 4%, which often is much greater than the total economic growth, resulting in a serious decline in per capita production. It is a basic fact that many areas in Latin America produced more per capita in the past than they do at present. The problem is widely recognized, but in only a few selected areas has it been remedied.

Puerto Rico is most unusual in that it has experienced a rapid economic growth probably unequaled in any other area of similar size in the world. This has been accomplished in spite of a limited base. Only about one-fourth of the island is arable. Close to 70% of the island is mountainous or hilly with extremely high slope topography, and perhaps an additional 5% is arid or semiarid. There is no exploitable mineral wealth except local building materials. The population of almost 2½ million occupies 3435 square miles, resulting in a very high population density of over 750 persons per square mile. There are close to 3000 persons per arable square mile, making the island one of the most densely populated areas in the world.

* The basic data for this paper were derived from the following: *Indices Seleccionados de Progreso Economico y Social: Años Economicos 1939-40, 1946-47 al 1961-62,* Negociado de Planificacion Economica y Social, Junta de Planificacion; and Rafael Picó, *Economic Growth in Puerto Rico,* an address to the Annual Convention of Bankers' Association for Foreign Trade, Edgewater Park, Mississippi, May 24, 1959.

In spite of the large and expanding population and lack of resources, an analysis of key economic indices reveals remarkable economic progress. (Table I.) In 1940, the income per capita was $121 and was typical of most Latin-American countries. Puerto Rico's per capita income increased to $682 in 1962 and now ranks first in all Latin America. Total gross production has reached almost $2 billion in 1962; this represents a seven-fold increase since

TABLE I

SELECTED ECONOMIC INDICES

	1940	1950	1954	1958	1962
Gross product *	287	755	1104	1384	1985
Income per capita	121	279	423	500	682
Domestic investment	23	111	173	280	433
Income by origin *					
Agriculture	71	149	169	161	208
Manufacturing	27	89	155	223	381
Trade	27	102	155	199	266
Insular Government	19	70	103	147	221
Federal Government	25	46	116	81	85
Employment (thousands)					
Agriculture	230	216	174	151	139
Manufacturing	56	55	67	79	91
Trade	54	90	83	97	100
Government	19	45	50	62	57
Industries (promoted)	0	82	278	539	863
External trade *					
Exports	92	235	348	467	740
Imports	107	345	525	728	1084

* In millions of dollars.

1940 and a 43% increase over the brief period of the last four years. Gross fixed domestic investment of 433 million dollars in 1962 reveals an increase of 55% since 1958. In 1961, private long-term investments in Puerto Rico from the continental U. S. were appreciably larger than similar U. S. investments in all other countries in Latin America combined. External trade has increased over three-fold since 1950. Today, Puerto Rico is the largest market of the U. S. in Latin America and fifth in the world, following Canada, Great Britain, Germany, and Japan.

The spectacular economic growth, however, has not been apparent in the agricultural sector of the economy. In 1952, net income from agricultural sources was about $195 million, the highest in the history of the island to that date. It had increased only 6% by 1962. This indicates that agricultural development has about reached the limits of the physical resource base. In 1952, almost two-thirds of the net income was derived from agricultural sources, while in 1962 agriculture accounted for only about one-third of the total net income. In the 1940's, Puerto Rico's economy was basically a one-crop agriculture economy based on the production of sugarcane. By 1962, beef, poultry, and dairy products together accounted for 38% of the total agricultural income, while sugarcane had declined to less than 36% of the total. The number of people employed in agriculture has decreased about 40% since 1940. All indications imply a more diversified agriculture in the future, but the agricultural sector of the economy as a whole will become appreciably less important.

The rapid growth of Puerto Rico's economy has been the result of remarkable increases in manufacturing and the development of the tourist industry. Manufacturing has been increasing at the rate of 14 to 15% annually over the last few years, representing a total gain in net income of 336% between 1950 and 1962. In 1962, the number of people employed in manufacturing was almost double that of 1950. Over 860 new industries have been established since 1950, and they are increasing at the rate of two per week. Tourist expenditures in 1962, in excess of $66 million, represent an increase of 18% over the previous year and an increase over 450% in a decade.

The remarkable economic growth and the diversification of the economy are reflected in social progress. (Table II.) The death rate has consistently declined; it was over 18 per thousand population in 1940 and was less than 7 in 1962. Life expectancy has risen from 46 years to 70 years during the same period. It is comparable now to the U. S. and Europe, and the present death rate in Puerto Rico is lower than that of the continental U. S. The rapid industrialization has resulted in a steady decline of the birthrate. From an all-time high of 44 per thousand population per year in 1947, there has been a steady decrease in the birthrate to about 30 in 1962. In the 1940's, Puerto Rico's pattern of population growth was one of high fertility and high mortality. In the 1960's, it is one of low mortality and medium fertility. Increasing employment opportunities have resulted in less migration to the mainland. In 1953, the net migration was at an all-time high of over 74,000, and by 1962, the net migration had declined to about 11,000. In 1962, over 30,000 students were enrolled in Puerto Rico's University and colleges, an increase of 100% from 1954. The University of Puerto Rico alone now spends more annually than the entire government did in 1940. The illiteracy rate has decreased from 32% to 12% in the last two decades.

TABLE II

SELECTED SOCIAL INDICES

	1940	1950	1954	1958	1962
Population *	1869	2211	2211	2284	2482
Birthrate (per 1,000)	39.0	39.6	34.6	31.8	30.2
Death rate (per 1,000)	18.2	10.5	7.7	7.2	6.6
Life expectancy (years)	46	61	61	68	70
Net migration *	+ .5	— 34	— 44	— 26	— 11
Literacy (% of population 10 years old and over)	68	75	80	85	88
College enrollment *	5	13	15	21	30
Number of teachers *	6	9	10	12	17

* In thousands.

Regardless of what economic and social yardsticks are employed, a remarkable progress is revealed. The economic and social growth has been appreciably greater than that of the U. S. as a whole. During the 1957-58 recession, for example, gross product, plant investments, construction, and industrial output expanded rapidly in Puerto Rico, while in the U. S. these indices declined significantly or showed no advance.

Conversion of a one-crop agricultural economy to a highly diversified one along with substantial growth of the entire economy was made possible by the establishment of several key governmental agencies in 1942. The Planning Board of Puerto Rico was established to determine the economic resources of the island and to create a general plan to insure orderly and balanced development of the total economy. The Planning Board coordinated the activities of various governmental agencies and served as the policy-making commission in formulating priorities and provided direction for the total development program. The Economic Development Administration, commonly referred to as "Fomento," was another key agency established in the early 1940's. This agency's functions included the development, promotion, and encouragement of industry, tourism, and trade. Its operation was similar to an industrial development organization in that it made economic studies, located plant sites, and arranged for the construction or alteration of industrial buildings. It used governmental funds for development purposes to encourage investments from private sources. In 1962, 863 manufacturing plants, producing a great diversity of products, have successfully been established, altering the island's entire economy. Tax exemption for a 10- or 15-year period has proved to be a powerful incentive in attracting private industry. A third key agency has been the Development Bank, which was established to make long- and medium-term loans to private business and to serve as a fiscal agent for public borrowers. Since its establishment in 1942 it has disbursed over $95 million. About 70% of the total loan disbursements has been made to manufacturing firms or for construction of industrial buildings, 23% for commercial and related purposes, and 6% for tourist facilities. The Development Bank has played an essential role in making capital available both to private and public groups for development purposes.

In addition to the three key governmental development groups, other agencies have also been essential. The Water Resources Authority, established in 1941, has developed the island's power system. Almost all potential hydropower sites have been fully developed, and thermal and nuclear power capacities are being built. Electrical output in 1960 had increased twelve-fold over that of 1940. It is anticipated that electrical output will double every four years during the 1960's and early 1970's. The Puerto Rico Aqueduct and Sewer Authority has brought sanitation to all sections of the island, and the Puerto Rico Ports Authority has developed the transportation network. There is no country in Latin America that has as dense a road network as presently exists in the island.

It is fully recognized that Puerto Rico is unique among the islands of the Caribbean and countries of Central and South America. Its Commonwealth status and close economic ties with continental U. S. are not duplicated anywhere else in Latin America.

U. S. currency and postal services are used, as well as other facilities of the Federal Government. It is a compact island well serviced by automotive roads and is accessible to the Eastern Seaboard and the Midwest, one of the greatest consuming markets in the world. However, these advantages alone do not fully explain the rapid economic and social progress. For over 40 years Puerto Rico enjoyed similar privileges, and little progress was made until the halfway mark in the century. Other areas, including states and sections of the continental U. S., have also enjoyed similar or superior advantages and have not progressed nearly so rapidly.

Recognizing that other underdeveloped Latin-American countries do not have the precise geographic, political, and economic conditions as those in Puerto Rico, it may be noted, however, that their problems are similar to those that confronted the island in 1940. The Puerto Rican system could well serve as a model or a desirable format. A planning board to determine resources, coordinate action, and formulate the general program; a development bank to provide available capital for public and private uses; and an economic development administration to effectively promote nonagricultural sectors of the economy are essential. This framework may well be effectively employed in many areas of Latin America, as well as in underdeveloped countries of Africa and Asia.

PAPERS OF THE MICHIGAN ACADEMY OF SCIENCE, ARTS, AND LETTERS
VOL. XLIX, 1964 (1963 MEETING)

POPULATION CHANGES IN MEXICO, 1950-60

PAUL CROSS MORRISON
Michigan State University

MEXICO had a population of 34,923,129 in 1960.[1] This was 9,132,112 over the 1950 total or an increase of 35.4%. The growth surprised even the demographers. Not only did annual additions considerably exceed those estimated by Mexican officials, but the 1960 population was some 800,000 more than predicted for that year by United Nations personnel using "high assumption" projections in 1954.[2] The rate of growth was among the highest in the world.

The 1950's were also a period of remarkable economic expansion in Mexico, but at a slower pace than during the 1940's. Use of underestimated annual population totals resulted in some overstatement of income and gross national product per capita during the intercensus years. One reason for writing this paper, therefore, was to briefly reexamine how economic development compared with population growth, but the primary purpose was to analyze national and regional changes in the numbers of people between 1950 and 1960.

POPULATION VERSUS ECONOMIC GROWTH

Mexico's population has been growing at an increasing rate. It grew 18.7%, 31.2%, and 35.4% respectively during successive decades, 1930-60. Although there was a net gain of over 500,000 persons through immigration,[3] the 1950-60 rise was mainly due to a high rate of natural increase. The annual average excess of births over deaths 1950-54 was 29.5 per 1000 population and was even

[1] *Población urbana y rural, de las cabeceras municipales y del resto de las localidades por entidades federativas*, Dirección General de Estadística, Secretaría de Industria y Comercio, México, D. F., January 1962 (release from 8o. Censo General de Población, 1960).

[2] United Nations, Department of Economic and Social Affairs, Population Division, *The Population of Central America (including Mexico), 1950-1980* (ST/SOA. Series A, Population Studies, No. 16), New York, 1954, 42.

[3] Gilberto Loyo, *La población de México: estado actual y tendencias, 1950-1980* (Editorial Logos: México, D. F., 1960), 17; originally published in *Investigación Económica*, 1st quarter, 1960.

greater, 33.4 per 1000, during the next five years (Table I). Both averages were larger than any preceding one for a like period.

Reported births reached a historic high of 46.9 per 1000 population in 1959. The rise was probably partly due to more accurate reporting than in the past, but it no doubt also reflected the better health of mothers which resulted in a higher proportion of pregnancies ending in live births. The death rate, on the other hand, dropped from 16.2 per 1000 in 1950, to an all-time low of 11.4 in 1960.

TABLE I

CRUDE BIRTH, DEATH, AND NATURAL INCREASE RATES PER THOUSAND POPULATION
Five-Year Averages, 1920-59; Annual, 1950-60

Time	Births	Deaths	Increase *
1920-24	31.4	25.1	6.3
1925-29	33.3	25.5	7.8
1930-34	44.5	25.6	18.9
1935-39	43.5	23.3	20.2
1940-44	44.6 *	22.1 *	22.5
1945-49	44.4	17.8	26.6
1950-54	44.9	15.4	29.5
1955-59	45.9	12.5	33.4
1950	45.5	16.2	29.3
1951	44.6	17.3	27.3
1952	43.6	14.9	28.7
1953	44.7	15.8	28.9
1954	46.0	13.0	33.0
1955	45.9	13.6	32.3
1956	46.1	11.9	34.2
1957	46.6	13.0	33.6
1958	44.0	12.3	31.7
1959	46.9	11.7	35.2
1960	45.0	11.4	33.6

* Calculated.

Sources: United Nations, *Demographic Yearbook, 1953, 1957, and 1961.*

The natural increase in 1959 was the largest on record—35.2 per 1000 population—and it exceeded 3% every year between 1954 and 1960 inclusive (Table I). Since the birthrate is not likely to drop much soon, and the death rate will no doubt decline some more, the prospect is for a continued rapid expansion of numbers over the near future. A major question is whether Mexico can keep on expanding her economy at a more rapid rate than the population grows.

Some are of the opinion that the cumulative momentum of economic development "has reached a self-propelling stage . . . . [and] high rates of population growth will be absorbed and will provide additional thrust for the upward ascent." [4] Others are dubious, to say the least, that this is true and that development can be sustained at a high enough rate to materially raise the level of living of the burgeoning population.

Expansion of Mexico's economy has run ahead of population growth most years during the last two decades. This has resulted in a fairly substantial gain in both income and gross national product per capita over the period. In monetary terms, per capita income rose from 304 pesos in 1939, to 1278 in 1949, and to 3,237 in 1959. Income adjusted to constant prices, however, only increased from 304 pesos, to 476, and to 544 respectively. Thus, in monetary terms income per inhabitant in 1959 was over ten times that of the 1939 base, but in real terms it was less than twice as large.[5] Significantly, the rise in real income per person during the last decade was less than half as large as during the previous one. Average income in Mexico is obviously still very low and appears to be increasing at a decreasing rate.

The picture looks much the same if gross national product per capita is analyzed.[6] Calculated at 1950 prices, this rose from 1056 pesos in 1939 to 1873 in 1958. During the five-year periods 1940-45, 1946-51, and 1952-57 the average of the annual percent of increase in the value of gross national product progressively declined. The averages were 7.8, 5.7, and 4.6% respectively. During the same periods, on the other hand, the rise in population averaged 2.6, 2.7, and 2.9% annually. With gross national product expanding

[4] Edmundo Flores, "The Significance of Land-Use Changes in the Economic Development of Mexico," *Land Economics*, 35 (May 1959), 124. For opinion nearer the opposite pole see John M. Ball, "Some Comments on Mexico's Population," *Journal of Geography*, 61 (October 1962), 296-301.

[5] Banco Nacional de Comercio Exterior, S. A., *Mexico* (Mexico, D. F., 1960), 99.

[6] See James G. Maddox, "Economic Growth and Revolution in Mexico," *Land Economics*, 36 (August 1960), 266-278, particularly Table 1, "Mexico—Gross National Product, Population, and Gross National Product per Capita: 1939-58," on 267.

at a decreasing rate and inhabitants at an increasing one, there has been a noticeable slowing in the rise of gross national product per capita. The average change from the previous year was 5.2%, 1940-45; 2.9%, 1946-51; and only 1.5%, 1952-57. Also dimming somewhat the luster of 1950-60 accomplishments is the fact that the statistics quoted were based on annual population totals which were underestimated.

Regional and Intraregional Changes

Comparisons of the populations of the 29 states, 2 territories, and the Federal District of Mexico[7] during the census years 1950 and 1960, along with the proportions of these totals that were rural and urban, are presented in Fig. 1. It is obvious that there was a wide range of population size, of density, and of relative numbers of country and city people each of these years; also large differences in the absolute and relative gains of population during the intervening decade.

For statistical comparisons Mexico is commonly divided into five regions, namely, the North Pacific, the North, the South Pacific, the Central, and the Gulf Coast (Fig. 1). In 1960 the Central Region, the plateau heart of Mexico, had approximately half of the nation's people and 14% of the land. It was the area of greatest population density—160.4 per square mile (Table II). The desert North Pacific was the most sparsely inhabited region, with 16.4 per square mile. Density in the arid, semiarid, and mountainous North was only a little higher, but in the wetter, more tropical South Pacific and Gulf Coast it was almost exactly the national average of 46.0 per square mile. These latter two divisions had nearly equal shares of Mexico's population and surface.

As would be expected, considering the large 1950 base, the 1950-60 gain in number of inhabitants was largest in the Central Region. It slightly exceeded that in the other regions combined and amounted to 4,649,959 persons. The relative gain of 37.3%, however, was only a little more than the 35.4% increase for all Mexico (Table II). Lacking statistics, but assuming that rates of birth and mortality were approximately the same in each division, it can be concluded that any excess increase over the national average was due to in-migration. The North Pacific, with the

[7] Hereafter in this study all will be referred to as states.

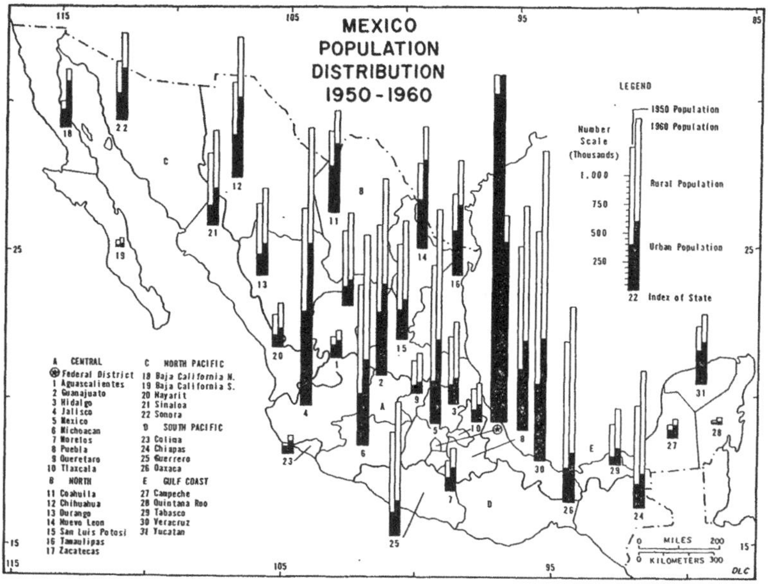

FIG. 1.

TABLE II

MEXICO: POPULATION AND AREA OF DIVISIONS *

Division	Population, in hundred thousands 1960	Increase 1950-60, percent	National population, percent	Approximate area, sq. miles	National area, percent	Population per sq. mile
Central	17.1	37.3	49.0	106,586	14	160.4
Gulf Coast	4.1	32.2	11.7	91,713	12	44.2
North	6.9	32.6	19.7	309,002	41	22.2
North Pacific	2.6	51.6	7.4	159,247	21	16.4
South Pacific	4.3	27.7	12.2	92,003	12	46.6
Nation	34.9	35.4	100.0	758,551	100	46.0

* Based on preliminary figures of the 8°. Censo General de Población, 1960, provided by Banco Nacional de Comercio Exterior, S. A., Mexico 1, D. F.

smallest base population, had the least increment—889,229 persons—but the relative gain of 51.6% made it the most rapidly growing region. The North Pacific was thus the most important destination of migrants, although not many more moved there than moved into the Central Region. Relative increases in the North, the Gulf Coast, and the South Pacific were all less than the national average, indicating these were sources of net out-migration. Movement appears to have been greatest, some 500,000 persons, from the South Pacific. There, the intercensus growth was only 27.7%.

Percentages of population increase in the different states between 1950 and 1960 are shown in Fig. 2; those for the previous decade in Fig. 3. Although the natural growth rate varied from one state to another, statistics concerning this were not obtained. It was assumed that in most cases the differences would be small. Consequently, in drawing the two maps, population gains within approximately 2½% of the national average were classed as about average, those within another 5% each way as moderately above or below average, and those still greater as much above or below average. In general, it can be concluded that the higher a state's percentage of increase above average, the greater the proportion of its residents who had moved in from elsewhere. Conversely, the lower the percentage below average, the more important out-migration was relative to those who remained.

Between 1950 and 1960 these states had population increases much above the average (Fig. 2): Baja California Notre and Sonora in the North Pacific, Chihuahua and Nuevo León in the North, Colima in the South Pacific, Quintana Roo in the Gulf Coast, and the Federal District. The percentage of growth was larger in both Baja California Norte and Quintana Roo than in the Federal District, but the absolute gain was much less (Figs. 1 and 2). States with growth rates much below the average were Coahuila, Durango, Zacatecas, and San Luis Potosí in the North, Querétaro, Hidalgo, Tlaxcala, and Puebla in the Central, Oaxaca in the South Pacific, and Yucatán in the Gulf Coast. No state of the North Pacific was in this category. Thus, some of the more obvious streams of internal migration during the decade appear to have been from Oaxaca, and the eastern states of the Central Region, into the Federal District; from the states of the North into the northeastern and northwestern parts of that region and into the North Pacific; and from Yucatán into Quintana Roo.

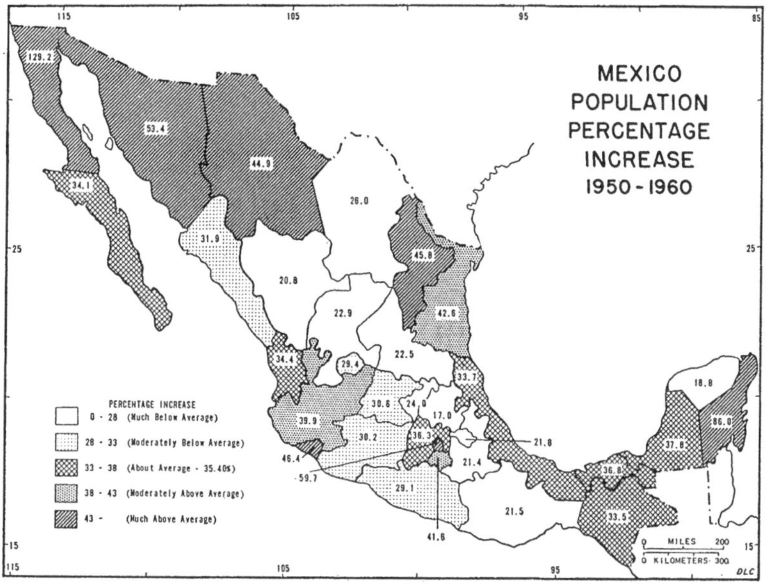
MEXICO
POPULATION
PERCENTAGE
INCREASE
1950 - 1960
PERCENTAGE INCREASE
0 - 28 (Much Below Average)
28 - 33 (Moderately Below Average)
33 - 38 (About Average - 35.40%)
38 - 43 (Moderately Above Average)
43 - (Much Above Average)
MILES 200
KILOMETERS 300
DLC

FIG. 2.

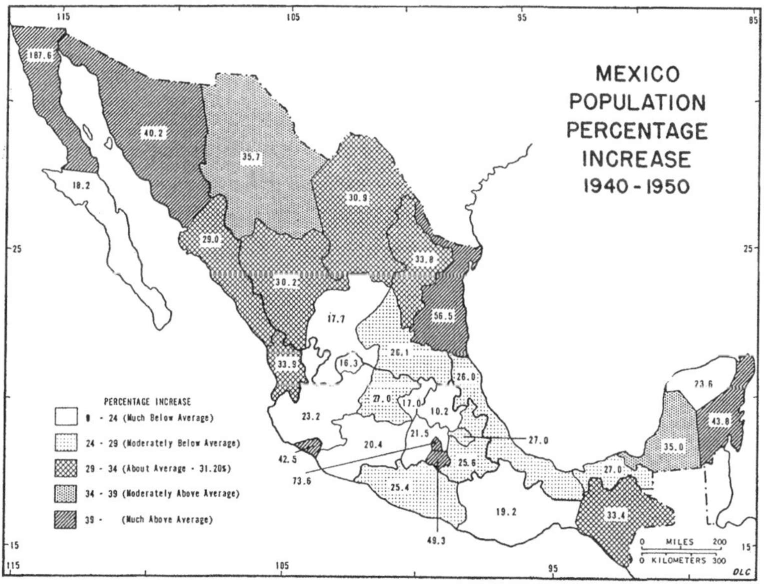
MEXICO
POPULATION
PERCENTAGE
INCREASE
1940 - 1950
PERCENTAGE INCREASE
24 - 29 (Moderately Below Average)
29 - 34 (About Average - 31.20%)
34 - 39 (Moderately Above Average)
39 - (Much Above Average)
MILES 200
KILOMETERS 300
DLC

FIG. 3.

Comparison of Figs. 2 and 3 shows that although the gross patterns of relative growth during the two successive decades illustrated were broadly similar, there were considerable differences in detail. The most apparent changes, i. e. changes of more than one category of growth, were the raised positions of Baja California Sur in the North Pacific, Nuevo León in the North, and Jalisco and México in the Central Region, and the lowered positions of Coahuila and Durango in the North. Also significant in the Central Region, because of influence on number and direction of flow of migrants to the Federal District, were the upping of Michiocán and Aguascalientes west of Mexico City from a much below average to a moderately below average growth, and the dropping of Puebla and Tlaxcala east of the capital from a moderately below average to a much below average increase.

States with a much above average 1940-50 growth rate were Baja California Norte, the Federal District, Tamaulipas, Morelos, Quintana Roo, Colima, and Sonora (Fig. 3). In each, part of the increment was obviously due to migration, since it was greater than could have occurred through the excess of births over deaths alone. These states, with the exception of Sonora, were among the eight which Whetten and Burnight found in their study of state of birth data " obtained more than a fifth of their population increase for the decade from migration." [8] Sonora, however, did get 17.7% of its increase this way.

Much more significant than Sonora's absence is the appearance of Nuevo León and Aguascalientes on the Whetten-Burnight list. Both had approximately 28.5% of their gain due to migration, although Nuevo León had an about average and Aguascalientes a much below average growth during the period. These apparent anomalies were the result of two-way movement. In the case of Aguascalientes the number leaving was much larger than that arriving, with the result that although there was a much below average population growth, the proportion of this made up of persons who were not native-born was high. Since neither differences in net migration or in rates of natural increase were considered

[8] Nathan L. Whetten, and Robert G. Burnight, "Internal Migration in Mexico," *Estadística*, 16 (March 1958), 65-77. This article appeared originally in *Rural Sociology*, 21 (June 1956), 140-151. It provides an informative analysis of state of birth data obtained by the Mexican census and a model for future studies.

in preparing the present paper, only broad conclusions concerning migration in Mexico could be reached. These are useful, however, pending arrival at more exact ones from study of 1960 census data not yet available.

Rural-to-Urban Migration

During the last 30 years there has been an expanding movement of people cityward in Mexico. This urban movement has been the most important factor in the differential growth of population. (Urban is defined here as a place with over 2,500 inhabitants.) In 1930 only about one-third the people lived in urban areas; by 1960 just over half (Table III). Percentages in 1940 and 1950 were 35.1 and 42.6 respectively. In 30 years the number of urban dwellers increased from 5.5 million to 17.7 million, or 222%, while the rural population grew only 56%, from 11 million to 17.2 million.

The Central Region was the most urbanized in 1960, with 58.4% of its people so classed. Both the North Pacific and the North had slightly over half of their totals in cities (Table III). The South Pacific was the least urbanized. Nearly 74% of its inhabitants was rural. The Gulf Coast Region was also somewhat more rural (58.1% is rural) than urban.

13 states had more people in cities than in the country. In the Federal District, 95.8% were urban. Other highly urbanized states, each with over 60% of the people living in cities, were: Baja California Norte in the North Pacific; Nuevo León and Coahuila in the North; Campeche in the Gulf Coast; and Colima in the South Pacific. On the other extreme, inhabitants were 70% or more rural in seven states, namely, Chiapas, Guerrero, Oaxaca, Tabasco, Querétaro, Hidalgo, and Zacatecas. Three of these states were in the South Pacific; none was in the North Pacific.

Between 1950 and 1960 nearly three-fourths of Mexico's population gain was in urban residents. The ratio was highest, 81.1%, in the Central Region, but it was almost as high in both the North and the North Pacific (Table III). It was lowest, 38.1%, in the South Pacific, the only region which had a higher rural than urban increase. In 22 states more than half the population growth was urban. The share was over 100% in Coahuila and Nuevo León, because of small losses of rural dwellers. Chiapas, in contrast, had only a 28.5% increase in urban population, ranking lowest in this regard.

An even clearer picture of the impact of rural-to-urban migration on 1950-60 growth rates can be obtained by examining the relative gains in urban and rural populations (Table III). If, as assumed, differences in the rate of natural increase were small from state to state and from city to country, and if there was no migration, the growth rate everywhere would be close to the national figure of 35.4%. Since, in Mexico as a whole, the urban population grew

TABLE III

POPULATION OF MEXICO, 1960, AND GAIN, 1950-60

	Population 1960		Gain 1950-60		Relative Gain 1950-60	
	Total	Percent Urban *	Total †	Percent Urban *	Urban ‡	Rural ‡
North Pacific...	2,613,470	52.5	889,229	78.4	103.3	18.3
North	6,864,521	51.3	1,687,666	79.1	61.0	11.8
Central	17,099,160	58.4	4,649,959	81.1	60.6	14.1
Gulf Coast.....	4,056,676	41.9	987,765	58.6	51.6	21.0
South Pacific...	4,289,302	26.2	929,256	38.1	46.0	22.2
Total Mexico...	34,923,129	50.7	9,132,112	73.6	61.2	16.3

* Percent rural can be obtained by subtracting percent urban from 100.
† Relative gain 1950-60 is shown in Fig. 2.
‡ Indicates the 1950-60 gain as a percent of the 1950 number.
Source: Based on preliminary figures from 8° Censo General de Población, 1960, released January 1962 and provided by Banco Nacional de Comercio Exterior, S. A., Mexico 1, D. F.

61.2% and the rural only 16.3%, it is clear there was much movement to the cities during the decade.

In every state the rate of urban gain was larger, usually much larger, than the rural. In only three states—Zacatecas, Hidalgo, and Yucatán—was it smaller than the national average. In all but three states—Baja California Norte, Quintana Roo, and Colima—on the other hand, the rate of gain of rural people was below, usually well below, that which would be expected through natural increase. In two states, as noted previously, there were small losses.

In Baja California Norte, where the urban and rural inhabitants increased 176 and 44% respectively, there was a net movement from outside into both the urban and the rural parts of the state. The same was true of Quintana Roo and probably also of Colima, but in no other state was there a net gain of country population from migration. It is also clear that states such as Yucatán and Hidalgo, where both rural and urban populations grew much less than in the national average, exported people from their city as well as country areas. Although it can be concluded that the great bulk of net gains due to migration, either interstate or intrastate, resulted from rural-to-urban movement, there was much movement from place to place in the rural areas. These migrants replaced others who had left for the city, and many eventually moved on to the city themselves. Some cities, including large ones, grew even less rapidly than would be expected. They were losing people to other cities.

The Larger Cities

There were 37 cities of over 50,000 population in Mexico in 1960, as compared to 24 in 1950, and 13 in 1940. Percentages of the total population living in these centers were 22.8, 19.1, and 13.6 respectively. Cities of over 100,000 population numbered 17 in 1960, 10 in 1950, and 4 in 1940. In order they had 18.6, 15.1, and 10.2% of the nation's people. Thus, between 1950 and 1960, there was a smaller share of Mexico's total population added to the bigger cities than from 1940 to 1950, although the number of such cities increased more than during the earlier decade.

As can be seen from Table IV, the growth rate of the cities over 50,000 in 1960 had varied greatly during the 1950-60 period. It ranged from 166.9% in Mexicali, to only 10.1% in Pachuca. Northern border cities comprised four of the five places that grew most rapidly. All more than doubled in size. In addition to Mexicali, these were Tijuana, Ciudad Juárez, and Matamoros. In each a high level of increase from 1940 to 1950 was continued from 1950 to 1960, but at a lower rate, although the number of people added was much larger the last decade. Hermosillo, in the North Pacific, was the other center among the first five.

Guadalajara, second city of the Republic, stood sixth with a relative increase of 94.8%. It was the only one among the ten fastest-growing centers that was outside the North Pacific and the North. Monterrey, the third largest city of the nation, was the ninth most rapidly growing one. Its relative increase was 79%, while its absolute increase of 263,571 persons was third biggest after the 497,337 in Mexico City and 397,330 in Guadalajara.

The absence of Mexico City among the ten cities with the highest percentages of growth from 1950 to 1960, and the fact that its growth of 21.3% was much below what would be expected by

TABLE IV

CITIES OF MEXICO OVER 50,000 POPULATION, 1960

City	Population 1940	Population 1950	Percent Gain 1940-50	Population 1960	Percent Gain 1950-60
North Pacific:					
Mexicali	18,775	64,658	244.4	172,554	166.9
Tijuana	16,486	59,950	263.6	151,939	153.4
Hermosillo	(A)	43,519	?	96,122	120.9
Culiacán	22,025	48,936	122.2	84,602	72.9
Mazatlán	32,117	41,754	30.0	74,934	79.5
North:					
Monterrey	186,092	333,422	79.2	596,993	79.0
Ciudad Juárez	48,881	122,726	151.1	261,683	113.2
Torreón	75,796	128,971	70.2	179,955	39.5
San Luis Potosí	77,161	125,612	62.9	159,640	27.0
Chihuahua	56,805	86,961	53.1	149,437	71.8
Saltillo	49,430	69,842	41.3	99,101	41.9
Durango	33,412	59,407	77.8	97,520	64.2
Nuevo Laredo	28,872	57,669	99.7	93,787	62.6
Matamoros	7,961	45,737	474.5	93,334	104.1
Gómez Palacio	25,558	45,842	79.4	60,765	32.5
Ciudad Madero	28,075	41,047	46.2	53,526	30.4
Ciudad Victoria	(A)	31,815	?	50,727	59.4
Central:					
Mexico City	1,448,422	2,334,795	61.2	2,832,133	21.3
Guadalajara	229,235	337,016	64.5	734,346	94.8
Puebla	138,491	211,331	52.6	287,952	36.3
León	74,155	122,566	65.3	209,469	70.9
Aguascalientes	82,234	93,363	13.5	126,229	35.2
Gustavo A. Madero	25,929	60,239	132.3	102,602	70.3
Morelia	44,304	63,245	42.7	100,258	58.5
Irapuato	32,377	49,445	52.7	83,505	68.9
Toluca	43,429	52,983	22.0	76,871	45.1
Querétaro	33,629	49,160	46.2	67,277	36.8
Pachuca	53,354	58,658	9.9	64,564	10.1
Azcapotzalco	(A)	49,617	?	63,303	27.6
Coyoacán	(A)	46,031	?	54,866	19.2

CITIES OF MEXICO OVER 50,000 POPULATION, 1960 (Cont.)

City	Population 1940	Population 1950	Percent Gain 1940-50	Population 1960	percent Gain 1950-60
Gulf Coast:					
Mérida	98,852	142,858	44.5	170,513	19.3
Veracruz	71,720	101,220	41.1	144,232	42.5
Tampico	82,475	94,342	14.4	122,197	29.5
Orizaba	47,910	55,531	15.9	69,672	25.5
Jalapa	39,530	51,169	29.4	66,317	29.6
Villahermosa	25,114	33,578	33.7	51,611	53.7
South Pacific:					
Oaxaca	29,309	46,632	59.1	72,313	55.1

(A) Less than 20,000.

Sources: Table 85 in Durán Ochoa, Julio, *Población* (Fonda de Cultura Económica: Mexico, D. F., 1955) , 210, which used census data; and preliminary figures from 8°. *Censo General de Población, 1960,* released March 15, 1962 and provided by Banco Nacional de Comercio Exterior, S. A., Mexico, D. F.

natural increase alone, while the numerical increase was less than half that of the previous decade, is at first glance surprising. The figures given are for the city proper, however, a city which had nearly filled up before 1950; hence much of the subsequent growth was outside the corporate boundaries in suburbs and satellite centers. Statistics for urban growth in the Federal District more accurately reflect the conditions for metropolitan Mexico City. In 1950 the Federal District had an urban population of 2,884,133, of which 2,334,795 were in Mexico City proper; in 1960, 4,666,028, with 2,832,133 inside the city boundaries. Thus, Mexico City in the larger sense can be said to have grown nearly 1,750,000 persons during the 1950's. This was a relative gain of almost 60%, not large enough to place the capital among the ten centers with the highest rates of increase, but certainly one worthy of the nation's "primate city." Mexico City proper had nearly four times, and the metropolitan area over six times, the population of the next largest place.

In most Mexican cities, however, a rate of increase less than the national average is related to movement to more distant places. Pachuca, with a 1950-60 relative increase of 10.1%, and Mérida with one of 19.3%, are good examples of larger cities which experienced reduction of gain due to net migration. Both are in states which had below-average coefficients of rural and urban growth during the period.

It is noteworthy that the South Pacific had only one city of more than 50,000 inhabitants in 1960. Also noteworthy, the North Pacific, and to a lesser extent the North, had a disproportionally large number of centers of this class compared to their total populations (Tables II and IV). The Central Region had fewer cities of over 50,000 than did the North if suburban parts of Mexico City are excluded, but it did have the nation's primate city and its second, fourth, and sixth largest ones. A greater share of its population was urban than that of any other division.

Reasons for Differences in Population Growth

In large part these differences can be attributed to variations in economic opportunity, the increased awareness among the people of these distinctions, and their growing willingness and ability to travel to take advantage of them. Before the 1910 Revolution the great majority of Mexicans were immobile, bound to the soil by the "hacienda" system. Today, because of a complex of factors including land reform, more education, the radio, automobiles, improved highways and railroads, expanded cash-crop agriculture, industrialization, attraction of urban life, and the push from over-populated rural areas, the people are surprisingly mobile.[9]

That this is true is evidenced not only by the recent movement to the U. S. and return of over 400,000 *braceros* annually, but by published studies. Whetten and Burnight, running a correlation analysis of measures of economic change and measures of migration, concluded that because of migration, "the Mexican population in 1950 was probably better distributed in terms of economic opportunities than it has been a decade earlier."[10] Laura Randall, studying minimum rural and urban wages in individual states, found it fairly easy to show that there was an economic motive to migration. Although there was rapid in-migration in the North Pacific area, for example, investment there was so great that relative urban

[9] For additional discussion see articles by Flores and Maddox, previously cited. See also Stanley R. Ross, "Mexico: Golden Anniversary of the Revolution," *Current History*, 38 (March 1960), 150-154; and David H. Shelton, "Mexico's Economic Growth: Success of Diversified Development," *Southwestern Social Science Quarterly*, 41 (December 1960), 304-319.

[10] Whetten and Burnight, *op. cit.*, 77.

and agricultural wages both rose, continuing the attraction of the area to migrants. In the South Pacific, on the other hand, investment and out-migration were insufficient to keep both relative rural and urban wages from falling, thus increasing motives for leaving the region.[11]

Some of the more obvious reasons for growth differentials in Mexico include the relationship between the rapid increase in the northern border towns and their rising importance as ports of entry and destinations for vacationers from the U. S. Three-quarters of a million tourists entered Mexico in 1959 and spent nearly $637 million.[12] The border towns too, particularly Mexicali and Matamoros, have been stimulated by the great increase of irrigation and cash crop agriculture in their vicinities. The northwestern states of Sonora and Sinaloa have also been leading beneficiaries of new irrigation facilities, with cities like Culiacán, Mazatlán, and Hermosillo experiencing vigorous growth as fields of wheat, cotton, rice, sugarcane, and winter vegetables replaced the desert and as large quantities of hydroelectric power become available.[13] Irrigation programs absorbed some 15% of Mexican public expenditures during the last decade.[14] Resulting expansion of cash-crop agriculture was such that its products came to make up some 60% of the nation's exports by value, with cotton the leading item, while domestic processing of these goods increased greatly.

The growing emphasis on development of industries producing iron and steel, oil, chemicals, cement, paper, and related products is reflected in the high numerical and relative growth of Monterrey, the nation's leading heavy industry city, of Guadalajara, a political, commercial, and manufacturing focus in a rich mining and agricultural region, and of Mexico City, which is not only the cultural, political, and economic heart of the Republic, but the center with the largest concentration of industries of all types. Thus, in part at least, is explained the growth of some cities, and indirectly the rapid population gain of states in which these centers are situated.

[11] Laura Randall, "Labour Migration and Mexican Economic Development," *Social and Economic Studies,* 11 (March 1962), 73-81.

[12] Banco Nacional de Comercio Exterior, S. A., *Mexico* (Mexico, D. F., 1960), 238.

[13] Willis B. Merriam, "Irrigation Progress in the Mexican Northwest, *Journal of Geography,* 56 (December 1957), 429-441.

[14] Banco Nacional de Comercio Exterior, S. A., *op. cit.,* 108-9, 216-217.

Extension of the remarkable expansion and balanced development of the Mexican economy of the past 20 years indefinitely into the future, if accomplished, will not affect equally all rural and urban areas in the years to come any more than it has up to now. Many of the growth differentials and migration trends which have developed since 1940 will continue, especially movement to the cities. Realization of fuller benefits per capita from additional economic development, however, would seem to depend most on a sizeable reduction of the rate of population increase. This is not now in prospect.

PAPERS OF THE MICHIGAN ACADEMY OF SCIENCE, ARTS, AND LETTERS
VOL. XLIX, 1964 (1963 MEETING)

RECENT CHANGES IN THE DISTRIBUTION OF POPULATION IN ARGENTINA

ROSS N. PEARSON
The University of Michigan

THE results of the Fifth National Census of Argentina,[1] published in 1961, provide the first opportunity in 13 years to study the changing patterns of population distribution in that country. The period from 1947 to 1960 was marked by political unrest, social change, industrialization, expanded irrigation agriculture, greater mineral production, and some increase in immigration. Most of these conditions contributed to significant changes in the distribution of population within Argentina.

The arrangement of people on the land during the 1940's (Fig. 1) is familiar to most students of geography. The major concentration of population centered upon the city of Buenos Aires and the shores of the Rio de la Plata. From the capital areas of high density extended north and west along the Paraná and across the plains to the cities of Santa Fé and Córdoba. Other areas with densities in excess of 50 persons per square mile focused upon the oases along the Andean Front (Fig. 2) and upon the city of Posadas in the northern Mesopotamian region. Lands with moderate densities (10 to 25 persons per square mile) extended south and west across the Pampa from Buenos Aires, covered the southern half of Mesopotamia, and could be found in the eastern part of the Chaco. The western mountains and foothills, most of the Chaco, and all of Patagonia were sparsely inhabited.

The map showing population density in Argentina for 1960 (Fig. 3) displays the same general features as the one for 13 years earlier. Areas with 100 persons or more per square mile are somewhat larger; sections with moderate densities have decreased in extent in the western Pampa and southern Chaco; and several departments[2] in the south show a noticeable increase. The altered patterns in the provinces of Chaco and Misiones are in part the result of

[1] *Censo Nacional de 1960, Población, Resultados Provisionales,* Dirección Nacional de Estadistica y Censos, Republica Argentina (Buenos Aires, 1961).

[2] The major subdivisions of most Argentine provinces are called *departmentos,* in a few they are termed *partidos.*

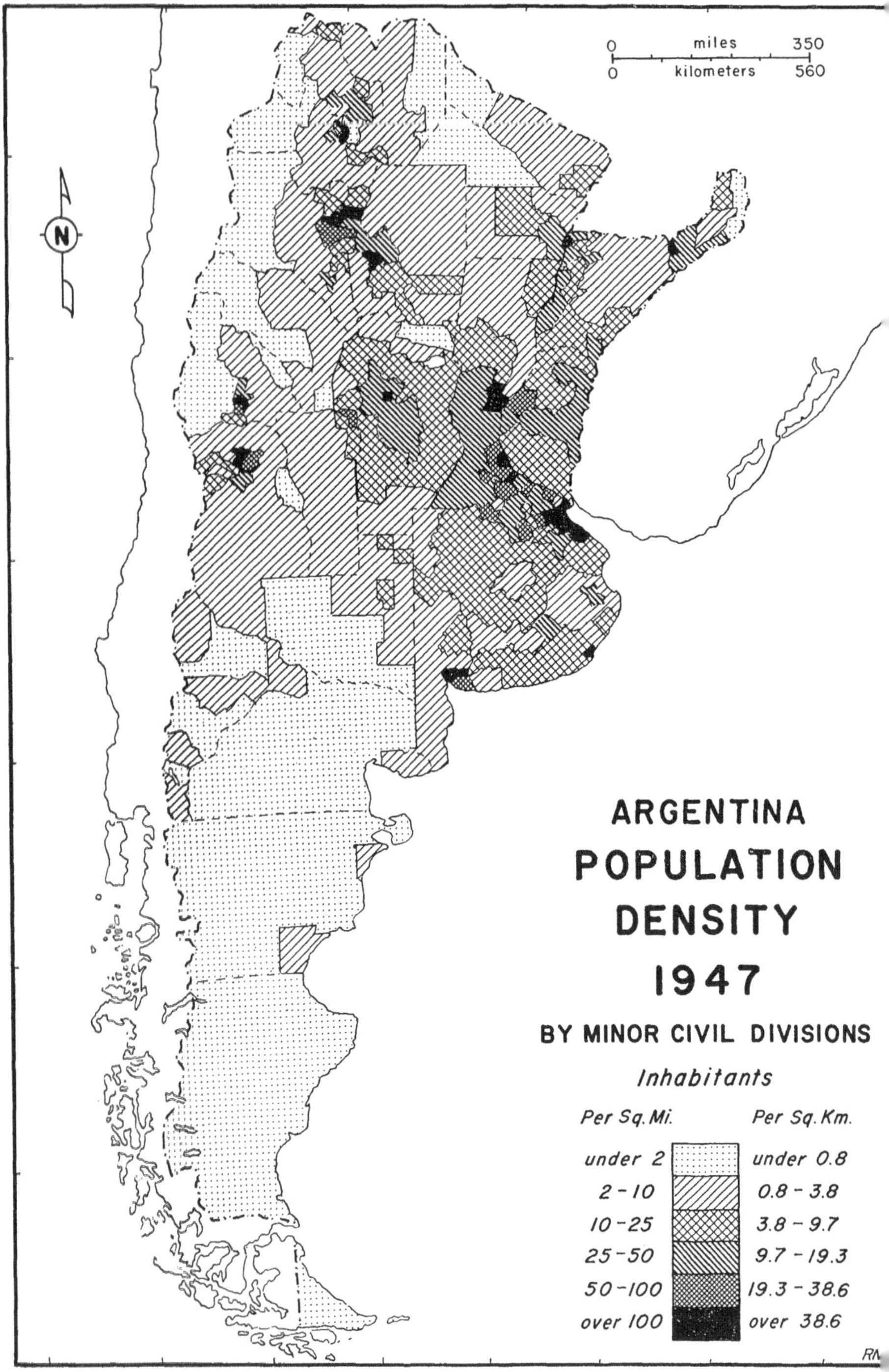
0 miles 350
0 kilometers 560
N
ARGENTINA
POPULATION
DENSITY
1947
BY MINOR CIVIL DIVISIONS
Inhabitants
Per Sq. Mi.
Per Sq. Km.
under 2
under 0.8
2 - 10
0.8 - 3.8
10 - 25
3.8 - 9.7
25 - 50
9.7 - 19.3
50 - 100
19.3 - 38.6
over 100
over 38.6

FIG. 1.

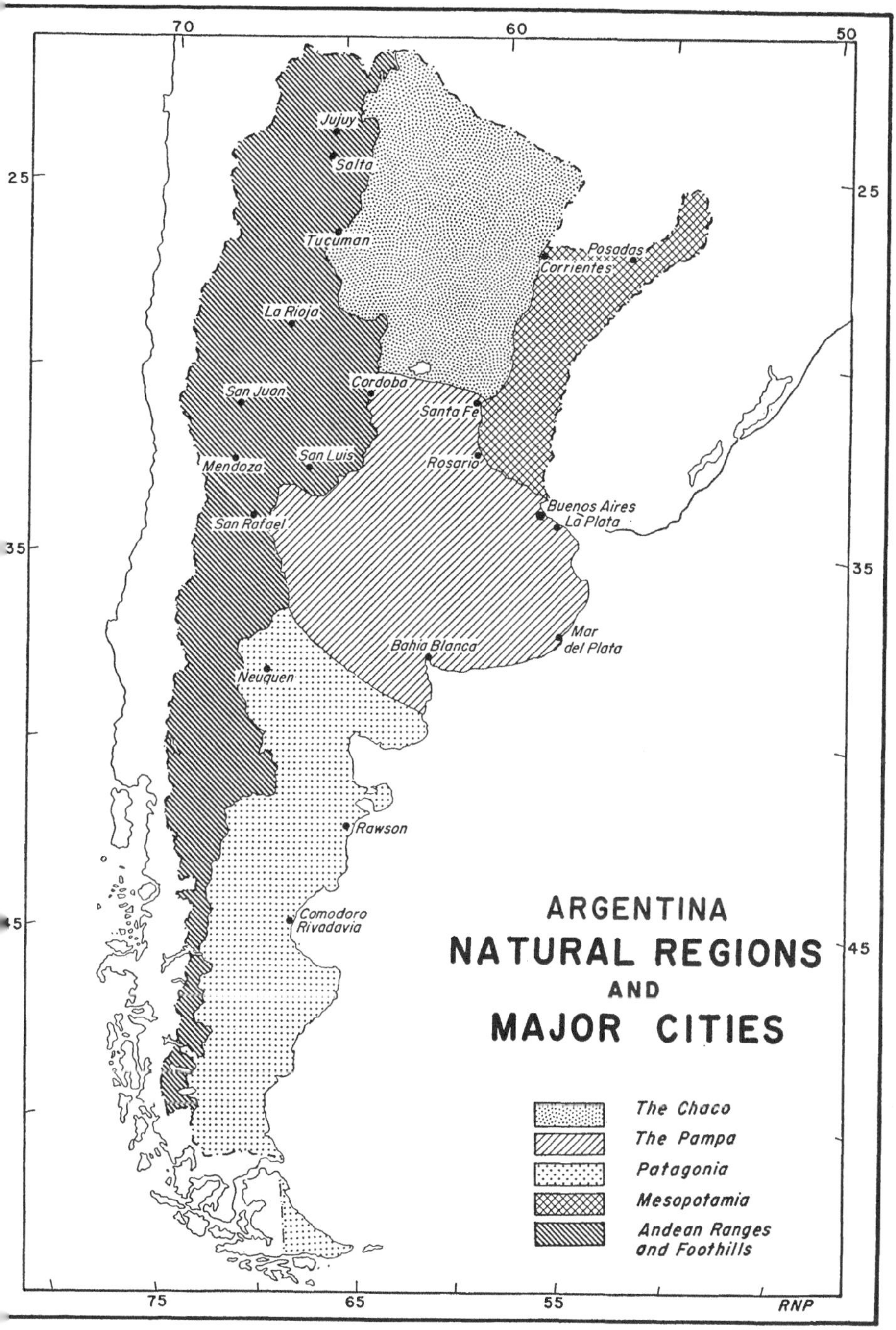

ARGENTINA
NATURAL REGIONS
AND
MAJOR CITIES
Jujuy
Salta
Tucuman
La Rioja
Cordoba
San Juan
Santa Fe
Rosario
San Luis
Mendoza
San Rafael
Posadas
Corrientes
Buenos Aires
La Plata
Mar del Plata
Bahia Blanca
Neuquen
Rawson
Comodoro Rivadavia
The Chaco
The Pampa
Patagonia
Mesopotamia
Andean Ranges and Foothills
70
60
50
25
35
45
75
65
55
RNP

Fig. 2.

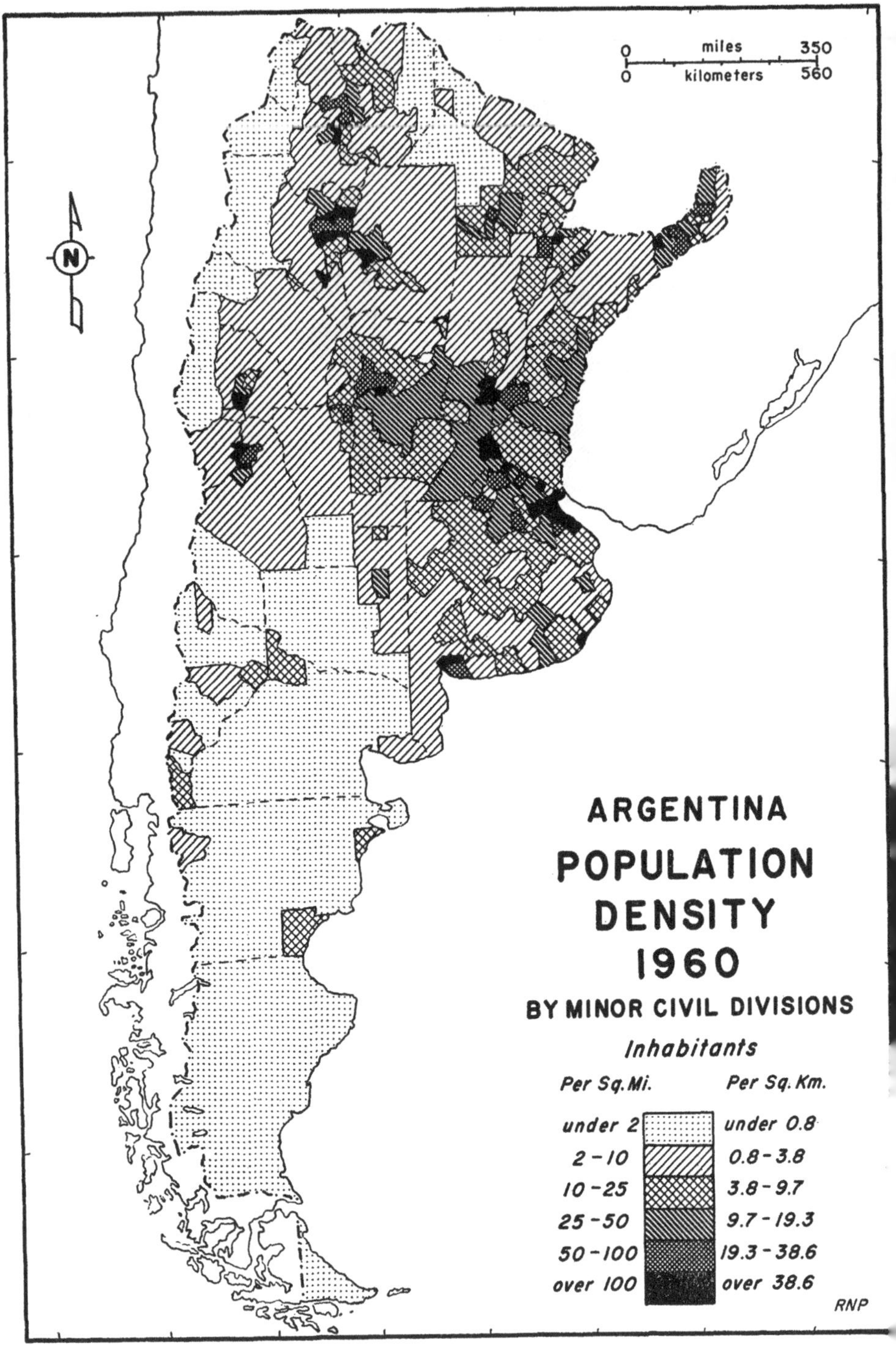
0 miles 350
0 kilometers 560
N
ARGENTINA
POPULATION
DENSITY
1960
BY MINOR CIVIL DIVISIONS
Inhabitants
Per Sq. Mi.
Per Sq. Km.
under 2
under 0.8
2 - 10
0.8 - 3.8
10 - 25
3.8 - 9.7
25 - 50
9.7 - 19.3
50 - 100
19.3 - 38.6
over 100
over 38.6
RNP

FIG. 3.

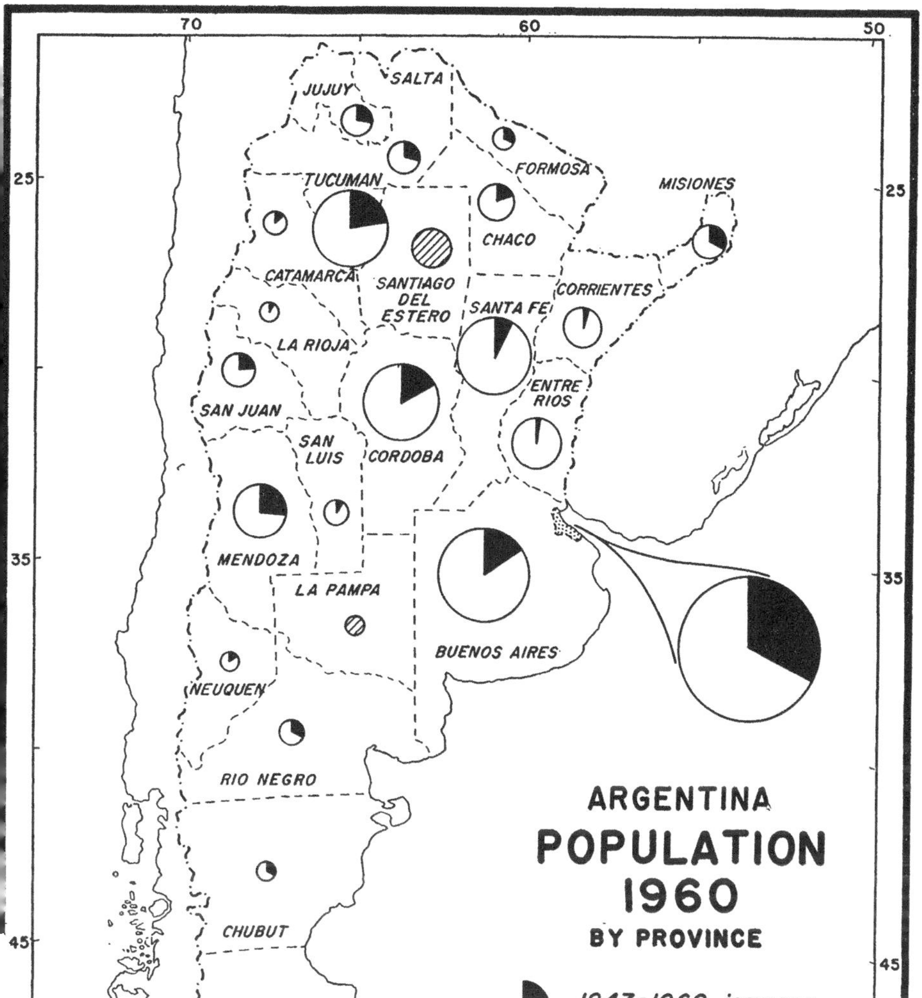
ARGENTINA
POPULATION
1960
BY PROVINCE
1947-1960 increase
Provinces with 1947-1960 decrease
Greater Buenos Aires
6,250,000
2,500,000
810,000
640,000
160,000
50,000
JUJUY
SALTA
FORMOSA
MISIONES
TUCUMAN
CHACO
CATAMARCA
SANTIAGO DEL ESTERO
SANTA FE
CORRIENTES
LA RIOJA
SAN JUAN
ENTRE RIOS
SAN LUIS
CORDOBA
MENDOZA
LA PAMPA
BUENOS AIRES
NEUQUEN
RIO NEGRO
CHUBUT
SANTA CRUZ
TIERRA DEL FUEGO

FIG. 4.

the reorganization of administrative boundaries. It is apparent that maps of this type and degree of detail do not show all the changes which have been inferred.

From 1947 to 1960 the population of Argentina increased from 15,897,127 to 20,008,945. This represents a growth of 25.9% and an annual increase of 17.6 per 1000 inhabitants. 11 of the 24 provinces and territories (Fig. 4) exceeded the national average and only 3—La Pampa, Santiago del Estero, and the Federal Capital—showed a decline for the intercensus period. If the area known as Greater Buenos Aires is included with the capital district the combined increase for the metropolitan region was 32%. This urbanized zone added 1,727,111 inhabitants and the other provinces containing Argentina's major cities also experienced large gains: Buenos Aires province (excluding the Federal Capital and Greater Buenos Aires) —406,199, Córdoba—262,010, Mendoza—237,304, Tucumán—186,977, and Santa Fé—162,563. The additions which made up the largest percentages of the 1960 totals tended to be in the outlying provinces of Misiones (37%), Formosa (36%), Chubut (35%), Río Negro (30%), Jujuy (30%), Salta (29%), and Mendoza (28%).

Within major political subdivisions there were great variations in percentages of population gain or loss (Fig. 5). Among the most conspicuous of the areas showing population decline are departments in the two provinces (La Pampa and Santiago del Estero), which earlier were shown to have had a 13-year loss. Both are located in regions of low and uncertain rainfall. The area of population loss in La Pampa, however, extends into southwestern Santa Fé and from there on southeasterly across Buenos Aires province. Much of the Patagonian region declined in population as did some of the Andean departments and the interior part of south and central Mesopotamia.

Regions with high percentages of population gain (Fig. 6) tend to be somewhat more scattered than those showing decline. This is particularly related to the distribution of cities, most of which increased by 35% or more. Widespread areas showing population increase are found in the far north, in the valleys of the Rio Colorado and the Rio Negro, along the foothills of the central and northern parts of the Andes, and in the far south. Most of the latter are in the parts of Argentina which have traditionally been sparsely inhabited.

Maps showing percentage increase or decrease in total population

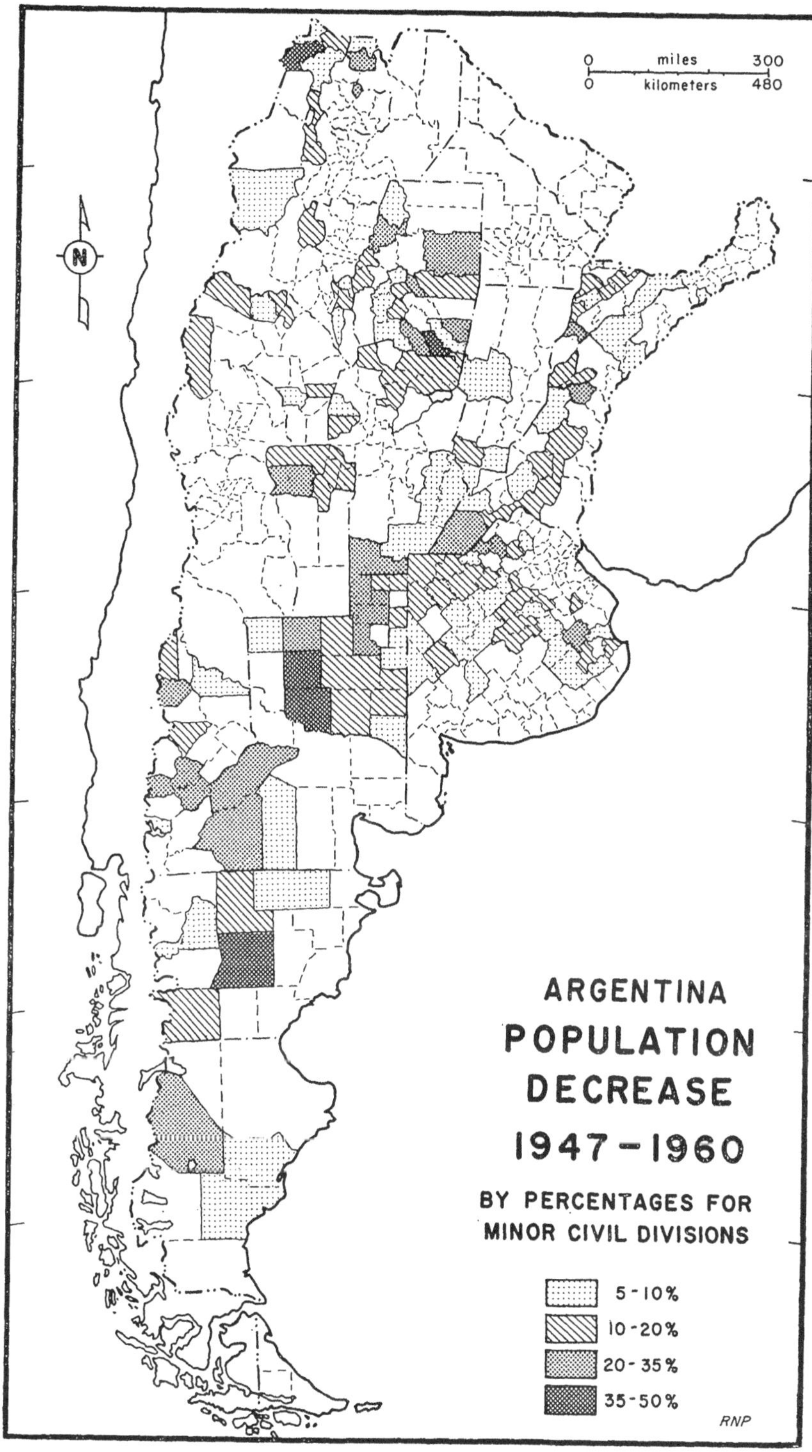
0 miles 300
0 kilometers 480
N
ARGENTINA
POPULATION
DECREASE
1947-1960
BY PERCENTAGES FOR
MINOR CIVIL DIVISIONS
5-10%
10-20%
20-35%
35-50%
RNP

FIG. 5.

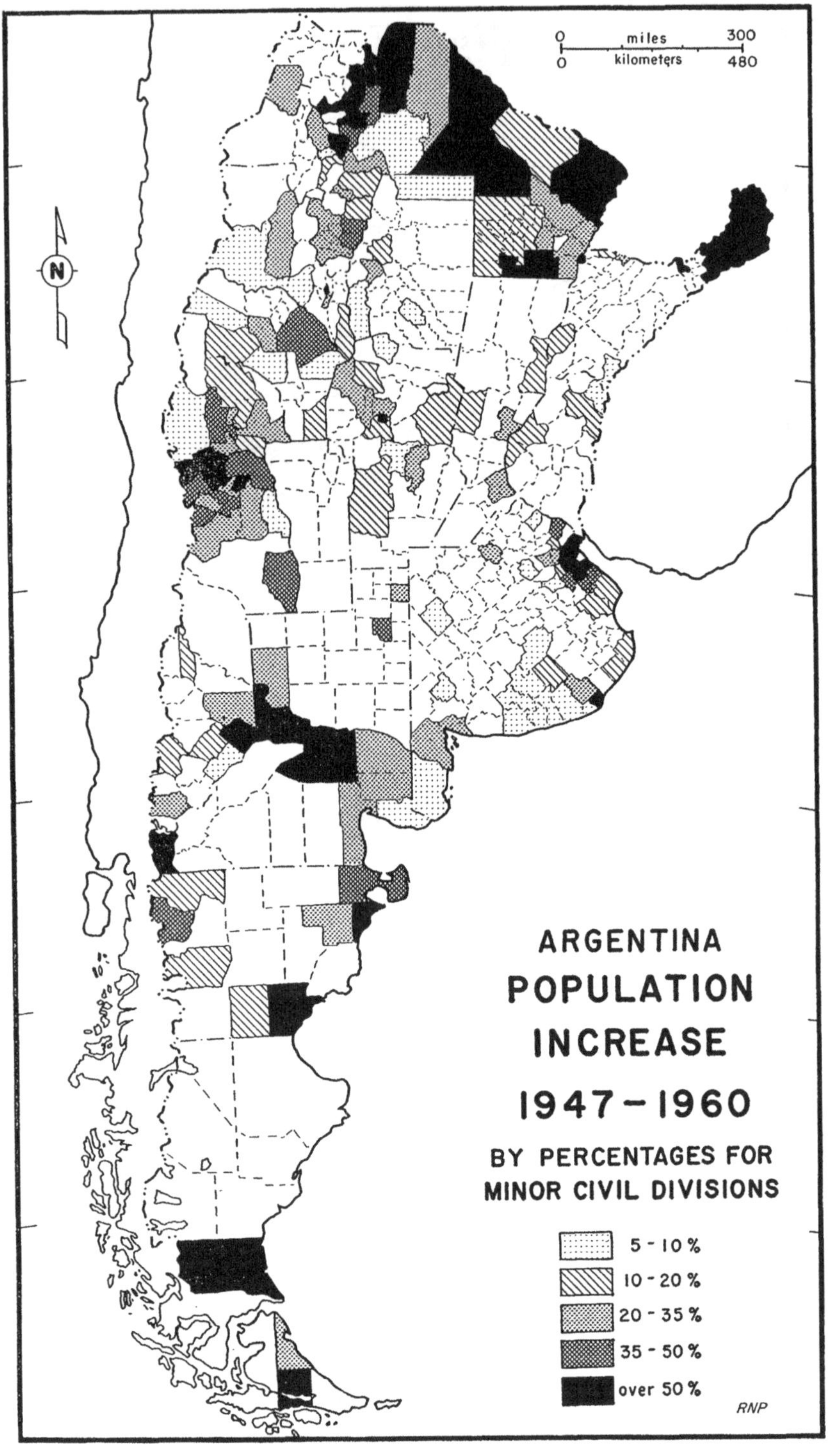
0 miles 300
0 kilometers 480
N
ARGENTINA
POPULATION
INCREASE
1947 - 1960
BY PERCENTAGES FOR
MINOR CIVIL DIVISIONS
5 - 10 %
10 - 20 %
20 - 35 %
35 - 50 %
over 50 %
RNP

FIG. 6.

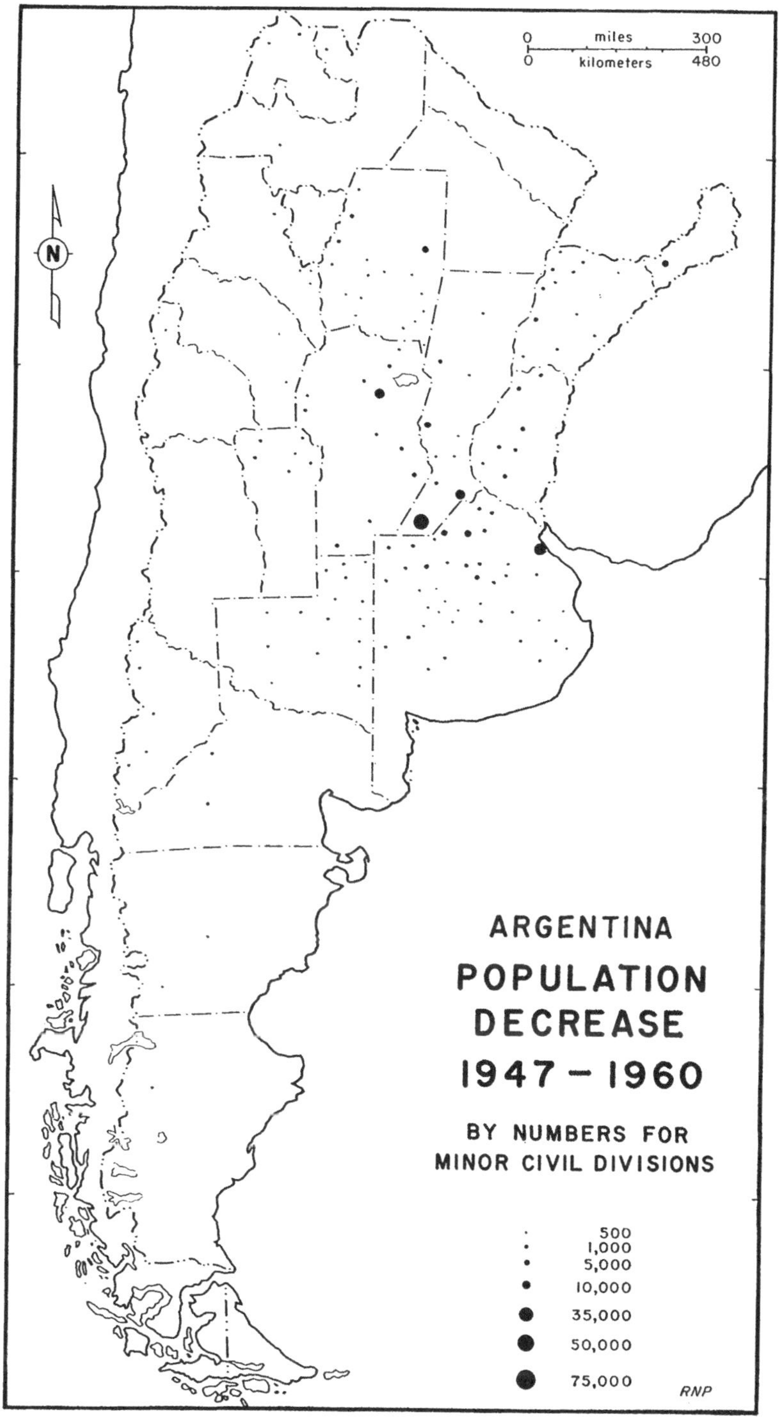
0 miles 300
0 kilometers 480
N
ARGENTINA
POPULATION DECREASE 1947 - 1960
BY NUMBERS FOR MINOR CIVIL DIVISIONS
500
1,000
5,000
10,000
35,000
50,000
75,000
RNP

FIG. 7.

by minor civil divisions, although useful in illustrating trends, do not fully indicate the nature of the changes that have taken place. The losses (Fig. 7) in Patagonia, the dry margins of the Pampa, and the southern Chaco are in the nature of 1 or 2 thousand inhabitants per department and are of small significance considering other changes which have occurred. Of far greater importance in terms of numbers are the areas of decline lying to the west of the cities of Buenos Aires, Rosario, and Santa Fé. Over 80,000 persons, in 9 minor civil divisions, presumably have migrated to other parts of the country. The total loss of departments in Mesopotamian provinces amounted to approximately 70,000 people. The Federal Capital (the city of Buenos Aires proper) showed a decline of 14,277 for the period between the two enumerations. This rather surprising, although small, decrease may be related to the fact that the census was taken on a weekend when many of the residents of the city were taking a holiday at Mar del Plata or some other resort.[3] Whatever the case may be, this loss was more than compensated for by the huge gains in other parts of the Buenos Aires metropolitan district.

One of the most striking features of population trends in Argentina has been the increasing urbanization (Fig. 8). Metropolitan Buenos Aires (the city and surrounding area) added almost 2 million to a population which 13 years earlier had numbered in excess of 4.7 million. If the city of La Plata is included, this one urbanized zone, which has less than 0.2% (0.16) of the area of the country, now contains 7,092,939 people or over 37% of the national total. Although the growth of the other urban areas is not so pronounced as that of Buenos Aires, most of them reflect the same situation. The population of the department made up largely of the city of Córdoba increased by 202,325 inhabitants, Rosario by 142,051, Mar del Plata by 123,811, Mendoza by 85,856, and Tucumán by 83,449. The gain of Mar del Plata may actually be somewhat smaller than indicated by the statistics because of the time at which the census was taken and the nature of the city. However, by the simple process of adding the population of the urban areas mentioned and 6 others[4] together it is certain that

[3] Private conversation with Elena Margarita Chiozza, Instituto de Geografia, Facultad de Filosofia y Letras, Universidad Nacional de Buenos Aires (August 1961).

[4] Bahia Blanca, Corrientes, Salta, Santa Fe, San Juan, and San Rafael.

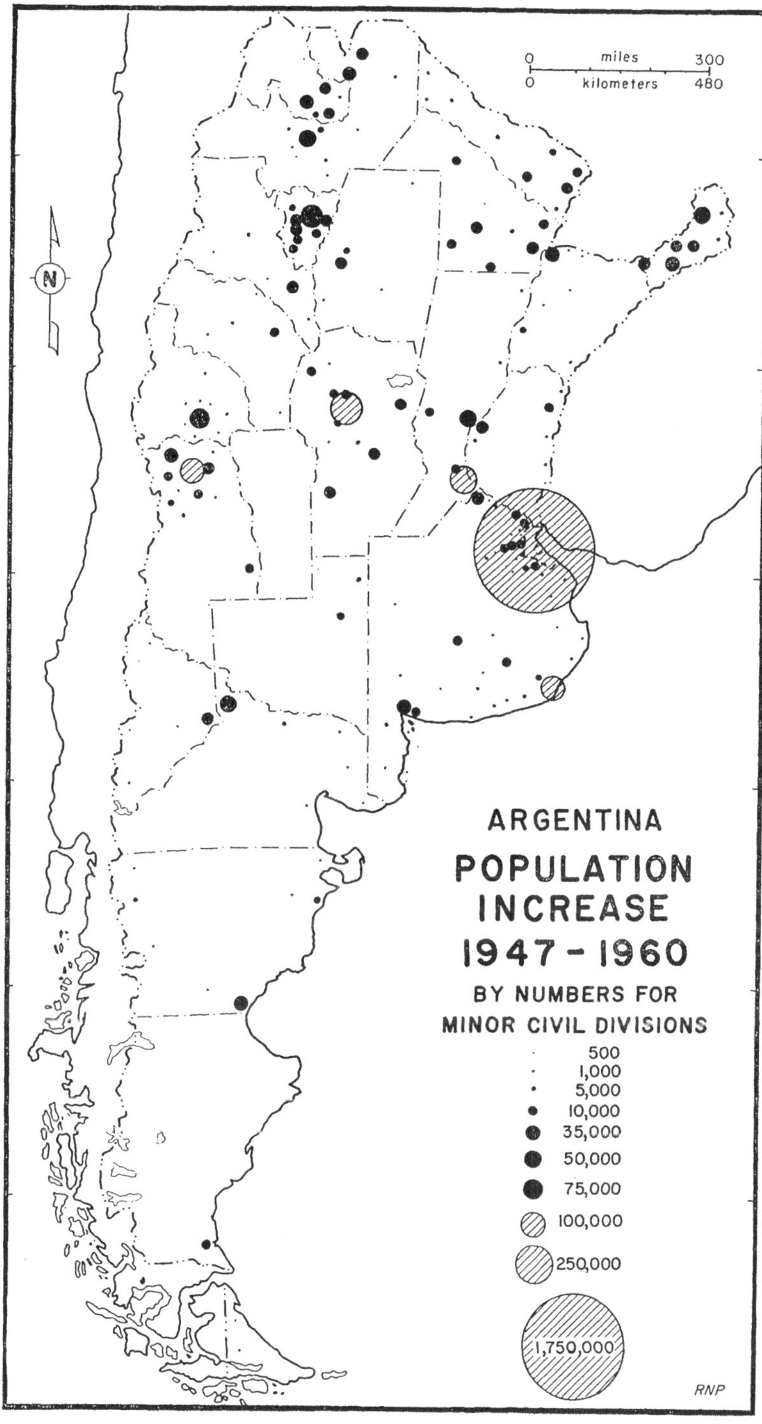
0 miles 300
0 kilometers 480
N
ARGENTINA
POPULATION INCREASE 1947-1960
BY NUMBERS FOR MINOR CIVIL DIVISIONS
500
1,000
5,000
10,000
35,000
50,000
75,000
100,000
250,000
1,750,000
RNP

FIG. 8.

well over half of the people in Argentina live in cities of 100,000 or more inhabitants.

The other regions with significant population growth tend to be located in the outlying portions of the country. Two provinces in the northern Chaco increased by a total of 169,556 persons; Misiones, in the extreme northeast, added 144,698; and the Andean Front zone of Salta and Jujuy increased 190,376. In the south significant gains occurred in the vicinity of the Neuquen oases, where the population is 70,232 more than in 1947, and in the areas around Rawson and Comodoro Rivadavia, where the increases were 7,499 and 34,356 respectively.

The shortage of statistical information regarding birth and death rates, internal migration, occupational structure, and economic activity makes it difficult to understand or to account for the changes that have taken place. It has been frequently stated that the social and economic programs of the Peron administration set off a widespread movement of people from rural to urban areas. The common assumption is that the wave of industrial development, the programs designed to benefit urban workers, and the handicaps imposed upon meat and grain producers caused large-scale cityward migration. This supposedly was accompanied by a nationwide decline in the production of corn, wheat, and meat. The data which are available do not, however, support these suppositions.

If one considers the numbers involved among the areas which lost or gained population (Figs. 7 and 8) it is hard to discover how the relatively small decreases in certain areas could result in such large increases in others. The birthrate in Argentina stands rather low among the South American nations (22.6 per 1,000 in 1959),[5] and there is no reason to assume that it would be higher in the cities than in the rural areas. It is certain that a part of the urban growth is the result of rural to city migration, but this cannot be the only factor involved. One can deduce that many nonurban regions have a birthrate that is high enough to maintain existing population levels, or to show some increase and at the same time provide a surplus which moves to the cities.

It is apparent that the rural-urban trend has been overly emphasized when one examines the agricultural statistics for the period under consideration. A recent edition of the *United Nations Statis-*

[5] *Demographic Yearbook 1960.* United Nations (New York, 1961).

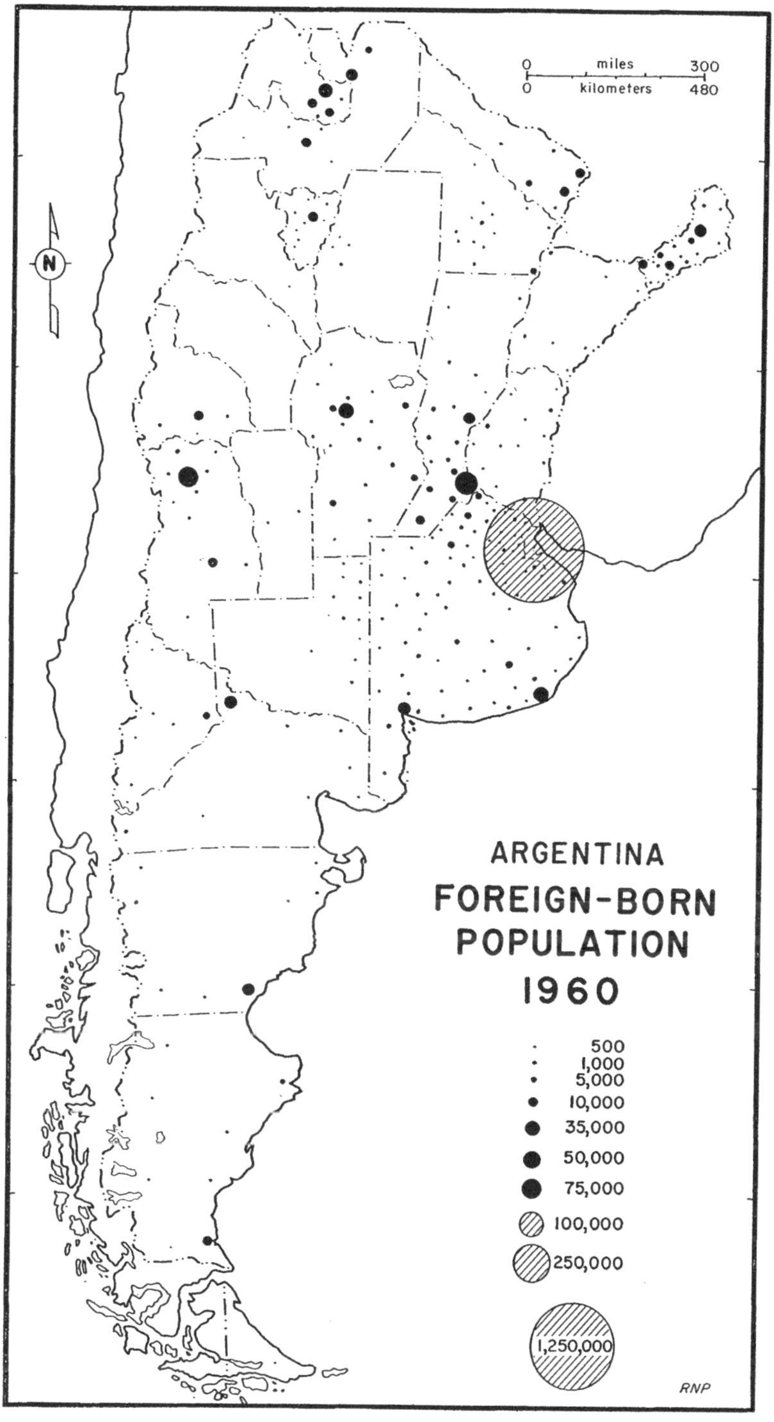
0 miles 300
0 kilometers 480
N
ARGENTINA
FOREIGN-BORN
POPULATION
1960
500
1,000
5,000
10,000
35,000
50,000
75,000
100,000
250,000
1,250,000
RNP

FIG. 9.

tical Yearbook[6] indicates that wheat and corn production actually expanded and that there were only small decreases in the number of livestock. Argentina's wheat crop averaged 5.2 million metric tons a year from 1948 to 1952, reached over 7 million in 1954 and 1956, and stood at 5.8 million tons in 1959. Corn production from 1948 to 1952 averaged 2.5 million tons and was 4.1 million in 1959, but in three of the intervening years exceeded 4.5 million. The situation with respect to major grains appears to be toward greater efficiency in production and a concentration of such crops on the lands that are best suited for them, rather than widespread abandonment and migration. While these changes were taking place the number of cattle in the country declined from 45 to 40 million and the number of sheep from 51 to 48 million.

A more important source of people, than internal migration, for the parts of Argentina that have shown population increase has been immigration from other countries. Informal movements across Argentina's northern and western borders are well known and the government statistical office recorded some 998,000 incoming migrants during the years from 1947 to 1958.[7] The map (Fig. 9) showing the distribution of the nation's foreign-born population displays a strong resemblance to the one for numerical population increase. The provinces which had the greatest percentages of population growth (Fig. 4) also have the highest percentages of residents of alien birth: Misiones 24.1, Formosa 22.4, Rió Negro 19.5, Chubut 19.1, and Jujuy 18.4. The foreign-born population of metropolitan Buenos Aires made up 23% of the total in 1960. We should note, of course, that all of the people of foreign birth in Argentina did not immigrate to the country between 1947 and 1960. However, the peak year of immigration, 1949, had over 150 times the official number for 1945 and the other years in the period had from 35 to 120 times the 1945 figure. We hasten to add that these observations are not intended as the final explanation for the origins of the new residents of Argentina's rapidly growing cities and pioneer zones. What we are saying is that foreigners constitute a more important element in the areas of expanding population than has otherwise been noted.

[6] *Statistical Yearbook 1960*. United Nations (New York, 1961).

[7] *Statesman's Yearbook, 1954-1957 Editions*. The number of emigrants for the same period was 425,200. The excess of immigrants over emigrants was 573,500.

The objective of migration is usually greater economic opportunity. In Argentina this is provided by the expansion of job possibilities in the manufacturing plants and commercial establishments of urban centers; in the new agricultural lands of northern Mesopotamia, the eastern Chaco, and northern Patagonia; and in the expanding mineral industries of southern Patagonia and the Andean region.

PAPERS OF THE MICHIGAN ACADEMY OF SCIENCE, ARTS, AND LETTERS
VOL. XLIX, 1964 (1963 MEETING)

C. GEOGRAPHY OF MICHIGAN

THE LAKE DWELLERS OF DELTON, MICHIGAN *

DUNCAN M. ANDERSON
Michigan State University

ONE of the social phenomena of the 20th century, and particularly of the postwar years, has been the areal spread of urban influence far beyond the actual political confines of urban municipalities. This trend has been manifest in urban sprawl, in dormitory and satellite communities, and by increasing pressures on outdoor recreational resources, particularly those areas close to large urban centers. The impact of urban population pressures upon outdoor recreational space has also exerted many influences on the economic life of smaller municipalities within the recreational areas. The purpose of this paper is to explore one type of urban influence that has affected the character of a rural Michigan village.

The small village of Delton is situated in Barry County, Michigan, in the southwest part of the state, as indicated in Fig. 1. This is one of the moraine areas of Southern Michigan, in which many kettle and depressional lakes have been utilized for recreational purposes. The village is within 20 miles of Battle Creek and Kalamazoo and is thus within their spheres of influence. It has a population, according to the Michigan State Highway Department, of 250 persons in 1962.[1] Delton is on State Highway M-43, but it is not on a major traffic artery connecting either Battle Creek or Kalamazoo with other equivalent urban centers. The major traffic movement between Kalamazoo and Grand Rapids passes to the west, and while the community is located on one of the two routes connecting Battle Creek and Grand Rapids, it is on the longer and less direct route. The apparent traffic-generating factors of its location, aside from its function as a rural village, are the general recreational activity within the region, the existence of the Gun Lake recreational area seven miles to the north, and the fact that

* This paper is an outgrowth of field study during the Michigan State University field course in geography conducted in the summer of 1962.

[1] Michigan 1962 Official Highway Map (1962).

Highway 43 is the most direct route between Kalamazoo and Hastings, the county seat of Barry County.

During field study in this area during the summer of 1962, it was noted that this rural village of 250 persons possessed a much

TABLE I

COMPARISON OF DELTON AND BARRY COUNTY THRESHOLD POPULATIONS [2]

	No. of Establishments in Delton	Delton Threshold Population	Barry County Threshold Population
Gasoline service stations	3	83	69
Food stores	3	83	71
Restaurants	3	83	93
Hardware stores	1	250	102
Beauty shops	3	83	110
Physicians	1	250	115
Auto repairs	3	83	120
Apparel shops	1	250	130
Appliances	3	83	131
Insurance agents	3	83	140
Car dealers	2	125	143
Elementary schools	1	250	155
Lumberyards	1	250	181
Health practitioners	2	125	202
Dry-cleaning establishments	1	250	207
Variety stores	1	250	212
Drugstores	1	250	232
Drinking establishments	2	125	296
Sporting goods	2	125	325
Fuel dealers	2	125	332
Department stores	–	–	350
Florists	1	250	362
Hay, grain, and feed stores	1	250	375
Laundry establishments	1	250	404
Real estate agents	3	83	404
Secondary schools	1	250	714
Funeral homes	1	250	771
Banks	1	250	814

[2] The threshold population for a specific retail function is the ratio of the number of retail establishments of the specific type to the total population of the urban area. It is expressed as the number of persons per individual establishment. Threshold values for Barry County are derived from an unpublished central place study conducted by Arthur Getis, assisted by students of the 1962 Michigan State University field course in geography.

higher proportion of retail establishments in relation to its population than would normally be found in a rural village of similar size in this agricultural region. The village contained, for example, one bank, three beauty parlors, three real estate offices, and three grocery stores. The full list of retail outlets is contained in Table I. Most of these establishments represented a rather high ratio for the particular retail function in relation to the village population. In a central place study in this area (see Table I) during the same period, it was established that 814 persons was the average threshold population in Barry County for a banking establishment, 404 for a real estate office, and 83 for a beauty parlor. Conspicuous among the retail establishments were three bait shops and a boat-and-sporting-goods store, indicative of nearby recreational activity. There are seven moraine lakes of over 200 acres each within the immediate vicinity of Delton. These lakes appeared to explain the recreationally oriented retail outlets but hardly seemed to offer adequate explanation for the high proportion of other retail outlets.

Productivity of the community's agricultural hinterland did not explain the seeming excess of retail services, for this part of Barry County is an area of predominantly moraine and irregular outwash topography. The village of Delton is located in the narrow angle of a wedge between two moraines, and thus the agricultural potential of the area is severely limited. In the publication *Types of Farming in Michigan,* the area is said to have "rolling to rough terrain with lakes, swamps and marshes in the basin-like associated areas. . . . The farms range from 1st class on the smoother areas to sub-marginal or abandoned on the steeply sloping sandier soils. Many hilly areas are unsuitable for farming and the best use is for forestry and recreation." [3] Thus, since agricultural productivity is low, the question of why the high level of retail outlets remained unanswered. Did shopping by Kalamazoo or Battle Creek residents, seasonal recreational business, or other factors account for this anomaly?

Interviews with merchants in Delton eliminated the question of shoppers from the nearby urban centers, but regarding recreational business, reports were conflicting. The owner of the local hardware store reported that in the three summer months the store

[3] Elton B. Hill and Russel G. Mawby, *Types of Farming in Michigan,* Special Bulletin 206, Michigan State College, Agricultural Experimental Station, Department of Agricultural Economics (September, 1954), p. 79.

accounted for over half its annual volume. The antique store owner reported that it was customary to close the store for the three winter months. However, other merchants did not concur with this. The local bank manager, who evidenced considerable knowledge of the local economy, considered that Delton's chief source of revenue was its function as service center for the surrounding area and that seasonal recreational business had a very small influence on the total annual retail volume of the community. Opinions expressed as to the number of cottages or the number of permanent homes on nearby lakes varied greatly. It was not possible to obtain from the community any specific information on the number of dwellings, either seasonal or permanent, on the nearby lakes.

Thus the following questions remained unanswered:

a) What were the number and location of lake dwellings close to Delton?

b) What proportion of these lake dwellings was summer cottages and thus purely seasonal in influence; and what proportion was permanent residences and thus an unseen part of the Delton hinterland?

The local post office represented one source of information as to numbers of seasonal dwellings, but this information was not readily accessible, nor would it enable precise location of the lake shore dwellings. Township assessment records would permit location of the lake dwellings but would not entirely answer the question of permanence of residence, nor would they provide information as to the retail purchase orientation of the lake dwellers. The most recent topographic maps (1946) indicated some cottage locations, but spot checks indicated that they were too out of date to be more than a rough guide to lake dwelling locations. Further, the period of time available for study was limited to three days.

Thus the method adopted for obtaining the information was that of conducting interviews with a sample of the lake shore occupants. A reconnaissance to the two known major cottage concentrations indicated that there were many more dwellings than could be covered individually; thus an arbitrary sample size of 10% was decided upon. The method of sampling was that of consistently selecting every 10th dwelling for interview. This method did not provide a statistically random sample, but was considered functional for an exploratory study of the nature planned. In cases where there was no occupant present, the procedure followed was to visit the adjacent residence in the direction of movement. This procedure was followed throughout, in order to retain consistency in this rudimentary attempt to approach randomness in sampling.

The following specific questions were asked:

a) Is this a seasonal (summer use) residence or a permanent year-round home?

b) If a seasonal residence, where is the family's permanent residence?

c) If a permanent residence, where was the family's previous home?

d) Where do you shop for the bulk of your day-to-day household requirements? (This was qualified to be clearly understood to be groceries and household supplies.)

As a delineation of the precise boundary of the Delton hinterland was not an objective of this explanatory study, the limit of the area investigated was initially taken as a five-mile radius from the village of Delton. This approximate distance was indicated by the location of the neighboring communities. As the interviewing progressed, the boundary zone in terms of shopping for household supplies was considered to be lake locations where more than half of the families interviewed stated that they did their basic shopping at locations other than Delton.

Observations

The 10% sample of the lake dweller population was then projected as an indicator of the approximate percentage of the total number of dwellings. The survey data and the percentages derived are shown in Tables II and III below.

Within the area in which the lake-shore households appeared to be oriented toward Delton for basic services, there was a total, by actual count, of 963 lake dwellings. Based upon the sample projections, slightly over half (56%) of the dwellings appeared to be permanent homes, and slightly less than half (44%) were seasonal dwellings, summer cottages in the true sense. Of the seasonal dwellings, the sampling projection indicated that less than half (43%) did their basic shopping in Delton, and approximately one-third (33%) of the permanent households shopped there.

TABLE II

LAKE SHORE HOUSEHOLD INTERVIEW SAMPLE DATA

	Shopping Orientation						Origin			
	Total	Delton	Kalamazoo	Battle Creek	Hastings	Other	Kalamazoo	Battle Creek	Hastings	Other
Seasonal Dwellings	42	18	8	5	8	3	17	14	5	6
Permanent Dwellings	54	18	19	6	8	3	25	12	8	9
Total	96	36	27	11	16	6	42	26	13	15

TABLE III

LAKE SHORE PROJECTION OF SAMPLE TO PERCENTAGES

	Permanent Lake Shore Dwellings					Seasonal Lake Shore Dwellings				
	Delton	Kalamazoo & Battle Creek	Hastings	Other	Total	Delton	Kalamazoo & Battle Creek	Hastings	Other	Total
Shopping Orientation	33%	46%	15%	6%	100%	43%	31%	20%	6%	100%
Origin	–	74	12	14	100%	–	69	15	16	100%

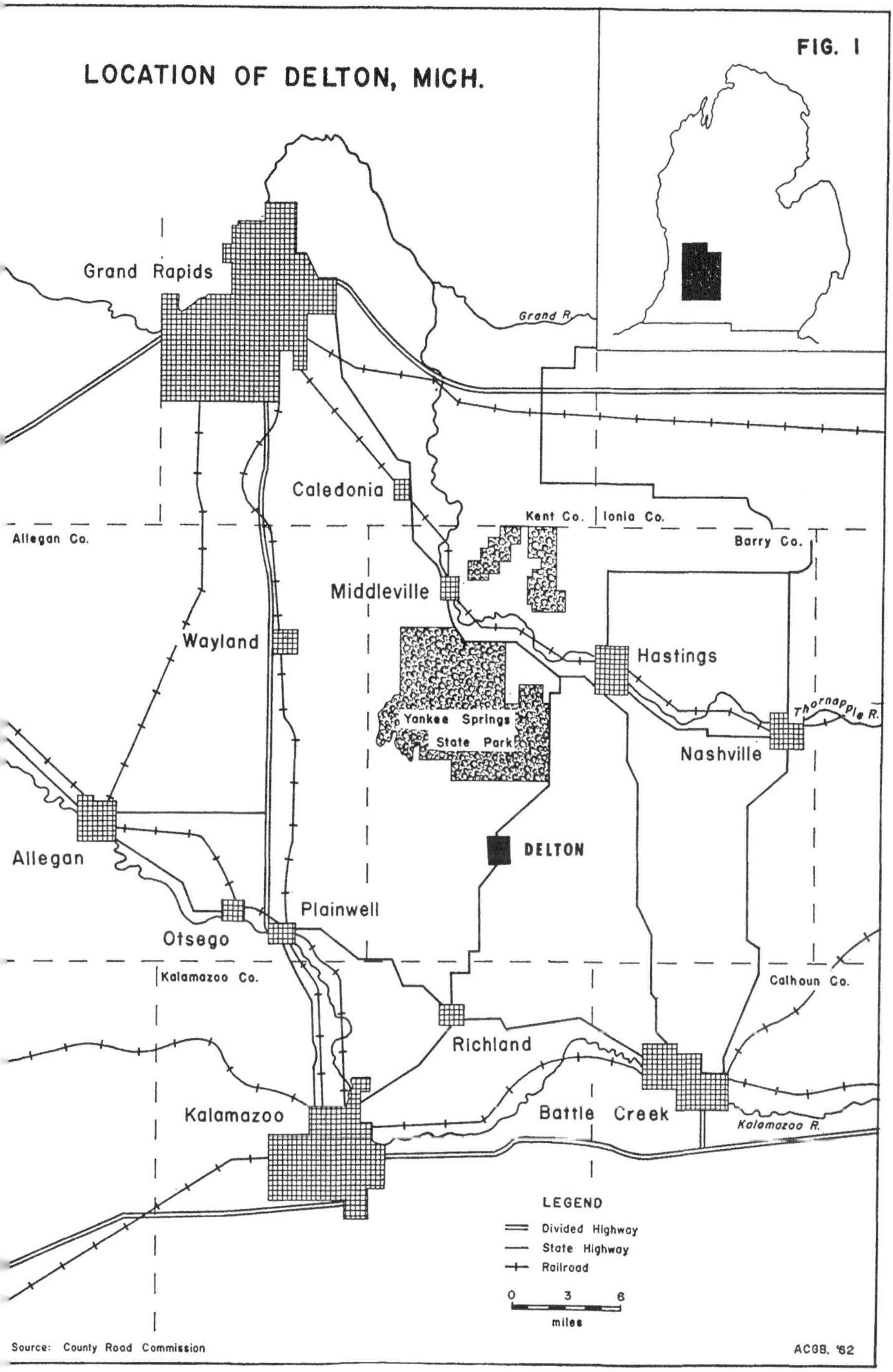
FIG. 1
LOCATION OF DELTON, MICH.
Grand Rapids
Grand R.
Caledonia
Kent Co.
Ionia Co.
Allegan Co.
Barry Co.
Middleville
Wayland
Hastings
Yankee Springs
State Park
Thornapple R.
Nashville
Allegan
DELTON
Plainwell
Otsego
Kalamazoo Co.
Calhoun Co.
Richland
Kalamazoo
Battle Creek
Kalamazoo R.
LEGEND
Divided Highway
State Highway
Railroad
0 3 6
miles
Source: County Road Commission
ACGB. '62

FIG. 1.

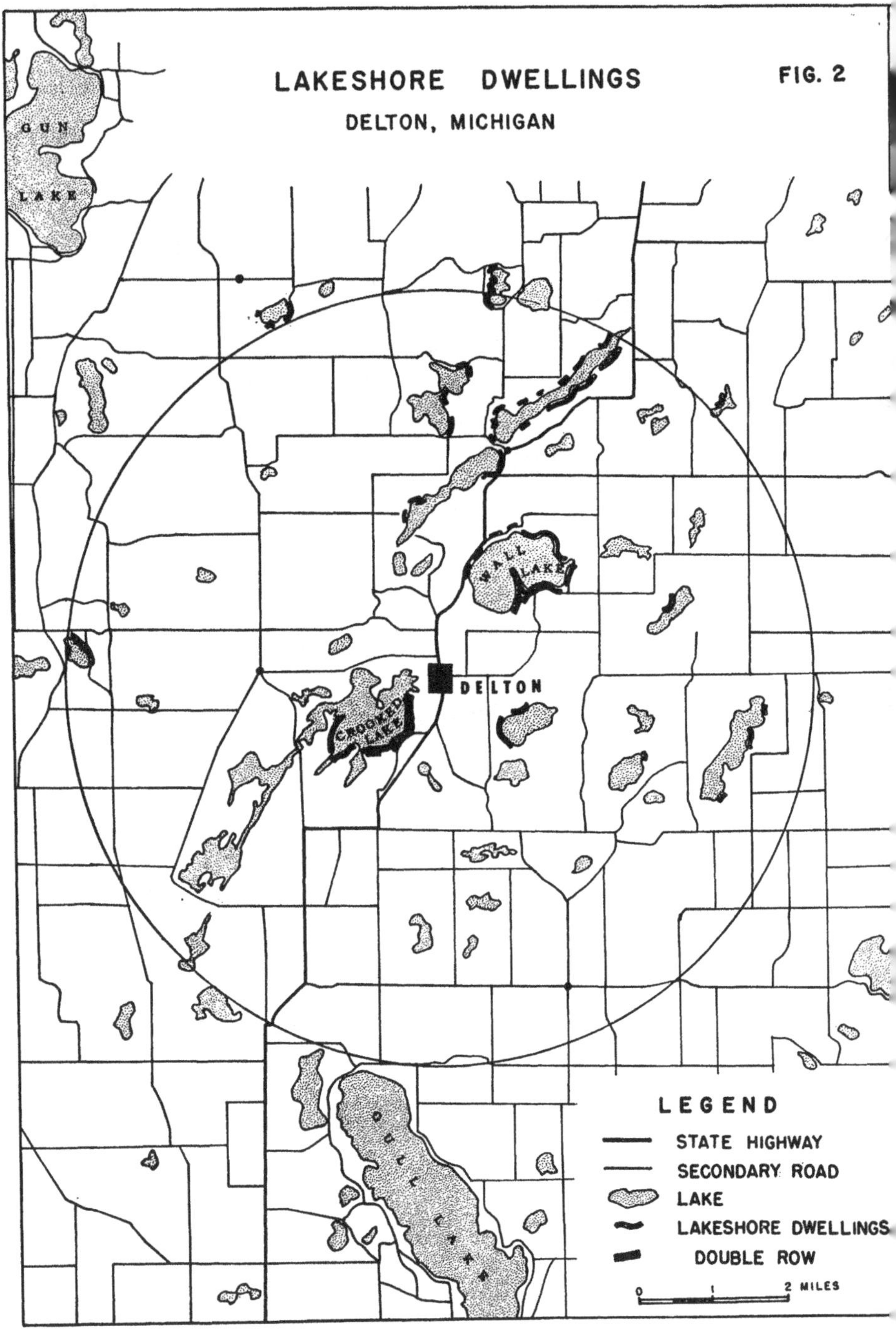
LAKESHORE DWELLINGS
DELTON, MICHIGAN
FIG. 2
GUN LAKE
WALL LAKE
CROOKED LAKE
DELTON
GULL LAKE
LEGEND
STATE HIGHWAY
SECONDARY ROAD
LAKE
LAKESHORE DWELLINGS
DOUBLE ROW
0
1
2 MILES

Fig. 2.

Of the total group of households, over half indicated that they did their basic shopping in Battle Creek, Kalamazoo, or Hastings, all over 15 miles from the village of Delton.

As far as the origin of the lake dwellers was concerned, it was not surprising to find that close to three-fourths of both seasonal and permanent lake dwellers originated from either Battle Creek or Kalamazoo. 12% to 15% of both seasonal and permanent lake dwellers originated from Hastings, and the remaining proportion came largely from the surrounding villages or rural areas.

In addition to the information on the lake-shore occupance, some of the most interesting information derived from the study came from observation and unsolicited information during the course of the interviews. This additional information represents a tangible benefit of the interview method for an exploratory study.

A large proportion of the lake dwellings appeared to be converted cottages, rather than homes built originally for year-round occupance. In many cases the occupants were older persons who had upon retirement sold their urban homes and winterized their cottages for permanent residence. In a proportion of the permanent homes, some member of the family, frequently a son or daughter, worked in either Battle Creek or Kalamazoo. This was put forward in many cases as the reason for grocery shopping in the larger centers. The family member working in the city would pick up the week's supply of groceries at the city supermarket and thus benefit from the lower prices of the large supermarkets. This factor of lower prices was advanced by a number of people as a reason for shopping in the city, but it may simply be rationalization of a desire for a weekly trip to the city.

With the exception of Crooked and Cloverdale lakes, most of the lake-shore dwellings appeared to be of postwar origin. Crooked Lake has been a recreational area for over 75 years. The lake was served by a line, now abandoned, of the Chicago, Kalamazoo, and Saginaw Railroad, and in the pre-automobile area, Crooked Lake possessed a railway station and a number of resort hotels and recreational establishments. The major postwar developments were located at Wall, Big Gilkey, and Long Lake.

Delton is the location of a public school and a new centralized high school. This is a major asset of the community, and according to both lake residents and the real estate people in Delton, it is an important consideration to people with families who are considering establishment of residences outside of Battle Creek or Kalamazoo.

A major problem of the predominantly shallow moraine lakes in the area was that of relatively rapid encroachment by vegetation. Large areas of some lakes, notably Crooked Lake, which appear on the topographic map as of 1946 as open water, were completely overgrown with aquatic vegetation and an encroaching bog type of vegetation. This development was reflected in the complete lack of cottages at the south end of Crooked Lake, and the lack of new cottages and the number of For Sale signs at the shallow north end of the lake. During the summer of 1962 the lake-shore property owners at Crooked Lake were attempting to organize and raise funds for a study on the feasibility of dredging and clearing work on the lake. This is an acute problem for many small lakes in Michigan and a problem which offers a major challenge for geographic and ecological study.

Conclusions

It is apparent that the village of Delton, in addition to a seasonal influx of some 400 cottage households, possesses a "hidden hinterland" of over 560 households in the form of permanent lake-shore residences. This hidden community is approximately three times the size, in terms of households, of the village of Delton itself. However, this population does not offer an automatic market for Delton retail services. Although within less than five miles of Delton, the lakes are largely populated by urbanites or ex-urbanites who are still considerably oriented toward the urban centers, for income, purchasing, and perhaps social reasons.

For various reasons, only a third of the permanent lake-shore households appear to be oriented toward Delton for their basic household needs. The addition of these families to the Delton population brings the threshold values more into line with the averages for Barry County. However, it is apparent that the influence of the larger urban centers is so pronounced that the lake dwellers do not contribute as much to the community economy as their numbers might suggest.

Observation and interviews with lake-shore residents revealed a number of specific problems concerning their future and their relations with the municipality and adjacent communities. The vegetation encroachment on the lakes, the question of the variation of urban influence with distance, the age of the residents, and the needs for schools and municipal services by the lake dwellers who both earn and spend outside the community: these constitute very real problems.

With the postwar increase in leisure time and mobility, many lake areas in Michigan have become developed in like fashion. These areas represent a rather nebulous, but certainly new, type of community. Answers to these problems would be of significance to many municipalities in urban-influenced recreational areas, not only in Michigan but in much of the populated Great Lakes region.

PAPERS OF THE MICHIGAN ACADEMY OF SCIENCE, ARTS, AND LETTERS
VOL. XLIX, 1964 (1963 MEETING)

A CLIMATIC SURVEY OF THE GRAND TRAVERSE BAY AREA

DIETER BRUNSCHWEILER AND ROBERT VOGEL
Michigan State University

INTRODUCTION

THE Great Lakes have been mentioned as climate modifiers ever since the white man began to settle along their shores. For instance, before it was officially opened for settlement, a rather observing white "squatter" in the Grand Traverse Bay region wrote:

> Some persons think the statement, that in this region the thermometer never sinks so low as it does two or three hundred miles south, in the interior, incredible; but the Philosopher finds no difficulty in accounting for it. When cold weather commences, the water first cools on the surface and then sinks, being heavier . . . than warm water, which rises from below and sinks in turn when it has cooled. Thus a continual circulation goes on . . . till the whole body of water cools to the freezing point. . . . In the act of freezing, a very great quantity of latent heat is set free, and warms both the air and water. . . . Now this bay being from 600 to 900 feet deep, the whole body of water does not cool to the freezing point till the middle of February, and must, therefore, constantly radiate heat to the atmosphere, and continue to do so during the formation of ice; while Lake Michigan on the west never cools to the freezing point, and continues to warm the air all winter. This explanation applies, however, only to countries lying on the east side of large and deep waters, and extends inland in proportion to the extent of contiguous bodies of water. . . . The only reason for this difference that science can at present assign, is the prevalence of wind from the west. South winds do not necessarily warm the atmosphere in this place, as they do in southern Michigan.[1]

The most conspicuous landscape elements along the eastern shore of Lake Michigan are the orchards which occupy a narrow zone a few miles wide (Fig. 1a). The northernmost portion of this fruit belt lies in the vicinity of Grand Traverse Bay where probably the greatest density of cherry trees on earth is found. Every farmer in this area is convinced that his forebears selected the fruit sites not so much because of relief, soil, or market conditions but as a consequence of their judgment as to the beneficial influence of Lake Michigan and its bays upon fruit-growing. It is well known that locations near the lake have 50 or more frost-

[1] *Michigan Farmer*, XIII (July 1855), 202.

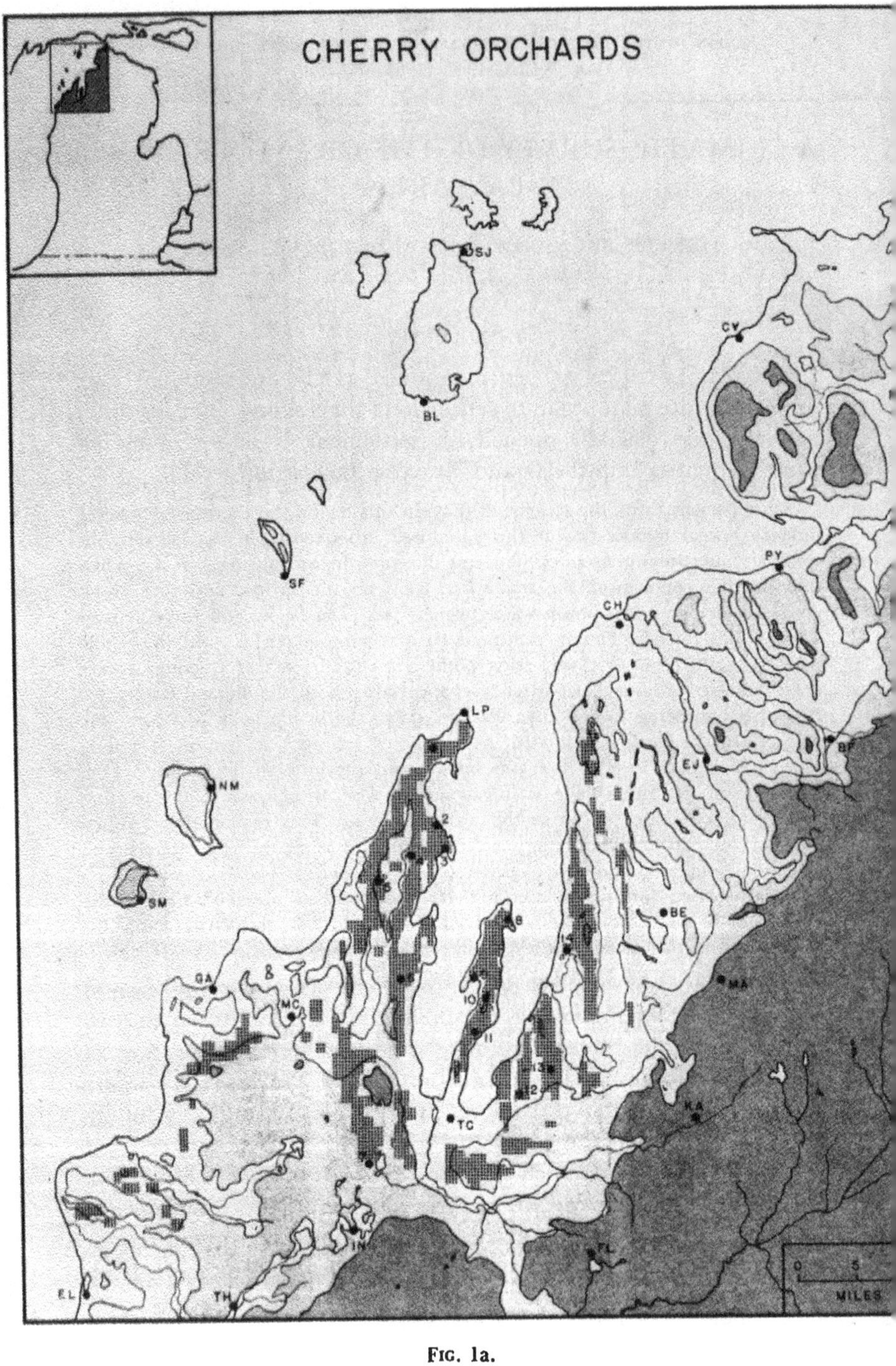
CHERRY ORCHARDS
SJ
BL
CV
SF
PY
CH
LP
EJ
NM
SM
BE
GA
MC
MA
KA
TC
IN
TL
EL
TH
MILES

FIG. 1a.

free days than inland locations. From the sharply defined distribution of the fruit farms it may be inferred that the climatic influence is very limited. It has been believed that late frosts during the blossoming period of the trees are less frequent near the Lake since the cold water has enough of a chilling effect to delay blossoming until the danger of a cold wave has passed. Whether this is the main reason for the high concentration of the orchards in the immediate vicinity of the Lake has never been demonstrated, nor do we know what the landward limits of this so-called lake effect are. Are the less severe thermal conditions, compared to inland stations, and the high amounts of snowfall in winter, or are the comparatively cool summers the major causal factors for this very specialized type of land use? To shed some light on these questions is the main objective of this investigation.

Organization of Project

In order to isolate the lake effect it was felt that the existing official U. S. Weather Bureau network was not dense enough. This network, consisting of 24 stations (including inland stations used for comparisons), was augmented by 14 stations operated by cherry farmers who cooperated enthusiastically. The Weather Bureau at Lansing provided three self-recording rain and snow gauges, in addition to granting permission to use unpublished data. The project was financed by Michigan State University through an All-University Research Grant.

The cooperating observers made one observation daily of the maximum and minimum temperatures, the 8:00 A. M. temperature, and the precipitation for the preceding 24 hours. Also, they were asked to observe the depth of the snow on the ground, wind direction and force, and to remark on any meteorological phenomena that they cared to. Although they pose some intriguing questions, these latter observations were inconsistent and any analysis of the derived data would be suspect.

So far only data for one complete year are available, but these, when plotted, show significant modifications of the pattern developed solely from the less dense Weather Bureau network.

Results of the Investigation

Winter conditions.—Maps of the temperatures for January and February 1962 give some indication of the magnitude of the lake effect (Figs. 2a & 2b). Only about a 4° gradient (or about 1° F. per five miles) is observed on the map of means, but with the first cold wave the gradient is sharply steepened to 25° or about 1° per mile, as is demonstrated by the map of minimum temperatures. When the mean temperature map is compared with the one constructed only from Weather Bureau statistics, it becomes clear that without the observations of the cooperating cherry farmers, the effect of the Bay is hardly demonstrable. The tongue of colder temperatures in the Leelanau Peninsula indicates the narrow band that is affected by the lake to a very marked extent. On the January snowfall map, the windward side of the Leelanau Peninsula shows nearly twice the snowfall that is received in its lee and on the Old Mission Peninsula (Fig. 1b). Local farmers in this lee area have indicated that one of their greatest climatic problems is to receive sufficient winter precipitation to prepare their orchards for the spring. This becomes especially acute on the Old Mission Peninsula where snowfall is at a minimum and where much is lost to the bays by wind action.

Spring and summer conditions.—Spring is the most critical season for the orchardists. The phenological diagram for May 1962 shows some of the critical dates for the cherry industry (Fig. 3a). The first sweet cherry blossoms were reported on the fourth, or exactly five days before the last frost of the season. Full blossoms were reported for both the sour and the sweet cherries at a steadily earlier date with increased distance from Lake Michigan. The darkened area on the diagram indicates the first five consecutive days in which the temperature maximum exceeded 80°. The appearance of the first sour cherries exhibits an even more pronounced delay near Lake Michigan than did the first blossoms.

The map of absolute minima during May still shows the warming influence of Lake and Bay, but to a less marked degree than in the winter map (Fig. 4a). However, the map of maxima shows a direct reversal of the winter condition (Fig. 4b). These temperatures occurred from one to two weeks after the minimum and indicate that the water then had a cooling rather than a warming influence. The recording of the absolute maximum on the Old Mission Peninsula again indicates the very narrow lake shore margin that receives a significant lacustrine influence. This cooling power of the lake on the immediate shore is even more pronounced in July when the range of observed maximum temperatures on the Leelanau Peninsula was 12° (Fig. 5a).

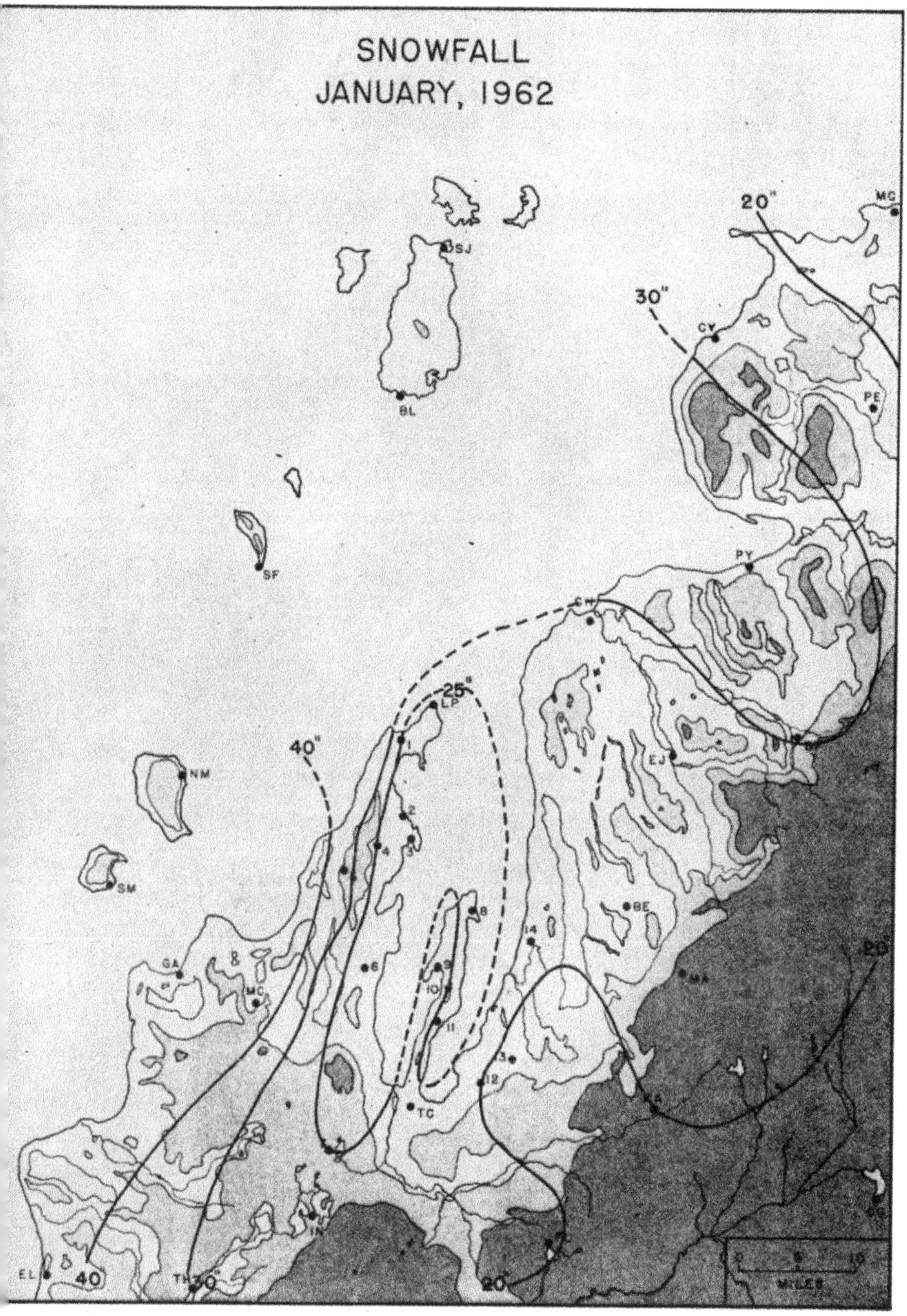

SNOWFALL
JANUARY, 1962
20"
30"
25"
40"
40
30"
20"
20
SJ
BL
SF
NM
SM
GA
MC
CY
PE
PY
CH
LP
EJ
BE
MA
TC
IN
EL
TH
MILES

FIG. 1b.

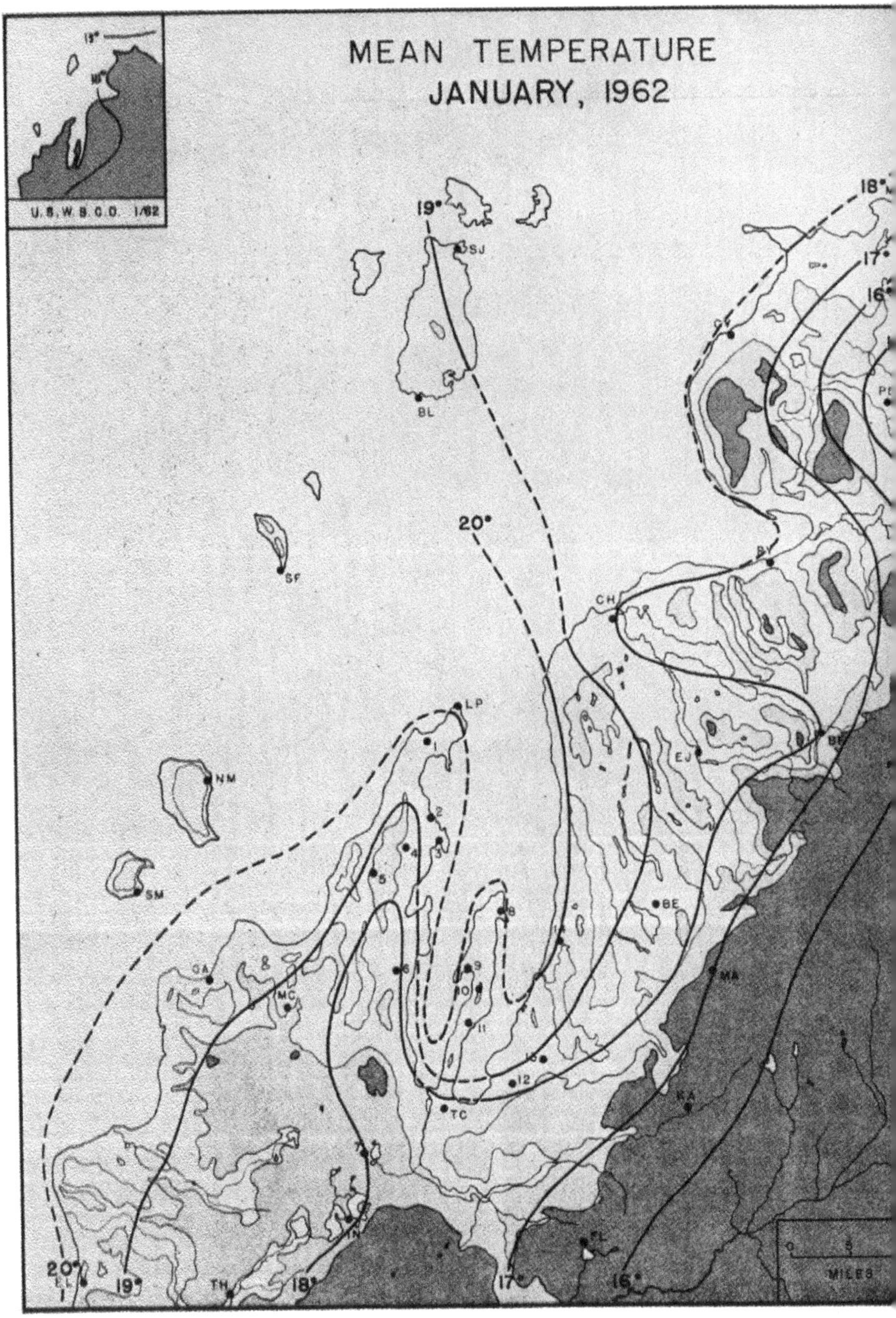
MEAN TEMPERATURE
JANUARY, 1962
U.S.W.B.C.O. 1/62
19°
20°
18°
17°
16°
SJ
BL
SF
NM
SM
LP
CH
BY
EJ
BE
MA
KA
GA
MC
TC
IN
TH
FL
MILES

Fig. 2a.

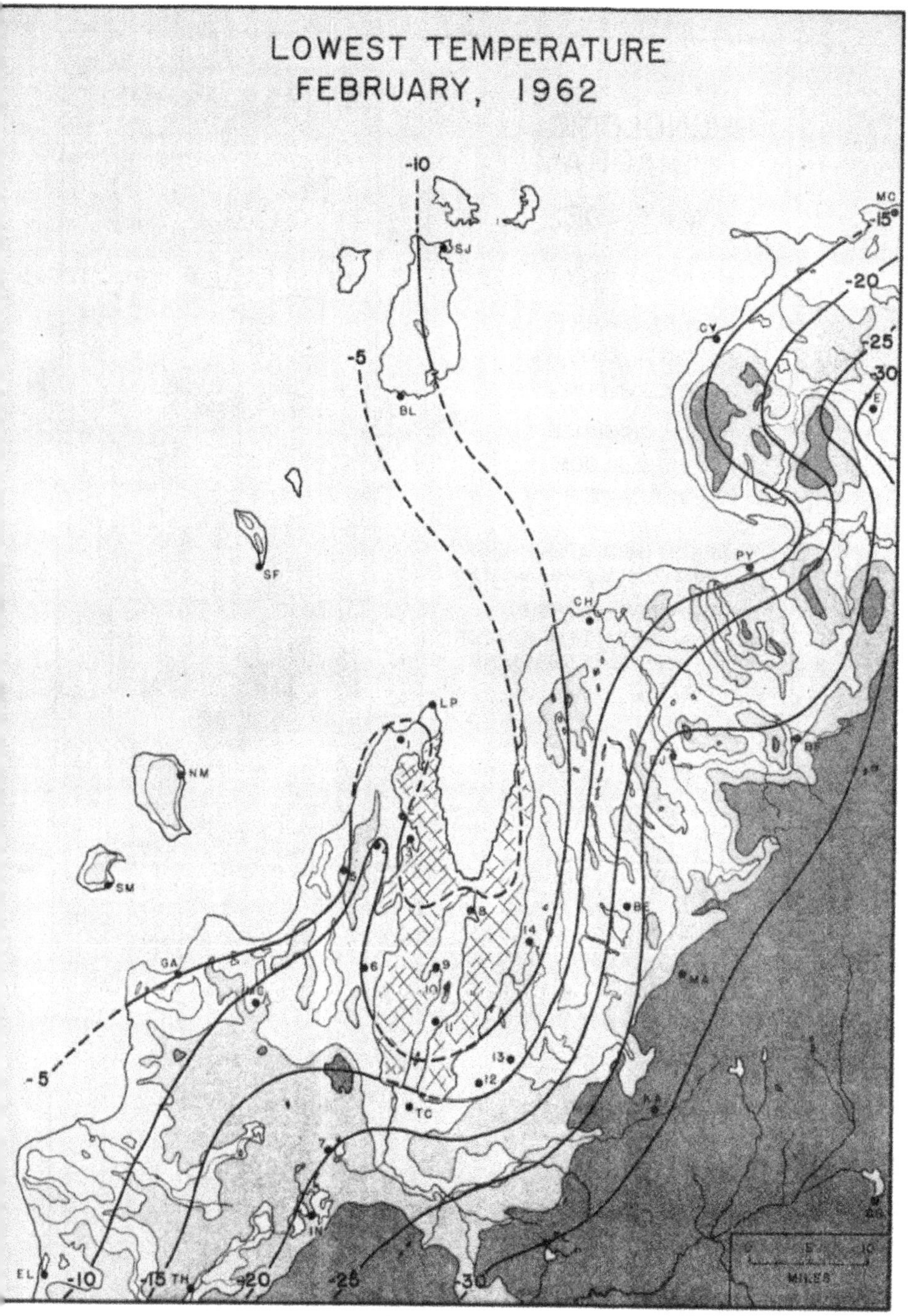

LOWEST TEMPERATURE
FEBRUARY, 1962
-10
-5
-15
-20
-25
-30
MC
SJ
BL
CY
SF
PY
CH
LP
NM
SM
BE
GA
MA
TC
IN
EL
TH
MILES

FIG. 2b.

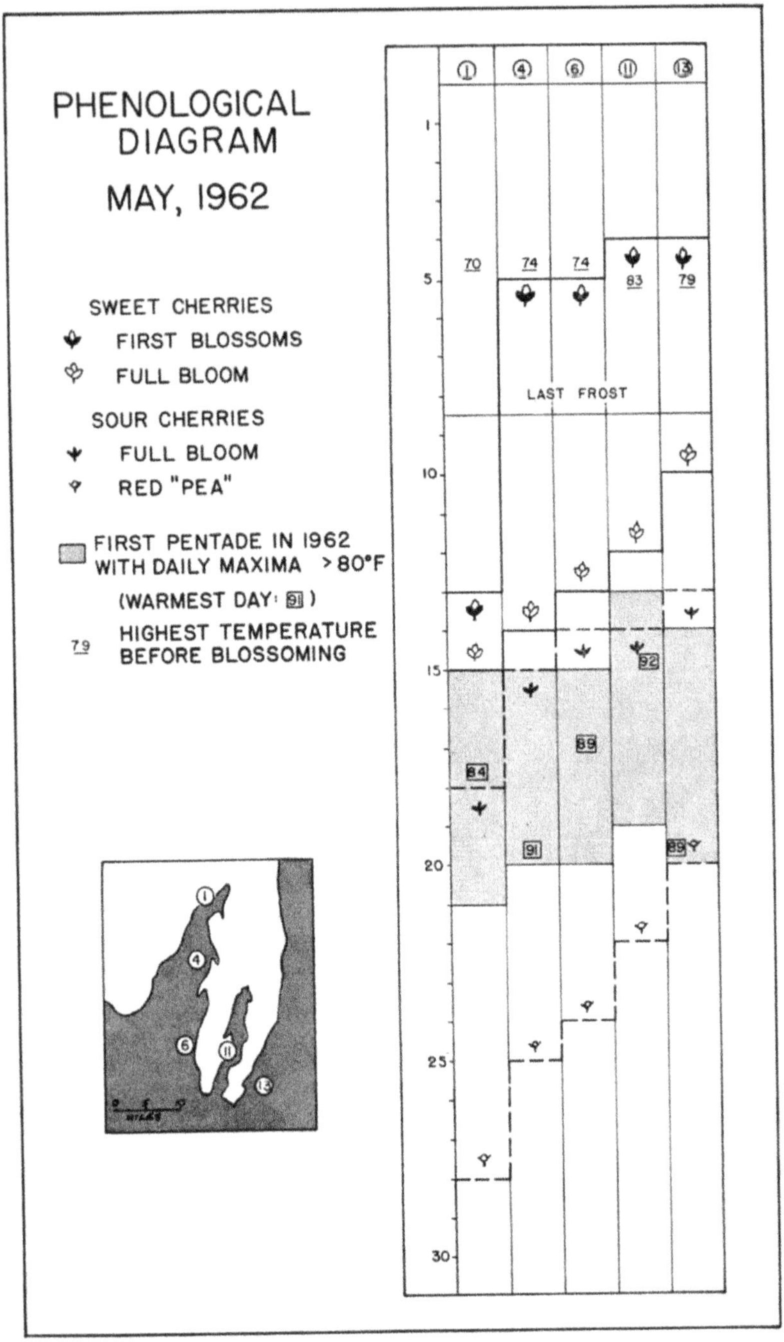
PHENOLOGICAL
DIAGRAM
MAY, 1962
SWEET CHERRIES
FIRST BLOSSOMS
FULL BLOOM
SOUR CHERRIES
FULL BLOOM
RED "PEA"
FIRST PENTADE IN 1962
WITH DAILY MAXIMA >80°F
(WARMEST DAY: 91)
79
HIGHEST TEMPERATURE
BEFORE BLOSSOMING
1
4
6
11
13
1
5
10
15
20
25
30
70
74
74
83
79
LAST FROST
84
91
89
92
89

Fig. 3a.

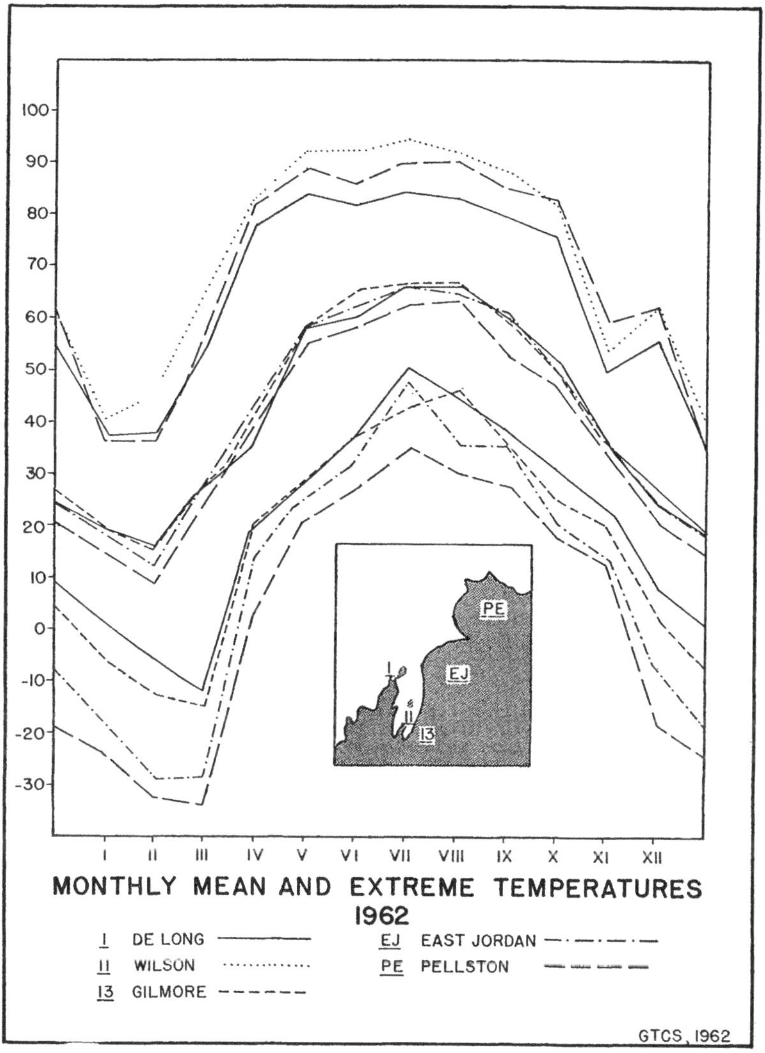
100
90
80
70
60
50
40
30
20
10
0
-10
-20
-30
I II III IV V VI VII VIII IX X XI XII
PE
EJ
I
II
13
MONTHLY MEAN AND EXTREME TEMPERATURES
1962
I DE LONG
II WILSON
13 GILMORE
EJ EAST JORDAN
PE PELLSTON
GTCS, 1962

FIG. 3b.

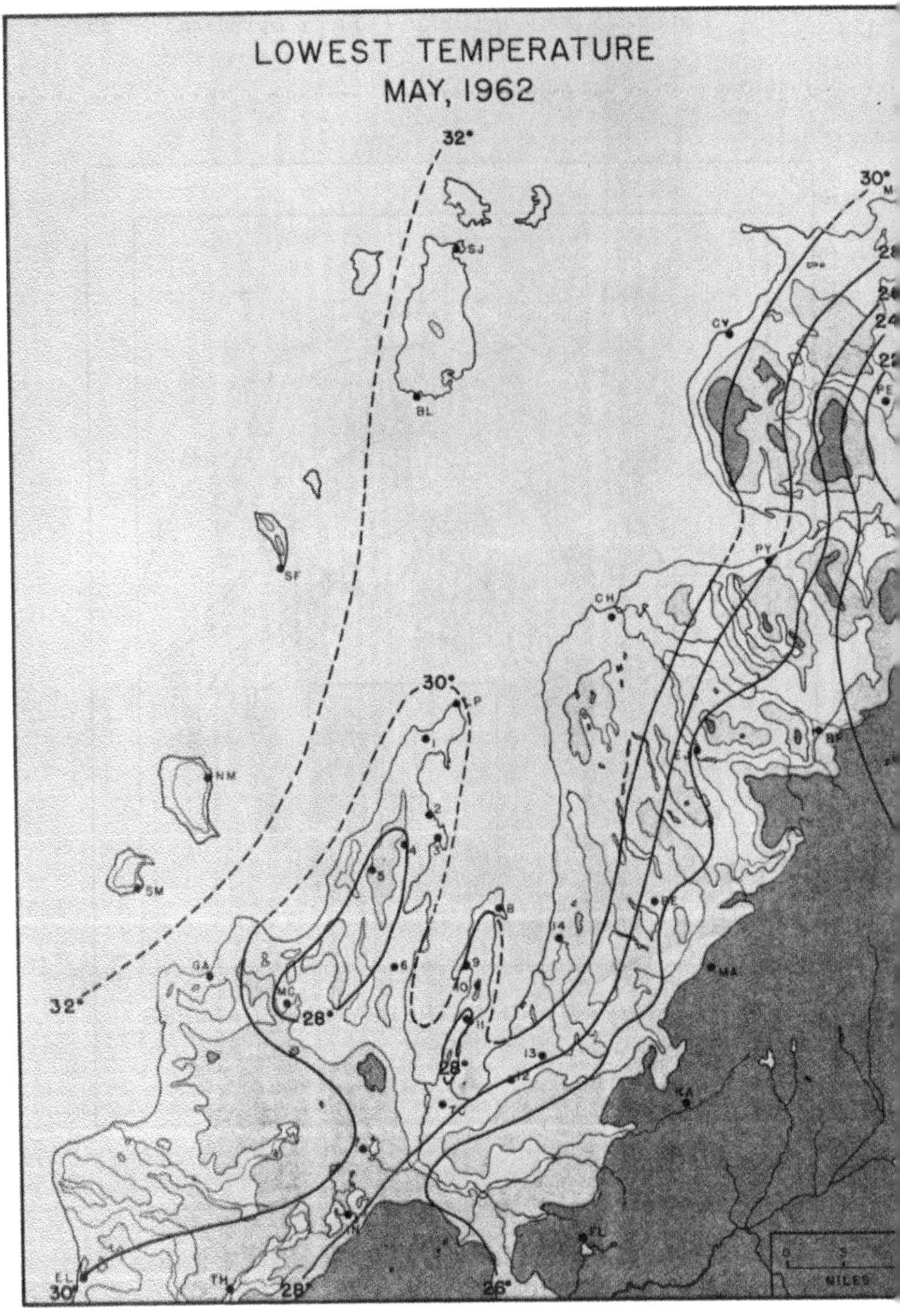

FIG. 4a.

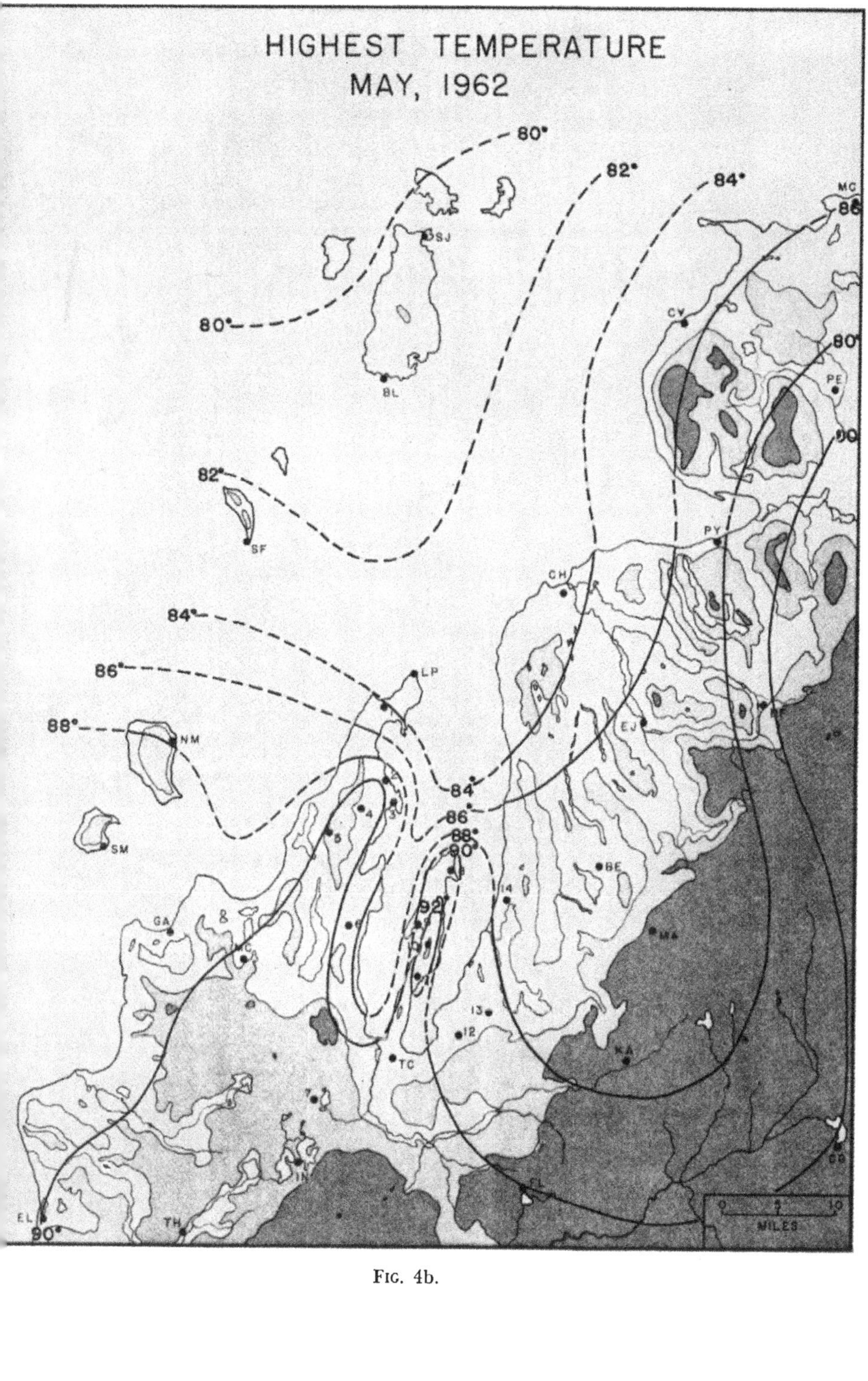
HIGHEST TEMPERATURE
MAY, 1962
80°
82°
84°
86°
88°
90°
92°
SJ
BL
SF
LP
NM
SM
GA
CV
PE
PY
CH
EJ
BE
MA
KA
TC
TH
IN
EL
MC
CB
MILES

FIG. 4b.

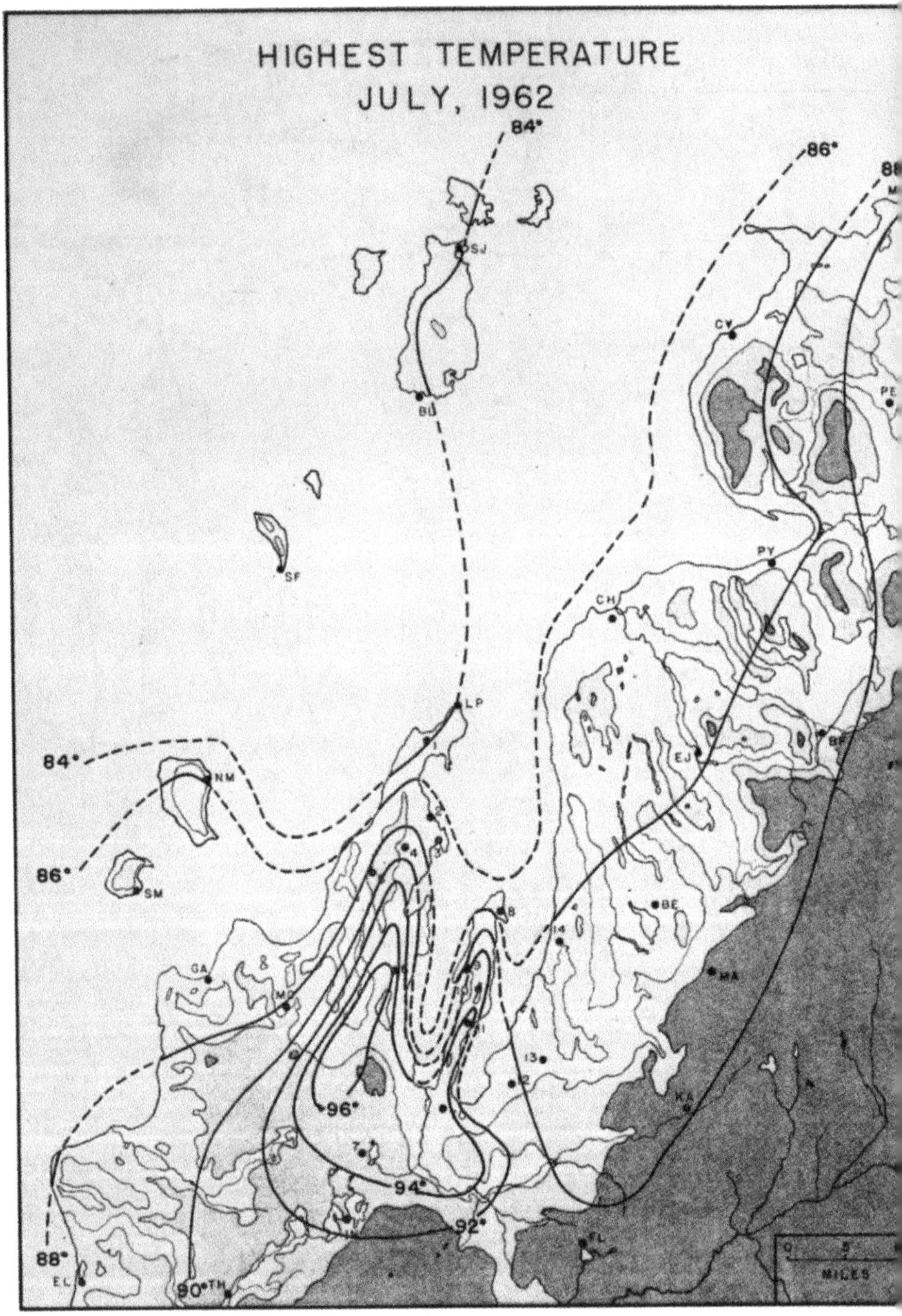
HIGHEST TEMPERATURE
JULY, 1962
84°
86°
88°
90°
92°
94°
96°
SJ
BU
SF
LP
NM
SM
GA
CY
PE
PY
CH
EJ
BE
MA
KA
FL
EL
TH
MILES

FIG. 5a.

The extremely variable rainfall noted on the June map leads to the conclusion that one year's data is insufficient to make valid generalizations about consistent rainfall patterns in the area (Fig. 5b). However, the stabilizing influence of the lake is marked by the very low figures in its immediate vicinity. One is led to suspect by the high figures on the Old Mission Peninsula and in the Boardman Valley that the stability is short-lived once the eastward moving air passes over the warmer land surface. Temperature and precipitation conditions observed during the fall of 1962 showed the lake effect to be mainly a retardation of the beginning of the winter season in the immediate vicinity of the shore.

The monthly march of the mean and extreme temperatures indicates that while there is a diminished winter temperature at the inland stations, the general pattern remains the same throughout the region (Fig. 3b). The winter is delayed until well into February, and absolute minima were not recorded until March. The summer part of the curve demonstrates the cooling power of the Lake. Maximum temperatures were fairly warm, but the broad, peakless curve represents a marked lacustrine situation. The section of the high temperatures of fall are no more than a confirmation of the same phenomenon. The upward sweep of the December maxima may only represent a freak year, or it may be a manifestation of the locally vaunted "Indian Summer."

Isopleth diagrams were constructed which show the behavior of temperature and precipitation against the variables of time and distance from the lake (Fig. 6a). While the mean temperature isopleths exhibit little lake-inland contrast for 1962, those for the monthly minima clearly express the negative gradient landward from January through March. Most conspicuous is the sharp change from the winter thermal conditions to those of summer during the month of April. The isopleths for precipitation illustrate the reversal of the lake effect from winter to summer (Fig. 6b). A double peak is clearly visible in the precipitation regime on the Leelanau side. The central portion of the area has a "normal" continental (single maximum) regime, the interior uplands a strong summer maximum with a secondary peak in winter.

Conclusion

The single year's observations leave a great deal to be desired in relation to sound generalization, but they do appear to afford

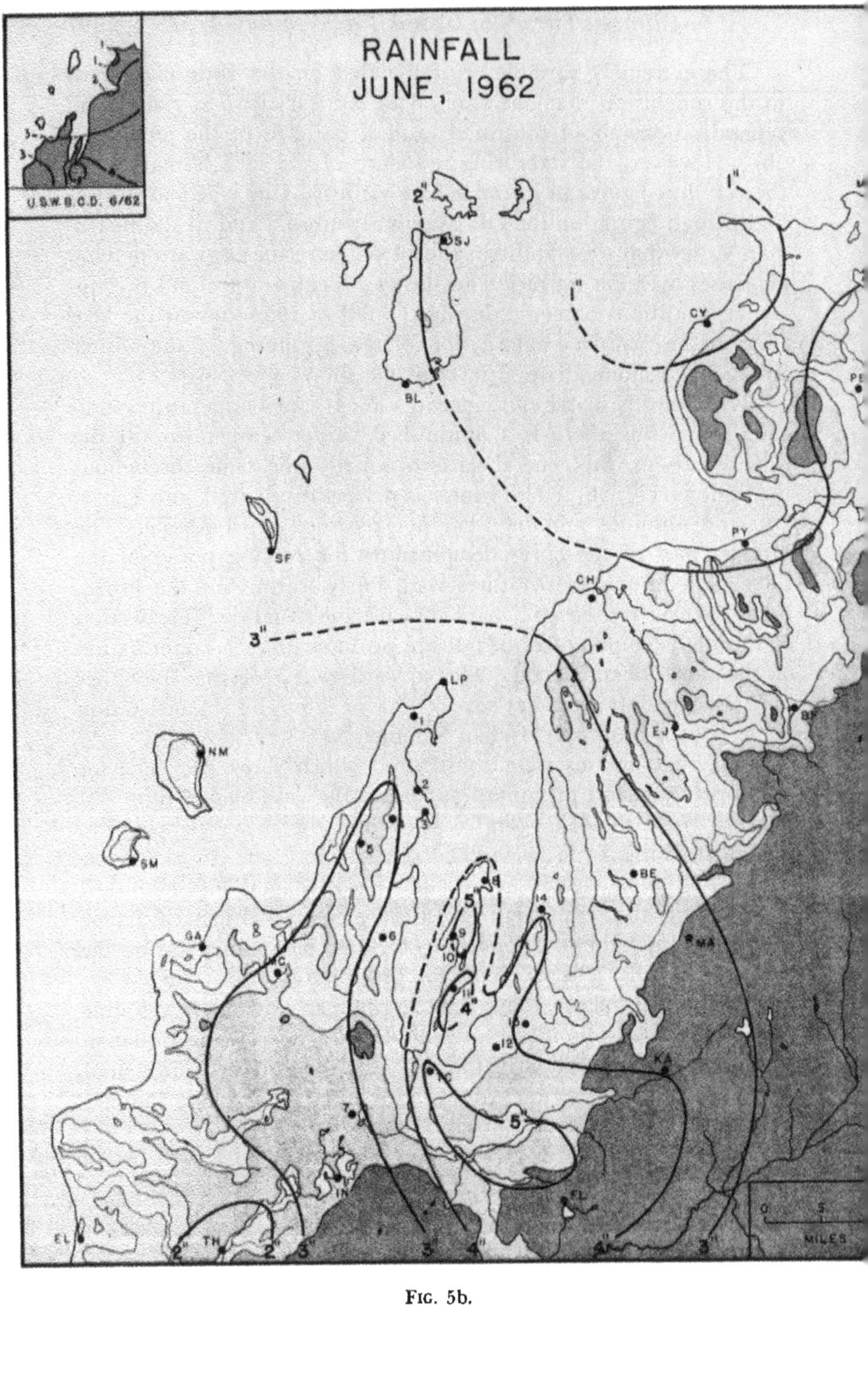
RAINFALL
JUNE, 1962
U.S.W.B.C.D. 6/62
1"
2"
3"
4"
5"
SJ
BL
CY
PY
CH
SF
LP
NM
SM
EJ
BE
GA
MA
KA
IN
FL
TH
EL
MILES

Fig. 5b.

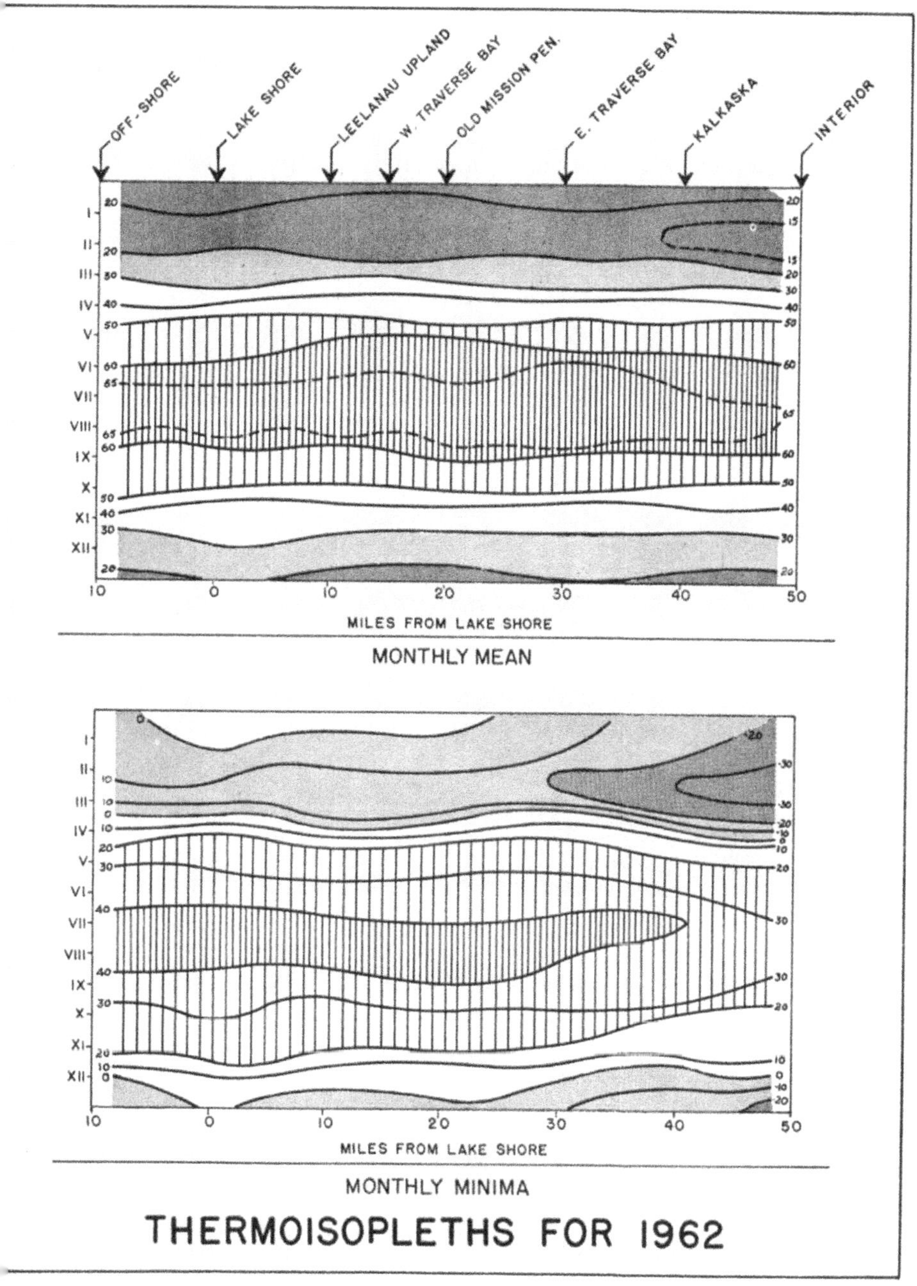
OFF-SHORE
LAKE SHORE
LEELANAU UPLAND
W. TRAVERSE BAY
OLD MISSION PEN.
E. TRAVERSE BAY
KALKASKA
INTERIOR
I
II
III
IV
V
VI
VII
VIII
IX
X
XI
XII
10
0
10
20
30
40
50
MILES FROM LAKE SHORE
MONTHLY MEAN
MILES FROM LAKE SHORE
MONTHLY MINIMA
THERMOISOPLETHS FOR 1962

FIG. 6a.

new insights into the pattern of the lake effect and, to some extent, of the area most stringently affected. Before really confident conclusions can be reached, a greater time period of observation is necessary, and some obvious hiatuses in the network need to be filled. If we were to start the project anew, we would be very much tempted to restrict the area of study and try to increase the

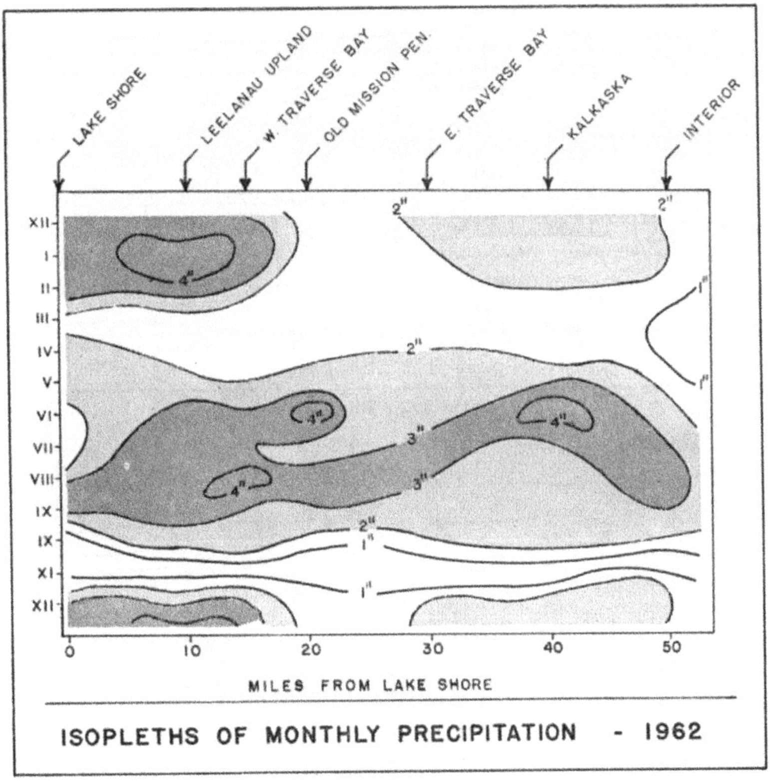

ISOPLETHS OF MONTHLY PRECIPITATION - 1962

FIG. 6b.

density of the observation network. Of all the conclusions reached, the one that strikes us most profoundly is that the climate of this region is extremely complex and that it can only be imperfectly understood until a detailed microclimatic study over a considerable period of time is completed.

ACKNOWLEDGMENTS

The authors wish to express their gratitude to the following persons: A. H. Eichmeier, head, and Stanley L. Baldwin, East Lansing Office, U. S. Weather Bureau, for loaning and servicing rain and snow gauges; the cooperators in the Grand Traverse Bay area for their conscientious recording of daily weather data; Wayne T. Strand and Miss Angie J. Rini, Michigan State University, for their cartographic and research assistance.

PAPERS OF THE MICHIGAN ACADEMY OF SCIENCE, ARTS, AND LETTERS
VOL. XLIX, 1964 (1963 MEETING)

VERMONTVILLE AND AUGUSTA: A STUDY OF TWO MICHIGAN VILLAGES

JOHN FRASER HART
Indiana University

THE American village is one of our least studied and most poorly understood settlement forms. Most Americans, including those who live in them, believe that our villages are dying, if they are not actually dead.[1] The casual observer who drives through the village may note its vacant stores and empty houses standing in unkempt, weed-grown lots. The death of the village excited considerable interest in the period between the two World Wars, but since 1940 scholars appear to have taken its demise more or less for granted and have turned their attention to other matters.

This has been unfortunate for, contrary to popular opinion, villages—in the Middle West, at least—have not been dying, if one defines a village as an incorporated place of less than 1000 persons which is outside an urbanized area, and if one defines the decline of the village as a decrease in the number of people who live in it. Two-fifths of the nation's 10,000 villages in 1960 were in a nine-state area in the Middle West (Table I). No data were published on the 1950 population of the 219 villages which had been incorporated since 1950. Just over half of the remaining villages had more people in 1960 than they had had in 1950, while just under half had lost population. In many villages, however, the gain or loss was so slight that it appears unreasonable to assume that the population actually changed at all, and I finally decided, quite arbitrarily, that a gain or loss of 10% or less represented essentially no change in the population of the village. It would appear, then, that about three-tenths of the villages gained population between 1950 and 1960, about one-fourth lost, and almost half remained stable in population.

The fact of village growth was much more impressive in Michigan, where two-fifths of the villages grew by 10% or more and

[1] One of the most recent and widely disseminated statements of this notion is Lauren Soth's "Iowa, or What's Happening to Farming," *New York Times Magazine* (December 2, 1962), p. 62. The same notion was dramatically presented on the nation's television screens by Chet Huntley in "The Land," Westinghouse Special, 3, March 13, 1962.

less than one-tenth lost 10% or more of their people (Table I). More than three-fourths of the villages in the state experienced some population growth during the decade 1950-60.

When viewed over a long-time period, the pattern of village population change is extraordinarily complicated. In general, it

TABLE I

INCORPORATED PLACES OF LESS THAN 1,000 PERSONS, OUTSIDE URBANIZED AREAS, BY RATE OF GROWTH, 1950-60

	Nine Middle Western States		Michigan	
	No.	Percent	No.	Percent
Total number in 1960	3,943		196	
Incorporated since 1950	219		4	
Remainder	3,724	100.0	192	100.0
Plus 10% or more	1,106	29.8	81	42.2
Plus 0.1-9.9%	880	23.6	66	34.4
Minus 0.1-9.9%	839	22.5	31	16.1
Minus 10% or more	899	24.1	14	7.3

appears that the population of many villages rose to a peak around 1900 to 1920, dropped in 1930, and subsequently remained stable or increased slowly and rather erratically (Fig. 1). But village population growth warrants a separate study in itself; suffice it to say that the population of many villages, contrary to popular belief, has been increasing over the last three decades, at least.

The continued, and generally unsuspected, growth of the village necessitates reexamination of a number of stereotypes which have been based on the assumption that the village has been fading away. Doesn't the village population consist largely of widows and retired farmers? Isn't the village merely a "fished-out pond" consisting largely of people who remain there simply because they have lacked the initiative to leave? Wouldn't any normal young couple prefer to live and raise a family in a place more lively than the village? And aren't the villagers simply "living off the welfare" because no jobs are available for them locally?

A preliminary search for answers to some of these questions was part of the first annual Big Ten Faculty Field Research Seminar in Geography, on "The Spatial Structure of Settlement," which was held at Michigan State University's Gull Lake Biological Station. Although I accept full responsibility for the use I have made of the data and the interpretations I place upon them, I must express my sincere appreciation to those colleagues who labored so hard and so faithfully to collect these data, and who engaged in

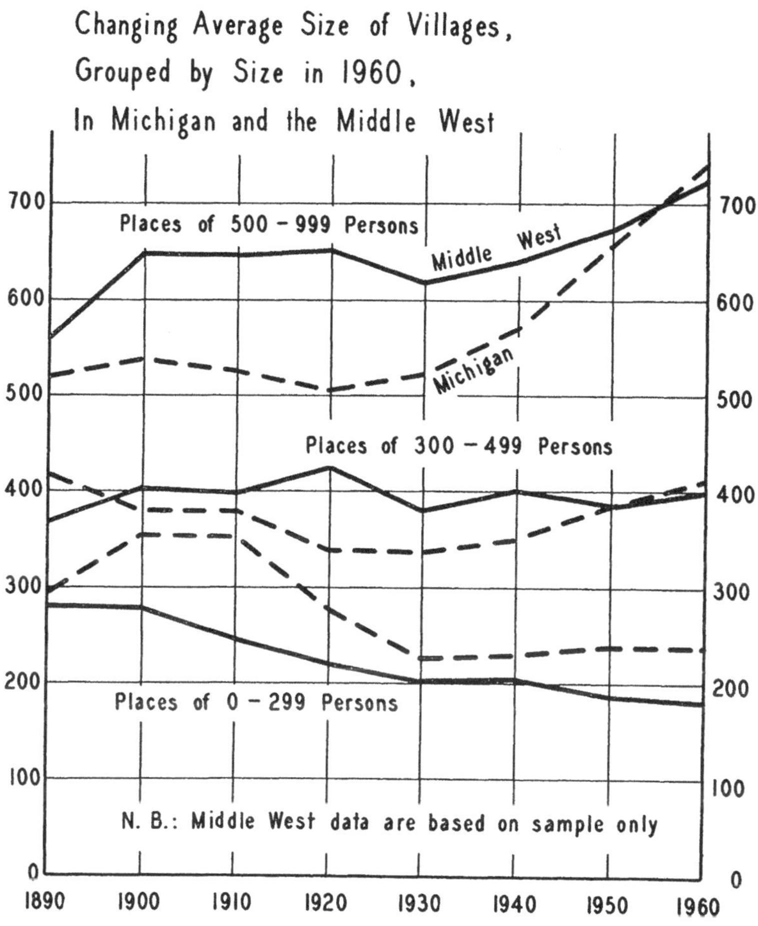

FIG. 1.

such stimulating discussions as to their significance: Fred W. Foster, Ronald Larson, and Robert McGregor of the University of Illinois; Wayne Kiefer, Chris Metcalfe, and Paul Whippo of Indiana University; Neil Salisbury and Paul Johnson of the State University of Iowa; K. C. McMurry and Don Lenick of The University of Michigan; Allen K. Philbrick, Stuart Denslow, Judy Getis, Robert Holz, and Edward Myles of Michigan State University; Eugene Mather of the University of Minnesota, and Edward Taaffe of Northwestern University. I owe a special debt of gratitude to Professor Salisbury for his wise counsel and advice throughout the project.

Two villages in the Gull Lake area were investigated in some depth during the late summer of 1961. Vermontville, about 27 miles southwest of Lansing and 23 miles north of Battle Creek, is not on a main highway; it was selected as an example of an isolated village. Augusta was selected because of its location 6 miles west of Battle Creek and 14 miles east of Kalamazoo, on the assumption that it had been subject to some degree of urban influence.

The population of both villages has been increasing recently. Vermontville grew from 544 persons in 1870 to 730 in 1890, then declined continuously to 564 in 1940, but rose once more to 707 in 1950 and to 768 in 1960 (Fig. 2). Augusta had 498 people in 1890, rose to 541 in 1900, dropped to 494 in 1910, and has grown continuously to its peak of 972 persons in 1960.

The two villages had a total of 507 houses when the fieldwork was done. Questionnaire schedules were collected from 417 (82%) of these houses, but one-fourth of these schedules was so incomplete as to be useless, and comparatively complete schedules are available for only 287 households or 57% of the total. The questionnaire, whose defects early became obvious, failed to instruct enumerators to make a distinction between "vacant" and "no response," and thus no information is available as to the number of houses which were vacant, although from casual observation I estimate it to be comparatively small, perhaps on the order of 5% or less.

A more serious bias was introduced by two groups of nonresponders: first, old ladies, at home by themselves, who refused to answer the doorbell for strangers; and second, childless couples with both spouses working and away from home in the daytime. The latter bias could be reduced by return calls in the evening, but those old ladies apparently will always pose a problem for geographers conducting village interviews. (Comments made by their neighbors indicate that geographers are not alone in having trouble with those old ladies!)

Those who did respond to our questions were not merely a

collection of old people, retired people, or childless people. When our village data are compared with national norms, it appears that the traditional stereotype of the village has been somewhat exaggerated, or even erroneous (Table II). Despite serious problems of data

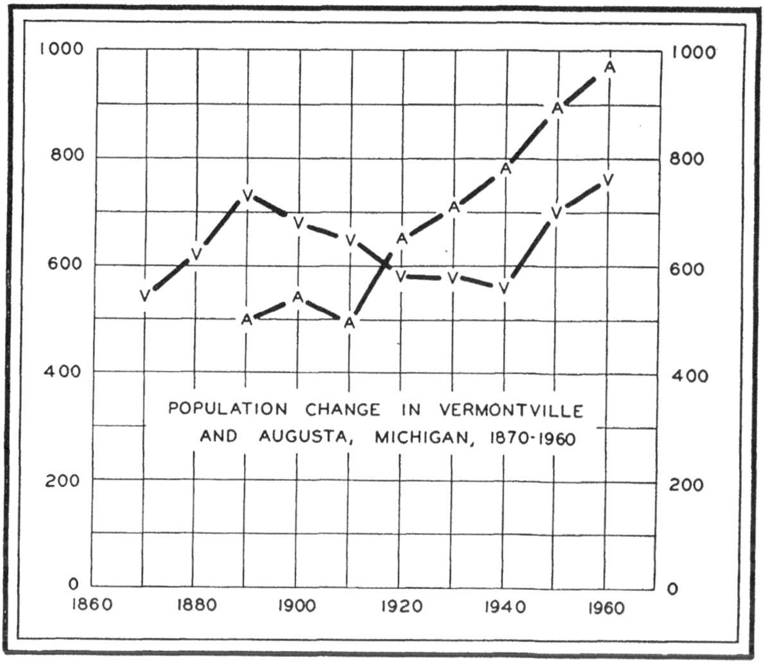

FIG. 2.

comparability, estimates of national norms have been included here to provide a rough standard of comparison. The villages do have a higher percentage of elderly heads of households, to be sure, but at the expense of the middle-age groups, for the percentage of heads of households in the younger, child-bearing age groups is the same as the national figure (Table II). And these men were doing their duty, for the average household in the two villages was slightly larger than in the nation, and the villages had a slightly higher percentage of large families, with fewer one- and two-person households (Table II).

One-fourth of the villagers were retired, but only one in five was employed in the professional, managerial, clerical, and sales activities which would presumably characterize a rural service center, whereas three-fourths of the employed workers were operatives or laborers (Table II). In other words, if one excludes the retired people and those few who operate the usual establishments which have not dried up, the village population consists of young skilled and semiskilled workers with relatively large numbers of children.

TABLE II

ESTIMATES OF SELECTED CHARACTERISTICS OF THE POPULATION IN TWO MICHIGAN VILLAGES AND THE UNITED STATES

	Vermont-ville %	Augusta %	Both Villages %	United States %
Percentage of heads of households				
Over 60 years old	33%	34%	34%	20%
40-60 years old	32%	26%	29%	43%
Under 40 years old	35%	40%	37%	37%
Average number of persons per household	3.47	3.34	3.40	3.38
Percentage of households consisting of				
1-2 persons	40%	39%	39%	42%
3-4 persons	31%	41%	36%	36%
5 or more persons	29%	20%	25%	22%
Percentage of heads of households who are retired	30%	24%	27%	9%
Percentage of employed persons employed as				
Operatives, laborers, or kindred workers	69%	75%	72%	36%
Professional, managerial, clerical or sales workers	30%	14%	22%	43%

One of the more unexpected results of our survey was the discovery of fairly large numbers of " Johnny-come-latelies " in the two villages. We found that 38% of our respondents had moved into their present houses within the preceding five years, and 71% had moved in since the Second World War, although we estimated that approximately the same percentage of the houses in the two villages had been constructed before the First World War.

Why did they move here? Only 6% mentioned any putative advantages of life in a village. Although the question was only tardily inserted in the questionnaire, a higher percentage of our respondents admitted that they detested village life! 15% gave a variety of reasons, such as wanting more room for the children, or "wanting a house we could call our own," or a cheaper place to live, or "better transportation," or simply, "we inherited the place."

29% (and the figure is higher in Augusta than in remoter Vermontville) said that they had moved to the village because they had a job in the area, and our data show that roughly half the breadwinners in each village were employed in the village itself. 45% of the wage earners of Augusta worked in Battle Creek (6 miles east) or in Kalamazoo (14 miles west), and 24% of the workers of Vermontville made the 27-mile daily trek to Lansing, while 23% worked in smaller communities closer at hand.

But only half of the people of these two villages moved here because of employment opportunities in adjacent cities or in the village itself, or because of the purported advantages of village life. The other half said they live in the village because one or both were born "here," or have family connections here, or had lived here before. Unfortunately, our questionnaire failed to request any specific information as to what was meant by "here," for we soon realized that "here," in an undetermined number of cases, included surrounding agricultural areas as well as the village proper. This raised the fascinating question, which we cannot yet answer, of the degree to which the village population is being replenished from the surrounding countryside, and the reasons therefor.

Three factors might account for a farm to village movement. One might be related to the traditional migration of surplus farm youths to urban areas.[2] If the farm lad were reluctant to try his wings in the metropolis, for whatever reason, he might well choose to live in a small community close to his friends and relations. If this were indeed the explanation, however, one might well inquire why village growth suffered such a setback in the 1920's, when off-farm migration was certainly no less than it is today.

A second, perhaps more likely, hypothesis might be based on the growing size of farms and the fact that entire families, rather than only sons and daughters, are becoming surplus in agricultural areas.[3] Whereas the single person might be attracted by the bright

[2] For example, see Gladys K. Bowles, *Farm Population: Net Migration from the Rural-Farm Population, 1940-50,* U. S. Department of Agriculture Statistical Bulletin No. 176 (1956).

[3] Some implications of changing farm size are treated in M. L. Mosher,

lights of the city, the family group might desire to remain closer to their friends. And housing is definitely cheaper in the village than it is in town.

A third hypothesis, which does not appear to be supported by our data, might be the growth of sidewalk farming in response to this self-same growth in farm sizes.[4] Smith's work in Minnesota has shown that farms grow larger by acquiring noncontiguous tracts of land, and he found some townships in which as much as 57% of the farm units consisted of noncontinguous tracts.[5] Under such conditions farmers might well choose to live in a central location—the village—and go out to their scattered acres as necessary. We have no evidence that this is actually the case, however, and quite possibly the farmer who decided to move in to town might really move in to *town,* and by-pass the village completely.

It is perfectly obvious, of course, that villages might grow on a "do it yourself" basis of natural replenishment, by maintaining a surplus of births over deaths. Unfortunately, our data are silent on this point, although casual observation indicates that village increase was the product of in-migration and not of natural increase.

Are these in-migrants city people? Certainly Philbrick's extraordinary work in southern Michigan has demonstrated that the great residential mobility associated with the automobile has produced an almost unbelievable scatter of nonfarm homes through the countryside of this area.[6] Mather, in an intensive traverse along a 10-mile strip of country roads here, found 102 dwellings, of which only 7 were occupied by full-time farm families; another 9 were occupied by part-time farmers, but the remaining 86, including 7 farmhouses, were occupied by nonfarm people.[7]

Farms Are Growing Larger, University of Illinois Agricultural Experiment Station Bulletin 613 (1957).

[4] For an extended discussion of sidewalk farming, see Walter M. Kollmorgen and George F. Jenks, "Sidewalk Farming in Toole County, Montana, and Traill County, North Dakota," *Annals of the Association of American Geographers,* 48 (1958), 209-231.

[5] Everett G. Smith, Jr., "Road Functions in a Changing Rural Environment" (unpublished Ph. D. dissertation, Department of Geography, University of Minnesota, 1962).

[6] Allen K. Philbrick, "Urban Dispersal in Metropolitan Lansing, Michigan" (unpublished field trip guide for the 1961 meeting of the Association of American Geographers).

[7] Eugene Cotton Mather, "One Hundred Houses West," *The Canadian Geographer* (in press, spring 1963).

But we found few urban " refugees " in our two villages. When city people move to the country, they seem to want to take a bit of the city with them still. Although cheaper housing in the older houses of the village may have a real appeal for former farm families, the urban family wants the latest fashion in boxes, with all modern conveniences, and it is not attracted by the roomy (and drafty?) older house.

When it does move to the village at all, the ex-city family commonly moves into the new subdivisions on its outskirts, which in many cases lie outside the incorporated area of the village, and thus are not included with it for Census purposes. Where such subdivisions exist—and presumably they are not uncommon on the fringes of villages near major urban centers—the growth of the village population may be sharply underestimated when reliance is placed wholly on Census data. It is tempting to postulate a movement of city people to the countryside, of farm people to the village, and of village youths to the city.

To summarize, it appears that the Tin Lizzie induced economic leukemia in the village after the First World War when the irregular (perhaps weekly) trip in search of goods could be made to a center offering a wider range than the village could provide. This car's improved descendents have made the daily trip to work much easier and have facilitated residential mobility of city workers. But they have also increased the employment mobility of the village resident, which permits him to use the capital invested in village housing while earning his living at some distance from the village. Quite clearly we need to take a longer look at the processes influencing the growth of the village population, and to do more fieldwork to test the ideas presented here, which are based on our preliminary investigations at Gull Lake. In other words, we need to find out why it is that the village, contrary to all expectations, simply refuses to curl up and die.

PAPERS OF THE MICHIGAN ACADEMY OF SCIENCE, ARTS, AND LETTERS
VOL. XLIX, 1964 (1963 MEETING)

PRECIPITATION ANOMALIES FOR MICHIGAN

BENJAMIN MOULTON
Indiana State College

INTRODUCTION

THIS short paper is prepared as the third of a series on various aspects of the climatology of Michigan. The first was presented in the spring of 1957 as a map of the climatic regions of Michigan. A second phase on degree days was presented last year and this, on precipitation anomalies, represents the third. No attempt is made to present all the data worked with, but selections are made of some deviations from average of precipitation statistics for Michigan.

Three phases of the precipitation are to be discussed in this paper: (1) duration of precipitation in terms of consecutive days of precipitation; (2) extremely wet and dry years; (3) wet and dry months within average and extreme years.

LIMITATIONS OF THE PROBLEM

In order to work with the phases just mentioned several major premises were accepted as part of the use of the material, and their acceptance does not necessarily indicate that absolute extremes in the climatic record are used here.

1. 50 climatic stations are used that show a history of observation from 1960 to 1873 as maximum for some to others that represent only 42 years. A few of the shorter history stations were used because they represented critical points in the state and wide distribution was desired. In general, only one station in a county was used.

2. The statistics for some portions of the study reflect use of random extremes rather than the purposeful selection of absolute extremes. A study carried on aside from the material here revealed a correlation of .93 between the use of random selection climatic statistics and total group when 25 out of 50 available items were used. Thus it would appear that even though not all stations are used, a correlation high enough to be conclusive would exist.

3. Precipitation effectiveness is not dealt with.

Consecutive Days of Precipitation

As a pilot study to determine what values might be obtained, periods of precipitation were analyzed, and, on a strictly random

TABLE I

Consecutive Days of Precipitation

Interval	Frequency of Period
5 to 9 days	715
10 to 14 days	136
15 to 19 days	24
20 to 24 days	5
25 to 29 days	8
30 to 34 days	4
35 to 39 days	1
40 to 44 days	1
45 to 49 days	2

TABLE II

Monthly Distribution of Consecutive Days of Precipitation

Month	Frequency of Period								
	5–9	10–14	15–19	20–24	25–29	30–34	35–39	40–44	45–49
January	92	35	6	1	2	–	–	–	–
February	80	20	2	–	–	–	–	–	–
March	78	18	3	–	2	–	1	–	–
April	27	4	–	–	–	–	–	–	–
May	20	1	–	–	–	–	–	–	–
June	79	1	–	–	–	–	–	–	–
July	7	–	–	–	–	–	–	–	–
August	51	–	–	–	–	–	–	–	–
September	6	1	–	–	–	–	–	–	–
October	85	7	–	–	–	–	–	–	–
November	129	27	4	1	–	1	–	1	–
December	61	22	9	3	4	3	–	2	2
Total	715	136	24	5	8	4	1	3	2

sample of years, the year 1955 was chosen as the base for analysis. Consecutive days in which at least a trace of precipitation or more fell were tallied into groupings from 5 to 49 days. Table I breaks down this information. Intervals were based upon the availabilities and apparent usefulness of the precipitation statistics. No less than 5 consecutive days were recorded since the study was in extremes of precipitation.

In cases where rainy periods began or ended at either end of the year they are tallied in by going back to the 1954 record or ahead to the 1956 record far enough to measure the period. The period was included in 1955 if more than half the days fell in that year.

Consecutive days of precipitation of rain or snow play their role in the activities of man. Two stations in 1955 recorded 45 days or more where at least some precipitation fell. Although 715 periods, 5 to 9 days in length, occurred, this dropped rapidly to 136 10-to-14-day periods and 24 15-to-19-day periods.

A grouping of the statistics by months, as shown in Table II, reveals that the longer periods are in the cooler months, and in July and August no period of precipitation over 9 days existed and only 7 periods of 5 to 9 days existed in July. May, June, August, and September are all characterized by short precipitation periods, although May and September are by far the wettest months of the year in precipitation received for most stations in Michigan.

The location of each station is not noted here, but the perusal of the data would show that the longer periods of precipitation exist for stations in the Upper Peninsula. Certainly, the extreme stations, Kenton, Kinross, Grand Marais, Houghton, Marquette, Stambaugh, and Beechwood support this conclusion.

One can have little doubt of the impact of sustained periods of precipitation on agriculture, tourism, and other activities of man.

An Average Year, a Wet Year, and a Dry Year

Fig. 1 illustrates average distribution of precipitation in Michigan. Two small areas, western Upper Peninsula and Southwestern Michigan receive more than 36″ a year, but the drier portions receive between 27″ and 30″. The range of the state is approximately 10″. Although no attempt is made here to reduce this to effective precipitation factors, it is quite evident that there is basically enough precipitation in average years to support tree growth. This map is not basically different in form from many maps that have been drawn on average annual precipitation for Michigan. Supporting evidence in the form of monthly precipitation maps and

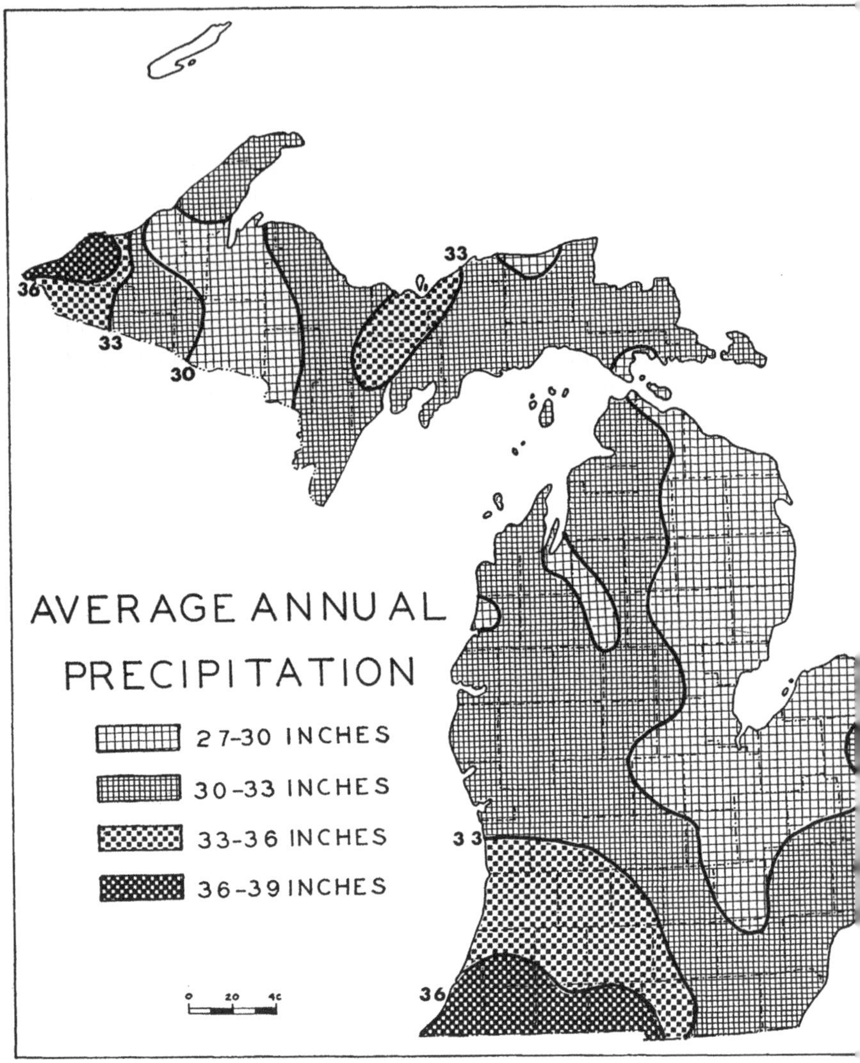
AVERAGE ANNUAL
PRECIPITATION
27-30 INCHES
30-33 INCHES
33-36 INCHES
36-39 INCHES
36
33
30
33
33
36

FIG. 1.

seasonal maps were drawn, but are not here depicted. However, this information is used in making some of the general conclusions of this paper.

A Wet Year

There was difficulty in defining a wet year for the state. To facilitate some understanding of what a wet year is, the 4 wettest years for each station, 50 stations, were tabulated and the wet year as defined in Fig. 2 represents the most frequently occurring year in the study.

A wet year as shown by the map reveals that, for many stations, 10″ to 15″ of precipitation above average occur, and this is particularly true in the wetter parts of the state, namely Southwestern Michigan and the western Upper Peninsula. An analysis of the wet year statistics since 1873 reveals that in climatic history the 3 wettest years were 1942, 1945, and 1951, and each one of these years was separated from the earlier and later periods by drier years. A cluster of several years of above-average precipitation took place in the year groups 1881-83, 1902-5, and 1911-12 as well as 1949-51.

At this point it should be noted that although many stations experienced a wet year in 1951, for some stations it was not necessarily so, or it was for them an average year or a slightly wetter but not extreme year. The question also may be raised as to whether it was exceptionally wet throughout the year or did the above-average amount represent a seasonal distortion. The climatic charts leave the impression that the latter is the case, and it is an exceptional amount in a season or month that causes the extreme. This could also be interpreted in view of the problem that a cluster of years are wet separated by dry seasons of one or more years in length. Even in the predominantly wet years studied, at least one month is dry and drier than the average.

Dry Years

The third map, Fig. 3, illustrates the dry year for Michigan. Precipitation statistics vary from around 16″ for some stations in the southeast to somewhat more than 30″ in the western part of the Upper Peninsula. Again it is possible to say that a dry year means usually 10″ less than normal for any station experiencing an average dry year. The year here is 1930. This was by far the driest year for more stations than any other year. The 5 driest

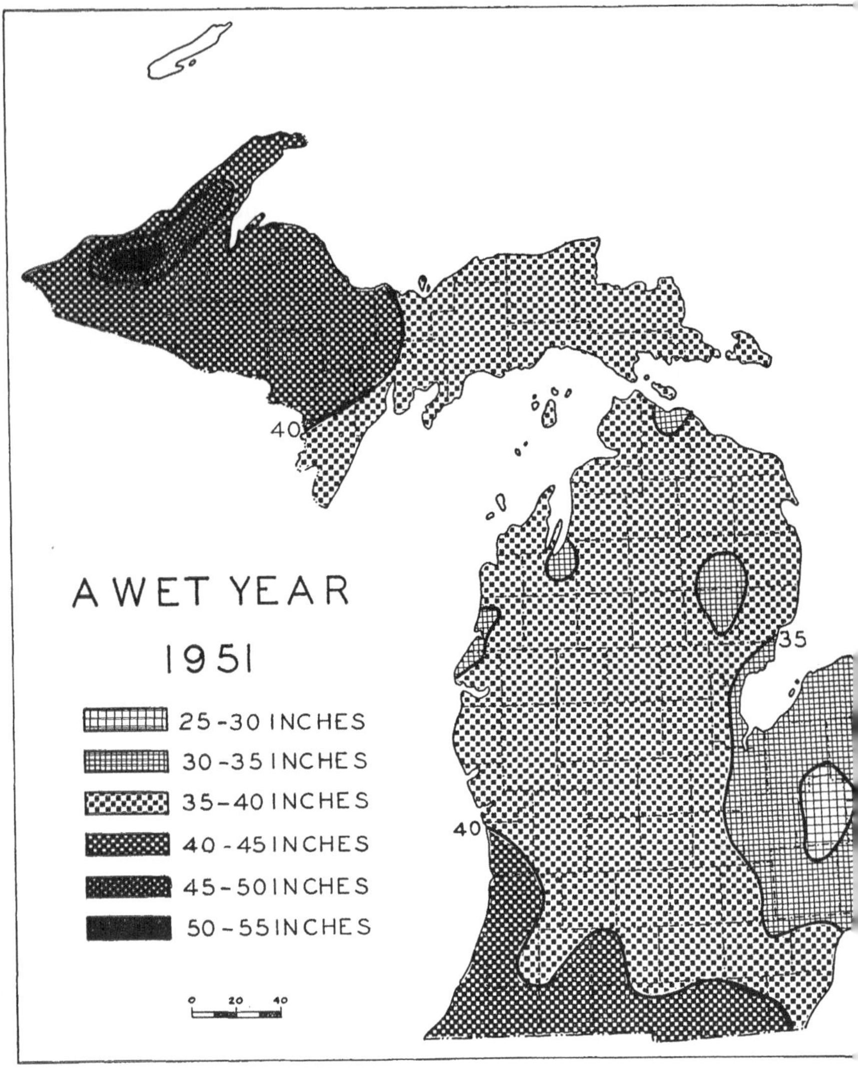
A WET YEAR
1951
25-30 INCHES
30-35 INCHES
35-40 INCHES
40-45 INCHES
45-50 INCHES
50-55 INCHES
0 20 40
40
35
40

Fig. 2.

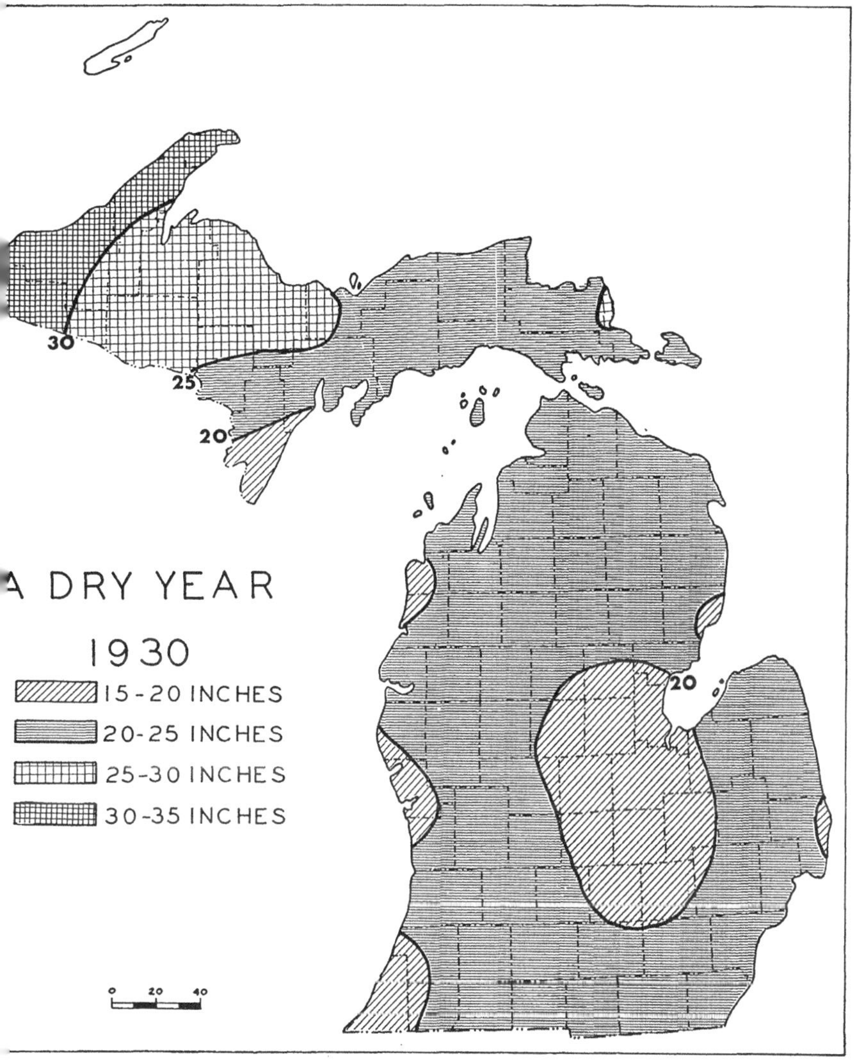

A DRY YEAR
1930
15-20 INCHES
20-25 INCHES
25-30 INCHES
30-35 INCHES
30
25
20
20
0 20 40

FIG. 3.

years in the record are 1930, 1934, 1936, 1925, and 1946. The four clusters of dry years are 1941-46, 1934-36, 1901-2, 1912-13. 1930 and 1925 were practically singularly dry years, preceded and succeeded by years of heavier rainfall.

Again it should be pointed out that this map represents a year in which a large number of stations had an annual amount of precipitation that was far below the average, occurring perhaps only once in 20, 40, 60, or 80 years, depending upon the individual station; whereas, for another station it was somewhat less than a dry year. This presents the problem that climatic statistics over a large state area have such a wide range that there is difficulty expressing a generalized theory for the entire state. The dry year as revealed by the map is nevertheless a static fact. The presentation of the maps to represent a dry year and a wet year are more realistic than trying to represent a composite of a wet year made up of records from many years, and likewise a dry year map based on the same technique.

Wet and Dry Months

A study of the average precipitation statistics shows that the most frequently occurring wet month in Michigan is September. This is true in nearly all the Lower Peninsula with the exception of a series of stations from lower Western Michigan to Detroit that have the wettest month in May and a few in June. In the Upper Peninsula the situation is more confusing with most of the western part having the wettest months in June, but in addition August, September, October, and November are recorded.

In the study of the 50 stations for wet and dry years the wettest month is May, but all summer months are represented, with February and November the driest in the list. Of the 3 extreme wettest months on record 2 were in June and 1 in October. The conclusion might be made that if it is to be a wet year it is going to be the summer that is the wet phase.

The dry month of the year is almost universally February for Michigan, with a few stations in the southeast experiencing a dry month in January. A few stations in the Upper Peninsula have their dry month in December, March, or April. The dry month is much more clearly defined than the wet month of the year. In the study of the 50 stations of the wettest and driest, this basic conclusion was maintained but accented by that in the wettest years the wettest months are summer months, May through September, and the driest months are December through April, although July and August can be dry. This last conclusion is part of a significant climatic pattern in Michigan.

SUMMARY

The study of the extremes of precipitation years and periods reveals a few patterns that could be further studied:

1. Long periods of consecutive days of precipitation are not uncommon. Most long periods occur in winter and in Upper Peninsula stations. The shorter periods occur in the Lower Peninsula and in summer.
2. Wet years occur both in clusters and in single years, with 1951 being an extreme year for more stations than any other.
3. Dry years occur both in clusters and single years, but 1930 was driest for more stations than any other.
4. The use of the term wet or dry year based on frequency of occurrence is an alternative technique to using absolute maximum, or absolute minimums, averages, medians, or modes. The basis of selection on a partial random basis perhaps could eventually lead to a more useful climatic statistic.
5. Most climatic stations experience annual distribution of a double maximum period of precipitation separated by drier period.

PAPERS OF THE MICHIGAN ACADEMY OF SCIENCE, ARTS, AND LETTERS
VOL. XLIX, 1964 (1963 MEETING)

A SERVICE CLASSIFICATION OF MICHIGAN CITIES [1]

EARL J. SENNINGER, JR.
Flint Community Junior College

THE statistical information revealed by the 1960 Census of Population indicated the continued urban growth of the U. S. This pattern of growth has affected all states, therefore Michigan is no exception. Unfortunately very little is known about the specific service functions of Michigan cities other than the visual evidence of such things as retail stores, doctors' and lawyers' offices, manufacturing plants, and assorted personal services. Almost everyone is aware that the functional character of the many cities and urban places of Michigan are different. It was apparent, therefore, that a need existed for defining and analyzing the significant service functions of this state's major cities.

The purpose of this paper is to describe the service functions of the 74 cities (and urban places) [2] in Michigan which had a population of 10,000 or more inhabitants at the time of the 1960 Census [3] (Fig. 1).

This study is patterned after the one published by Howard J. Nelson,[4] who developed a service classification used in classifying 897 individual urban concentrations in the U. S. which had 10,000 or more inhabitants according to the 1950 Census. Since this study is concerned with a smaller geographic area, fewer cities, and a different purpose it was necessary and also useful to modify Nelson's method of classification to fit the purpose of this study.

The Census of Population contains detailed occupational information for each of the 74 cities. Table I illustrates the variety of occupational information available in the Census and also the

[1] This study was made possible by a Faculty Research Grant from the Flint Community Junior College.

[2] Lakeview and Eastlawn were the only unincorporated cities in this study.

[3] *U. S. Census of Population*, 1960, General Social and Economic Characteristics, Michigan, Final Report PC (1) -24C., Table 75, Washington, D. C.: U. S. Government Printing Office, 1962.

[4] Howard J. Nelson: "A Service Classification of American Cities," *Economic Geography*, 31 (1955), 189-210.

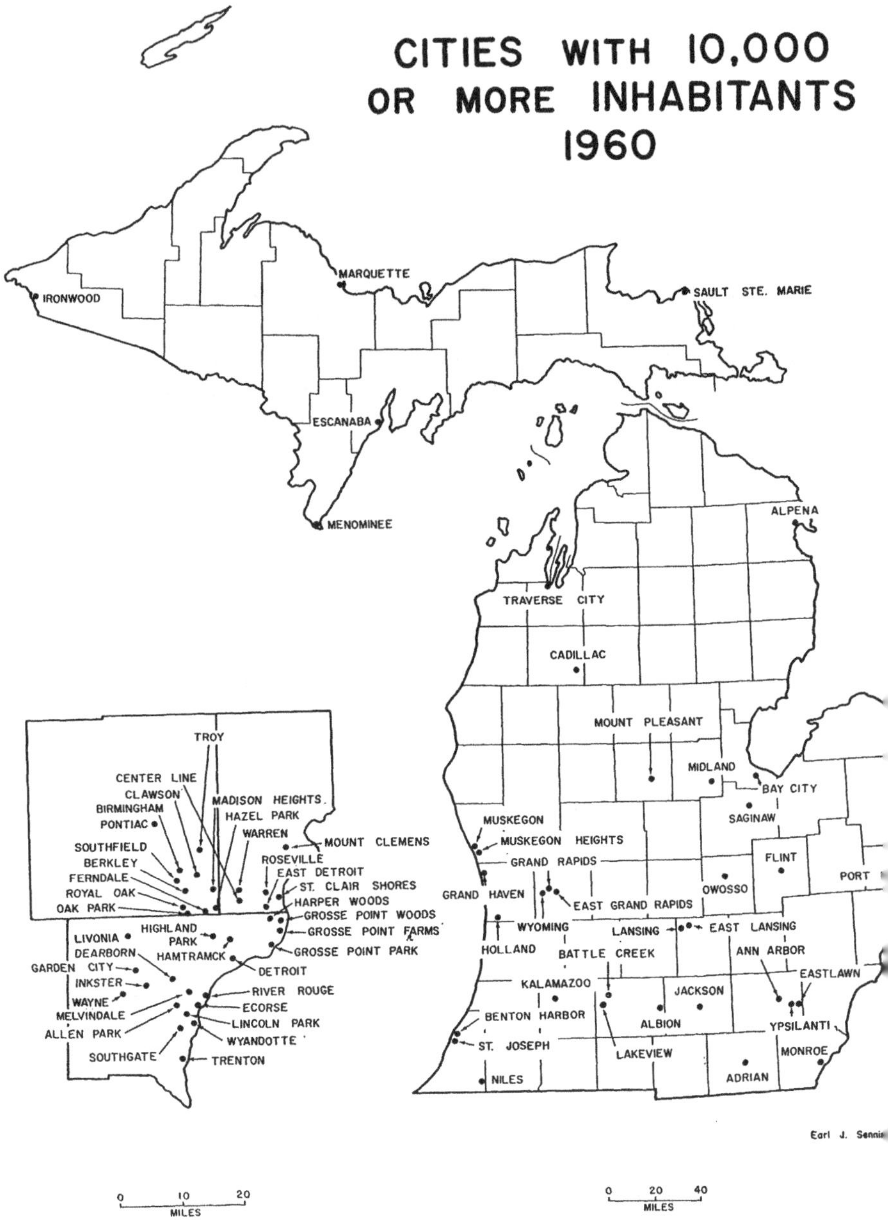
CITIES WITH 10,000
OR MORE INHABITANTS
1960
IRONWOOD
MARQUETTE
SAULT STE. MARIE
ESCANABA
MENOMINEE
ALPENA
TRAVERSE CITY
CADILLAC
MOUNT PLEASANT
MIDLAND
BAY CITY
SAGINAW
MUSKEGON
MUSKEGON HEIGHTS
GRAND RAPIDS
GRAND HAVEN
EAST GRAND RAPIDS
OWOSSO
FLINT
PORT
WYOMING
LANSING
EAST LANSING
HOLLAND
BATTLE CREEK
ANN ARBOR
EASTLAWN
KALAMAZOO
JACKSON
BENTON HARBOR
ALBION
YPSILANTI
ST. JOSEPH
LAKEVIEW
MONROE
NILES
ADRIAN
TROY
CENTER LINE
CLAWSON
MADISON HEIGHTS
BIRMINGHAM
HAZEL PARK
PONTIAC
WARREN
SOUTHFIELD
MOUNT CLEMENS
ROSEVILLE
BERKLEY
EAST DETROIT
FERNDALE
ST. CLAIR SHORES
ROYAL OAK
HARPER WOODS
OAK PARK
GROSSE POINT WOODS
HIGHLAND PARK
LIVONIA
GROSSE POINT FARMS
DEARBORN
GROSSE POINT PARK
HAMTRAMCK
GARDEN CITY
DETROIT
INKSTER
RIVER ROUGE
WAYNE
ECORSE
MELVINDALE
LINCOLN PARK
ALLEN PARK
WYANDOTTE
SOUTHGATE
TRENTON
0 10 20
MILES
0 20 40
MILES

Fig. 1.

categories which were combined and/or omitted in this study. I saw no value in including elements of the labor force from such categories as Agriculture, Forestry and Fisheries, Construction, and Industry Not Reported. The following 10 occupational service

TABLE I

MAJOR OCCUPATIONAL ACTIVITIES

Census Classification by Industry Groups *	Service Classification
Agriculture	Omitted
Forestry and fisheries	Omitted
Mining	Mining
Construction	Omitted
Manufacturing	Manufacturing
Railroads and railway express service	
Trucking service and warehousing	
Other transportation	
Communications	Transportation and communication
Utilities and sanitary services	Utilities
Wholesale trade	Wholesale trade
Food and dairy products stores	
Eating and drinking places	
Other personal services	Retail trade
Finance, insurance, and real estate	Finance, insurance, and real estate
Business services	
Repair services	
Private households	
Other personal services	Personal services
Entertainment and recreation services	
Hospitals	
Educational services: government	
private	
Welfare, religious, and nonprofit membership organizations	
Other professional and related services	Professional services
Public administration	Public administration
Industry not reported	Omitted

* *U. S. Census of Population, 1960,* General Social and Economic Characteristics, Michigan, Final Report PC (1) -24C., Table 75 (Washington, D. C.: U. S. Government Printing Office, 1962).

categories were chosen for investigation in this study: Mining; Manufacturing; Transportation and Communication; Wholesale Trade; Retail Trade; Finance, Insurance, and Real Estate; Pro-

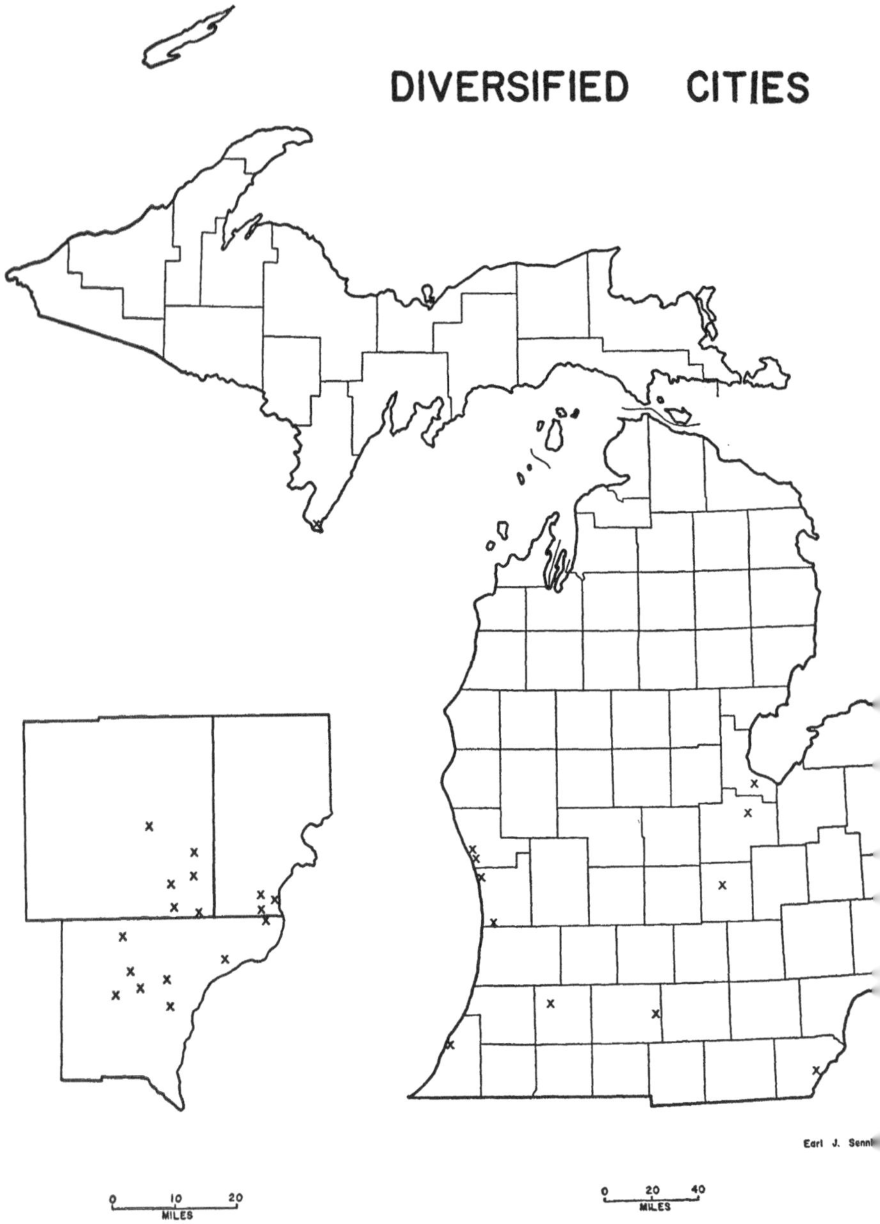

Fig. 2.

fessional Service; Personal Service; Public Administration; and Utilities. One of the basic problems encountered in this study was the fact that the cities varied greatly in population, with the largest being Detroit and the smallest Cadillac with 1,670,144 and 10,112 respectively.

PROCEDURE

The percentage of the labor force engaged in the 10 occupational categories was computed for each of the 74 cities and these in turn were used to determine the average percentage of persons employed in each of the 10 occupational categories described in this study (Table II). It should be noted that these are listed in rank order starting with manufacturing (38.31), the highest, and mining (.50), the lowest.

In addition to computing the averages for each of the selected occupational categories an additional calculation was made to determine the variation from the average for each category for the 74 cities. The degree of variation, therefore, was the method by which the significance of each city's occupational activities could be ranked. This relatively simple measure of variation is known as the Standard Deviation and, hereafter, will be referred to as the SD.

The computation of the percentage of the labor force of each of the cities in the various categories, the average for each of the categories, and the SD's from the averages were obtained and analyzed with the aid of an IBM 1620 Data Processing System.[5]

Three degrees of variation from the average were established for each of the categories and used to classify the relative significance of each occupational category for each of the 74 cities. To use an example, cities that ranked 1 SD above the average in Retail Trade were given an R rating, those over 2 SD's an R2 rating, and those over 3 or more SD's an R3 (see Tables II and III for a complete illustration of this procedure).

By use of the above rating procedure each of the 74 cities was ranked according to its degree of significance in each service category. Table III shows the results of this rating procedure, and it also illustrates that the 74 cities exhibit great variations in their occupational service activities to the extent that some rate high in as many as four categories.

[5] I am indebted to Miss Patricia Graff, IBM Systems engineer, for her valuable assistance in developing the program used in this study and to the Flint Branch for the use of their 1620 Data Processing System.

TABLE II

AVERAGES AND STANDARD DEVIATIONS FOR SELECTED ACTIVITY GROUPS

	Manu-facturing	Retail Trade	Profes-sional Service	Personal Service	Transpor-tation and Communi-cation	Finance, Insurance, and Real Estate	Public Adminis-tration	Wholesale Trade	Utilities	Mining
Average	38.31	15.44	14.79	7.29	4.60	3.65	3.64	3.06	1.45	.50
S.D.	11.39	2.64	8.56	1.99	2.20	1.67	1.44	1.59	.80	3.05
Average plus 1 S.D.	49.70	18.08	23.35	9.28	6.80	5.32	5.08	4.65	2.25	3.55
Average plus 2 S.D.	61.09	20.72	31.91	11.27	9.00	6.99	6.52	6.24	3.05	6.60
Average plus 3 S.D.	72.48	23.36	40.47	13.26	11.20	8.66	7.96	7.83	3.85	9.65

TABLE III

A SERVICE CLASSIFICATION OF MICHIGAN CITIES

Key

	Plus 1SD	Plus 2SD	Plus 3SD
Manufacturing	Mf	Mf2	Mf3
Retail trade	R	R2	R3
Professional service	Pf	Pf2	Pf3
Transportation and communication	T	T2	T3
Personal service	Ps	Ps2	Ps3
Public administration	Pb	Pb2	Pb3
Wholesale trade	W	W2	W3
Finance, insurance, and real estate	F	F2	F3
Mining	Mi	Mi2	Mi3
Utilities	U	U2	U3
Diversified	D		

Adrian	R
Albion	D
Allen Park	D
Alpena	R2
Ann Arbor	Pf3
Battle Creek	PsPb
Bay City	D
Benton Harbor	Ps2
Berkley	R
Birmingham	W2F
Cadillac	R2Ps
Center Line	Mf
Clawson	D
Dearborn	D
Detroit	D
East Detroit	D
East Grand Rapids	W3F2
East Lansing	Pf3Pb
Eastlawn	T2
Ecorse	Mf
Escanaba	T3RPsPb
Ferndale	D
Flint	Mf
Garden City	D
Grand Haven	D
Grand Rapids	W
Grosse Pointe Farms	F3Ps3
Grosse Pointe Park	WF2
Grosse Pointe Woods	WF2
Hamtramck	Mf
Harper Woods	D
Hazel Park	Mf
Highland Park	Ps
Holland	D
Inkster	D
Ironwood	Mi3RPb
Jackson	U3
Kalamazoo	D
Lakeview	F2Pb
Lansing	Pb3
Lincoln Park	T
Livonia	D
Madison Hgts.	Mf
Marquette	T3PfPb3
Melvindale	T2
Menominee	D
Midland	Mf
Monroe	D
Mount Clemens	RPb2U2
Mount Pleasant	MiPf2
Muskegon	D
Muskegon Hgts.	D
Niles	T
Oak Park	W2R3F

TABLE III (Continued)

A SERVICE CLASSIFICATION OF MICHIGAN CITIES

Owosso	D	Southfield	D
Pontiac	D	Southgate	MfTU
Port Huron	TRU2	Traverse City	RPfPsU2
River Rouge	Mf	Trenton	U3
Roseville	D	Troy	D
Royal Oak	D	Warren	Mf
Saginaw	D	Wayne	D
St. Clair Shores	D	Wyandotte	Mf
St. Joseph	D	Wyoming	TW
Sault Ste Marie	TRPsPb2	Ypsilanti	Pf2Ps

As would be expected there are a number of cities that do not rank high in any category to obtain a rank of plus 1 SD. These cities have been designated as "Diversified" cities, which simply means that they did not rank significantly high in any one category and, therefore, may be considered as "average" cities.

A discussion and brief explanation of each of the categories follow:

Manufacturing.—As one might expect this category is the most outstanding in terms of numbers of employed persons and it also has the highest average for any of the categories (38.31). There were only 11 cities that received a plus 1 rating; however, this is understandable since a plus 2 rating would require nearly 50% of the work force engaged in this category. Considering the fact that the average for this category is the highest in the study it is significant to point out that 47 cities had an average employment in this category which equaled or exceeded the average. The cities that ranked highest and lowest in this category were Midland (53.55) and East Lansing (6.99) respectively.

The concentration of manufacturing in Michigan cities is well known to most persons and obviously is greatest in the southeastern part of the Lower Peninsula. Most of the cities within the three-county Detroit Metropolitan area have achieved the average or above-average scores in this category. Further evidence of the Detroit area's concentration of manufacturing is borne out by the fact that 9 of the 11 cities with a plus 1 rating are located in the Detroit Metropolitan area. The only exceptions are Flint and Midland. The percentage of employment in this category drops off very sharply north of the Muskegon-Bay City line; however, Menominee (43.73) is the only exception. Generally speaking, the employment is below average in this category in the cities of the northern part of the Lower Peninsula and the Upper Peninsula.

Retail trade.—The geographic distribution of cities rating high in retail trade is relatively large and the centers are located in the northern, eastern, and southern parts of the state. There are 11 cities which received at least a plus 1 rating in this category, and 6 of these are located in the northern part of the Lower Peninsula and Upper Peninsula. These cities should be considered as the key retail trade centers for those areas. In addition, there is an understandable concentration of this activity in the Detroit Metropolitan area. It should also be noted that no city on the western side of the Lower Peninsula other than Traverse City is in this group.

Professional service.—Only 6 cities are outstanding in this category and as would be expected all but 1 of them are locations of large State-supported colleges and universities. In addition, a number of hospitals are also located in several of these cities, which adds to their significance as centers of professional service. East Lansing (56.06) and Ann Arbor (51.92) have the highest averages of the 6 cities.

Personal service.—The 9 cities in this category tend to fall into two distinct classes: those having a stable year-round population and those with a significant seasonal population increase due primarily to resorts or other tourist attractions. Grosse Pointe Farms (15.81) has the highest percentage of its labor force engaged in this activity and therefore received a rating of plus 3; Benton Harbor (11.88), a resort city, is the only one to receive a plus 2.

Transportation and communication.—The significance of population distribution, port facilities, and railroad routes are brought out in the cities that have reached at least a plus 1 rating in this category. 4 of the 10 cities are located in the Detroit area, 3 in the Upper Peninsula, and the others are scattered. The port cities are Marquette, Port Huron, Escanaba, and Sault Ste Marie. In addition to the port functions of these cities they would also have significant railroad facilities since they are situated at relatively strategic locations in the state. The railroad function would also be extremely important for such places as Eastlawn, Melvindale, Niles, and

Wyoming. Truck and air transportation facilities would also be of great significance in the Detroit Metropolitan area. It is interesting to note that Escanaba and Marquette, both in the Upper Peninsula, and port cities, have achieved a rating of plus 3.

Finance, insurance, and real estate.—This activity appears to be concentrated in three distinct areas of the state, the Detroit Metropolitan area, Grand Rapids, and Battle Creek. Of the 7 cities in this category, 5 are located in the Detroit Metropolitan area, and these, together with the other 2, would be generally regarded as high income communities.

Public administration.—Most of the 9 cities in this category are government administrative centers such as the state capitol of Lansing or county seats. 4 of these 9 cities are county seats. Lansing (9.14) and Marquette (8.48) have plus 3's; the latter has this rank because it is a county seat and a significant center of military activity. Sault Ste Marie (7.95), a county seat, center of military activity, and also the location of the Soo Locks, received a plus 2. It also seems significant to point out that 25 of the 74 cities are county seats.

Wholesale trade.—A relatively small percentage of the work force of the 74 cities is employed in this category (3.06). Two rather interesting concentrations of this activity occur in the state; they are in the Detroit Metropolitan and Grand Rapids areas. East Grand Rapids has a plus 3 rating while Birmingham and Oak Park have attained plus 2's. It should also be noted that none of the cities which are high in this category has achieved a rating of plus 1 in manufacturing.

Utilities.—This category has the second lowest average (1.45) and, with the exception of Jackson, all are located close to deep water. Two cities, Jackson and Trenton, received plus 3 ratings; the former is the headquarters of Consumers Power Company and the latter has a Detroit Edison generating plant. The other cities also have electric generating facilities.

Mining.—This category attained the smallest average in the study (.50). Only 2 cities achieved a rating; Ironwood (plus 3) and Mount Pleasant (plus 1). Ironwood, an iron mining city, had 26.10% of its work force engaged in this category while Mount Pleasant, a petroleum-producing center, had 3.78%. Only 2 other cities exceeded the average: they were Marquette (.79), an iron-producing city and Albion (1.76), a petroleum center. In addition, there were 18 cities that had no employment in this category.

Diversified cities.—All of the cities which did not meet the minimum requirement of the average plus 1 SD in any one category have been designated as diversified cities. 29 of the 74 cities analyzed in this study fall within this group. All but 1 of these cities fall south of the line running from Muskegon to Bay City (Fig. 2). The only city north of that line is Menominee. Most of the cities in this designation are located in the Detroit Metropolitan area. It is also interesting to note that Detroit, which is often regarded as a very significant manufacturing city, has been defined as diversified.

SUMMARY

This classification of Michigan cities has attempted to point out those cities which are significant for selected occupational service categories. One of the difficulties or shortcomings in a statistical study of this nature is the fact that Detroit and the Detroit Metropolitan area cities definitely influence the averages and standard deviations for each of the categories. However, I feel justified in using all 74 cities since each contributes to the urban character of the state.

In conclusion, it is hoped that this study has led to a further understanding of the service functions of the cities that have been analyzed. It seems feasible that other studies of this type would be valuable for the smaller cities and communities within the state.

PAPERS OF THE MICHIGAN ACADEMY OF SCIENCE, ARTS, AND LETTERS
VOL. XLIX, 1964 (1963 MEETING)

A POLYNOMIAL REPRESENTATION OF MICHIGAN POPULATION

W. R. TOBLER
The University of Michigan

THERE are many purposes for which specific and detailed information concerning the distribution of population is required. There are other purposes, however, for which the entire mass of information as represented by, say, the detailed tract data available from the Bureau of the Census, is not necessary. This very great detail is in fact often a handicap. The following represents an experimental attempt to use a technique which summarizes and filters information concerning geographical distributions.

The underlying concept is quite simple. It consists in describing an empirically observed distribution by a mathematical function. Suppose one has a set of observations which may be represented in the form of dots on a graph or frequency diagram. The procedure is then to find a mathematical expression for a curve which passes through all of the points on the graph. This equation may of course be quite complicated and difficult to obtain. In practice one does not look for a perfect description, but one which approximates the given empirical observations fairly well.

For geographical distributions a similar procedure is available. In this instance the equation no longer represents a line curve, but a "surface" which assigns a value to every location. Such an equation has been referred to as a *numerical map*.[1] The technique, incidentally, has also been employed in geology, meteorology, and geophysics; an extensive literature is available.[2]

[1] S. M. Ostrow, *Handbook for CRPL Ionospheric Predictions Based on Numerical Methods of Mapping*, National Bureau of Standards Handbook 90 (Washington: Government Printing Office, Dec. 1962), 40 pp.

[2] A small sample:

R. A. Bryson, "Fourier Analysis of Spatial Series," in *Quantitative Geography*, W. L. Garrison (ed.), forthcoming.

W. E. Byerly, *Fourier Series and Spherical, Cylindrical, and Ellipsoidal Harmonics* (Boston: Ginn, 1893), pp. 144-218.

B. Haurwitz and R. Craig, *Atmospheric Flow Patterns and Their Representation by Spherical Surface Harmonics*, Geophysical Research Paper No. 14 (Cambridge: Cambridge Research Center, 1952), 80 pp.

The detailed procedures required to obtain the descriptive equations may become quite complicated. One of the difficulties is that there are possibly several equations which will do the job. The advantages and disadvantages of each may rest on complicated and obtuse mathematical theorems, a discussion of which goes beyond the purposes of the present paper.[3]

An Example Experiment

The geographical distribution of population density in the state of Michigan has been chosen for use as an example experiment. Any other phenomenon or characteristic which varies from one part of the state to another could, of course, have been chosen.

The process began by a recording of the population density of each of the 83 Michigan counties. These densities were then assigned to the latitude and longitude of the centers of the counties[4] and recorded on punched cards. The specific form of the equation to be fitted was taken to be a polynomial of the type $P_n = (aX + bY + c)^n$. For example if $n = 2$, one obtains the quadratic polynomial,

$$P_2 = A_{00} + A_{20}X^2 + A_{10}X + A_{11}XY + A_{01}Y + A_{02}Y^2,$$

W. B. Jones and R. M. Gallet, "Methods for Applying Numerical Maps of Ionospheric Characteristics," *Journal of Research*, National Bureau of Standards, D, 66D (6) (Dec. 1962), 649-662.

W. C. Krumbein, "Trend Surface Analysis of Contour-type Maps with Irregular Control-Point Spacing," *Journal of Geophysical Research*, 67 (7) (July 1959), 823-834.

A. Prey, "Darstellung der Höhen- und Tiefenverhältnisse der Erde durch eine Entwicklung nach Kugelfunktionen bis zur 16. Ordnung," *Abhandlung der Königlichen Gesellschaft der Wissenschaften zu Göttingen*, Mathematische Physikalische Klasse, Neue Folge, Bd. XI, 1 (Berlin: Weidmannsche Buchhandlung, 1922), pp. 1-42.

S. M. Simpson, Jr., "Least Squares Polynomial Fitting to Gravitation Data and Density Plotting by Digital Computers," *Geophysics*, XIX (1954), 250-257.

[3] See: H. Mandelbaum, "Statistical and Geological Implications of Trend Mapping with Non-orthogonal Polynomials," *Journal of Geophysical Research*, 68 (2) (Jan. 1963), 505-519; or W. B. Jones and R. M. Gallet, "The Representation of Diurnal and Geographic Variations of Ionospheric Data by Numerical Methods," *Telecommunication Journal*, 5 (May 1962), 3-23; also W. C. Krumbein, *op. cit.*

[4] Since this project was initiated the U. S. Bureau of the Census has announced the availability of latitude and longitude information for the centers of population of small areas (National Location Code Areas). That an error is introduced by assigning densities to point locations is recognized. The pragmatic objective, however, justifies the procedure.

where X represents the latitude and Y represents longitude. The coefficients (A_{ij}) are numbers whose values are to be obtained and which depend on the particular geographical distribution. The polynomial form has the advantage of being extremely simple; in particular, the polynomial is easily integrated.[5]

The next step is to estimate the numerical values of the coefficients for several degrees of polynomial. The coefficients are obtained by standard least-square techniques. In the present instance the University of Michigan IBM 7090 digital computer performed the arithmetical work, using a stepwise multiple regression program with transformed variables.[6] The entire procedure should terminate when the geographical distribution has been reasonably well described. One test for this is to calculate the population density value which the equation obtained assigns to the centers of the counties. These values can then be compared with the actual (assigned) values at each of these points (see Appendix). If the equation fits well, there will be only a small difference in the two values at each point. The regression program provides an additional measure of the goodness of fit in the coefficient of determination (R^2), which can be interpreted as the proportion of the distribution satisfactorily described.

The results of the computations for $n = 1$ through $n = 3$ are given in the accompanying table (Table I). It can be seen that the fit is not particularly good, since the cubic is only 18% successful in describing the geographical variation of population density in the state of Michigan.[7] Casual observation (and the large variance),

[5] Polynomials also have disadvantages. The procedures employed here yield terms which are not independent. The constant term (A_{00}) therefore varies for each polynomial. See the references cited above (note 3). An excellent discussion and description of polynomial fitting also can be found in B. Arden, *An Introduction to Digital Computing* (Addison-Wesley, 1963).

[6] The author wishes to express his appreciation to the University of Michigan Computing Center for access to the 7090, and particularly to Professor Westervelt, who provided advice regarding the use of the regression program.

[7] Polynomials of higher degree ($n = 4$ through $n = 7$) were also computed but did not improve the description, apparently because of numerical difficulties. Rounding problems appear in the matrix inversion required for the least-squares solution. The 36×36 matrix for P_7 contains elements ranging in absolute value from 3×10^{-12} to 2×10^{12}, for example. It becomes necessary to eliminate certain variables from the analysis because of the computational problems. Conversion of the values to rectangular coordinates on Lambert's cylindrical equal area projection was of no apparent benefit, contrary to expectation.

however, indicate that the population density fluctuates very considerably. It is suggested that some other form of equation (e.g., Fourier series) might provide a more adequate description of the population density.

TABLE I

MICHIGAN, POPULATION DENSITY, 1960
Results of Polynomial Fitting

NOTE—The first subscript on the coefficient refers to the exponent of the latitude ($41 < X < 48$), the second to the exponent of the longitude ($-91 < Y < -82$, west longitudes taken as negative). Coefficients not listed are zero. The mean density is 147.8 persons per square mile, with a variance of 496.0 persons per square mile.

Linear equation (P_1 ; $R^2 = 0.0799$)

$A_{00} = 6314.4$ $A_{10} = -68.674$
$A_{01} = 36.903$

Quadratic equation (P_2 ; $R^2 = 0.1523$)

$A_{00} = 207{,}528.7$ $A_{01} = 2389.3$
$A_{10} = -4547.1$ $A_{11} = -52.336$

Cubic equation (P_3 ; $R^2 = 0.1811$)

$A_{00} = 126{,}895.7$ $A_{11} = 38.076$
$A_{10} = -2214.3$ $A_{30} = -.73229$
$A_{01} = 171.58$ $A_{21} = -1.2831$
$A_{03} = 0.034469$

A second attempt was made, using a somewhat "smoother" distribution, since a simpler distribution should be easier to describe. The rural population density is less variable than the total density, and the same procedure was attempted for this class of information. In this case the attempt was, as expected, considerably more successful; 65% of the geographical distribution is described by five numerical coefficients (Table II).

APPLICATIONS

(NOTE—All of the applications given here assume that the equations obtained provide an adequate [for the purpose contemplated] description of the phenomenon in question. In the experiment discussed here, only 65% of the geographical variation of the rural population density in Michigan has been represented, which [for my purpose] does not appear adequate. It is anticipated that a more complete description can be obtained by a more complicated function, and further work in these directions is contemplated.)

The contemplated use of analytical representations of geographical distributions on my part is primarily cartographic. For example, a number of map projections have recently been derived which require a description of population density in a form which allows integration of certain partial differential equations.[8] The polynomial representation is more satisfactory for this purpose than are census tables, though less accurate. For the machine production of population density contour maps,[9] the polynomial representation is again more convenient than are census tables. A very simple computer program has been prepared which makes such maps by evaluating the descriptive equation some 5000 times per page and then assigning appropriate symbols to class intervals. The polynomial, of course, should not be employed for areas outside Michigan, since no observations were taken from these locations.

TABLE II

MICHIGAN, RURAL DENSITY, 1960

Mean rural density, 38.76 persons per square mile, with variance of 33.08 persons per square mile.

Sextic equation coefficients (P_6 ; $R^2 = 0.6463$)

$A_{00} = 5513.6$	$A_{01} = 28.718$
$A_{10} = -89.134$	$A_{60} = .44669 \times 10^{-7}$
$A_{24} = .55034 \times 10^{-8}$	

The description by powers of latitude and longitude also cannot be said to "explain" the distribution of population, since there is no direct theoretical structure which would lead one to accept (or expect) such an explanation. This remark does not imply that one cannot assign interpretations to the coefficients; it is intended to emphasize that the present objective is one of obtaining convenient and specific characterizations of the manner in which the

[8] W. R. Tobler, "Geographic Area and Map Projections," *Geographical Review*, LIII (1) (Jan. 1963), 59-78.

[9] Tobler, "Automation and Cartography," *Geographical Review*, XLIX (4) (1959), 526-534, and the references therein.

incidence of phenomena differs from place to place on the surface of the earth. Certainly the anomalies (or residuals) can be examined to make statements concerning the departures of certain counties from the statewide trends, but this is not the primary purpose; furthermore, if the intent is to " explain " the distribution of population density, then a model should be employed which has some theoretical justification; such a model should also give better results, but it might be based on information which is more difficult to obtain empirically than latitude and longitude.

Within-city variations might also be described by the procedures presented here. Equally apparent should be the application of a descriptive equation which includes the temporal variation of some subject of study.[10] In this case the generating polynomial might be of the form $P_n = (aX + bY + cT + d)^n$, where X and Y are the locational and T the temporal label.

[10] W. B. Jones and R. M. Gallet, "The Representation of Diurnal and Geographic Variations . . . ," *loc. cit.*

APPENDIX

MICHIGAN, RURAL POPULATION DENSITY, 1960

County	Location *		Actual Density †	Computed Density ‡
Alcona	44.62	–83.55	9.000	23.48
Alger	46.50	–86.58	5.430	2.684
Allegan	42.58	–85.87	51.41	60.98
Alpena	45.00	–83.58	24.30	17.06
Antrim	45.00	–85.12	21.00	14.04
Arenac	44.03	–83.92	26.00	33.62
Baraga	46.50	–88.42	7.000	8.522
Barry	42.58	–85.33	45.60	62.97
Bay	43.83	–84.00	76.80	37.59
Benzie	44.62	–86.00	24.00	18.55
Berrien	42.00	–86.37	130.0	75.00
Branch	41.88	–85.07	50.73	84.16
Calhoun	42.23	–85.00	68.45	74.11
Cass	41.88	–86.00	60.38	79.93
Charlevoix	45.08	–85.00	18.78	13.11
Cheboygan	45.50	–84.75	11.94	8.263
Chippewa	46.33	–84.75	8.540	1.329
Clare	44.00	–84.83	20.00	31.41
Clinton	42.92	–84.58	51.61	57.04
Crawford	44.62	–84.67	8.000	20.52
Delta	45.92	–87.00	11.54	5.730
Dickinson	46.00	–87.92	8.246	7.565
Eaton	42.58	–84.83	53.24	65.03
Emmet	45.50	–84.92	20.88	8.111
Genesee	43.00	–83.59	130.4	59.51
Gladwin	44.00	–84.33	21.00	32.88
Gogebic	46.42	–89.75	6.531	14.92
Grand Traverse	44.62	–85.50	32.40	19.10
Gratiot	43.25	–84.72	37.96	48.24
Hillsdale	41.85	–84.53	44.46	87.89
Houghton	46.97	–88.67	23.05	10.74
Huron	43.75	–83.05	4.920	43.31
Ingham	42.58	–84.37	67.48	67.13
Ionia	42.92	–85.08	48.98	55.06
Iosco	44.37	–83.67	30.00	27.70
Iron	46.28	–88.67	10.95	9.601
Isabella	43.67	–84.75	35.32	38.52
Jackson	42.25	–84.40	79.29	76.44
Kalamazoo	42.23	–85.52	89.71	71.83
Kalkaska	44.62	–85.17	7.000	19.59
Kent	43.00	–85.50	77.88	51.60
Keweenaw	47.42	–88.12	4.000	10.32
Lake	44.00	–85.75	9.000	29.32
Lapeer	43.12	–83.25	53.74	57.99
Leelanau	44.92	–85.75	12.00	14.51
Lenawee	41.85	–84.12	59.84	90.19
Livingston	42.58	–83.87	57.62	69.61
Luce	46.50	–85.58	5.328	1.041
Mackinac	46.12	–85.00	6.930	2.634
Macomb	42.63	–82.98	106.2	73.08

APPENDIX (Continued)

MICHIGAN, RURAL POPULATION DENSITY, 1960

County	Location *		Actual Density †	Computed Density ‡
Manistee	44.37	-86.00	19.14	22.45
Marquette	46.50	-87.58	11.40	5.398
Mason	44.00	-85.25	25.08	30.36
Mecosta	43.67	-85.55	21.72	36.28
Menominee	45.50	-87.58	12.49	9.521
Midland	43.67	-84.33	45.08	39.93
Missaukee	44.33	-85.08	12.00	24.62
Monroe	41.85	-83.43	129.4	94.36
Montcalm	43.25	-85.25	39.60	46.42
Montmorency	45.00	-84.08	7.000	15.83
Muskegon	43.25	-86.25	98.01	43.67
Newaygo	43.55	-85.75	24.08	38.32
Oakland	42.63	-83.37	92.87	70.86
Oceana	43.58	-86.25	30.00	36.62
Ogemaw	44.37	-84.08	16.00	26.41
Ontonagon	46.67	-89.42	8.000	13.69
Osceola	44.00	-85.33	23.00	30.18
Oscoda	44.62	-84.12	6.000	21.83
Otsego	45.00	-84.58	9.240	14.84
Ottawa	42.92	-86.00	93.98	51.99
Presque Isle	45.33	-83.92	12.80	11.55
Roscommon	44.37	-84.67	13.00	24.83
Saginaw	43.25	-84.00	72.07	51.12
St. Clair	43.00	-82.75	72.72	63.91
St. Joseph	41.88	-85.50	51.63	82.11
Sanilac	43.50	-82.67	33.00	51.20
Schoolcraft	46.25	-86.25	3.185	2.874
Shiawassee	42.92	-84.12	55.66	59.06
Tuscola	43.50	-83.33	45.42	47.97
Van Buren	42.25	-86.00	64.15	69.38
Washtenaw	42.25	-83.77	70.80	79.82
Wayne	42.25	-83.25	109.8	82.85
Wexford	44.37	-85.50	14.46	23.15

* Latitude and longitude of the assigned centers of the counties, in decimal degrees.

† Rural density defined as the nonurban population per county divided by the area of the county (in miles). The population data were taken from preliminary census information; slight discrepancies exist between these data and the published census materials.

‡ Computed from the sextic polynomial given in the body of the paper, at the latitudes and longitudes given. All numbers rounded to four digits. The detailed computational results can be obtained from the author.

PAPERS OF THE MICHIGAN ACADEMY OF SCIENCE, ARTS, AND LETTERS
VOL. XLIX, 1964 (1963 MEETING)

HISTORY

PHILIP OF HESSE AND THE ESTABLISHMENT OF THE UNIVERSITY OF MARBURG

PAUL G. FRIED
Hope College

THE figure of Philip of Hesse (1504-67) looms large in the history of the Reformation and undoubtedly deserves more attention than it has received recently. Though a good deal has been written about various aspects of Philip's career in the last 50 years,[1] the last full-scale biography of the Landgrave dates back to 1830.[2] There is a real need, then, for a study which will bring that work up to date by incorporating the results of 130 years of scholarship. While a short paper obviously cannot meet this need, it may serve to call attention to the importance of the topic and to some of the work which could be done in this area.

This study will focus primarily on a period of less than a year which, to me, appears central to the career of Philip as a leader of the German Reformation. Between August 1526, when the Diet of Speyer adjourned, and May of the following year, Philip introduced the Reformation in Hesse and established the first Protestant university at Marburg. As we shall see, the two events are closely related to each other and they illustrate Philip's bold decision to take advantage of the political situation created by the Diet of Speyer.

Philip began his long rule in Hesse in 1518, when he was not quite 14 years old. Three years later, he entered the larger stage of German politics when he attended the Diet of Worms. Here he met Luther for the first time, and it appears that the reformer made a great impression on the 17-year-old prince.[3] Though it is

[1] Most useful was F. H. Littell, *Landgraf Philipp und die Toleranz* (Bad Nauheim, 1957).

[2] Dietrich Christoph von Rommel, *Philipp der Grossmuetige* (Giessen 1830) Two volumes. Also Wigand Lauze, *Leben und Thaten des Durchleuchtigsten Fursten und Herren Philippi Magnanimi.* Contemporary 16th century biography republished in 1841.

[3] Friedrich Kuech, "Landgraf Philipp auf dem Wormser Reichstage des

not clear just when and under what circumstances Philip made his decision for the Reformation, he showed increasing interest in the new faith in the two or three years which followed his first encounter with Luther.

Possibly a chance meeting with Melanchthon in July 1524 provided the final impetus. By this time Philip may already have been considering the possibility of reforming the church in his territory. In any case, he discussed the matter with Melanchthon, who promised to send him a summary of Luther's teachings as well as some practical recommendations for ecclesiastical reforms in Hesse. Melanchthon kept his word and also sent some of his own writings as well as Luther's German translation of the New Testament,[4] which Philip studied carefully.

By March of 1525 the young Landgrave freely discussed the new faith in letters to his mother and to his father-in-law. Rather dramatically he wrote, according to F. W. Hassencamp, that "he would rather lose his life and his country than to depart from the Word of God." [5] In the meantime he had also become acquainted with Adam Krafft, an able preacher, who had been accused of inciting the people of Hesse to revolt. Philip investigated the charge by listening to Krafft preach and promptly appointed him as his own court chaplain.[6]

Philip may not have been fully aware of the political implications of his conversion. In any case, he made no immediate effort to impose his new religious beliefs on his people. But, in the spring of the following year, when the Catholic princes formed the League of Dessau as a means of fighting against the reform movement, Philip concluded that if it were to survive, the Reformation would have to become a political movement as well. His energetic efforts to create a similar league among the Lutheran princes resulted in the Torgau compact. Thus, when the Imperial Diet met at Speyer that summer, the dividing lines were clearly drawn and the 22-year-

Jahres 1521" in *Festschrift zum Gedaechtniss Philipps des Grossmuetigen* (Kassel, 1904), p. 207.

[4] Johann B. Rady, *Geschichte der katholischen Kirche in Hessen* (Mainz, 1904), pp. 634-636.

[5] *Hessische Kirchengeschichte* (Marburg, 1852), pp. 45-46.

[6] F. W. Schaefer, "Adam Krafft, der Reformator Hessens," *Archiv fuer Hessische Geschichte*, VIII (Darmstadt, 1912), 1-46 and 67-110. The letter of appointment appears on p. 75.

old Philip had emerged as one of the principal backers of the reform movement.

The extended negotiations and particular political circumstances which contributed to the rather tolerant formula adopted by the Diet of Speyer have received considerable attention and need not be reviewed here.[7] The Diet agreed that each prince should conduct himself according to the dictates of his conscience before God and the Emperor until a general council could be called. Philip had already recognized that no general solution to the religious crisis was likely to follow and that the responsibility for reforming the Church would have to be assumed by the territorial rulers.[8]

Apparently it was also during the Diet at Speyer that Philip first met Francis Lambert of Avignon, who was to become his principal adviser in preparing the reform of the Church in Hesse. Lambert was a former member of the Franciscan order who had come under the influence of Zwingli and had then gone on to Wittenberg. He had remained there for about two years, lecturing and writing. During this time he had considerable contact with Luther, but, since he could not master German, he returned to France and eventually settled in Strasbourg.[9] Philip took an immediate liking to the fiery Frenchman and asked him to come to Hesse. Apparently Philip had earlier hoped that either Luther or Melanchthon would be willing to come to Hesse to help him introduce the reforms they advocated, and he appointed Lambert only after he learned that neither of the Wittenberg reformers could accept his invitation.

Philip left Speyer early in August of 1526. It is not clear whether Lambert accompanied him or followed a few weeks later. In any case, by the end of September, Lambert was in Hesse. He had, in the meantime, prepared the theological statements as well as a detailed plan with which the Reformation was to be introduced in Hesse. By October 5 invitations were sent out, in the name of the

[7] Theodor Brieger, *Der Speirer Reichstag von 1526 und die religioese Frage der Zeit* (Leipzig, 1909). For opposite view see Walter Friedensburg, *Der Reichstag zu Speier 1526 im Zusammenhang der politischen und kirchlichen Entwicklung Deutschlands im Reformations Zeitalter* (Berlin, 1887).

[8] Ernst Benz, "*Die Gruendung der Universitaet Marburg und die Reformation*" in *Pastoralblatt* fuer Kurhessen-Waldeck (Marburg, Aug. 1952), p. 8. I am indebted to Frau Dr. Doehner of Marburg for locating this most useful recent article.

[9] Gerhard Mueller, *Franz Lambert von Avignon und die Reformation in Hessen* (Marburg, 1958).

Landgrave, to mayors, city councillors, pastors and other secular and ecclesiastical leaders in Hesse to come to Homberg on October 20 for a discussion of the great religious issues which were disturbing the tranquillity of the land. This meeting, generally referred to as the "Synod of Homberg," in reality represented an important innovation; for in its composition it combined the functions of a meeting of the estates and of a provincial synod.

The synod convened in the early morning of October 21, 1526. Johann Feige, Philip's chancellor, opened the proceedings and briefly outlined the purpose for which the assembly had been called. He explained that Philip, charged by the Diet of Speyer with responsibility for the temporary settlement of the religious question, had decided to follow the example of early Christian rulers by calling a synod; and that, in order to provide a basis for discussion of the needed reforms, he had commissioned Lambert with the preparation of the articles, now posted on the doors of the church.[10]

Presently Lambert was introduced and he read, explained, and justified the 158 theses contained in his *Paradoxa*. Since Lambert did not know German, he presented his ideas first in Latin. The most important ones were then explained in German by Adam Krafft, Philip's court chaplain, for the benefit of those who did not understand Latin. Lambert's program listed the familiar abuses which had crept into the Church and the steps which had to be taken to achieve reform. His first paradox, the church is deformed—it must be reformed—was followed by the explanation that only God's Word teaches what is deformed and what has to be reformed.[11]

Only two of the assembled clerics took up Lambert's challenge and rose in defense of the old faith. One was Nicolaus Ferber, the guardian of the Franciscan monastery in Marburg.[12] Interestingly enough he did not attempt to reply to Lambert's charges of needed reform or try to refute them on the basis of the Bible. Rather, he questioned the right of the assembly to discuss religious questions or to make any changes in the established order. Philip himself, as well as his chancellor, answered Ferber and demanded that he dispute with Lambert on the basis of the Bible. This Ferber refused

[10] Rady, *op. cit.*, pp. 667-668, also Mueller, *op. cit.*, pp. 33-43.

[11] *Ibid.*

[12] Franz Emil Weber, *Die Einfuehrung der Reformation in Hessen und der Franziskaner Nikolaus Herborn* (Kassel, 1907). Also Otto Clemen, "Zu Landgraf Philipps reformatorischen Anfaengen," *Zeitschrift des Vereins fuer hessische Geschichte*, XXXIV (Kassel, 1910), 109-114.

to do, and, knowing himself in dangerous minority, he departed before the end of the meeting.[13]

The only other, somewhat pathetic, defender of the old order was Johannes Sperber. He was pastor of a small village near Kassel. He appeared both confused and ignorant, and his question about continued worship of the Virgin Mary was refuted so decisively by Lambert that everyone felt sorry for him.[14] Since no one else rose to debate the *Paradox,* the synod closed on the third day after a committee had been elected which was to work out the practical details of reform and the new church order.

Not much is known of the members of this committee except that they met in Marburg and that they completed the draft of the new reformation order by about the middle of December. This new church "constitution" provided for a democratic, congregational structure in which the individual congregations were to be almost completely independent. They were to choose and support their own preachers, elect deacons and elders, and only had to send a delegation to an annual synod at Marburg. There, an executive committee of 13 members was to be elected to serve for one year and three additional members were to be appointed as visiting superintendents. The Landgrave was to have a seat and a vote in the synod, but was not accorded any other special rights.[15]

With the calling of the Homberg synod, Philip had created a revolutionary *fait accompli.* He was careful enough, however, to submit the resulting organizational plan to Luther for his opinion.[16] Luther's reply, dated January 7, 1527, expressed strong disapproval and recommended that Philip should as much as possible avoid hasty changes.[17] Philip accepted Luther's counsel of caution, and Lambert's draft of the new reformation order was abandoned in favor of one which followed the Wittenberg model more closely.

The part of the Lambert program which was not dropped was the plan for the opening of a university in Marburg.[18] The synod

[13] Mueller, *op. cit.,* pp. 37-38.

[14] *Ibid.,* p. 39. The interpretation of this incident depends on the religious point of view of the author. For a Catholic report see Rady, *op. cit.,* pp. 669 ff.; for Protestant view see Hassencamp, *op. cit.,* pp. 86 ff.

[15] Gustav Conrad, *Die Reformationsordnung* (Halle, 1897).

[16] Benz, *op. cit.,* p. 11.

[17] Friedrich Kuech, "Landgraf Philipp und die Einfuehrung der Reformation in Hessen" in *Festschrift, op. cit.,* p. 228; also Mueller, *op. cit.,* p. 46 and Benz, *op. cit.,* p. 10.

[18] Benz, *op. cit.,* p. 10.

at Homberg had recommended the establishment of a university primarily in order to provide a school where a new generation of theologians could be trained. At least this is the purpose which Philip outlines in his first letter to Melanchthon dealing with this subject. However, as plans for the new university began to take shape, several factors combined to minimize the central position of theology in the new institution.

Since neither Luther nor Melanchthon had been willing to come to Marburg and particularly since Luther disapproved of the whole Homberg plan, Philip had to rely on his own advisers. Of these Lambert was seriously handicapped by his inability to speak German. Also, Luther's repudiation of his proposals had obviously undermined his position. This left Krafft, who was more than fully occupied, however, with the practical details of reform in individual churches throughout Hesse.

Thus the major responsibility for the establishment of the university and for the selection of a suitable faculty fell to Johann Feige, Philip's chancellor. Feige had studied at Erfurt and had been greatly influenced by the humanist circle there. Assured of his young master's full support, Feige pushed forward plans for the opening of the university with such vigor that the new institution was actually ready to begin operations by the spring of 1527.[19]

Throughout the early months of 1527 Feige sent invitations to scholars in all parts of the Empire in order to recruit a distinguished faculty.[20] By Easter he was ready to advertise the new university at Kassel and at the Spring Fair in Frankfurt. The announcement listed as prospective teachers five theologians, five jurists, three physicians, one teacher each for Latin, Greek, Hebrew, philosophy, and mathematics, and four instructors for the preparatory school which was to be set up in conjunction with the university.[21]

The faculty of the new university appears to have been surprisingly cosmopolitan and reflects the broad humanistic approach of both the Landgrave and his Chancellor. In selecting men to teach theology, for example, Philip had balanced the rather dogmatic Lambert, who was more or less identified with the Swiss

[19] Georg Heer, *Marburger Studentenleben, 1527-1927* (Marburg, 1927), p. 3.

[20] F. Kuech, " Beitraege zur aeltesten Geschichte der Marburger Universitaet," *Zeitschrift des Vereins fuer hessische Geschichte*, Neue Folge, 46 (Kassel, 1927), 15-16.

[21] Rady, *op. cit.*, pp. 753-755.

and Strasbourg reform movements, with Erhard Schnepf, recommended to him by Luther [22] and Adam Krafft, who combined the ideas of the Wittenberg school with a broad humanistic outlook. The distribution of the rest of the faculty followed somewhat the same pattern. Sebastianus Nouzensus had been called from Wittenberg to become professor of Hebrew; the Greek professor Johannes Lonicerus had taught in Strasbourg and Freiburg; Euricius Cordus, appointed professor of medicine, had studied in Erfurt and Ferrara.[23]

The most widely known of the new professors who assembled at Marburg in the spring of 1527 was undoubtedly Hermannus Buschius (Hermann von dem Busche), who was appointed to teach history and poetry. He had studied in Erfurt, where he had formed a friendship with Feige, and had later traveled widely in Italy, France, Holland, and England. He had taught at many of the major universities and left a professorship at Heidelberg to accept Feige's invitation. Apparently he also helped considerably in the preparations for the opening of the new university. In any case, the suggestion that a preparatory school should be added seems to have come from him.[24]

It is significant that the man whom Philip and Feige selected as rector of the university was a jurist and not a theologian. Johann Eisermann (Ferrarius Montanus) had studied at the University of Wittenberg and had later become dean of the faculty of arts there. In 1523 he had married into a Marburg family and had settled there to serve as judge on the Hessian tribunal.[25] Eisermann's long residence in Wittenberg had undoubtedly brought him into contact with Luther as well as Lambert, both of whom had lectured at the university there. Nonetheless, as a judge and a high magistrate in Hesse, he was undoubtedly interested in making the new university a training ground for future jurists and civil servants. The University was the obvious place to prepare the future administrators for the Hessian state. Perhaps this is why, from the very beginning, students were promised that upon successful completion of their studies, they would be given preference in appointments to official positions in the territory.

The location of the new university in Marburg, the geographical

[22] Franz Grundlach, ed. *Catalogus Professorum Academiae Marburgensis* (Marburg, 1927), pp. 4 ff.

[23] *Ibid.*, passim.

[24] Rady, *op. cit.*, p. 753.

[25] Grundlach, *op. cit.*, pp. 84 85.

center of Hesse, which already housed the highest tribunal of the state, which had been designated as the meeting place of the annual synods, and to which Philip transferred his residence early in 1527, further underscored the close relationship between the university and the total concerns of Philip's territory. As a result, the university could utilize the services of judges for instruction in law, while theology professors could also be entrusted with the practical tasks of church administration and visitation.

An additional factor in the selection of Marburg was the availability of a number of monasteries and other church institutions which could be secularized and used to house the university. The Dominican monastery became the university's first home, and remains the center of academic life in Marburg to the present day. Altogether there were some thirteen monasteries and convents which, with their outlying sources of income, formed an important economic base for the new university.[26] The confiscated church properties thus provided the necessary physical facilities as well as a steady income for the university, which even covered faculty salaries and scholarships for needy students.

In addition to a faculty, administration, and physical facilities, the university also had to have students. Philip realized that, at least at the outset, Marburg could hardly expect to attract many students from outside of Hesse. They therefore had to come from Hesse and since the nobility showed little interest in the project,[27] the majority of the students would have to come from the middle and lower classes. Philip and his advisers therefore devised an interesting system which enabled poor students to attend the university. This made the young Landgrave a champion of the common people.[28] In order to utilize church property secularized as a result of the Reformation for the common good, much of it was turned over to the individual towns and villages. These, in turn, were asked to set aside the income from this property for the support of a specified number of poor and upright boys from their own or neighboring communities who wanted to study in Marburg.[29]

[26] H. Hermelink, und S. A. Kaehler, *Die Philipps-Universitaet zu Marburg 1527-1927* (Marburg, 1927), pp. 31-38.

[27] Wilhelm Falkenheiner, *Personen und Ortsregister zu der Matrikel und den Analen der Universitaet Marburg 1527-1652*, p. 279.

[28] Hermann Haupt, "Sozialistische und religoese Volksbewegungen in hessischen Staedten 1525-1526," *Festschrift, op. cit.*, pp. 444-459.

[29] Ludwig Zimmermann, "Das hessische Stipendiatenwesen im Zeitalter der

On May 30, 1527 the University of Marburg formally opened its doors with 88 students and a faculty of 11.[30] A few years later, when the charter of the university was confirmed by Emperor Charles V, enrollment had grown to 338.[31] The opening phrases of the university's charter clearly express the hopes which Philip placed in this new institution:

> In order to aid the sciences, threatened with extinction in this age, we have established in our city of Marburg a university for all general studies together with a school for three languages which shall produce men of good counsel (as Homer calls his heroes) who shall distinguish themselves through scholarship as well as through intelligence and Christian piety, so that they may become leaders in the Christian Church and in the administration of the state. For learning can no more disappear than the sun could cease to shine upon this world. No human society, nor any other form of life, can exist without learning. Right indeed is he who calls that state a happy one in which philosophers reign and princes study philosophy.[32]

In view of his later career we could hardly call Philip of Hesse a philosopher-prince. Yet his decisive actions leading to the reform of the church in Hesse and the founding of the first Protestant university indicate that he clearly recognized the interrelation of the political and religious questions of his day.

The establishment of the University of Marburg without the endorsement of either pope or emperor underscores Philip's claim to territorial sovereignty and independence from both. But the importance of this first Protestant university transcends the immediate political and religious purposes which led to its formation.

As we have noted, the new university had three distinctive features which may make it appear modern, even 400 years later:

1. Although this was to be an institution which would serve the cause of reform, Philip was eager to see that the university would not become narrowly identified with one particular creed. He extended invitations to Melanchthon, as well as to Zwingli,[33] to join

Gruendung der Universitaet Marburg 1527-1560," *Zeitschrift des Vereins fuer hessische Geschichte*, 46 (Kassel, 1927), 72-123.

[30] Heer, *op. cit.*, p. 3.

[31] *Marburg*: Die Philipps-Universitaet und ihre Stadt, ed. Dr. Kurt Goldammer (Marburg, 1952), p. 37.

[32] Hermelink, *op. cit.*, p. 55, my translation

[33] Mueller, *op. cit.*, p. 120. The best brief statement concerning Philip's religious tolerance is given in George H. Williams, *The Radical Reformation* (Philadelphia, 1962), pp. 436-443.

its faculty. The calling of the famous Marburg conference of 1529 illustrated the nonsectarian character with which Philip wanted to endow the university which bears his name.

2. It was to be true to the Word of God, and responsible for teaching the truth as it is found in all fields of learning, not only theology. In effect, Philip hoped that the new university would reflect both the spirit of the Reformation and that of Christian Humanism.

3. The task of the University in the teaching of both theology and other subjects was to be clearly related to the needs of the territory, and it was to provide education for the people of Hesse regardless of economic or social standing.

Given these objectives, the establishment of the University of Marburg may be taken as a reflection of Philip's recognition of the social and intellectual demands of his day and of his skill in bringing into a meaningful synthesis the great religious and political forces at work in Germany during the Reformation.

PAPERS OF THE MICHIGAN ACADEMY OF SCIENCE, ARTS, AND LETTERS
VOL. XLIX, 1964 (1963 MEETING)

POLITICAL SCIENCE

THE POLITICAL PARTIES AND THE MICHIGAN CONSTITUTIONAL CONVENTION OF 1961

ROBERT W. KAUFMAN
Western Michigan University

IT is almost axiomatic among writers on political affairs today to assume a proper and beneficial, if informal, role for the political parties in the governmental systems of modern democracies. In spite of their absence in the Federal Constitution, they were implicitly anticipated by James Madison in the Federalist papers, became a living entity with the election of Thomas Jefferson as President, and have assumed a rightful place in the political system ever since, exerting no little influence on its operation. Thus, the American public is conditioned to having the political parties participate in every phase of the political process from nominations to patronage appointments and major policy decisions—at least at the state and national levels of government.

This paper raises the question of the role of partisan politics in a state constitutional convention. More specifically, it addresses itself to the impact of the parties on the product of the 1961 Michigan Constitutional Convention and contrasts the role of party in the normal political process with the role of party in any constitutional convention.

MICHIGAN PARTIES AND GOVERNMENTAL STALEMATE

Politics in Michigan perhaps are best understood in terms of the changes, social, cultural, and economic, which have been brought about chiefly through the industrialization of the southeast corner of the state. These changes have divided Michigan into various factions, but the basic cleavages may be described in terms of rural vs. urban and suburban, outstate vs. Detroit, and labor vs. business. Such cleavages are not unique with Michigan; to the contrary, they are a phenomenon characteristic of every industrialized state in the nation. What is peculiar to Michigan, however, is, first, the highly polarized ideologies of the parties as they reflect these cleavages;

and second, the regional patterns supporting these cleavages. Each party has within it highly militant minorities with deeply committed economic interests and articulate leaders. The parties, thus aligned, have become hardheaded spokesmen for these minorities: they are in conflict in ideology and in program and they disagree almost always on what should be done, by what method, and by whom. For if zealous Republicans see the Democratic party as composed of decadent city dwellers who reject the past, inject constant change and turmoil into the present, and embrace the future as the millenium, there are equally ardent Democrats who view the Republicans as country hicks or Neanderthal business barons living in yesterday, refusing to recognize the need for change, and resisting the future with every policy decision. But these are the reciprocal party images and we expect them to be exaggerated. Nonetheless, there is a deep commitment by many adherents of each party to a way of life, and neither group is psychologically disposed to give in to the other. The Republicans historically have found their strength in the rural areas—the farmers and small townsmen, who shared complementary economic interests and social viewpoints. The selection of local leadership through the Republican party became a way of life. To this cadre of freedom-loving ruralists was added, as the state grew, increasing numbers of equally freedom-loving enterprisers of the urban areas. Not the least of the participants was the automobile industry. Thus, with business contributing most of the money and organizational know-how, and with the rural areas providing most of the votes, prior to the election of G. Mennen Williams in 1948 these two forces ruled Michigan under the Republican banner for 97 years, interspersed with 7 brief Democratic intervals totaling 16 years.

During the early years of the Williams era, the objectives of the coalition participants were compatible. The rural interests opposed any change in public policy that in any way recognized the needs of urban areas, particularly when those needs meant spending rural tax money. The industrialist and the Main Street merchant were concerned with keeping to a minimum the influence of organized labor in the development of state policy. Both objectives were satisfied if no major decisions were made during the period of Democratic rule. But as the 1950's came to a close, this position became untenable. Stalemate was an acceptable policy for a short period; for the long haul it would lead to disaster for the entire state. The compelling demands for change inherent in an industrial and urbanizing society clearly precluded the continuation of reciprocal vetoes on the part of the governor and the State Senate. In recent years some urban, suburban, and industrial leaders have been working to develop a new Republican leadership which would come to terms with positive government. The effects of their work became apparent in the 1960 elections of the state legislature, and the trend toward moderation continued in the state Republican primary in 1962.

The Democratic party, by contrast, has been politically significant in the state only in the last two decades. Its gubernatorial victories in the first third of the century were mere quirks resulting from Republican overconfidence and factional disputes. The Democratic party's rise to power in statewide elections was the result of a coalition of liberals with industrial unionism, especially the Political Action Committee (PAC) of the former Congress of Industrial Organizations (CIO), which in 1948 publicly declared itself a part of the Democratic party. Thus, the mold of the Democratic platform was cast: since 1948 it has been strongly labor, and it has been strongly urban. With this program, with a dedicated leadership, and with continuing and well-financed campaigns, the Democrats converted Michigan into a two-party state. They first gained control of the Executive Office and slowly emerged with a majority on the State Supreme Court through gubernatorial appointment. But while they approached control of the State House of Representatives, their numbers rarely exceeded more than one half of the Republican membership in the State Senate.[1]

By 1960, the legislative-executive stalemate forced remedial action by independent groups not identified with the political parties per se. The calling of the Constitutional Convention (known colloquially in the state as "Con-Con") largely was due to the work

[1] Various publications give similar and substantiating reports concerning recent development in Michigan politics. See for example, Stephen B. and Vera H. Sarasohn, *Political Party Patterns in Michigan* (Detroit: Wayne State University Press, 1957); Fay Calkins, *The CIO and the Democratic Party* (Chicago: University of Chicago Press, 1952); Joseph LaPalombara, *Guide to Michigan Politics* (East Lansing: Bureau of Social and Political Research, Michigan State University, 1960); and Robert Lee Sawyer, Jr., *The Democratic State Central Committee in Michigan, 1949-1959: The Rise of the New Politics and the New Leadership* (Ann Arbor: Institute of Public Administration, The University of Michigan, 1960).

of the League of Women Voters of Michigan, whose members are drawn mainly from urban and suburban areas, the Junior Chamber of Commerce of Michigan, made up of urban and suburban shopping center merchants, and George Romney's Citizens for Michigan, whose 19 chapters by and large were found in the urban communities of Southern Michigan. It is worthwhile noting that overwhelming majorities from the 4 large metropolitan counties in the Detroit area voting in favor of calling a Con-Con overshadowed the majorities of the 79 other counties in the state which voted against calling a Con-Con. In the election of delegates to the Convention, the outstate urban and rural interests elected 99 Republicans, while the Democrats won only 45 seats, almost all of them from the Detroit Metropolitan area.[2]

CON-CON LEADERSHIP

Republican factions.—The two-to-one majority of the Republican party assured that one from its ranks would become president of the Convention. Not unexpectedly the outstate wing of the party prevailed upon former State Senator Edward Hutchinson of Fennville (population 705) in Southwestern Michigan to announce and actively campaign for the position. Mr. Romney, president of American Motors Corporation, was described as an "unannounced but active" candidate. On the Saturday prior to the scheduled October 3, 1961 convening date, the Republican delegation gathered in Lansing to discuss preliminary and procedural arrangements and to select a president. The expected short session lasted two days and the ensuing intraparty fight indicated as much as any action following that the same factionalism present in Michigan Republicanism also had infected the Convention delegation. The party split was revealed on the first ballot. The rural conservatives supporting Hutchinson had 35 votes; George Romney had 29 from suburban and urban areas; and an independent group split their 31 votes between two academicians, John A. Hannah, president of Michigan State University and James K. Pollock, chairman of the University of Michigan Political Science Department. On the 17th ballot, two days later, the nomination went to Stephen S. Nisbet, a compromise

[2] *Michigan Manual, Michigan Department of State,* 1961-1962 (Lansing: 1962) Voting on the gateway amendment is reported on page 462; voting in the election of delegates is on 636 ff.

choice. Nisbet had served as a member of the State Board of Education, thereby avoiding much of the intraparty strife which made most available candidates unacceptable to one faction or the other. As a face-saving gesture, the Convention elected Romney and Hutchinson, along with one Democrat, to serve as vice-presidents.[3]

The intraparty squabbles multiplied in organizing the Convention. Committee chairmanships were sought with equal vigor by both moderate and conservative Republican delegates, with the actual assignments being almost equally divided. The urbane John C. Martin, Jr., Republican National committeeman, was selected to head the Executive Branch Committee; D. Hale Brake, a local government-small town spokesman, the Tax and Finance Committee; Arthur Elliott, a suburban Oakland (Detroit area) County chairman, the Local Government Committee; and the conservative Alvin Bentley, a former U. S. representative, the Education Committee. Then, too, there were men of independence like Hannah and Pollock heading the Legislative Organization and the Declaration of Rights committees respectively. Further, the Republicans found themselves with not one, but two caucuses. The caucus of the entire Republican delegation selected Arthur T. Iverson, a lawyer, as chairman. Iverson freely admitted that he was elected only because of his early experience with the outstate party and his later experience with the Detroit party. He did not pretend to act as floor leader, nor did anyone else in the party. In fact, the urban moderates consented to the caucus only on the condition that delegates would not be bound by caucus votes on substantive matters. Thus the caucus was limited to deciding procedural matters and serving as a sounding board to measure the acceptance of proposals coming out of the committees.[4]

A second Republican caucus materialized chiefly through the efforts of the Michigan Farm Bureau. Long a political power in Michigan, the group had perceived that the political power of the farmer was being eroded by the rising forces of urbanism. Having lost in its effort to defeat the proposal to call the Convention, the Farm Bureau mobilized support for and elected 23 of its members to

[3] *State Journal* (Lansing), October 1, 2 and 3, 1961.

[4] Michigan Constitutional Convention, *Handbook, 1961-1962* (Lansing: 1961). Data concerning the role of the Republican caucus was obtained from interviews with delegates. Arthur T. Iverson on May 11, 1962 and Walter DeVries on May 1, 1962.

Con-Con.[5] The goal of these delegates, almost to a man, was to support only those constitutional measures which would lessen urban control over the executive and judicial branches of government or strengthen the role of rural-local government interests in the legislative branch.[6] Because these delegates belonged to the Republican majority, because they were so well agreed on basic issues, and because their leader, D. Hale Brake, proved to be an astute political tactician, the views of this group became more influential than those of any other in the Convention. Brake, a lawyer from Stanton, Michigan (population 1139), was a former State treasurer and a conservative. In recent years he had served as director of the Institute of Local Government, sponsored by the Michigan County Supervisors Association.[7] Many outstate Republicans consider him an elder statesman; to others, he is a knowledgeable and able politician. But whatever epithet is used, there is considerable respect for his judgment among those who believe in him and respect for his power among those who don't. He gathered delegates around him and organized what in reality was a second Republican caucus numbering between 30 and 35 delegates. The significance of the group was not its size, which when compared to a total of 144 delegates is an unimpressive minority; its strength lay rather in the cohesion demonstrated in voting and in the partisan political circumstances which permitted it to hold the balance of power between the moderate Republicans and the Democrats. Estimates vary that between 30 and 40 delegates from urban and suburban areas usually voted with the Romney leadership; however, there was no attempt to mobilize these votes into a bloc.[8]

Democratic cohesion.—The Democrats made an interesting contrast. Since almost all of their outstate delegate races had been lost

[5] *Michigan Farm News*, October 1961. (Official monthly publication of the Michigan Farm Bureau.)

[6] Robert W. Kaufman, "The Michigan Constitutional Convention: An Interpretation Based on Social and Economic Change and the Control of Political Institutions," a paper presented before the Michigan Academy of Science, Arts, and Letters, March 23, 1962.

[7] The Institute primarily is concerned with directing "Government Days" in Michigan high schools. The program permits high school seniors to discuss problems with local and state government officials face to face and to hear Brake speak on the importance of local government.

[8] The sizes of the Brake and Romney groups were estimated by various convention delegates within the limits suggested in the text.

to Republicans, no real faction of significance was present within their delegation at the Convention. Furthermore, through their leadership they were tied to the party in power outside the Convention. The Democratic selection for vice-president of the Convention was Tom Downs, counsel for the Michigan State AFL-CIO. Downs was assisted by William C. Marshall, executive vice-president of the same labor group. Adelaide Hart, a former State Democratic vice-chairman, served as caucus chairman. It was widely acknowledged that these leaders shared the views of and communicated with Governor John B. Swainson, the Democratic National committeeman Neil Staebler, and Gus Scholle, president of the Michigan State AFL-CIO. The Democratic delegation recognized its minority position outside the inner circle which was to operate the Convention and design the new Constitution. The alternatives seemed obvious: either they could work with one or the other of the Republican factions and play a minor role in writing the new document, in which case they would be obligated to support it, or they could vote as a minority and exercise little influence, in which case they could take the new document or leave it. Actually, subsequent events left them no choice.[9]

Convention Politics

The bipartisan facade.—Even after the initial display of political bickering within the Republican party and in spite of a long history of conflict and stalemate between the parties, newspapers began to discuss what they called the "bipartisan" atmosphere of the Convention. Indeed, during the first two or three months, which were taken up with questions of organization and procedure, there was little else to discuss. For wasn't it the combined efforts of three "nonpartisan" groups in calling the convention that laid the groundwork for this bipartisan spirit? And certainly the election of Nisbet as president took off some of the partisan edge. Another factor was the apparent harmony with which the party delegations settled some of the convention's organizational and procedural problems. And, on the eve of the Convention, Romney called for a continuation of the harmonious spirit by saying to the first meeting of the Republican delegation that "it is fundamental to

[9] This data is based on interviews with Democratic delegates Adelaide Hart and William C. Marshall on May 8, 1962 and Robert G. Hodges on April 24, 1962.

put partisanship behind us . . . ," a statement which must have shaken some of his Republican supporters to the core.[10]

While it is true that partisan considerations could not be eliminated from the Convention simply by eliminating the terms Republican and Democrat, the addition of these labels gave an emphasis to partisan interests which was hardly desirable. The parties, which are supposedly the solution-seekers to the conundrums of society, had become so identified with these problems that they were helplessly neutralized. But when the state legislature voted to stitch partisan labels on the aspiring delegates, it thereby tied a noose around the neck of the Convention and severely limited the possibility that the basic problems facing the State of Michigan would be met squarely and honestly.[11]

The Romney candidacy.—The "bipartisan" spirit described in the press only served to sharpen the disappointment when the facade was crumbled by the thunder of partisan moves. As it turned out, committee reports for action reached the Convention floor about the same time in December that Romney informally stated that he was considering running for the Republican nomination for Governor.[12] It was of course not unexpected that sooner or later he would indicate an interest in a high state office. Mr. Romney had turned his attention toward public affairs more and more during the previous two years through his Citizens for Michigan group, and most people assumed an announcement would be the next step. But the announcement placed him in a new political perspective. He had entered the partisan arena and was fair game for Democratic sniping.[13] And snipe they did, using all the party leaders, delegates, and public officials they could muster, including the Governor.

The Democratic Party line was set two days after Romney's statement. Tom Downs and Adelaide Hart sponsored a resolution in the Convention which would have forced the resignation of any delegate announcing his candidacy for state office. The resolution, referred routinely to the Rules Committee, was routinely disposed of by a heavy Republican majority. Miss Hart stated on the floor of the Convention that any delegate who became a political candi-

[10] *State Journal*, October 3, 1961.

[11] *1962 Public and Local Acts of Michigan* 125.

[12] *State Journal*, December 5, 1961.

[13] Some of this paragraph is based on interviews with Adelaide Hart, Robert G. Hodges and William C. Marshall.

date while still retaining his delegate status "cannot escape the public appearance of using the Convention to write a platform for partisan advantage." [14] Later, the Democratic State Chairman, Joe Collins, speaking to members of the State Central Committee, stated that the Republicans no longer had a "serious concern" for writing a good Constitution but rather were determined "to write a platform on which to run a gubernatorial nominee in 1962, and a presidential candidate in 1964." [15]

The concern of the Democrats was a practical one, for they saw Romney, already a vice-president of the Convention, being placed in a position which would draw more and more public attention. Their fears were more than justified, for it wasn't long before he was being mentioned by syndicated columnists as a possible Republican contender for the Presidency in 1964. After that, as far as the Democrats were concerned, things got worse and never got better. Not only was the Convention hall overrun by correspondents from the wire services and news magazines, but Romney was asked to appear on several nationwide television programs. Miss Hart recalls with exasperation that one network sent word into the Education Committee hearing room requesting that witnesses be quieter so as not to interfere with the televising of the Romney interview in the next room.[16] The high volume of nationwide publicity continued through December, January, and into February when Romney made a formal announcement of candidacy. He emphasized that he was an "inactive" candidate for the Republican nomination and stated that he would "not take an active, personal part in the campaign" until his work as a convention delegate had been completed.[17]

Democratic criticism increased both in volume and in strength. James M. Hare, the Democratic secretary of state, wrote Nisbet that Romney's candidacy would "sacrifice the long-term needs of the State of Michigan in return for the hope of short-term political rewards." He expressed his fear that the Republicans were using the Convention in a "desperate" attempt to "fit a political schedule," to produce a document that would be "geared to and motivated by the needs of a particular candidate for a particular office

[14] *State Journal*, December 5, 1961 and December 19, 1961.

[15] *Ibid.*, January 29, 1962.

[16] Interview with Miss Adelaide Hart.

[17] *State Journal*, February 11, 1962.

at a particular time . . . ," rather than for the benefit of the entire state.[18] In reply, Nisbet said Hare's "unfortunate" choice of words reflected on the "integrity of all the Republican delegates," by implying they would follow Romney like sheep. Nisbet pointedly added that it was "not a bossed convention" and suggested that the Romney candidacy would not upset the Convention unless he campaigned actively. He urged the Democrats to wait and see.[19] They waited in a state of agonized frustration as they watched what their partisan hearts told them was a two-million dollar expenditure of public money for launching a dynamic personality into orbit for the governorship.

Earlier, however, Nisbet appeared to have held some reservations about the Romney candidacy. When he spoke in January to a St. Joseph (Michigan) Chamber of Commerce meeting, Nisbet stated that Romney's role as a potential candidate for Governor had "thrown some political implications into the Convention that were not there before."[20] Other than his reply to Hare, Nisbet apparently made no further public statement on the matter after Romney's formal announcement in February.

Toward the end of March, the situation had, at least in the view of some delegates, deteriorated. A Republican delegate of some prominence, John A. Hannah, suggested a seven-month recess in order to ease the political feuding which had come to dominate the Convention. Governor Swainson was quick to reply, noting that Hannah's plan recognized the "political atmosphere that had pervaded the Convention since the announcement by a delegate of his intention to run as a Republican candidate for Governor."[21] Hannah's plan was not accepted, partly because of the technical complications involved, such as staffing and housing after an extended interim. But the headlines across the state made a point the Democrats had been trying to emphasize for months.

The Republican compromise.—By mid-March 1962 the voting in Convention was being interpretated by the Romney leadership and by the rural conservatives. For if the latter found the moderate Republicans voting with the urban Democrats to remove the 15-mill tax limitations from the Constitution,[22] the former saw the rural

[18] *Ibid.*, February 12, 1962.

[19] *Ibid.*, February 13, 1962.

[20] *Ibid.*, January 31, 1962.

[21] *Ibid.*, March 24, 1962.

[22] Michigan Constitutional Convention, *Journal of the Constitutional Convention*, No. 76: 573-577 (February 9, 1962).

Republicans voting with the Democrats to preserve local county offices from destruction by home rule legislation,[23] and again to help disperse the executive power of the Governor with respect to his Administrative Board.[24] The rural conservatives obviously had most to gain by siding with the other faction of their own party. For there they were in a position of strength: first, they were well organized and the resulting cohesion contrasted with the loosely affiliated Romney group of moderates; second, they were led by seasoned state politicians, whereas the moderate group contained relative newcomers to state politics; and third, they were satisfied with the status quo and therefore anything the moderates wanted changed would have to be obtained at a price.

But the alternatives open to the moderates were extremely limited. They could: (1) do nothing; (2) compromise with the Democrats; or (3) compromise with the conservative Republicans. To follow the first course would mean that the conservative Republicans and the Democrats would write every crucial provision of the Constitution, which not only would destroy the moderate Republican delegation as a force within the Con-Con, but would also ruin Romney's opportunity to develop stature as a party leader. The second alternative, at least theoretically, might have been the best solution for Michigan because the representation from the urban aud suburban areas of the state thereby could have been harmonized to cope with a determined and entrenched ruralism. On the practical political level, however, this became impossible because of Romney's candidacy and the partisan labels worn by the delegates. Except for the proposals concerned with the Executive officers, and legislative representation, Romney's group and the Democrats agreed on many of the so-called gut issues of the Convention. However, it was unthinkable that the Democrats would agree to bolster the Romney image before the eyes of the Michigan voting public. Thus, the third course of action became the only realistic choice. Therefore Mr. Brake, speaking for the rural conservative Republicans, and Mr. Romney, speaking for the urban and suburban moderate Republicans, worked out a compromise.

The compromise settled the outcome of most major issues to come before the Convention. The contents of the package were never released to the public, but according to newspaper reports,

[23] *State Journal*, January 28, 30, and 31, 1961.
[24] *Ibid.*, February 22, 1962.

such issues as property taxes, property assessment, sales tax diversion, retention of earmarked funds, the Administrative Board, and legislative apportionment were issues that were settled between these two factions of the Republican party.[25]

Politics and the Proposed Constitution

It is with the foregoing political factors in mind that we now undertake an analysis of the proposed Constitution, attempting to describe the impact of these forces upon the final document produced by the Convention. Our analysis will commence with the Executive, continue with the Judicial, and conclude with the Legislative Branch.

The Executive Branch.[26]

The functions of the Governor may be discussed in terms of his political power, that is, his relationship to the Administrative Board[27] and to other branches of state government, and in terms of his administrative power, that is, the degree of direction the Governor may exercise within his own branch in executing the laws of the state.

The political power of the Governor.—In the Constitutional Convention the debate over the degree of political power to be vested in the Governor may be described as a three-dimensional tug-of-war. The Democrats elected to the Convention assumed at first that their party's entrenchment in the Governor's office was

[25] *Kalamazoo Gazette*, March 16, 17, and 24, 1962.

[26] Several analyses of the 1962 Proposed Michigan Constitution have been utilized in this paper. Credit should be given to those analyses which proved helpful in preparing this paper. The papers include those of Robert S. Friedman on the Executive Branch, Norman C. Thomas, on apportionment, Dorothee S. Pealy and Daniel R. Fusfeld, on finance and taxation, and Arthur W. Bromage on local government, all of The University of Michigan; Milton Greenberg on civil liberties and civil rights, Leo C. Stine on education, both from Western Michigan University; Harold J. Spaeth on the judicial branch, University of Detroit; and Edgar W. Waugh on the Highway Commission, Eastern Michigan University.

[27] Michigan statutes have formalized the Governor's cabinet under the title of Administrative Board. Created in 1920, the board has "supervisory powers" over the Executive Branch of the Government. In 1927 the legislature permitted any five board members to override a veto by the Governor on board actions. Membership on the board is ex officio and consists of the Governor, lieutenant governor, and six elected department heads.

secure. Thus, it appeared that they would accept an Administrative Board appointed by the Governor. However, the more astute leadership of the Democratic party viewed the election of Mr. Romney as a delegate to the Constitutional Convention as his political debut, and they became aware of the threat to Democratic hegemony over the Governor's office and its perquisites. Thus, when the Democratic State Central Committee took a stand recommending that the Administrative Board continue to be elected, the Democratic delegates at the Convention followed suit.[28]

The position taken by Mr. Romney and his followers would provide for a politically strong Governor. Romney's experience as an executive in a large business corporation and perhaps his ambitions toward the state house prompted him to recommend authority for the Governor commensurate with responsibility. The Governor, he argued, should have the power to appoint the Administrative Board, just as the President has the power to appoint his cabinet.[29]

Ironically, the rural conservatives found themselves holding a position identical to the Democrats, although in response to a different sequence of political logic. This group had learned that no matter which party controlled the state executive machinery, the Governor would never be rurally oriented. Their strategy therefore was to weaken the political power of the Governor and to disperse as widely as possible the authority of the Administrative Board.[30] As the outstate politician viewed the matter, participation in the appointment of Administrative Board members with the chief executive officer of the state in his Governor's suite after the election was far less desirable than participation in the selection of party nominees for Administrative Board positions with the gubernatorial nominee prior to his election and in the milieu of the party convention.

This three-way split resulted in the compromise written into the proposed Constitution which permitted the Governor to appoint only the State Treasurer. The Secretary of State and the Attorney General were to be elected as before, and the Auditor General was

[28] Karl A. Lamb, "Michigan Legislative Apportionment: Key to Constitutional Change," in Malcolm E. Jewell, ed., *The Politics of Reapportionment* (New York: Atherton Press, 1962), p. 285.

[29] *State Journal*, December 7, 1961.

[30] *Michigan Farm News*, October, 1961.

to be appointed by the State Legislature. The Highway Commissioner and the Superintendent of Public Instruction were to be appointed by their respective boards.[31] This mélange was in keeping with the dispersive objectives of the conservative forces. If the 1908 provisions made it possible to frustrate the efforts of the Governor to lead, the new provisions appear to have further removed him from the center of influence and control of his own Administrative Board.

Other provisions of the proposed Constitution would further erode the political power of the Governor. First, the elimination of his control over the interim appointments of justices of the State Courts of Record removed the only significant check the Executive has upon the Judicial Branch of State government.[32] Since the 1908 Constitution permitted the Governor to fill vacancies in the courts, one must admit a net decrease in the political power of the Governor as a result of the new provision.

Second, two new restrictions on the Governor's political power are found among the fiscal provisions of the proposed Constitution. To understand the strategy of the rural conservatives in this area one must again look at the Governor's office as they do. For rural Michigan sees that office as a center of power that has been removed from rural influence and the Governor as the spokesman for all of those agencies and forces within the Executive Branch which represent urban views. These urban views are seen as a network of expensive schemes and programs to assist in solving urban problems, provide for urban needs, and in general contribute to additional urban sloth. If the Executive were not held in check, the result would be higher taxes for the farmers; their objective, therefore, was to limit the initiative of the Governor and to hold him

[31] 1962 *Proposed Michigan Constitution*, Art. V, sec. 3 (State Treasurer); Art. V, sec. 21 (Attorney General and Secretary of State); Art. IV, sec. 53 (Auditor General); Art. V, sec. 28 (State Highway Director); and Art. VIII, sec. 3 (Superintendent of Public Instruction).

[32] 1962 *Proposed Michigan Constitution*, Art. VI, sec. 23. The 1908 *Michigan Constitution* permitted gubernatorial appointment without confirmation to serve until the next general election. See Art. VII, sec. 20. Courts of record are defined in the 1908 Constitution as the Supreme Court, Circuit Courts, Probate Courts. See Art. VII, sec. 17. The proposed Constitution added the court of appeals and other courts designated as such by the legislature. See Art. VI, sec. 19.

in check through these provisions on state finance.[33] First, new short term borrowing authority was limited to 15% of the undedicated revenue of the preceding year.[34] This would mean a legal limit of approximately 70 million dollars. However, this authority is limited to one year and the principal must be repaid as taxes are collected but in any event no later than the end of the current fiscal year. The provision would thus eliminate "payless-paydays" for state employees and would prevent the recurrence of delinquent school aid payments from the state, but would do little to assist the state during a prolonged and severe financial crisis. Making the financial arrangements even more stringent is the provision which requires the Governor to submit a balanced budget.[35] This has significant consequences, because ultimately that balanced budget must meet with the approval of the appropriations committees of the State Legislature.[36] The provision in practice will require him to adjust his budget recommendations to meet any shortage of revenue from any cause whatever. This would include the refusal of the legislature to appropriate adequate funds or its refusal to vote increases in taxes consistent with the Governor's program.

These new provisions effectively curtail the leadership that might otherwise be generated in the Governor's office. During a period of financial crisis the limitations on short term debt and the requirement for an annual balanced budget would suppress the Governor's initiative and limit his alternatives in developing those programs he believed necessary to meet the crisis. Furthermore, the constitutional authority given directly to the appropriations committees would deny the Governor an appeal to the entire State Legislature in seeking support for his programs. This system assumes that the judgment of the rural legislature is always correct and that the judgment of the urban Governor is always wrong.

The administrative power of the Governor.—Although the political power of the Governor has been weakened, the proposed Constitution does give new strength to the strictly administrative aspects of the governor's job. First we should note the enlargement of the Governor's domain as he is transferred from a mere "chief execu-

[33] Robert W. Kaufman, *op. cit.* (see note 6), pp. 16-20. See also the discussion below under the Legislative Branch.

[34] 1962 *Proposed Michigan Constitution*, Art. IX, sec. 14.

[35] *Ibid.*, Art. V, secs. 18 and 20.

[36] *Ibid.*, Art. V, sec. 20.

tive" to one holding the state's "executive power." [37] This is a broader delegation of power, and other provisions reenforce this extension. For example, the Governor may inquire into the condition and administration of any public office at any time and may remove any state officer for cause.[38] Also, the provision for initial reorganization of the Executive Branch agencies gives additional administrative prestige and control to the Governor, a control which is enhanced by permanent authorization to the Governor to submit reorganization proposals which would go into effect unless disapproved by a majority of both houses of the state legislature.[39] Yet, even here an important check on the executive agencies by the legislature has been included in the proposed Constitution that was not in the 1908 Constitution. A new provision would endow a joint interim committee of the State Legislature with power to "suspend any rule or regulation promulgated by an administrative agency" until the close of the next regular legislative session.[40]

Thus, in summary, the political powers of the Governor were severely restricted and circumscribed with respect to the State Legislature, completely eroded with respect to the Judicial Branch, and all but dispersed with respect to the Administrative Board. By contrast, the administrative power of the Governor was improved in several important aspects, giving him greater managerial control over executive administrative units. The proposed Constitution would appear to be saying to the Governor: "You may not exercise political power to any significant degree. However, you may have all the authority you need to organize the Executive Branch and to manage its operations. But should the executive agencies make decisions at variance with the views of the Legislature, such decisions will be suspended summarily by one of its committees."

The Judicial Branch

For much the same reasons, the political logic underlying decisions on the Judicial Branch paralleled that of the Executive Branch. The Democrats were aware that their twelve-year tenure

[37] *Ibid.*, Art. V, sec. 1. The comparable provision in the 1908 *Michigan Constitution* is in Art. VI, sec. 2.

[38] *Ibid.*, Art. V, sec. 8 and 10. In the 1908 Constitution this power was limited to times when the legislature was not in session. See Art. IX, sec. 7.

[39] *Ibid.*, Art. V, sec. 2.

[40] *Ibid.*, Art. IV, sec. 37.

in the office of Governor had enabled them to place on the State Supreme Court a majority of their own kind by 1955 and to maintain control until 1962.[41] The decisions of the Court after 1955 became marked with a distinctly urban and liberal tinge, replacing the rural and conservative outlook of earlier decades.[42] The rural conservatives, again drawing from their recent experiences in both political parties, could hardly expect judicial appointments by a Governor of either party to reflect a rural viewpoint. Their strategy, therefore, was to find some other arrangement for the selection of State Supreme Court justices. The system placed in the proposed Constitution provided that the Supreme Court would appoint retired judges to vacancies on a temporary basis. Such appointees would be ineligible for election.[43] This would force an election contest between two candidates, neither of which would be an incumbent. Having thus removed the urban influence of appointment by the Governor, the next step was to create an election scheme which would not only reflect some outstate views but build in some outstate, preferably rural, influence. The suggested medium for achieving both of these objectives was the election of justices on a district basis. From a strong majority position within the Committee on the Judicial Branch the rural conservatives launched a concerted effort to construct a Supreme Court that would be elected from districts.[44] This was the most controversial provision of the Committee's report. As finally adopted, the district system was stricken, but the voting was extremely close: In one instance, the statewide election scheme was substituted for the district system by four votes; in another, the district system failed to win readoption

[41] The 1962 election of Republican Michael D. O'Hara over the Democratic incumbent Paul L. Adams, returned the Court to a 4-4 division between parties.

[42] See for example, Park v. Employment Security Commissioner (355 Michigan 103) 1956—The Ford-Canton case; Lockwood v. Commissioner of Revenue (357 Michigan 517) 1958—Constitutionality of the Sales Tax, and Scholle v. Secretary of State (360 Michigan 1) 1961—Constitutionality of Michigan's apportionment system.

[43] 1962 *Proposed Michigan Constitution*, Art. VI, sec. 23.

[44] 1962 Michigan Constitutional Convention, *Journal of the Constitutional Convention*, No. 71: 467-468. The report of the Committee on the Judicial Branch stated that on January 16, 1962 the committee opposed the appointment of justices as a general principle by a vote of 1-18; on the same date the "Missouri" and "ABA" plans were opposed by a vote of 3-15. On January 31 the committee voted to adopt the district system as written in its report by 15-5.

by two votes.[45] It should be noted, however, that the scheme for statewide election of justices to the Supreme Court did not necessarily prevail. The 1908 Constitution stated that the justices would be "chosen by the electors of the state."[46] This language was replaced in the proposed Constitution by the phrase "as provided by law."[47] The district election system, therefore, remains a possibility under the proposed document.

The same strategy was applied in the establishment of a new State Court of Appeals. Such an intermediate level of courts had been urged for Michigan by professionals for many years. The proposed Constitution provided for the nomination and election of the justices for this court on a district basis using nonpartisan elections.[48] Seven judicial districts were described in the report of the Judicial Branch Committee, but the report also contemplated a Court of Appeals sitting in three divisions of three judges each. Two divisions would serve southeastern Michigan and a third would serve the remainder of the state including the Upper Peninsula.[49] Since most appeals go no further than the intermediate level, this arrangement makes provision for outstate controversies to be decided without influence of justices elected in the metropolitan areas surrounding Detroit. If we consider, too, that the Circuit Court justices are also elected on a district basis, it becomes evident that only outstate justices, elected by outstate voters, will preside over cases which develop in the outstate areas. Stated this way the judicial pattern as proposed by the Committee on the Judicial Branch promised to develop two distinct lines of precedent: one reflecting outstate views would tend to be conservative, rural, or small-town, and probably antilabor; a second, reflecting southeastern Michigan, would tend to be liberal, cosmopolitan, urban, suburban, and probably pro-

[45] *Ibid.* The floor action begins in No. 85, p. 669 (February 22, 1962) and continues through No. 98, p. 759 (March 13, 1963). Two of the crucial votes came on amendments, the first by Mr. Downs, which authorized a nonpartisan statewide ballot, adopted 56-52, No. 97, p. 741 (March 12, 1962); and a second amendment to reinstate a district system by Messrs. Dehnke, Iverson, Hoxie, McAllister, Rood, Pugsley and Prettie, which was not adopted on a division vote of 62-64 No. 97, p. 745 (March 12, 1962).

[46] *1908 Michigan Constitution,* Art. VIII, sec. 2.

[47] 1962 *Proposed Michigan Constitution,* Art. VI, sec. 23.

[48] *Ibid.,* Art. VI, sec. 12.

[49] 1962 Michigan Constitutional Convention, *Journal of the Constitutional Convention,* No. 71, p. 467 (February 2, 1962).

labor. Two important elements of this ingenious system were lost to the rural conservatives at the Convention. The proposed Constitution permits the State Legislature to describe the districts from which the Appeals Court and Supreme Court justices will be elected. But should they choose to follow the pattern of the Committee on the Judicial Branch in dovetailing the districts at the intermediate level with those of the Supreme Court in a manner designed to isolate the metropolitan areas of southeastern Michigan, it is very likely that the battle for control of the state judiciary would become as heated and perhaps as prolonged as the battle for control of the State Senate.

The Legislative Branch

If the political history of the state of Michigan has shown the Executive and the Judicial branches to be strong seats of power through which urban interests have found expression, the Legislature has just as clearly asserted itself as the voice of rural and conservative outstate Michigan. What would happen to the political power of the Legislature under the proposed Constitution? In considering this question, it might be advantageous to concern ourselves first with those provisions which effected power exchanges between the Legislative and the Executive branches, followed by a consideration of other power adjustments to the Legislative Branch.

Power exchanges between the legislative and executive branches.—An important power exchange was that of the selection of the Auditor General. His exclusive appointment by the State Legislature[50] contrasts with the procedure of the federal government where the Controller General is appointed by the President and confirmed by the Senate. Under the proposed Constitution, there will be no check on the part of the Governor with respect to the Auditor General. The legislature gained a staff agency designed to scrutinize the operations of the Governor, which conceivably could be used to reduce his effectiveness.

Two other provisions, already discussed, tend to limit the power of the Executive, or to strengthen that of the Legislature, or both. The so-called "executive budget," which requires the Governor to meet the demands of the legislative appropriations committees,[51]

[50] 1962 *Proposed Michigan Constitution*, Art. IV, sec. 53.

[51] *Ibid.*, Art. V, secs. 18 and 20. See also discussion above under Executive Branch.

and the new provision in the proposed Constitution for a joint legislative committee to suspend the rule of an executive agency [52] both tend to move political power toward the legislature and away from the Governor in selected but crucial areas of public policy.

Additional power increments to the Legislature.—Several other provisions which accrue power to the legislature may appear of little importance when considered singly; yet, their accumulated impact effectively and significantly augments the political power of the Legislative Branch. Thus, the taxing power of the State Legislature is increased by giving new authority to provide for a nongraduated income tax,[53] and the proposed Constitution would augment legislative power with respect to the sales tax by removing Constitutional restrictions and permitting the legislature additional authority over rate and base.[54] Also, the limitation on short-term borrowing was raised from $250,000 to 15% of the undedicated revenues of the previous year,[55] an amount approximating $70 million.

Other provisions increasing the power of the Legislature are found in the local government article. Thus, the conditions under which a county might adopt, amend, or repeal a home rule charter must be set by the Legislature.[56] Also, the Legislature gains additional authority to write the laws under which new forms of multipurpose governments might be created in metropolitan areas.[57] These are questions vital to the growing and increasingly complex urban and suburban areas in southeastern Michigan, and it is significant to note that a legislature with a historic rural orientation is now endowed with an increased political power to set conditions under which the urban areas of Michigan may attempt to solve their pressing and difficult governmental problems.

This is by no means an exhaustive listing, yet it does contain a number of the more important increments added to the legislative power which were not previously there. One could argue that at least some of these power increments could be marked up to an accidental dovetailing of rural conservative objectives with professional recommendations from many sources directing a more power-

[52] *Ibid.*, Art. IV, sec. 37.

[53] *Ibid.*, Art. IX, sec. 7.

[54] *Ibid.*, Art. IX, sec. 8. See also 1908 Michigan Constitution, Art. X, sec. 23.

[55] *Ibid.*, Art. IX, sec. 14.

[56] *Ibid.*, Art. VII, sec. 2.

[57] *Ibid.*. Art. VII. sec. 27.

ful and a more responsible State Legislature. But whatever the reasons for the shifting of power, it cannot be denied that the outcome was first, to vastly increase the political power of the legislature vis-a-vis the executive; secondly, to emasculate the only significant check the Executive had with respect to the Judicial Branch; and thirdly, to provide for legislative controls which are virtually without check from either of the other two branches of government. One cannot avoid the conclusion that if the rurally oriented State Legislature was able to control to some degree the political system under the 1908 Constitution, it will accede to a new dominance with the adoption of the proposed Constitution.

On the other hand, the Legislature did lose some power, and in one instance, at least, it was a noticeable loss. The power to redistrict the House of Representatives, along with a new power to redistrict the State Senate, was given to an apportionment commission or to the courts.[58] But most of the other limitations placed on the Legislature reflect merely a discontent with procedural actions within and about the legislature, rather than being an attempt to shift political power. Among the provisions which express discontent with procedures are: (1) the removal of the legislative incumbency designation on the ballot; [59] (2) a provision to eliminate secret hearings by requiring public notice and announcing subjects to be considered; (3) a requirement that a public record be made of all committee actions by name and vote;[60] (4) a provision insuring a record by name and vote be made of floor actions including legislative elections, confirmation of appointments by the Governor,[61] and on final passage of all bills; [62] and (5) a provision requiring action with regard to gubernatorial appointments. The legislature can no longer withhold action indefinitely, but must reject the appointment within 60 days or the individual in question is considered confirmed.[63]

The apportionment provisions.—It is only with some understanding of the shift of power within the three branches of government that we can fully appreciate the significance of the new apportionment formula as viewed [64] by the rural conservatives. Having

[58] *Ibid.*, Art. IV, sec. 6.

[59] *Ibid.*, Art. II, sec. 4.

[60] *Ibid.*, Art. IV, sec. 17.

[61] *Ibid.*, Art. IV, sec. 18.

[62] *Ibid.*, Art. IV, sec. 26.

[63] *Ibid.*, Art. V, sec. 6.

[64] *Ibid.*, Art. IV, sec. 2 describes Senate apportionment and sec. 3 House apportionment.

removed the urban influence from the Judicial Branch and further injected therein a worthy outstate influence, having dispersed the Governor's authority with respect to his Administrative Board, and isolating him from the Judicial Branch, and having established financial brackets for his budget, and having insured the "proper" behavior of his administrative agencies, their next major problem was to insure that these gains would not be eroded by the ever-increasing pressure of an acknowledged urban majority.

In the light of these achievements, the proposed Constitution as it looked at the close of the third reading reflected the real hopes of the rural minority. At that point, the document provided that the changes with respect to the State Senate would not go into effect until "after the official publication of the total population count of the 1970 decennial federal census.[65] This would put off any possible threat to rural control until the year 1973. However, at the final session of the Convention, the section delaying the Senate apportionment was removed. The action, taken in the light of *Baker v Carr*,[66] perhaps was intended to render moot the court test of the Michigan apportionment as challenged by *Scholle v Hare*[67] which was pending before the U. S. Supreme Court. Thus, once again, a key provision insuring rural dominance was lost at the Convention. But the new apportionment formulas did not release the State Senate from its rural conservative majority, although the House of Representatives, which would be districted on an almost straight population basis, would probably contain an urban majority. But the Senate, apportioned on a unique system of 80% on population and 20% on land area, would remain under the dominance of a majority of members (52.7%) selected from outstate and rural Michigan, which have 42.5% of the population. By contrast, the four high population counties of southeastern Michigan plus Kent County (Grand Rapids) in western Michigan, which contain 57.5% of the state's population, would send 47% of the State Senate membership from their districts.[68]

[65] 1961 Michigan Constitutional Convention, *Journal of the Constitutional Convention.* No. 133A: 27. See Schedule of Temporary Provisions, sec. 6.

[66] 369 U. S. 186 (1962).

[67] *Brief (of Plaintiff), August Scholle v. James M. Hare, Secretary of State of the State of Michigan, Defendant and Frank D. Beadle, et al., Intervening Defendants. Filed with the Supreme Court of the United States, January 30, 1960.*

[68] The League of Women Voters of Michigan, *It's Your Choice:* The 1908

Conclusions

The social, economic, and cultural changes of the first half of the century created what might well be called classic urban-rural cleavages within the Michigan society. These cleavages found expression in the political institutions of the state in such a manner that divided and stalemated government was the result. And the political parties in Michigan, rather than performing their usual unifying function, lay helpless in their beds of regional geographic support and isolated in their cells of sharply contrasting party programs.

Perhaps most significant in this instance was the fact that the urban-suburban majority was divided by partisan labels. This division was forced upon the delegates of the Constitutional Convention when the state legislature voted partisan elections. But these two interests not only made up a majority of the state's population, and a majority of delegates at the Convention, they also suffered most under the 1908 Constitution, and since their interests were complementary, they stood to gain most by coalescing their forces in writing a new Constitution. But if the partisan labels made improbable any agreement between these two interests, the announcement of George Romney that he would become a candidate for the gubernatorial nomination on the Republican ticket made it impossible. His announcement set in motion a partisan psychology that galvanized the Democratic minority into stiff resistance and made it politically imperative that it oppose him and if possible crush him at every opportunity. Above all, the Democrats could never have placed themselves into a position which might in any way develop a pleasing image of Mr. Romney before the Michigan public. His announcement also made it imperative that he seek the support of the outstate interests within the Republican party. And he appeared to have staked his political life on his ability to encompass them in his plans. But the seasoned political spokesmen for the rural conservatives did not confuse political ideals with realistic political goals. When the rural spokesmen went into conference with the suburban moderates, the latter returned second best. The moderates made some gains in terms of administrative improvements in the Executive Branch, and they certainly gained procedural

or the 1963 Constitution (Detroit: the League of Women Voters of Michigan, 1962), p. 8.

reforms in the legislature; but they lost large and important portions of political power which had been used to the advantage of the urban and suburban interests. In terms of political power, the moderate Republicans must admit defeat before their more clever rural extractors.

The hope of the state of Michigan in October of 1961 was that a new basic document could be developed that would end governmental paralysis and give to the state a modern, that is, an urban Constitution. But the proposed Constitution is not one that would permit this acknowledged majority to play a role of dominance. It is more accurately a treaty of peace between two warring factions of the Republican party and even is more accurately described as a set of new rules under which the rural conservatives would consent to cooperate with the suburban Republicans in the quest for partisan control of the state political institutions. But the new rules have rearranged the power differentials to such an extent that the outstate interests, other factors being equal, would be capable of determining the direction of governmental policy for at least the next decade.

It is for this reason that one questions the wisdom of permitting partisan interests to be accentuated by partisan elections to a Constitutional Convention. For truly the Constitutional Convention is a political activity outside the normal political process. Its purpose is to develop the framework in which that process may be conducted. And to give control of such a political activity to partisan interests appears in light of the evidence here to be both unfair to minority party interests and defeating to the best interests of the state as a whole.

PAPERS OF THE MICHIGAN ACADEMY OF SCIENCE, ARTS, AND LETTERS
VOL. XLIX, 1964 (1963 MEETING)

RUSSIAN STUDIES

DECEMBRISTS IN SIBERIAN EXILE: 1826-56

JEANNE MEAKIN HASKETT
Central Michigan University

AN unsuccessful uprising at St. Petersburg occurred December 14, 1825; two weeks later a related uprising began in the southern part of Russia. This Decembrist Revolt lacked leadership, organization, well-defined objectives, and mass support. The leaders, members of several secret societies who had hoped to use the confusion following the death of Alexander I to liberalize the regime, were mostly noblemen and guards officers. They were not audacious revolutionary leaders, but liberals resorting impulsively to violence, men whose hastily contrived plans were quickly foiled and who themselves were easily captured and imprisoned.

The Imperial Investigating Commission presented its report in May 1826, indicting 579 alleged conspirators, of whom 121 were sentenced to death or exile. Emperor Nicholas commuted the death sentences in all but 5 cases, and 116 men went to Siberia.[1]

One Decembrist only, Paul Vygodovskii, was from the peasant class. 28 were colonels and generals; 31, majors and captains; 46, lieutenants; 6, ensigns; and 10 were civil servants of various ranks. Most had received broad liberal educations, supplemented often with foreign travel. Their ages ranged from 18 to 54, but most were in their 20's or early 30's. These political exiles lived a total of 2070 years in Siberia.[2]

The educational and cultural contributions of the Decembrists to Siberia were considerable. The 19th century historian Prizhov has well summed up their role in exile:

---

[1] The best work on the Decembrist uprising is Anatole Mazour, *The First Russian Revolution, 1825* (reprinted, Stanford, California, 1961). More popularized is *The Decembrists* by Mikhail Zetlin, trans. George Panin (New York, 1958).

[2] *Sibirskaia Sovetskaia Entsiklopediia*, vol. 1 (2 vols.: Novosibirsk, 1929), 799-802. V. N. Sokolov, *Dekabristy v Sibiri* (Novosibirsk, 1946), 228-239. Kirik Levin, *Dekabristy; istoriia vooruzhennogo vozstaniia 14 dekabria 1825 g.* (Moscow, 1918), 43.

> Despite the most unfortunate conditions of life [the political exiles] did as much good for Siberia as might have been accomplished in a hundred years or more. . . . They explored Siberia in the fields of anthropology, natural science, economics, and social and ethnographical conditions . . . they did incomparably more than had been done in the other Russian provinces. These people were the true benefactors of Siberia in a moral, social, and material sense.[3]

80 Decembrists were sentenced to periods of *katorga,* or forced labor, before settling permanently in Siberia. All, however, lost their civil, social, and political rights and privileges; they no longer owned property of any kind and were legally dead, their wives free to remarry. Only a few wives did remarry; most remained loyal, and 12 went to Siberia to share exile, renouncing their titles and property and leaving their children at home. Despite the remarkable efforts of these women during the period of *katorga* and later in settlement, life in exile became one of depressing routine that for some lasted two or three decades. Yet they remained cheerful, loyal, and proud, and added considerably to the cultural tone of the region. The 19th century Russian poet Nicholas Nekrasov has immortalized in verse the sacrifices of these women.[4] About 20 of the Decembrists eventually married in Siberia; most of these marriages were relatively happy and long-lasting.[5]

For each Decembrist at *katorga,* first at Chita and then at Petrovsk Zavod, the government provided four kopeks per day. As this sum was entirely insufficient for these exiles, they obtained whenever possible additional funds from home.[6] Because of the great disparity in wealth, the Decembrists organized two cooperatives or artels. The large artel took care of providing their food and

[3] I. G. Prizhov (1827-1885), quoted in *Literaturnoe nasledstvo,* ed. A. M. Egolin (Moscow, 1954), LIX-LX, 635.

[4] Maria Volkonskaia, *Memoirs* (St. Petersburg, 1904), 37, 39. Polina E. Annenkova, *Vospominaniia,* ed. S. Ia. Gessen and A. V. Predtechenskii (Moscow, 1929), 19. Pavel E. Shchegolev, *Nikolai I i dekabristy* (Petrograd, 1919), 37. Baron A. E. Rozen, *Russian Conspirators in Siberia*; a personal narrative, trans. Evelyn St. John Mildmay (London, 1872), 118, 134-135. *Tainye obshchestva v Rossii v nachale XIX stoletiia*; sbornik materialov, statei, i vospominaniia (Moscow, 1926), VIII, 121. Ivan D. Iakushkin, *Zapiski, stat'i, pis'ma,* ed. S. Ia. Streich (Moscow, 1951), 96, 108, 133, 599. *Vosstanie dekabristov materialy*; dela, sledstvennoi komissii o zloumyshlennykh obshchestvakh, ed. B. L. Modzalevskii and A. A. Sivers (Leningrad, 1925), VIII, 311.

[5] This figure is based on sources quoted in J. M. Haskett, *The Decembrists in Siberian Exile* (unpublished dissertation, Ohio State University, 1962), 72.

[6] Sokolov, 65. *Vospominaniia Bestuzhevykh,* ed. M. K. Azadovskii (Moscow, 1951), 160.

other necessities. The subscription rate was 500 rubles yearly; the wealthy, who received thousands of rubles from home, gave much more; the others gave what they could. The men elected officers of manager and buyer, who were permitted to leave the prison on business matters. At the year's end any surplus funds went into the second or small artel, established to assist needy Decembrists and their families in settlement. When a man left *katorga* the artel might supply money for needed furniture or farm tools. Impoverished comrades were helped, as well as widows and children, and the artel continued to function into the 1850's.[7]

Beginning in 1835 the state yearly paid up to 200 rubles to any Decembrist in settlement who had no income and asked for aid. The men were permitted to receive as much as 2000 rubles yearly from any source—family, friends, or their earnings—and most of the exiles avoided asking for state aid. An average of only 6 men a year received government assistance between 1835 and 1856.[8]

According to regulations set forth by the Third Division, all letters, money, and packages were required to pass through the offices of that department. The attempt was made to maintain thorough censorship, but the exiles had little difficulty bypassing official channels. Occasion to circumvent the Third Division came whenever Siberian friends traveled or when visitors came from Russia; the men also used the Cossacks, local officials, and Siberian nuns as couriers.[9]

For the Decembrists who went first to *katorga*, life was only nominally one of hard labor. In 1828 at Chita they formed an "Academy" in which specialists among them gave lectures on a broad range of subjects, including history, physics, chemistry, anatomy, mathematics, literature, philosophy, astronomy, geography,

[7] S. V. Maksimov, *Sobranie sochinenii* (St. Petersburg 1896), III, 276-277. N. V. Basargin, *Zapiski*, ed. P. E. Shchegolev (Petrograd 1917), 138-156 for description of the artel and its operation. *Vospominaniia Bestuzhevykh*, 159. Ivan I. Pushchin, *Zapiski o Pushkine; pis'ma iz Sibiri*, ed. S. Ia. Streich (Moscow, 1956), 191.

[8] S. Ia. Streich, "Dekabristy na katorge i v ssylke," *Katorga i ssylka*, XXI (August 1926), 126.

[9] M. N. Gernet, *Istoriia tsarskoi tiur'my* (Moscow, 1946), II, 152-154. E. E. Iakushkin, *Dekabristy na poselenii* (Leningrad, 1926), 64. N. A. Belogolovyi, *Vospominaniia i drugie stat'i* (St. Petersburg, 1897), 2. Pushchin, 111. N. K. Piksanov, "Iz arkhiva dekabrista Vasiliia L'vovicha Davydova; neizdannye pis'ma," *Istorik-marksist*, I (Moscow, 1926), 179. *Dekabristy*; letopisi gosudarstvennogo literaturnogo muzeia, ed. N. P. Chulkov (Moscow, 1938), III, 93, 319.

natural sciences, and military strategy and tactics.[10] There eventually were adequate books, journals, and newspapers to pursue almost any subject or interest. The exiles' library came to number in the hundreds of thousands, as the wealthy sent to Russia for their entire libraries and the wives constantly ordered books. The important journals of France, Germany, Italy, Spain, Poland, and England arrived regularly.[11] Apparently the government knew about this and posed few restrictions to the receipt of publications, so that the exiles had access to works denied free citizens of the Empire.[12]

The *katorga*-Academy also supervised the study of foreign languages. The few men unfamiliar with French mastered that language, while Baron Rozen, a German from the Baltic provinces, studied Russian. Others learned Greek, Latin, Hebrew, Arabic, Italian, Spanish, or English. English provoked the most comment with what was felt to be its weird pronunciations. An undetermined number of the Decembrists learned several languages; for example, Dmitri Zavalishin studied Greek, Hebrew, Arabic, and Spanish.[13] To perfect newly acquired skills they translated foreign works; translations were made of Benjamin Franklin's *Autobiography*, Pascal's *Thoughts*, Caesar's *Commentaries*, the Bible, and works of Cicero, James Fenimore Cooper, Montesquieu, and many others.[14]

One evening each week was devoted to concerts of chamber and choral music; occasionally the men performed the original works of their comrades, but unfortunately, most of these musical compositions have been lost. Another evening each week was devoted to

[10] S. S. Volk, *Istoricheskie vzgliady dekabristy* (Moscow, 1958), 7. *Dekabristy*; poeziia, dramaturgiia, proza, publitsistika, literaturnaiia kritika, ed. V. N. Orlov (Moscow, 1951), 579. (Referred to as *Dekabristy*; poeziia.) S. V. Maksimov, *Sibir i katorga* (St. Petersburg, 1900), 407. D. I. Zavalishin, *Zapiski dekabrista* (Munich, 1904), II, 95.

[11] I. D. Iakushkin, 131. Zavalishin, II, 94 mentions half a million volumes, but he perhaps exaggerates. *Vospominaniia Bestuzhevykh*, 176.

[12] A. P. Beliaev, *Vospominaniia dekabrista o perezhitom i perechuvstvovannom 1805-1850* (St. Petersburg, 1882), 213.

[13] Rozen, 131. D. I. Zavalishin, " Sukhorikov i Kornilovich, *Drevniaia i novaia Rossiia*, IV (June 1878), 170. *Vospominaniia Bestuzhevykh*, 279. Zavalishin, II, 95.

[14] Pushchin, 110. Beliaev, 222, 298. *Poeziia dekabristov*, ed. B. Meilakh (Leningrad, 1950), 660. V. Bogucharskii, V. I. Semevskii, and P. E. Shchegolev, *Obshchestvennoe dvizhenie v Rossii v pervuiu polovinu XIX veka* (" M. A. Fonvizin, E. P. Obolenskii and V. I. Shteingel," Vol. I, St. Petersburg, 1905), 309. Volk, 209.

reading original essays, poems, short stories, and biographical sketches. Alexander Odoevsky was the best known poet among the Decembrists, and it is his line which Lenin appropriated for *Iskra*—"from the spark shall come the flame." There are extant poems or fragments from about 20 of the exiles. Michael Bestuzhev, a man who had never seriously put pen to paper before exile, composed one of the most popular revolutionary verses sung during *katorga*.[15]

The exiles wrote little of the revolt at *katorga*, except for Michael Lunin, who collected statements from the participants and composed a "Review of the Report of the Investigating Commission," which gives what might be considered the Decembrists' interpretation of the uprising and the interrogation. This report circulated widely and secretly among the Decembrists throughout their period of exile. Lunin also made a study of *Russian Secret Organizations*, 1816-1826.[16]

Memoirs by members of the group, written in middle or old age, lack fervor and preciseness and generally give an impression that the authors were somewhat abashed at their earlier escapades. Yet there is nothing in these memoirs that indicates they disavowed the ideals that dominate earlier thinking: belief in the educability of the masses, in the efficacy of liberalism, in the duty to develop one's mind to its furthest limits and constantly to strive for a better world. These ideals permeated their Siberian experience, and the Decembrists were able to avoid the total nothingness that so very often is ascribed to the period after 1825.

Three Decembrists knew Siberia before exile.[17] The rest were almost completely ignorant of their new homeland but, utilizing education and past experience, they were able eventually to become expert upon some aspect of Siberia—its administration, climate, resources, plants, or animals. Recognizing the value of their training, Count Mordvinov in 1826 suggested the government use the exiles for agricultural development and scientific study of the region. Where else, he asked, would the government ever locate over 100 educated men with sufficient leisure and the occasion to

[15] Basargin, 118. *Literaturnoe nasledstvo*, LX, 176, 587. Rozen, 131. Beliaev, 224. Sokolov, 72.

[16] *Dekabristy*; poeziia, 515, 648. M. V. Nechkina, *Dvizhenie dekabristov* (2 vols. Moscow, 1955), II, 446.

[17] *Dekabristy*; neizdannye materialy i stat'i, ed. B. L. Modzalevskii and Iu. G. Oksman (Moscow, 1925), 168.

study and develop Siberia?[18] Though his suggestion was ignored, and imperial policy was designed to guard against this occurrence, the subsequent work of the exiles emphasizes the cogency of the idea.

Few if any of the scientific expeditions exploring Siberia during the second quarter of the last century did not have the services of at least one Decembrist.[19] At the request of Alexander von Middendorf, Decembrist A. I. Iakubovich was granted permission by local officials to participate in that scientist's 1828 expedition, on condition, however, that none of Iakubovich's work ever be attributed to him. Middendorf maintained secrecy, and only when the state archives were opened did it become known that Iakubovich had helped the expedition in the very northern Siberian district of Nazimov.[20]

At the suggestion of a Siberian official another Decembrist, Stephen Semenov, worked with Alexander Humboldt. Semenov's services were so unexpectedly competent that Humboldt spoke to the Emperor of his surprise at finding such educated persons in Siberia. This testimony to the cultural dispersion within the Empire pleased Nicholas until he realized to whom the great scientist referred. And then, since Nicholas always hated to have the Decembrists brought to public attention, he had Semenov transferred to a very remote village. The Siberian official who had first called attention to Seminov was severely reprimanded.[21]

Decembrists participated in six other expeditions, occasionally without the knowledge of officialdom, but most often with its complicity. A few exiles acted as translators for foreign scientists, while those who had scientific training performed mathematical computations or made astronomical observations. Many of the men turned over to scientists their collections of plants, fossils, minerals, and meteorological data.[22]

Some remarkable early studies of eastern Siberian flora and fauna were made by Peter and Andre Borisov, who were trained

[18] *Arkhiv Grafov N. S. Mordvinovykh*, ed. V. A. Bilbasov (St. Petersburg, 1903), VIII, 41-42.

[19] Lidiia K. Chukovskaia, *Dekabristy issledovateli Sibiri* (Moscow, 1951), 99-100. This writer concludes there was not one scientific expedition of the late 1820's and the 1830's in which at least one Decembrist did not participate.

[20] Chukovskaia, 102-105.

[21] Chukovskaia, 101.

[22] This statement is based on sources quoted in Haskett, 151-153, 154.

biologists and avid collectors. In addition, Peter was a fine artist whose drawings are excellent supplements to their collections. His sketches of numerous types of Siberian birds are especially valuable. These brothers sent their collections of plants and insects to friends for transferral to museums; but few of their materials went beyond the files of the Third Division. After the Bolshevik Revolution they were found and were finally transferred to museums in Moscow and Leningrad.[23]

Some Decembrist studies of climate and weather provide the earliest data of this kind on Siberia. Over more than 10 years at *katorga,* the exiles kept meteorological records which were later sent to the Berlin Academy of Science. Heinrich von Wild used data of three Decembrists—Alexander and Peter Beliaev and M. F. Mitkov —in his work on Russian meteorology.[24] Studies of the climate, of meteors, and of seismic disturbances in the Transbaikal, made by Nicholas Bestuzhev, were turned over to the Academy of Science in Petersburg by his friend, Admiral M. F. Reineke.[25]

At a time when there was growing interest in Lake Baikal, Michael Kuchelbeker was commissioned to make depth soundings along the eastern shores. Ferdinand Wolf, who had formerly been a staff doctor with the second army, was especially interested in the minerals and the mineral springs of Siberia and during his travels was able to chart many of the Siberian mineral springs. A map of the Transbaikal used for some time by local authorities was constructed by Dmitri Zavalishin, whose knowledge of the region was based on several sources. As a youth he had crossed Siberia en route from America to Petersburg; during *katorga* he sent scouts along the Amur and its tributaries in preparation for an escape. These records were later supplemented by his own travels during settlement.[26] Both Vladimir Shteingel and Ivan Iakushkin made geographic studies of western Siberia; Shteingel's study of Ishim district was published in 1843 in the *Journal* of the Ministry of Internal Affairs.[27]

[23] Chukovskaia, 82-86. Fedor A. Kudriavtsev, *Irkutsk;* ocherki po istorii goroda (Irkutsk, 1958), 108. Bogucharskii et al. (eds.), I, 276.

[24] *Sibir i dekabristy;* stat'i, materialy, neizdannye pis'ma, bibliografiia, ed. M. K. Azadovskii (Irkutsk, 1925), 98. Chukovskaia, 96.

[25] M. Iu. Baranovskaia, *Dekabrist Nikolai Bestuzhev* (Moscow, 1954), 185.

[26] S. Ia. Streich, *Moriaki dekabristy* (Moscow, 1946), 256. Zavalishin, II, 102, 104.

[27] "Statisticheskoe opisanie ishimskogo okruga tobol'skoi gubernii," *Zhurnal*

Nicholas Bestuzhev contributed one of the important ethnographic studies of Siberia in the early 19th century—*Guzinoe Ozero* (Goose Lake)—published anonymously in 1854. Bestuzhev's studies convinced him the lake had been formed following seismic disturbances in the 18th century.[28] Along the shores he found evidences of an early Buriat civilization which he judged of a high order on the basis of artifacts and an extant irrigation system. Bestuzhev also described at considerable length the contemporary life and culture of Transbaikal Buriats.[29]

Because the exiles could travel with relative freedom within the confines of Siberia, scholarly activities such as these were possible. Many moved from one village to another for various reasons—health, cost of living, climate—and some lived in three or four Siberian villages. Quite frequently they were permitted to travel to Tobolsk or Irkutsk for medical treatment, and if a man merely wanted to get away for awhile and see old comrades, the illness was more fictitious than real. Always the trips were leisurely and included many villages where Decembrists were living. Officials in the provincial capitals sometimes sent for their Decembrist friends, providing them with a brief holiday in the population centers. Nicholas Bestuzhev's last trip, for example, was in response to the governor's request to visit him; on this trip Bestuzhev contracted a fatal illness.[30] Over the years almost every exile had some occasion to travel.

Within a radius of 15 versts the men could roam freely, so long as absences from the village were registered. Some exiles were even allowed to hunt and carry firearms. Many farmed plots of land outside their villages.

The government granted each Decembrist in settlement 42 acres

ministerstva vnutrennikh del, II, 1843. N. P. Nikitin, "Ekonomiko-geograficheskie raboty dekabristov," *Voprosy geografii*, XLI (Moscow, 1957), 259.

[28] *Vestnik estestvennykh nauk* (1854). This article was later published in *Rasskazy i povesti starogo moriaka N. Bestuzheva* (Moscow, 1860) and is the source quoted here; 493-495, 502-503.

[29] Bestuzhev, 508-575 passim.

[30] Pushchin, 33, 227, 418. Beliaev, 250. *Dekabristy v zabaikal'e;* neizdannye materialy, ed. A. V. Kharchevnikii (Chita, 1925), 33. Sergei G. Volkonskii, *Zapiski*, ed. M. S. Volkonskii (St. Petersburg, 1901), 477. *Literaturnoe nasledstvo*, LIX, 398, 472. Nikolai V. Basargin, *Zapiski*, ed. P. E. Shchegolev (Petrograd, 1917), 205. N. I. Lorer, *Zapiski dekabrista*, ed. M. N. Pokrovskii (Moscow, 1931), 178 ff. I. D. Iakushkin, 390, 408, 418. E. E. Iakushkin, 29.

of land from which he was to support himself and family, and which was expected to keep him preoccupied and aloof from the populace. Few men knew anything practical about Siberian farming, so adversely influenced at that time by conditions of soil, climate, transportation, and markets; except in the western part, Siberia was inhospitable for large scale farming. At Selenginsk the Bestuzhev brothers made ambitious attempts to farm but were able merely to break even in these projects. However, further west, the Beliaev brothers became quite prosperous, raising cattle and grain and operating a mill for the peasants.[31]

Except for the Old Believers and some Buriat tribes, the Siberians were inefficient and ignorant farmers. They produced chiefly onions and cabbage, sufficient only for their own needs; they never experimented with new crops or better methods. Therefore, the example over several decades of exiles raising tobacco, melons, asparagus, cucumbers, rye, and other unusual crops astounded the Siberians. At *katorga* the exiles grew most of their food, as it was the only way to obtain a balanced diet. In settlement they raised favorite vegetables and fruits, often with notable success. A Soviet scholar maintains the Decembrists first introduced maize into Siberia and perhaps other crops, such as tobacco.[32]

As farmers, the exiles formed close ties with the peasants, providing much more than the example of ingenuity and industry. They liberally distributed new kinds of seeds, medical care, advice, and, when they could, money; they acted as referees between quarreling peasants and hired the natives to help on their farms. Such contacts were disturbing to those officials who saw in everything the possible subversion of authority. A few Decembrists were even moved once or twice in order to satisfy officials the suspects were sufficiently

[31] A. P. Golubkov, "Pervye zemliaki," *Eniseiskaia ssylka*; sbornik eniseiskogo zemliachestva, ed. V. N. Sokolov, III (Moscow, 1933), 17. Dmitriev-Mamonov, *Dekabristy v zapadnoi Sibiri*, 191. Pis'ma iz Sibiri dekabristov M. i N. Bestuzhevykh, ed. M. K. Azadovskii and I. M. Trotskii (Irkutsk, 1929), 87, 72. Beliaev, 329.

[32] P. Voitik, "Rol dekabristov v razvitii sel'skogo khoziaistva Sibiri," *Sel'skoe khoziaistvo Sibiri* (No. 1, 1957), 103. *Vospominaniia i rasskazy deiatelei tainykh obshchestv 1820-kh godov*, ed. S. Ia. Streich (Petrograd, 1922), 63. F. G. Safronov, *Dekabristy v iakutskoi oblasti* (Iakutsk, 1950), 26. N. G. Zalkind, "Nekotorye voprosy antropologii v rabotakh dekabristov," *Sovetskaia antropologiia*, II (No. 3, 1958), 173. Belogolovyi, 17. V. Maksimenko, "Poiavlenia kukuruzy v Sibiri," *Sel'skoe khoziaistvo Sibiri* (No. 6, 1957), 101.

isolated.[33] Local authorities generally, however, accepted the fact that political exiles could not be completely segregated from the populace. Prince Volkonskii spent most of his time with the peasants, either working in the fields or meeting with them at the market place. At harvest time each year Baron Rozen hired dozens of peasants to bring in the crop. After work they all gathered for Madame Rozen's delicious meals, with dancing until dawn.[34]

State ministers, senators, and officials from Petersburg made frequent tours of inspection in Siberia. Seldom did they miss an opportunity to visit with the exiles in order to renew old friendships or, more significantly, to obtain information on the Siberian administration. In the consistency with which they visited the Decembrists, we may read their low impression of the local administrators, who were often considered dishonest sources of information. The Decembrists, on the other hand, had every opportunity to learn of maladministration and corruption, and to assess the results of state policies.[35]

It was not at all rare for the Siberian officials themselves to seek help of the exiles. For example, the *gorodnichii* (mayor) of Chita daily visited Zavalishin with correspondence to be answered or problems that seemed beyond his competency. The *gorodnichii* even left Zavalishin in de facto charge of city affairs while he was out of town. During this period the governor of Eastern Siberia, V. I. Rupert (1837-47), consulted the exile on numerous occasions.[36]

Rupert's successor, Nicholas Muravev—later Amurskii—changed that routine, however; he never consulted Zavalishin nor did he approve of his subordinates' doing so. This made the exile a bitter foe of Muravev; but despite the new policy, Zavalishin in 1852 was able to help with the reorganization of Chita as an incorporated city, and in the next year to perform relief work among the peasants during the typhus epidemic. His constant criticism of Muravev, especially of the colonizing expeditions along the Amur, plus the publication of one caustic article in a Russian journal, prompted Muravev to try moving the exile to western Siberia. Muravev's request was granted but Zavalishin absolutely refused to leave Chita.

[33] Lorer, 171. Chukovskaia, 127. Sokolov, 159. *Sibir i dekabristy*, 88. M. D. Frantseva, "Vospominaniia," *Istoricheskii vestnik* (May 1888), XXXII, 754.

[34] Rozen, 202.

[35] Rozen, 206. Pushchin, 377. Zavalishin, II, 291.

[36] Zavalishin, II, 291, 311-320 *passim*, 341, 387.

Only in 1863 did the Siberian administration get rid of him, returning him to Russia, where he quickly adjusted to the "new exile" and began to write voluminously and polemically for journals until his death in 1892. Zavalishin was the last Decembrist.[37]

In 1842 authorities in Tobolsk province went to Vladimir Shteingel for help in averting an imminent rebellion among the famine-stricken peasants. Shteingel was a native of Siberia, a man extremely well liked by the populace, and whether or not his efforts were decisive, Tobolsk was at least spared the disorders of the neighboring province. Once the threat was over, Shteingel was sent to a remote village where he would not again be able to mix in public affairs.[38]

Michael Kuchelbeker was pressed into government service several times in eastern Siberia. Serving as translator, he accompanied an investigating commission to the Tungus villages around Lake Baikal. Later, he made a study of the peasant economy in that region and suggested that the government might partially overcome chronic depression by providing the peasants with potatoes for food, as well as for planting, so that they might diversify their crops. The report was ignored and its sole effect was to rouse official fears of Kuchelbeker's influence with the peasants.[39]

Either for financial or altruistic reasons, or from boredom, a number of exiles entered state service in Siberia. But none of these men, except possibly Alexander N. Muravev, who later became governor of Tobolsk province, ever achieved rank or responsibility commensurate with his education and ability. And Muravev, despite his position, was not trusted by the government, which continued to maintain surveillance on his activities.[40] These exiles in state service worked mainly as clerks and were often harassed by their ignorant and jealous superiors or by the government's obscurantist policies.

At this time education in Siberia was in a deplorable condition.

[37] Streich, *Moriaki dekabristy*, 251.

[38] Bogucharskii *et al* (eds.), I, 313-315. Chukovskaia, 127.

[39] Streich, *Moriaki dekabristy*, 255.

[40] *Dekabristy*; novye materialy, ed. M. K. Azadovskii (Moscow, 1955), 150-151. B. G. Kubalov, "Sibirskoe obshchestvo i dekabristy," *Katorga i ssylka*, XXI (August, 1925), 170. *Pamiati dekabristov*; sbornik materialov (Akademiia nauk SSSR, Leningrad, 1926), III, 77. S. N. Brailovskii, "Iz zhizni odnogo dekabrista," *Russkaia starina*, CXIII (March 1903), 556.

There was no university until 1888. In 1850 there were but half a dozen high schools and about 20 district elementary schools for all of Siberia, inferior institutions receiving meager support and no encouragement from state ministries.[41] In 1830 there were 117 students in the Tobolsk high school and 150 at Irkutsk. With a population in 1853 of half a million, Tomsk province had 95 high school students. Girls' schools were practically nonexistent, the first being established in 1838.[42] Those families who could afford it brought governesses from Europe. Most of the populace, however, remained ignorant of even the rudiments of education. The political exiles during the 18th and 19th centuries did considerable educational work. Prisoners of war during the reign of Peter I had established private schools in Siberia, and since these prisoners were mostly Swedes, in later years Siberia was dotted with enclaves of Swedish-speaking inhabitants. Also, in the 18th century political exiles from Ukrainian and Polish territory, and, in the 19th century Poles from the rebellions of 1830 and 1863, tutored the children of local merchants and officials, chiefly in music and foreign languages.[43]

The Decembrists were important educators during exile, opening or operating about 10 schools.[44] In 1831 at Petrovsk Zavod the Decembrist *katorzhniki* organized a choir for the children of criminal prisoners and villagers. Commandant Leparskii was persuaded to permit instruction in reading and writing, and the exiles augmented the curriculum to include science, arithmetic, French, Latin, Greek, and numerous crafts.[45]

Leparskii never reported these activities to Petersburg, and the school operated until the last Decembrists left *katorga* in 1839. Ivan Gorbachevskii, remaining at Petrovsk as a settler, continued to teach until his death in 1869. A few graduates of the *katorga*-school subsequently went to higher schools in Russia where their surprisingly good performances might well have helped curious

[41] Nikolai M. Iadrintsev, *Sibir kak koloniia v geograficheskom, ethnograficheskom i istoricheskom otnoshenii* (St. Petersburg, 1892), 558, 604.

[42] Iadrintsev, 559. R. M. Kabo, *Goroda zapadnoi Sibiri*; ocherki istoriko-ekonomicheskoi geografii XVII-pervaia polovina XIX vv. (Moscow, 1959), 113.

[43] Iadrintsev, 563-564. V. G. Bazanov, *Vol'noe obshchestvo liubitelei rossiiskoi slovesnosti* (Petrozavodsk, 1949), 76.

[44] Chukovskaia, 62.

[45] Maksimov, *Sobranie sochinenii*, III, 330-332.

officials discover the school, had not a friend of the Decembrists succeeded in quashing their investigation.[46]

The most significant Decembrist schools were at Ialutorovsk, southwest of Tobolsk. There a parochial school, opened in 1842 under the clerical authority of Father Stephen Znamenskii, was actually operated by the Decembrist Ivan Iakushkin. The school began with 6 students, as the parents at first were suspicious. By the end of the first year Iakushkin had 42 pupils, and the number continued to increase. In 1846 he opened a girls school, inspired by the memory of his wife.[47] Prior to 1856 these two schools graduated about 800 students, while 1600 boys attended the parochial school.[48] Eventually boys from as far distant as Kurgan and Tobolsk came to the school.[49]

Besides the usual parochial program, the schools offered Latin, Greek, geography, physics, history, geometry, algebra, drawing, botany, and zoology. The boys also studied mechanics, while the girls learned French and needlework. The pupils attended all year, four days a week, two hours in winter and four hours daily in summer. All work was done at the schools and no materials were taken home.[50] Warm clothing and transportation were provided in winter.[51]

The girls school quickly became self-sustaining through the sale of needlework, but the boys school at first depended upon contributions from the Decembrists and some of their wealthy Siberian friends. In 1848 Tobolsk authorities appropriated 200 rubles yearly

[46] *Vospominaniia Bestuzhevykh*, 161. Zavalishin, II, 97. Ivan I. Gorbachevskii, *Zapiski dekabrista*, ed. B. V. Syroechkovskii (Moscow, 1916), xvi. Kubalov, *Katorga i ssylka*, XXI, 151.

[47] The most informative work on Iakushkin's schools, aside from his memoirs, is N. M. Druzhinin, "Dekabrist I. D. Iakushkin i ego lankasterskaia shkola," *Uchenye zapiski*, II (Moskovskii gorodskii pedagogicheskii institut, Moscow, 1941), 33 96. Druzhinin, 34-35, 45, 75, 87. I. D. Iakushkin, 314, 489, 659. A. G. Bazanov and N. G. Kazanskii, "Pedagogicheskaia deiatel'nost dekabristov v Sibiri," *Sovetskii sever*, I (Leningrad, 1938), 31.

[48] M. S. Znamenskii, "Ivan Dmitrievich Iakushkin; po neizdannym materialam," *Sibirskii sbornik*, III (St. Petersburg, 1886), 89. Kubalov, *Katorga i ssylka*, XXI, 152.

[49] V. N. Orlov and V. G. Verzhbitskii, *Dekabristy smoliane* (Smolensk, 1951), 112.

[50] I. D. Iakushkin, 490, 510. Druzhinin, 76, 89.

[51] Chukovskaia, 121.

for its operation from the Ialutorovsk land tax, thereby indicating their appreciation of the value of this school.[52]

The parochial school was far more popular than the district school, and this fact added considerably to the chagrin of Ialutorovsk officials already vexed that the state criminal had such freedom in exile. On one occasion the district inspector of schools entered the classroom denouncing Iakushkin, of whom he was extremely jealous, and driving him from the school. Decembrists in exile at Tobolsk were able to influence provincial authorities, so that Iakushkin continued to run the school despite the inspector's attitude.[53] In fact, the latter was reminded that the parochial school was not within his jurisdiction.[54]

Petersburg apparently remained oblivious to all this activity in western Siberia. Father Znamenskii never mentioned Iakushkin's association with the parochial school, and when the suspicious hierarchy questioned him the priest always replied that the two were close friends and the exile often visited the school but did not take active part in school affairs.[55] Because Znamenskii was well liked and respected, the church officials accepted his statements.

Periodically the flood of denunciations from Ialutorovsk officials did threaten to interrupt Iakushkin's work, but the governor was a friend of the exiles and for many years ignored the reports. In their regular reports to the Third Division the provincial authorities spoke vaguely of Iakushkin, mentioning his exemplary manner and never touching upon his teaching.[56]

The most severe crisis in Iakushkin's work came in the early 1850's. After Father Znamenskii's transfer from the city in 1853, the new priest was hostile to the exile and the parochial school.[57] At the same time state controls over political exiles became more stringent, and the Tobolsk administration ordered Iakushkin to sever all ties with the schools, a command the new priest eagerly executed. From 1853 to his departure in 1856 Iakushkin had little contact with either of the schools. Following the amnesty and his

[52] Druzhinin, 89-90. K. Golodnikov, "Gosudarstvennye prestupniki v Ialutorovske i Kurgane," *Istoricheskii vestnik*, XXXIV (December 1888), 93.

[53] Druzhinin, 90.

[54] M. S. Znamenskii, *Sibirskii sbornik*, III, 88.

[55] Druzhinin, 92.

[56] Druzhinin, 90. *Dekabristy*; novye materialy, 309.

[57] I. D. Iakushkin, 328, 672.

return to Russia, the boys school was turned into a one-year preparatory school, and the girls school was closed.[58]

The teaching at these schools employed the Lancaster method, familiar to many of the Decembrists who had been in the army. A monitor, always an advanced student, drilled the pupils until they mastered information on a given series of wall charts; Iakushkin then tested each student before advancing him to the next level. The beginner learned to write letters, first in specially moistened sand, then on slates; only after achieving considerable proficiency did he advance to pen and paper.[59]

Chief advantages of this method were that it required only one qualified teacher and a minimum of material. Recognizing the method's mechanical, impersonal, and inflexible qualities, Iakushkin varied the procedure whenever possible. He constructed maps and globes for the study of geography, built demonstration models for the classes in physics and mechanics, and in summer months took his pupils into the country to study drawing and biology.[60] The success of the schools drew visitors from throughout western Siberia to study their operation. Girls schools at Omsk and Tobolsk were modeled on that at Ialutorovsk.[61]

Not far from the China border, at Selenginsk, the Bestuzhev brothers and their friend Konstantin Torson operated a school for Buriats who came from throughout the Transbaikal to learn manual arts, reading, and writing. Each pupil left the school with newly made tools necessary for carpentry, blacksmithing, locksmithing, and other crafts.[62] Perhaps describing briefly the three men who operated it would best indicate the nature of this craft school.

Paintings made by Nicholas Bestuzhev in Siberia provide the best visual record we have of the exiled Decembrists, their wives, children, friends, and the local populace. In the pursuit of this avocation, Bestuzhev often received permission to travel, chiefly

[58] I. D. Iakushkin, 340, 358. Druzhinin, 91, 93.

[59] I. D. Iakushkin, 325. Druzhinin, 48, 73, 79. Bazanov and Kazanskii, *Sovetskii sever,* I, 33.

[60] I. D. Iakushkin, 657. Chukovskaia, 118. *Dekabristy;* novye materialy, 237, 302.

[61] Bazanov and Kazanskii, *Sovetskii sever,* I, 33. Druzhinin, 87, 90. I. D. Iakushkin, 484, 664. Znamenskii, *Sibirskii sbornik,* III, 88.

[62] F. A. Kudriavtsev, *Istoriia buriat-mongol'skogo naroda* (Moscow, 1940), 226. *Vospominaniia Bestuzhevykh,* 316.

to Kiakhta and Irkutsk, and toward the end of his life spent several months each year in these cities.[63]

At Petrovsk Zavod Nicholas was spoken of as the man with the "golden hands," referring to his remarkable ability to make or repair almost anything that was needed at *katorga.* He helped make their clothing and shoes; made jewelry for the wives and families out of their discarded fetters; built furniture, toys, cradles, and sometimes coffins; tuned pianos, devised sprinkling systems for their garden, and repaired broken or obsolete machinery in the Petrovsk foundry.[64] But all these activities were incidental to Bestuzhev's most serious interest, his determination to perfect a simple chronometer.[65] Other inventions of Bestuzhev were an ingenious stove requiring relatively little fuel, improvements for a ship's steering mechanism, and a simplified rifle lock. The rifle lock had been suggested by a remark of Governor-General Muravev at the outset of the Crimean War. According to Michael Bestuzhev, his brother's invention was the ne plus ultra of simplicity, and Muravev referred it to officials in Russia.[66]

The younger brother, Michael, was the inventor of a carriage, known in the Transbaikal as the *bestuzhevka.* The carriage, a sturdy, durable thing of simple design, could be easily drawn over the mountainous terrain. Michael's shop supplied these carriages within a wide radius of Selenginsk; elsewhere his comrades reported seeing some feeble imitations of the *bestuzhevka.*[67]

The third member of this trio of former naval officers was Konstantin Torson. At *katorga* Torson devised farm machinery and gave the blueprints to comrades leaving prison to take up farming. At Selenginsk Torson built various types of machines and had visions of operating a large plant which would fill the peasants' orders for mechanized farm equipment, an impractical dream in view of the terrain, the economy, and the climate, to say nothing of the peasant's obvious reluctance to adopt innovations. Torson's

[63] Baranovskaia, 160-163. *Vospominaniia Bestuzhevykh,* 192. Many of Bestuzhev's portraits and some other works of his are in *Dekabristy;* 86 portretov, vid petrovskogo zavoda i 2 bytovykh risunka togo vremeni, ed. and comp. I. S. Zilbershtein (Moscow, 1906).

[64] Maksimov, *Sobranie sochinenii,* III, 263. Lorer, 148. Baranovskaia, 112-113. *Vospominaniia Bestuzhevykh,* 185, 313.

[65] *Vospominaniia Bestuzhevykh,* 152, 277 and passim.

[66] Baranovskaia, 165. *Vospominaniia Bestuzhevykh,* 183, 517.

[67] *Vospominaniia Bestuzhevykh,* 281, 323

pride was a machine that would winnow, thresh, and grind the grain and which proved to be wonderfully efficient, though cumbersome.[68]

These three energetic, lively men trained hundreds of young Buriats in the Transbaikal, while the mother and sister of Torson and the Bestuzhev sisters taught the women of Selenginsk sewing and needlework.[69]

Once their suspicions dissipated, the Siberians were appreciative students. In the 1920's, when Soviet historians were preparing publications to celebrate the centennial, they located in Siberia men and women to write about the Decembrist teachers who obviously had inspired in their students gratitude, loyalty, and devotion.[70]

A rather sizable number of the Decembrists at one time or another were private tutors for the children of Siberian officials. The governors-general of eastern and western Siberia, provincial governors, village mayors, and police chiefs, all considered themselves fortunate to have nearby political exiles who could educate their sons or daughters in the fine manners of European society.[71]

The Decembrists also educated their own Siberian-born children, as the Emperor honored practically none of their petitions for funds and permission to send the children to state schools. The Tsar answered Maria Volkonskaia's petitions to the effect that since Siberia had no teachers, such petitions were foolish.[72] In the late 1840's, however, a few of the exiles with very important family connections did receive permission to send children to schools at Irkutsk and Tobolsk.[73]

In 1842 the state announced it would educate any child born

[68] Beliaev, 310, 312. Kudriavtsev, *Istoriia buriatmongol'skogo naroda*, 228. *Dekabristy v buriatii* (Verkhneudinsk, 1927), 83-85.

[69] *Vospominaniia Bestuzhevykh*, 193, 316 ff.

[70] Kudriavtsev, *Istoricheskie pamiatniki irkutskoi oblasti XVII-XX vv.* (Irkutsk, 1949), 32. Sokolov, 129. Chukovskaia, 121, 73. Baranovskaia, 179. Kudriavtsev, *Istoriia buriat-mongol'skogo naroda*, 226. *Dekabristy v buriatii*, 7. Sergei Mikhailovich Volkonskii, *O dekabristakh*; po semeinym vospominaniiam (Paris, 1921), 97. V. P. Girchenko, "Dekabristy brat'ia Bestuzhevy na poselenii v Selenginske," *Zhizn' buriatii*, II (September 1925), 18.

[71] *Dekabristy v buriatii*, 26. Sokolov, 128. *Dekabristy*; neizdannye materialy i stat'i, 160. Beliaev, 248.

[72] S. M. Volkonskii, 84-85.

[73] S. G. Volkonskii, 481.

in Siberia and would restore to him all rights denied the parents, on condition they renounce the family name. Although this would have assured the opportunities they desired for their children, only Decembrist Davydov accepted the offer, and in accepting brought upon himself and his wife the bitterest denunciations of his comrades for such a traitorous act.[74]

Educations were provided for these children by the parents and the comrades whom they pressed into tutorial service. For example, Michael Lunin supervised the Volkonskiis' education and also taught them English; other Decembrists tutored them in mathematics and history; their parents taught them literature and French.[75] Such efforts were taken seriously, and even at the terrible Akatui silver mines, where he was imprisoned in 1841 for subversive writing, Lunin found opportunities to smuggle out letters of instruction and advice to young Volkonskii about his education and future career.[76]

The Decembrists thought in terms of giving their children broad and liberal educations; they especially emphasized foreign languages, literature, and music. There was no attempt to provide practical training such as manual arts, except where the offspring of an alliance with a Buriat or Tungus girl was expected to remain in Siberia.

That their educational activity was not effectively restricted is obvious. That the political exiles could achieve so much reflects the weakness of Nicholas' bureaucratic controls as well as the great need and demand in Siberia for some enlightenment.

The Decembrist exiles were able to exert considerable influence upon their Siberian environment. A man whose early teachers had been the Decembrists, Dr. N. A. Belogolovyi, wrote that the Decembrists had " brought real enlightenment " to Siberia.[77]

[74] *Literaturnoe nasledstvo*, LX, 285. Basargin, 212. Streich, *Katorga i ssylka*, XXI, 138. E. E. Iakushkin, 82.

[75] S. Ia. Gessen, *Dekabrist Lunin i ego vremia* (Moscow, 1926), 132, 269-272.

[76] Gessen, 275.

[77] Belogolovyi, 47.

PAPERS OF THE
MICHIGAN ACADEMY OF SCIENCE,
ARTS, AND LETTERS

VOLUME XLIX

(1963 Meeting)

PART III. LITERATURE AND ART

1964

VOLUME XLIX CONSISTS OF THREE PARTS:

I. *Natural Science*

II. *Social Science*

III. *Literature and Art*

Papers of the

MICHIGAN ACADEMY

OF SCIENCE, ARTS, AND LETTERS

VOLUME XLIX

(1963 Meeting)

Editor

RALPH A. LOOMIS

PART III. LITERATURE AND ART

Published for the Michigan Academy of Science, Arts, and Letters by

THE UNIVERSITY OF MICHIGAN PRESS

ANN ARBOR

1964

MANUFACTURED IN UNITED STATES OF AMERICA

CONTENTS

Part III. Literature and Art

LITERATURE

ART

PART III

LITERATURE AND ART

PAPERS OF THE MICHIGAN ACADEMY OF SCIENCE, ARTS, AND LETTERS
VOL. XLIX, 1964 (1963 MEETING)

LITERATURE

THE IDEA OF ROMANCE IN THE EIGHTEENTH-CENTURY NOVEL

SHERIDAN BAKER
The University of Michigan

I take my title from Fielding's opening admonition in *Joseph Andrews,* that ". . . it is possible the mere English reader may have a different idea of romance with the author of these little volumes." And I propose to show that this standard idea of romance, which Fielding assumes, and opposes, was an active ingredient and conditioner of the 18th-century novel, and that for the 18th-century reader the word "romantic" had a fairly specific lading of the chivalric, or of the Arcadian-chivalric—points largely overlooked by literary historians.[1]

Fielding assumes his readers to have an idea of romance derived from those huge folios of neochivalric French romance he names, those of La Calprenède and Mlle. de Scudéry, still read, still fixing the idea of romance in Fielding's day, 80 years after they were written. And when we find Sir Walter Scott, another 80 years beyond Fielding, still naming among the books that set his head romancing "the ponderous folios of Cyrus and Cassandra" (*Waverley,* Gen. Pref.), we may well suspect that the idea of romance of Fielding's public persisted amazingly, that Scott rather fanned a live coal than struck a spark, and that the idea was rather specifically this: "a fabulous chivalric story, together with the noble, genteel, elevated wish-fulfilling quality in daily life, or the wild and verdant quality in landscape, that seemed Arcadian and chivalric, as in the French romances." [2]

[1] I read an earlier, shorter version of this paper before the Nippon Eibungakai, at their annual meeting, Aoyama Gakuin (University), Tokyo, June 2, 1962; that version will appear at some future date in their proceedings.

[2] Johnson's definition (*Dictionary,* 1755) is "1. A military fable of the middle ages; a tale of wild adventures in war and love . . . 2. A lie; a fiction. In common speech. ROMANTICK. adj. [from *romance*] 1. Resembling the tales of romances; wild. 2. Improbable; false. 3. Fanciful; full of wild scenery." Clara Reeve (*The Progress of Romance,* 1785, Facsimile Text ed., New York,

Of course, although the French romances are primary—particularly La Calprenède's *Cassandra*—the 18th century's courtly idea and ideals of romance were strongly augmented on home ground by Sidney, Spenser, and Shakespeare. In 1725, Sidney's *Arcadia* appeared in a modernized version as well as in its three-volumed 15th edition, which in turn was reprinted in Dublin in 1739, the year before Richardson named his servingmaid after Sidney's princess.[3] Spenser's influence is widely acknowledged, but not for strengthening the image of knighthood in the 18th-century consciousness: in 1785, Clara Reeve's *Progress of Romance*, crediting the genre for teaching manners and virtue, quotes Spenser in praise of "chivalrie" on its title page. Shakespeare's romantic influence, especially (with *Hamlet*) on the later "Gothic" sensibility, is wide, obvious, and difficult to calculate. In short, everything seems to have conspired to give the 18th-century public a constant romantic referent, rather specifically centered on chivalric conduct, as in the romances they knew.

And yet literary history conventionally takes the word "romance" in the 18th century only as an awkward generic term for the new realistic novel. The Romantic Period has blocked our view: "It is generally agreed," writes Arthur O. Lovejoy, "that the word 'romantic' . . . was launched upon its tempestuous career through 19th-century criticism and the philosophy of Friedrich Schlegel."[4] To be sure, Lovejoy writes about *German* romanticism. But he clearly, and rightly, finds Schlegel the fountainhead of English Romantic terminology: the power of Schlegel's generalization simply washed out, for us, and for Lovejoy, the 18th-century idea of romance, which extends steadily into the 19th century, where the kindred German current floods it out. The power of the Ro-

1930) notes that La Calprenède, Mlle. de Scudéry, D'Urfé (who, with Sidney, was the archetype for the pastoral mode of chivalry), and others "revived the taste for prose Romances" (p. v); she defines her term as "an *Heroic fable,*—a fabulous Story of such actions as are commonly ascribed to heroes, or men of extraordinary courage and abilities.—Or if you would allow it, I would say an Epic in prose" (p. 13). Her interlocutors, taking a dimmer and more popular view, define romance as "a wild, extravagant, fabulous Story . . . built upon fiction, [with] no foundation in truth" (p. 6). She is clearly drawing from Fielding.

[3] Ernest Baker, *The History of the English Novel* (London, 1929, repr. New York, 1957), II, 88.

[4] "The Meaning of 'Romantic' in Early German Romanticism," *MLN*, XXX (1916), repr., *Essays in the History of Ideas* (New York, 1955), p. 183.

mantic Period has led even Northrup Frye, who gives us the essentially mythic and wish-fulfilling idea of romance more clearly than anyone else, to think primarily of Scott and the Brontës, and to tend to skip from *Don Quixote* to *Northanger Abbey*, and *Madame Bovary*, and *Lord Jim*.[5]

Specialists in the 18th-century novel have done no better. Cross, Mackillop, and Watt stress the novel's realistic departure. Even Ernest Baker presents the tremendous impact of the 17th-century French romances as virtually dying with the end of the Restoration.[6] Watt beautifully describes how the new individualistic society evoked the fictional biography of the new Mr. Everybody, the new common individual in the society growing classless. Watt shows us how Defoe's autobiographical focus was "as defiant an assertion of the primacy of individual experience in the novel as Descartes's *cogito ergo sum* was in philosophy" (p. 15). And it is indeed surprising to see how Defoe's "I" has replaced the Ideal, as Defoe instinctively follows his culture's common assumptions.

But this is not so altogether new as Watt would have it. The "I" had for some centuries already marked the picaresque tale; and we may easily see that this tale of the nimble outsider, successfully making his way around the established order, perfectly suits the very viewpoint and mode of the new society, by Watt's own splendid demonstration. Defoe—by whatever means, perhaps by sheer journalistic rediscovery—had adopted the picaresque tale for the new society of outsiders. That he changed it from comic to serious does not, contrary to Watt's claim, obliterate the old pattern nor its old social implications. The new novel simply found new social significance in the older kind of storytelling. For the 18th-century novel is essentially a success story, suiting the new individualistic society in which each unknown person hoped to succeed to some high place of security and recognition.

But adaptions from romances were deeper and more extensive

[5] *Anatomy of Criticism* (Princeton, 1957), p. 306.

[6] Wilbur L. Cross, *The History of Henry Fielding* (New Haven, 1918), I, 321; Alan D. McKillop, *The Early Masters of English Fiction* (University of Kansas, 1956), p. 110; Ian Watt, *The Rise of the Novel* (Berkeley and Los Angeles, 1957), pp. 9-34 and *passim*; Ernest Baker, III, 12, 30-37, 130. Benjamin Boyce describes the "romancy" quality of Charles II's times and the corresponding taste for French romances ("The Effect of the Restoration on Prose Fiction," *Tennessee Stu. in Lit.*, VI, 1961, 79-81).

than those from the picaresque.[7] Even Fielding's burlesquing brought into the novel the form it attacked. And the central idea of romance was hardly attacked at all, even by Fielding. The central idea of romance, is, I think, also a success story, a wish-fulfillment. It is indeed the ancestor of the picaro's successes, and probably the oldest story in the world. In fairytale or myth or romance, it is always the same: a young nobody of unknown origin rides in from nowhere and turns out to be somebody after all.[8] This shadowy story of success is also the central story of the 18th-century novel.

Indeed, 18th-century fiction and 18th-century life were permeated with attitudes inherited from the romances. We can distinguish three slightly different inheritances. First is the central daydream of success I have just described, as man continued to imagine himself starting low and ending high in the old romantic way, and as society, in a plebeian way, began to make the old dream come true. Second, the romance's ideas of excellence in conduct, even the very phrases of romance, began to shape both the daily life and the fictions of a society reaching for elegance and noble ideals. Ernest Baker outlines the beginnings: the mid-17th-century salons that played seriously at romance, imitating the language and manners of their beloved romances, assuming the names of Orinda and Astrea and Silvander (III, 30-31). But this conditioning persisted far into the century, and became the very mode of 18th-century politeness: Pope's real Martha Blount signs herself "Parthenissa"; Fielding's fictional Mrs. Western claims to have been called "the cruel Parthenissa" (*Tom Jones* XVII. iv); Jane Austen's mother and sister are christened Cassandra—probably not after Priam's daughter but after the disguised heroine of La Calprenède's famous

[7] Professor George Sherburn points out that the 18th-century novel drew its picaresque primarily from *Gil Blas* (1715-35) and not until midcentury ("The Restoration and Eighteenth Century" in *A Literary History of England*, ed. Albert C. Baugh, New York and London, 1948, p. 855). Arnold Kettle notes, however, that the picaro, the feudal outcast, had become the potential bourgeois of Defoe's day, and hence the typical "picaresque" hero of Defoe's novels (*Introduction to the English Novel*, Vol. I, New York, 1960, p. 55).

[8] See Otto Rank, *The Myth of the Birth of the Hero: A Psychological Interpretation of Mythology*, trans. (New York, 1952, *passim*); see also my "Henry Fielding's Comic Romances," *Pap. Mich. Acad.* (Ann Arbor, 1960), XLV, 416; "*Humphry Clinker* as Comic Romance," *Pap. Mich. Acad.* (Ann Arbor, 1961), XLVI, 649; "Fielding's *Amelia* and the Materials of Romance," *PQ*, XLI (1962), 445.

romance of the same name. And Fielding can have his Jones (XVI. ix) and his James (*Amelia* VIII. vii) speak words that might "have become the mouth of Oroöndates himself," referring to the same famous romance without fear that his audience would miss the reference nor take, in either instance, his young men as anything but perfectly realistic and fashionable.[9]

Similarly, when Richardson's Lovelace and Belford agree to write in "the Roman style" (*Clarissa*, Let. XXXI)—whencesoever the term, the style is à la romance ("roman"?)—they too are probably acting just as young 18th-century bravos would, as naturally as when Lovelace (and Tom Jones too) speaks in the romantic mode: "Stay one moment—but one moment stay, O best beloved of my soul!" (*Clarissa*, Let. XCIV). And again, Fielding's Billy Booth and Amelia address one another at a time of highest stress, in lines derived from Congreve's *Mourning Bride* (1697), a play based on a romance and still popular in 1751: "Who calls the wretched thing that was Alphonso"—"O! it is indeed my Alphonso, my only Alphonso!"[10] In 1785, Clara Reeve testifies that the romances "produced a particular kind of affectation in speaking and writing, which is still called the Romantic" (*Progress of Romance*, p. 66), and her whole defense of romances turns on the fact that they had, in earlier times and in her own, improved the manners and morals of England:

> If it taught young women to deport themselves too much like Queens and Princesses, it taught them at the same time that virtue only could give lustre to every rank and degree.—It taught the young men to look upon themselves as the champions and protectors of the weaker sex;—to treat the object of their passion with the utmost respect;—to avoid all improper familiarities, and, in short, to expect from her the reward of their virtues. (pp. 67-68)

Throughout the 18th century, then, young men and women continued to fly high in the language of romances. In fine, the ideas of romance virtually poured into the 18th-century novel, even when it resisted them, and from three directions: (1) from the romances themselves, particularly the 17th-century French ones, still read far

[9] R. S. Crane ("The Plot of Tom Jones," *Journal of General Education*, IV, 1950, 129) and E. M. W. Tillyard (*The Epic Strain in the English Novel*, London, 1958, pp. 52-54) both see Jones's romantic clichés as mildly stylized and comic, as indeed they frequently are; but they as frequently seem absolutely straight and realistic, reflecting midcentury manners actually styled after the romances, as Fielding smiles (if smile he does) at his society rather than his hero.

[10] II. vi. See my "Fielding's *Amelia* . . . ," p. 439.

into the century;[11] (2) from the theater, which had borrowed heavily from them; and (3) from actual genteel life itself—the "real life" that was the new novel's subject—which had acquired its elegant manners and speeches from the romances and the romantic theater both.

The illusions of romance, the specific roles of romance, in other words, had so colored English life that they did not fade for another century and more, on through the Victorian world, indeed, and on into the 20th century. The Clarissas and Amelias are domesticated paragons of that female excellence which can be traced, in books and out of books, from the courtly ladies of medieval romance to the Elizabethan and later French romances, into actual 18th-century life, and on through the 18th century, on through Dickens and long after Dickens. The ideas of romance shaped both 18th-century life and the life in the 18th-century novel. And the literary pictures of perfection continued to perpetuate the ideas of romance in daily life as illusions of romance became part of daily reality.

Like the attitudes of romance, the literary devices of romance became standard in serious and comic novels alike. All (with the eminent exception of *Clarissa*) built on the ancient romantic story of the low unknown rising to recognition, usually, as in *Tom Jones*, with courtly love added, the low unknown discovering his high birth, marrying the high, unattainable lady, and living happily ever after. All used disguisings and mistakes of identity, with frequent lost letters and mixed assignations. But in general, we may identify three periods, each with a differing blend of romance.

The first period includes Aphra Behn, Defoe, and Richardson, who, in differing ways, find the qualities of romantic experience in

[11] Clara Reeve states that Charlotte Lennox's *Female Quixote*, which had "finely exposed and ridiculed" the "passion for the French Romances of the last Century, and the effect of them upon the manners," was, in 1752, "thirty or forty years too late," since "the taste for those Romances was extinct, and the books exploded" (p. 6). But one of her interlocutors reports that her aunts had told her that they and her mother gathered weekly to sew and to listen to one of them read "The *Astrea* of *D'Urfé,—Cyrus* and *Clelia* by *Mesdemoiselles Scudery, Cassandra* and *Cleopatra* by *Calprenede* . . . ," presumably, judging roughly from Miss Reeve's birth date (1729), in about the 1720's (pp. 68-69). Fielding assumes the knowledge of those books in 1742. By 1785, they are, says Miss Reeve, "the lumber of the bookseller's shop, and frequently seen to wrap a pound of sugar from the grocer's" (p. 70). Her interlocutress says of Sidney's *Arcadia*, "I loved the book in my youth" (p. 79): that is, again roughly, in the 1740's.

the realities of modern life. The second is the 30 midcentury years of the comic romance, founded on Cervantes by Fielding in 1742, including Sterne, and ending with Smollett's *Humphry Clinker* in 1771, with a pleasant afterglow in Richard Graves's *Spiritual Quixote . . . A Comic Romance* (1773), and a stout representation in Charlotte Lennox's *Female Quixote* (1752). The third is that of Walpole's Gothic revival, combining the nobility and thrill of romance with the new plebeian novel's tighter form, and of the kindred sentimental novel where the romance's noble benevolence and sensibility tremble to a climax, in transparent comic wrapper, as Mackenzie's "Man of Feeling" dies from the sheer delicious agony of courtly love (1771). From here, the old high virtue and high language, with all the old devices, drive right on through Scott to Dickens, virtually ignoring Godwin and Jane Austen, anti-romantics with whom the century ends.

The new realists like Behn, Defoe, and Richardson, denouncing fiction as pertaining to aristocrats and romancers, claiming actuality itself, nevertheless looked for qualities of romance in actual experience, and then further romanced their tales, in differing lights, with the devices and formulas of their romantic predecessors. Mrs. Behn names her Oroonoko after the great Oroöndates, hero of La Calprenède's *Cassandra,* as qualified by the neighboring Orinoco River,[12] itself fabulous and romantic from Sir Walter Raleigh's search for Eldorado, the land of solid gold. As she adapts episodes from *Cassandra,* so she adapts the romance's heroic excellence to her plea for the noble savage, so cruelly traduced by Christians.[13] Her royal slave, with his French tutor, is a paragon, with the typical "nose . . . rising and Roman" from the romances, and a mouth "the finest shaped that could be seen; far from those great turned lips, which are so natural to the rest of the Negroes." Yet, as J. A. Ramsaran points out, the Coromanten Negroes, whom Mrs. Behn identifies, actually had thin noses and lips, and were as stoic under torture as her Oroonoko.[14] She had found in whatever she knew of exotic South America the kind of hero and the quality of experience the romances had told her about. And she describes in the

[12] Ernest Baker, III, 92.

[13] For Mrs. Behn's humanitarian source in Thomas Tryon, see Ruthe T. Sheffey's excellent "Some Evidence for a New Source of Aphra Behn's *Oroonoko,*" *SP,* LIX (1962), 52-63.

[14] *N&Q,* VII (1960), 143-144.

phrases of romance the South American life she heard about, or read about (or, just possibly, saw), thereby transmuting reality back again into something like the old golden dream.

The case for Defoe is similar. He too found romance in far-off actuality, in Selkirk's experience on a distant island. Defoe's circumstantial realism lets us feel exactly how it would be to live the anxieties of a romantic adventure. But there is more coloration from conventional romance than we at first suspect.[15] Since Defoe's island is far away, he also moves his data a little toward the romantic long ago as well—as if the romantic syndrome could not be resisted—pushing Crusoe about a generation into the past (forty-four years from Selkirk, twenty-eight from himself). Furthermore, he moves Selkirk's island from the Pacific over into the Caribbean, onto the actual island of Tobago, it seems, but also right into Mrs. Behn's romantic territory, in the very conflux of the "Oroonooko" river, spelled with an insistent doubling of the two "oo's" Mrs. Behn had acquired from the French romance (Raleigh, Defoe's geographical source, consistently spelled it "Orenoque"). And the man Friday, whose "nose was small, not flat like the Negroes," certainly derives some of his noble savagery from Mrs. Behn's romantic description of the real slave.[16]

But *Robinson Crusoe* takes a great step forward from *The Royal Slave* in treating romance as one of the realities of life. Defoe has widely and subtly romanced the facts, propelled, as it were, by the romance *in* the facts; but he moves on to treat, as part of his wonderfully real memoir, man's propensity to romanticize. As Crusoe enters his second cave, for instance, he sees "two broad shining eyes," hears a groan and indistinct words, then, in a cold sweat, finds "a monstrous, frightful, old he-goat" dying. In this episode, Crusoe is making fun of his own romantic fancy, as any anecdotalist would, engaging the same romantic expectations of his audience to make his realistic joke. It is exactly the rationalized ghost story Fielding uses, for fun, on Partridge, and Ann Radcliffe rediscovers for pleasant terror. But Defoe, here as at the end of his book, offers his readers romantic consolation: the cave "reflected

[15] In his introductory remarks, Ernest Baker says, "Defoe seems new but accommodates old patterns" (I, 298), but Baker virtually sets all previous fiction aside when he comes to discuss Defoe in detail (III, 130).

[16] Edward D. Seeber, "Oroonoko and Crusoe's Man Friday," *MLQ*, XII (1951), 286-291.

a hundred thousand lights . . . whether diamonds, or any other precious stones, or gold, which I rather supposed it to be, I knew not." Then further: "I fancied myself now like one of the ancient giants, which were said to live in caves and holes in the rocks, where none could come at them."

Crusoe's mind, in other words, has the conventional coloring of boyhood fairy tales and romances, somewhat coloring the world in turn, sometimes recognizing its fancies, sometimes not. Likewise he wryly thinks of himself as king, absolute monarch of nothing, with his crazy court of cats, dogs, and parrot. But he also takes this fantasy seriously, with a saving touch of ironic awareness, in his "castle," and "fortress," and "country seat," and "bower." Friday and his friends become Crusoe's "subjects," and Defoe manages a remarkable actualization of romance in making Crusoe's fanciful fortified kingdom function as if it really were so, frightening the mutineers, making them think "they were got into an enchanted island," and allowing Crusoe to play public hero at last, acting out his fantasy of lordship and saving the ship that is to rescue him. Crusoe exchanges goatskin rags for the clothes of power and riches, disguising himself, true to romance, but in an inversion of the romantic pattern that has its princes disguised in rags. Crusoe does not become king, nor even governor, really, as he would have in a romance. He has acted out the romantic apotheosis to his own and our satisfaction but stays real Crusoe all the time. He merely plays seriously at romance and comes away with the realistic equivalent of the romantic rise to glory at the end of the tale: a new suit and a nest egg.

As Lionel Trilling has said in *The Liberal Imagination*,[17] and as others before him have said, *Don Quixote* is the archetypical novel because it is the prime story of illusion and reality, the novel's primary business. This is exactly what Defoe discovered for himself in *Robinson Crusoe*. But I should like to emphasize one point: it is specifically the stuff of romance that the novel, from *Don Quixote* on, has used to dramatize man's illusions. Crusoe has laughed at his illusions in the cave by seeing them in the garb of romance.

Samuel Richardson, writing Pamela's realistic palpitations, moment by moment, writing a novel taken as breathing reality itself and no illusion at all, nevertheless gives his illusion of reality a

[17] (New York, 1950), pp. 208-209.

strong glow of romance. He heightens his language to reflect his heroine's excellence, for she, true to romance, is a paragon. He domesticates the ancient romantic fairy tale of Beauty and the Beast—by sheer psychological intuition, it seems[18]—just as he builds on the ancient romantic success story, the low person ending high and rich. The unknown squire of low degree becomes the unknown person of the new society, as represented by the powerless serving girl; the discovery of noble birth, which authenticates the squire's intrinsic excellence and permits the impossible marriage to the high lady, becomes the triumph of intrinsic virtue and the reward of impossible elevation and marriage. Richardson deliberately invites the romantic comparison—his smokescreen of authenticity and editorship notwithstanding—by naming his servingmaid not Betty nor Moll, but after the princess Pamela in the still-popular *Arcadia,* and by naming her Swiss jailor "Colbrand" after the Danish giant and jailor in the popular chivalric romance of *Guy of Warwick.*[19]

The idea of romance, of the noble, elevated, ideal life, also shapes *Clarissa,* her romantic name tied to reality by a surname within the pages of the book, as Watt has noted. *Clarissa* seems much more nearly a fair copy of 18th-century life than *Pamela,* as it mirrors and projects the excellence of language and conduct that society had idealized and copied. And yet it is built on abduction and filled with disguisings, as Lovelace, the idealized libido (as Frye might have put it) works his wiles.

The comic romances of Fielding and Smollett similarly uphold the romantic ideal, as they qualify it with low reality and laughter. Even Protean Sterne is still within the genre: he mocks the high and chivalric "Tristram" with the low "Shandy," a perfect Cervantic mockery signifying something like "noble, sad Sir Tristram Crackpot,"[20] a formula for burlesquing romance he seems to have

[18] My "Introduction," *An Apology for the Life of Mrs. Shamela Andrews* (Berkeley and Los Angeles, 1953), p. xxix.

[19] See Ian Watt, "The Naming of Characters in Defoe, Richardson, and Fielding," *RES,* XXV (1949), 322-338. I am indebted to Professor Arthur Sherbo for "Colbrand."

[20] Wilbur L. Cross, *The Life and Times of Laurence Sterne* (New Haven, 1925), I, 183. I draw henceforth on points from my "Henry Fielding's Comic Romances" and "*Humphry Clinker* as Comic Romance." See also Maurice Johnson's *Fielding's Art of Fiction* (Philadelphia, 1961), pp. 47-60, and John S. Coolidge, "Fielding and 'Conservation of Character'," *MP,* LVII (1960), 248.

borrowed from Smollett (high " Sir Roderick " against low " Random "), just as Smollett borrows back from him the latecoming and virtually nonexisting hero in his (high Sir) *Humphry* (low) *Clinker*. From Cervantes, Sterne develops his quixotic Tristram, who tilts against the windmills of authorship with comic assurance and comic unsuccess. From Cervantes's Quixote and Sancho, Sterne also derives his opposing pair of brother comics, Walter the rationalist, Toby the benevolist, and his basic comedy of the mind bemused in its bookish assumptions yet confident of its universal vision.

Fielding's romantic paralleling is explicit, depending broadly on the Cervantic comedy of having modern men mimic the adventures of knighthood, of substituting reality for wonders, and hard knocks for grand illusions:

> This kind of writing may be termed not improperly the comedy of romance [writes Johnson in *Rambler* No. 4, a year after *Tom Jones,* in what seems a commentary on Fielding], and is to be conducted nearly by the rules of comic poetry. Its province is to bring about natural events by easy means, and to keep up curiosity without the help of wonder: it is therefore precluded from the machines and expedients of the heroic romance, and can neither employ giants to snatch away a lady from the nuptial rites, nor knights to bring her back from captivity; it can neither bewilder its personages in desarts, nor lodge them in imaginary castles.

But the plot of Fielding's two great novels is the oldest romantic story of all, the basic daydream of the dispossessed: the young man of unknown origin but of noble qualities ending truly noble and rich. And this romantic center of Fielding is not altogether comic. In *Joseph Andrews* it is playful, but not ridiculous. Fielding indulges our romantic wishes far enough to allow us to see Joseph happily identified and elevated after his trials. *Tom Jones* is somewhat more serious. Jones is of gentle blood; he is elevated to the position and the lady he deserves, in the true romantic style. But he remains illegitimate; like a true modern, his qualities must depend essentially upon himself. *Amelia,* really akin to the century's final period of high sentiment, takes its romantic plot and devices more seriously still, and illustrates how nearly real life had approached the romances in idea and tone.

In Smollett, too, the sentimentally serious romance is evident, alongside the burlesquing of it, with all the standard romantic ingredients: the success story from low to surprising high, the disguisings, the mock knight with alabaster skin. Smollett, too, had aligned himself with Cervantes, who, he said, had reformed a taste "infected with a spirit of knight-errantry, . . . representing chivalry in the right point of view, and converting romance to purposes far more useful and entertaining, by making it assume the sock, and point out the follies of ordinary life." (*Roderick Random*, Pref.). Cervantes had made romance comic, in other words, and had used it to burlesque ordinary reality. And Smollett, like Cervantes, and the other master practitioners of the genre discovered beneath the satiric romancing of everyday follies the essential display of man as a comically illusioned contradiction.

After Smollett, a new rush of serious romance takes over in the Gothic thriller and the sentimental novel. The only basic innovation is that the ancient story of recognition and success takes on a dissonant Gothic ending, not happiness but comforting melancholy ever after, or the eternal thrill of damnation. And the setting becomes pastoral and Arcadian, "romantick" like the grassy or wild landscapes in the romances, most liable, as in Mrs. Radcliffe, to urge a "pleasing melancholy" and an inexplicable tear. The Romantic Period is well under way, though we have forgotten how closely it derived from the romances that named it.[21] Otherwise, the essential story, the ideas of chivalry, the noble sentiments, the disguisings, the mysterious strangers, are all more evident than ever. The Clarissas and Amelias ride a new sweet flood of daydream that sweeps unchecked into the 19th century, in spite of Godwin and Jane Austen, the two notable and different attackers of romance in the last decade of the 18th.

Godwin and Jane Austen both illustrate what has happened throughout the century: that the attackers of romance adopt the enemy's tactics. Both also illustrate what Defoe discovered and the comic romancers knew: that they could illustrate human illusion most shrewdly for their society by the illusions of chivalric romance. Although Godwin designed *Caleb Williams* to expose *Things as*

[21] Walter Allen represents our characteristic view of "romance" in the 18th century, and of Mrs. Radcliffe: "with her, indeed, we are already plumb in the middle of the romantic movement; her half-ruined castles on perilous crags, with dungeons beneath, her mountains, seas, and forests, her monks and nuns, the embodiment of hooded superstition, her bandits and her villains, were to become in a matter of years part of the territory and inhabitants of poetry proper" (*The English Novel*, New York, 1957, p. 102). He has simply named the machinery the Gothic writers developed from the older romances, and hence called "romantick."

They Are (his primary title), we still find the high language, the high virtue, the high unkind lady that all go back to the 17th-century romances: "Madam! Amiable, exemplary Laura! whom in the midst of all your harshness and inflexibility I honour!" (Rinehart ed., p. 347). We still find the old abducting; and also the old disguisings, though these are now a fugitive's necessities.

In Godwin's Falkland we find again the familiar quixotic of the comic romance, his head turned by too many books of chivalry. But the amiable Quixote is now a tortured Gothic hero; the offending romances are Italian! He represents the decayed and tyrannical aristocracy of Godwin's political philosophy. The charm of romance, though certainly not its power, seems at last to have broken. But the mad Falkland, even like Quixote, ends with his author's admiration in spite of his errors: "It was necessary to make him, so to speak, the tenant of an atmosphere of romance, so that the reader should feel almost prompted to worship him for his high qualities" (p. xxvi). Godwin assumes in his public, as Fielding had assumed 50 years earlier, an active idea of chivalric romance.

Moreover, Caleb Williams, like Fielding's hero, is doubly cut to the romantic pattern: he is both the Quixote of the comic romance and the Hero of the serious. His curiosity, which leads him to Falkland's secret and his own downfall, has been fed by "an invincible attachment to books of narrative and romance" (p. 4). True to romance, he is an orphan; his origin is in no way mysterious, but it is properly low and distant: "I was born of humble parents in a remote county of England" (p. 3). True to the Richardsonian reading of the ancient romantic success story, Williams's heroic innocence and virtue triumph over the wicked aristocrat. But Williams is Gothically miserable for destroying the noble Falkland: "thou [who] imbibedst the poison of chivalry with thy earliest youth" (p. 377). Romance has sprung the plot, shaped the novel, and finally overpowered villain, hero, and reader alike.

While Godwin was writing, Jane Austen was returning to the comic romance as such, burlesquing the illusions of life in delicate clichés of chivalry. The female quixote, the usual romance-reading girl of the 18th-century novel, becomes her central heroine, becomes the representative, indeed, of all human illusion, of the way the mind will inevitably generate its cloudy expectations, and of how the event will inevitably differ. This is the center of Jane Austen's comedy: this self-containing, self-deceiving mind, her legacy from

Locke by Sterne. And the modes of romance are its embodiment. *Northanger Abbey* treats the comedy of romantic notions explicitly, and in *Emma* the comic romance reaches perfection. *Emma* is Jane Austen's richest novel, and its base is the same ancient romantic story that Fielding had modernized for *Joseph Andrews* and *Tom Jones,* that of the unknown person who turns out to be noble. But with this story Jane Austen illustrates the comedy of our self-delusion. Emma has read Ann Radcliffe's *Romance of the Forest.* Emma's whole error comes down to her conviction that, since Harriet is of unknown origin, she must therefore be a nobleman's foundling daughter. And this moves on, as a comic romance should, to a deeper disclosure of the comic contradiction between what we are and what we think we are. Emma, like a true comic character, never learns her lesson: her ultimate illusion is that she has learned it. She forswears matchmaking only to continue to matchmake, even for herself—when the match is already pretty well made. The final, delightful romance of it all is to find Emma rewarded by the unknown dream-man she did not recognize, her Knightley in fatherly armor. Like Fielding, Jane Austen perpetuates the happy ending of romance,[22] as her nice but unexalted girls rise by marriage to stations above them. Like Fielding, Jane Austen qualifies the unending happy ending with irony, satisfying, yet teasing and dispelling, our romantic illusions, and specifically though delicately in the accoutrements of chivalric romance. Surely, she means " Knightley " to suggest a knight.

And so the idea of romance is remarkably persistent in the 18th-century novel, just as it seems to have been in 18th-century life. Taken seriously, it produces the Clarissas and Amelias in a social flirtation that saw all gentlemen vaguely as knights and all ladies as princesses gazing from unobtainable windows. Taken comically, it represents all human illusion, and laughs at people taking themselves as knights and ladies. But comically or seriously, the romance's most indelible legacy to the novel was the happy ending. *Pamela* and *Tom Jones* win out over *Clarissa.* Aside from *The Man of Feeling* and the Gothic tremors, no novel before *The Old Curiosity Shop* (1840-41), I think, follows Richardson to over-

[22] For her using the " conventional happy ending " set for the novel by Richardson's *Pamela* (but acquired, as I hope I have shown, from traditional romances), see Henrietta Ten Harmsel's " The Villain-Hero in *Pamela* and *Pride and Prejudice,*" *CE,* XXII (1961), 104-108.

turn the romantic happy ending; and the happy ending dominates for another 20 years, until *The Mill on the Floss* (1860) demonstrates that the old romantic dream of the unknown nobody's marrying the aristocrat is no longer believable, nor bearable, before the deeper dream of drowning in a brother's arms. "Real life," somehow, can no longer fulfill the daydream with high life happily ever after.

In the 18th century, I must repeat, the word "romance" was not what we have taken it for: as a generic term roughly equivalent to the word "novel" that displaced it. Throughout the century, it had a much higher specific connotation of chivalry than usually conceded. The imprint of the 17th-century French romances is distinct on any and every 18th-century novel we may name. The ideas of romance provide a constant measure from which we may describe the innovations and life of any 18th-century novel. I do not, of course, insist that this romantic matter is the only matter in the 18th-century novel. Far from it. I am merely insisting that the iron mask and gauntlet were never very far from the 18th-century writer's central awareness, especially in days when men still rode horses, and wore swords on high occasions, and occasionally used them. The Romantic Movement—and the Neo-classic antithesis—have obscured the armory for us, in spite of Keats's knights-at-arms and magic casements. But in the 18th-century novel, the word "romance" and the idea of romance are active and formative, whether presenting ideals or laughing at illusions, with a distinct clinking of armor from the older romances in the background.

Indeed, the essential idea of romance is still with us, the idea of the unknown low ending known and high. It had remained a fairytale for societies with little hope of social rise. In chivalric times it had moved into formal literature as the story of the only socially mobile type around, the young man-at-arms with spurs and high-born ladies to win. The story comes into the 18th-century novel, first by way of the French romances and the Restoration stage, then forcefully and officially, as it were, through Richardson and Fielding, gathering seriousness as it becomes the summative graph of a society demanding of everyone the upward push for place and identity, or, later, the upward refinement of feeling. Even Defoe, whose characters know their origins, has traces of it. His nobodies rise to relative riches. And the knight in shining armor is still a cliché, and a story, and a state of mind, even though he may now ride a Thunderbird in *The Ladies' Home Journal*.[23]

The novel became the dominant form, then, and continues to be the dominant form, not merely, as Watt suggests (Johnson, in *Rambler* No. 4, had previously seen this), because it realistically presented a life the new individual could authenticate from his own experience, and could enjoy as if he were seeing himself in print, but also because it seemed to reward his own struggles with the status he needed to recognize himself in the lonely crowd. Recognition and discovery, the supreme fictive dessert of the centuries, copied by the novel from the romance, had become a social and psychological necessity. In the new novel, the new individual could imagine his own secret excellence, which seemed hopelessly lost in the crowd to all but himself, publicly acclaimed like that of the unknown knight he had known from childhood.

[23] Or a white charger on the silver screen. The hero of TV's popular *Have Gun Will Travel*, a rescuer of the innocent and helpless, always alone, always moving on, with a chess knight on his holster, is Paladin, obviously named after the *paladins*, the 12 knights and champions of the Roland cycle of romances. See also Robert Warshow, "Movie Chronicle: The Westerner," *Partisan Review*, XXI (1954), 190-203. With no reference to medieval romances, or to any others, Mr. Warshow nevertheless describes a kind of archetypical cowboy that fits the old romantic outline almost perfectly, "fixed in the dreamy clarity of a fairy tale" (p. 200).

PAPERS OF THE MICHIGAN ACADEMY OF SCIENCE, ARTS, AND LETTERS
VOL. XLIX, 1964 (1963 MEETING)

MALORY'S "TALE OF SIR LANCELOT" AND THE QUESTION OF UNITY IN THE *MORTE DARTHUR*

GILBERT R. DAVIS
State University College
Geneseo, New York

IN a long introduction to *The Works of Sir Thomas Malory*, Eugène Vinaver asserted that the *Morte Darthur*, hitherto regarded a continuous narrative, was in truth a collection of eight separate "tales." [1] Prior to this pronouncement, Vinaver had been an advocate of the conventional view [2] that Malory "makes of this vast assemblage of stories one story, and one book." [3] Vinaver's thesis was received with mixed reaction in scholarly circles,[4] but it was not until some years after the publication of this edition that the opposition began to present the results of its investigations.[5] Vinaver's hypothesis had forced scholars to reexamine the text of the *Morte Darthur* to find conclusive evidence that the sections are integrally related elements of a complete narrative. The burden of proof fell upon advocates of the once generally accepted position, largely because Vinaver's theory that the *Morte Darthur* is a collection of eight separate tales gained credence by appearing in an

[1] Eugène Vinaver, *The Works of Sir Thomas Malory*, 3 vols. (Oxford, 1947), pp. xxx-xxxv. Hereafter, all references to the text of the narrative, cited by volume and page, will be from this edition and will appear in the body of the paper.

[2] Vinaver, *Malory* (Oxford, 1929).

[3] George Saintsbury, *The English Novel* (London, 1913), p. 25.

[4] For the comments of those reviewers who accepted Vinaver's thesis, see: J. H. Houseman, *Erasmus*, Vol. I (1947), cols. 921-925; M. Schlauch, *New York Times Book Review* (Sept. 12, 1948), p. 28; E. C. Batho, *MLR*, XLIV (1949), 260; and J. A. W. Bennett, *RES*, XXV (1949), 61-67.

[5] Cf. M. E. Dichmann, "Characterization in Malory's *Tale of Arthur and Lucius*," *PMLA*, LXV (1950), 878 ff.; R. H. Wilson, "How Many Books Did Malory Write?" *University of Texas Studies in English*, XXX (1951), 1-23; D. S. Brewer, "Form in the *Morte Darthur*," *Medium Ævum*, XXI (1952), 14-24; R. M. Lumiansky, "The Question of Unity in Malory's *Morte Darthur*," *Tulane Studies in English*, V (1955), 32-33 ff.; T. C. Rumble, "The First *Explicit* in Malory's *Morte Darthur*," *MLN*, LXXI (1956), 564-566; and C. Moorman, "Internal Chronology in Malory's *Morte Darthur*," *JEGP*, LX (1961), 240-249.

edition which is excellent in so many ways. Thus, Vinaver's claim is more forceful than it ought to be, or would have been, had it been published separately.

In view of this, it is my purpose in this paper to undertake another examination of the problem of unity in the *Morte Darthur.* Limiting my consideration to the section of the work which Vinaver calls "The Noble Tale of Lancelot Du Lake," I will try to show that this section is a part of a continuous narrative, and that it appears in that continuous narrative specifically where it does because of certain structural necessities. One of the primary difficulties in establishing the links between the "Lancelot" and the rest of the *Morte Darthur* is that the tale contains very few specific echoes picked up from earlier sections or resounded in later sections of the narrative; therefore, its place in the *Morte Darthur* must rest on clear structural connections which mark it as an integral part of the larger narrative.[6]

The first link to be considered is the one connecting the "Lancelot" with the section immediately preceding it, "The Noble Tale of King Arthur and the Emperor Lucius":

> Sone aftir that kynge Arthure was com from Rome into Ingelonde, than all the knyghtys of the Rounde Table resorted unto the kynge and made many joustys and turnementes. And some there were that were but knyghtes encresed in armys and worshyp that passed all other of her felowys in prouesse and noble dedys, and that was well proved on many.
>
> But in especiall hit was prevyd on sir Launcelot de Lake, for in all turnementes, justys, and dedys of armys, both for lyff and deth, he passed all other knyghtes, and at no tyme was he ovircom but yf hit were by treson other inchauntement. So this sir Launcelot encresed so mervaylously in worship and honoure; therefore he is the fyrste knyght that the Frey[n]sh booke makyth men[n]cion of aftir kynge Arthure com frome Rome. (I, 253)

These paragraphs clearly indicate that the "Lancelot" was meant to follow the "Arthur and Lucius" section. Furthermore, these paragraphs are original with Malory; they do not appear in his known French sources. In dealing with this transitional passage in his commentary, Vinaver mistakes the scope and structure of the whole narrative when he says that Malory

[6] Along similar lines, Moorman ("The Relation of Books I and III of Malory's 'Morte Darthur,'" *Mediaeval Studies*, XXII [1960], 361-366) has recently shown that thematically "Malory's purpose in Book III ["Lancelot"] is to look both backwards and forward to contrast the courtly perfection of Lancelot with the barbarity of earlier knights" (p. 366).

. . . apparently imagines that in the French Arthurian romances the story of the Roman campaign was followed by the adventures of Lancelot. In none of the numerous versions of the French Arthurian Cycle is there anything resembling such an arrangement. The Roman campaign invariably occurs at the very end, in the *Mort Artu*, long after the *Lancelot* proper; the latter constitutes the middle branch of the Cycle, separated from the *Mort Artu* by the *Quest del Saint Graal.*[7]

Surely no author as eclectic as Malory could have mistaken the order of events as he found them in his sources. It is clear from Malory's addition of the above-quoted transitional paragraphs that although he was aware of the sequence of events in his sources, he altered the conventional order of events to make them conform to his design. Moreover, what he altered, he seems to have altered for artistic purposes.

To attempt to discover something of that design, we shall have to turn momentarily to the " Arthur and Lucius " section of the *Morte Darthur,*[8] for several features of this section point clearly to what is going to take shape in the " Lancelot." In the alliterative *Morte Arthure,* the source of Malory's " Arthur and Lucius," Lancelot is a minor character; indeed he is mentioned only six times. But Malory had apparently already decided that Lancelot was to be a major figure; thus, he introduced him at this early point in the narrative and departed from his source markedly to portray him as a noble warrior, one whose renown is rapidly increasing. Though he begins as " yong sir Launcelot de Laake with a lyght herte " (I, 189-190), by the close of the " Arthur and Lucius " section he is praised as a knight of great strength (I, 216-217) who has accomplished great deeds (I, 220). These repeated references indicate Malory's conscious effort to prepare the way for the " Lancelot " section and the fuller development of the character who is to become eventually the central figure of the rise and fall of the Round Table.

Lumiansky has pointed out another connection between the " Lancelot " and the " Merlin." Malory's alteration of the name of the castle held by the two giants (whom Lancelot kills) to Tyntagyll (I, 272), and the mention of Ygrayne and Uther Pendragon

[7] *Works,* III, 1398.

[8] Miss Dichmann, in her article " Characterization in Malory's *Arthur and Lucius,*" has done a thorough job of showing how and why this portion of the *Morte Darthur* is so well related to the whole narrative.

. . . carry the reader back to Arthur's being conceived before the marriage of his father and mother, a situation which in the book as a whole—like that between Arthur and his sister Morgawse, from which Mordred was born—has immensely important thematic connections with the adulterous relationship of Lancelot and Guenevere.[9]

Though I am not entirely certain the reader would very readily associate the mention of Tyntagyll with the adulterous relationship of Lancelot and Guinevere, it does seem likely that Malory was in some way attempting to relate the "Merlin" and the "Lancelot" sections through the reintroduction of this castle and the brief mention of the names associated with it.

Still another connection exists in the "Lancelot" which relates this section to the rest of the *Morte Darthur*. At the end of the introductory remarks of the "Lancelot," we are told that

. . . quene Gwenyvere had hym [Lancelot] in grete favoure aboven all other knyghtis, and so he loved the quene agayne aboven all other ladyes dayes of his lyff, and for hir he dud many dedys of armys and saved her from the fyre thorow his noble chevalry. (I, 253)

I will deal later with Lancelot's love for Guinevere and the "many dedys of armys" he did for her. But his saving her from the fire "thorow his noble chevalry" is an episode which occurs much later in the narrative—in "The Most Piteous Tale of the Morte Arthur Saunz Guerdon"—and which brings about the open conflict in Arthur's Court and ultimately the final collapse of the Round Table. This early anticipation of a much later event is particularly interesting because Malory's source introduces the episode of the burning of Guinevere *before* any of the events recorded in the "Lancelot" section of the *Morte Darthur*. Thus, Malory's shifting of this episode from its early position in the French prose *Lancelot* to its late position in his own narrative suggests a definite thematic purpose: he did not want to introduce the irrevocable knowledge of the lovers' guilt here in the middle stages of the history of the Round Table; rather, he wanted to depict chivalry on the rise, coming into its flourishing state, and this is one of the important thematic purposes of the "Lancelot" and the "Gareth" sections of the *Morte Darthur*. Moreover, the late introduction of the episode of Lancelot's rescue of Guinevere functions in Malory's version to emphasize the deteriorated state of Arthur's kingdom and the

[9] R. M. Lumiansky, "The Relationship of Lancelot and Guenevere in Malory's 'Tale of Lancelot,'" *MLN*, LXVIII (1953), 91.

Round Table. Indeed, this episode signals the breaking point—the point at which the inevitability of the collapse of Arthur's realm is recognized.[10]

The motif of the love between Lancelot and Guinevere throughout the various sections of the *Morte Darthur* is a rather obvious connection, but it deserves repeated analysis because of the deliberate and methodical way in which Malory prepares his readers for the growth and development of the adulterous affair which is so instrumental in the final destruction of the Round Table. The first mention of this love is in the very early " Torre and Pellinor " section of the *Morte Darthur*:

> . . . Marlyon warned the kyng covertly that Gwenyver was nat holsom for hym to take to wyff. For he warned hym that Launcelot scholde love hir, and sche hym agayne. . . . (I, 97)

Merlin's prophecy is meant to be regarded seriously, and indeed the reader knows that this love is an eventuality which cannot be deterred. Though the next mention of the Lancelot-Guinevere love is rather cryptic, the indication is clear that some kind of intrigue is going on between Lancelot and the queen, even if the " love " is as yet one-sided. This mention comes in the " Arthur and Lucius " section (I, 195) , and, significantly, it is not to be found in Malory's source. In fact, Malory's addition of it suggests quite clearly that he was interested in maintaining a progressive development of this illicit love; first seen as a prophecy, here the illicit love is a reality which has already taken some kind of shape. By the opening of the " Lancelot " section, there can be little doubt that the love affair is well under way and will continue to grow throughout the rest of the narrative.

The love affair motif is further emphasized because, as Lumiansky has pointed out, four of the five references to the Lancelot-Guinevere relationship in the " Lancelot " are Malory's additions to his source.[11] The first of these additions I have already quoted. It appears in the introductory remarks of the " Lancelot " and serves as a summary statement, not only of what has just been completed in the " Arthur and Lucius " section, but also of what is to occur in the " Lancelot " itself. Therefore, part of the function of the " Lancelot " is to show the growing love of Lancelot for

[10] See Vida D. Scudder, *Le Morte Darthur of Sir Thomas Malory and Its Sources* (New York, 1921) , pp. 374-375 for a fine discussion of this point.

[11] " The Relationship of Lancelot and Guenevere," p. 86.

Guinevere, and this love is indicated either in the direct statements of those who confront Lancelot with it (whether he denies it or not), or by deed (for example, Lancelot's directing those whom he has conquered to report to the queen on Pentecost).

All three of the episodes in which Lancelot is confronted with the knowledge of his love for Guinevere are Malory's additions to his source. The first of these occurs when the four queens recognize Lancelot and comment on his love for Guinevere as though it were at this point almost widely acknowledged rumor. Here Malory reworks and adds to his source, because in the French prose *Lancelot* the four queens do not even recognize him, much less front him with his love for Guinevere.[12] Similarly, after the Chapel Perilous episode (I, 280-281), Hallewes the Sorceress confronts Lancelot with his love for Guinevere; here we have another episode not found in the French prose *Lancelot*.[13] And finally, the damsel who leads Lancelot to Tarquin refers to his love for Guinevere and upbraids him for not loving an eligible woman of the land (I, 270-271). In the French prose *Lancelot,* this recrimination is absent.[14] Thus, there are three separate but similar instances of Malory's having added material to his source in an effort to make the developing relationship between Lancelot and Guinevere perfectly clear. In these three episodes Malory's purpose is to establish the Lancelot-Guinevere relationship as a fairly widely known fact by the time of the opening of the "Lancelot" section. That Lancelot denies this relationship (I, 258) is not to be taken too seriously because he constantly denies it throughout the remainder of the *Morte Darthur*. Indeed, the subterfuges which he is forced to employ in order to try to keep the affair with Guinevere hidden do almost as much to denigrate his character as his knightly deeds of arms do to heighten it. When he is upbraided by the damsel who leads him to Tarquin, he no longer denies his relationship with the queen; instead, he avoids the issue entirely in a long speech in which he rejects, in general, the virtues of marriage for a knight seeking adventure.

[12] See Oskar Sommer, ed., *The Vulgate Version of the Arthurian Romances* (Washington, 1912), V, 92-93.

[13] R. N. Wilson has made a good case for the *Perlesvaus* being the source of this episode ("Malory and the *Perlesvaus*," *MP*, XXX [1932], 14-16).

[14] Sommer, V, 204-205.

I may nat warne peple to speke of me what hit pleasyth hem. But for to be a weddyd man, I thynke hit nat, for than I muste couche with hir and leve armys and turnamentis, batellys and adventures. And as for to sey to take my pleasaunce with peramours, that woll I refuse: in prencipall for drede of God, for knyghtes that bene adventures sholde nat be advoutrers nothir lecherous, for than they be nat happy nother fortunate unto the werrys; for other they shall be overcom with a sympler knyght than they be hemself, other ellys they shall sle by unhappe and hir cursednesse bettir men than they be hemself. And so who that usyth peramours shall be unhappy, and all thynge unhappy that is aboute them. (I, 270-271)

Lancelot's commentary on himself and his own circumstances is, of course, too obvious to require comment, but the passage functions clearly as an adumbration of the events which are to befall him in the closing books of the entire narrative.

Thus, it becomes clear that Malory's intention was to open this section with the summary statement that Lancelot "loved the quene agayne aboven all other ladyes dayes of his lyff, and for hir he dud many dedys of armys" (I, 253) and then to add the series of adventures and tests of Lancelot's chivalry and love. He is, of course, victorious on both counts. The way is then clear for the Lancelot-Guinevere relationship to develop fully in the books following the "Gareth" section,[15] until it culminates in the final destruction of Arthur's kingdom and the Round Table.

Those who have argued that the *Morte Darthur* represents a unified narrative greatly emphasize the "Lancelot," for they see it as the section in which the real protagonist of the "whole boke" is introduced.[16] Though Lancelot plays a role in the action of the "Arthur and Lucius" section, and though this role has been augmented by Malory over and above what was in his source, it is still a minor role. The early references to Lancelot are meant to present a consistently developing character—one growing in strength and reputation—but it is the function of the "Lancelot" section to present this youthful knight in some of his many exploits to illustrate his knightly perfection.

This perfection is illustrated in his conquering Tarquin (I,

[15] Lumiansky (cf. "The Question of Unity in Malory's *Morte Darthur*," pp. 35 ff.) has shown that the "Gareth" section is a flashback to events preceding the "Lancelot"; thus the appearance of the "Gareth" after the "Lancelot" does not really interrupt the development of the Lancelot-Guinevere relationship.

[16] See Scudder, *Morte Darthur*, p. 212; Wilson, "Malory's Early Knowledge of Arthurian Romance," *University of Texas Studies in English*, XXIX (1950), 40 ff.; and Lumiansky, "The Relationship of Lancelot and Guenevere," p. 90.

265-267), his freeing the knights of the Round Table held prisoner by Tarquin (I, 268), his assisting King Bagdemagus against the King of North Galys (I, 261-263), his defeating the two giants of the Castle Tyntagyll (I, 271), his thwarting Sir Phelot's treachery (I, 282-284), and a host of other adventures. Throughout all of these exploits, Lancelot is presented as a perfect knight; indeed, we are told that "he passed all other knyghtes, and at no tyme was he ovircom but yf hit were by treson other inchauntement" (I, 253). It is here in the "Lancelot" (as well as in the "Gareth" section to follow) that the heights of chivalry are presented. After the introductory sections in which Arthur was seen coming to the throne, placing his kingdom in order, conquering his enemies on the Continent, and establishing the concept of chivalry defined by the high ideals of the Round Table, the real protagonist of the narrative is finally set in his proper place; and, except for the brief flashback in the "Gareth" and the early part of the "Tristram" sections, the narrative is ready to run its course through the quest of the Grail, the civil strife, and the ultimate downfall of the kingdom and Round Table. This structural unity argues strongly that Malory conceived of the *Morte Darthur* as a continuous whole.

PAPERS OF THE MICHIGAN ACADEMY OF SCIENCE, ARTS, AND LETTERS
VOL. XLIX, 1964 (1963 MEETING)

KING LEAR'S " POOR FOOL "

ARTHUR M. EASTMAN
The University of Michigan

MY concern is with the meaning of "poor fool" in Lear's final speech:

> And my poor fool is hanged! No, no, no life!
> Why should a dog, a horse, a rat, have life
> And thou no breath at all? . . .
> (V. iii. 305-307)

With rare exceptions every standard edition of Shakespeare since 1773 glosses "poor fool" as Cordelia. Through the turn of the 20th century most editors offered supporting evidence; today most do not.[1] The traditional gloss has shaded into a dogma. I propose to challenge it on the grounds of its inadequacy.

[1] Steevens inaugurated the traditional gloss in the Johnson-Steevens edition of 1773. The rare exceptions among editors are Charles Knight (1841), Collier (1842), Halliwell (1850-53: in his edition of 1865 Halliwell returned to the traditional gloss), Singer (1855-56), Phelps (1896), Porter and Clarke (*First Folio Shakespeare*: 1905), Gildersleeve (*Tudor Shakespeare*: 1912). A good many editions, of course, have no note on this line. There are probably more exceptions among critics than I have noted. Reynolds's long note challenging Steevens's gloss appeared in the Johnson-Steevens edition of 1785. Later dissidents include C. A. Brown, *Shakespeare's Autobiographical Poems* (1838), 292; Stopford A. Brooke, *Ten More Plays of Shakespeare* (London, 1913), 219-221; William Empson, "Fool in *Lear*," *Sewanee Review*, LVII (1949), 209; and Harold Goddard, *The Meaning of Shakespeare* (Chicago, 1951), 548.

The school texts generally reflect tradition without qualification. I have found no college text on the market today that admits any question about "poor fool." My investigations include the following: G. B. Harrison (1948, 1952), Hardin Craig (1951), Neilson and Hill (1942), Parrott (1938), Brooke, Cunliffe, and MacCracken (1935), Kittredge (1936), Craig (*The Oxford Shakespeare*: 1904), and a good many less prominent editions. The glossaries accord with the school texts. Dyce-Littledale (1902), Onions (1919), and Schmidt (1923) all refer to this line in presenting the traditional gloss; Cunliffe (1910) is the sole exception.

Line references are to the Globe numbering. To avoid undue annotation editions will be referred to simply by the editor's name and date. Most of the critical material prior to 1880 will be found collected in H. H. Furness, *A New Variorum Edition of Shakespeare*, Vol. V: *King Lear*, 2nd ed. (Phila., 1880), 345-347.

It has been argued that " fool " or " poor fool " is a common expression of tenderness or compassion, here applied to Cordelia. So Verplanck (1847): " a common phrase of affectionate kindness, applied to any person or thing, where some feeling for helplessness or misfortune was mixed with natural tenderness." So Kittredge (1940): " often used as a term of affection." By way of proof the following parallels are adduced from Shakespeare's own works:

(1) *Twelfth Night,* V. i. 377 (Olivia to Malvolio on his release from prison): " Alas, poor fool, how have they baffled thee! "
(2) *Winter's Tale,* II. i. 118 (Hermione to her ladies when Leontes commands her imprisonment for adultery): " Do not weep, good fools, There is no cause."
(3) *Two Gentlemen,* IV. iv. 98 (Julia, in disguise, of Proteus, who has entrusted her with a message to Silvia): " Alas, poor fool! Why do I pity him . . . ? "
(4) *Venus and Adonis,* 578 (the narrator of Adonis, who after Venus's amorous attack wishes to leave): " The poor fool prays her that he may depart."
(5) *As You Like It,* II. i. 22 (Duke Senior on the venison they are about to hunt): " And yet it irks me the poor dappled fools . . . Should in their own confines with forked heads Have their round haunches gor'd."
(6) *Much Ado,* II. i. 326 (Beatrice replying to Don Pedro's remark that she has a merry heart): " Yea, my lord, I thank it, poor fool—it keeps on the windy side of care."
(7) *Antony and Cleopatra,* V. ii. 308 (Cleopatra to the asp at her breast): " Poor venomous fool, Be angry, and dispatch."
(8) *Taming of the Shrew,* III. ii. 159 (Tranio comparing Kate to Petruchio): " Tut, she's a lamb, a dove, a fool to him! "
(9) *Henry VI, Part III,* II. v. 36 (Henry wishes he were a shepherd): " So many days my ewes have been with young, So many weeks ere the poor fools will ean."
(10) *Romeo and Juliet,* I. iii. 31 (Nurse on the weaning of Juliet): " But, as I said, When it did taste the wormwood on the nipple Of my dug, and felt it bitter, pretty fool, To see it tetchy, and fall out with the dug! "

Although " fool " connotes tenderness or compassion in these passages, the referents vary. In the first four passages " fool " applies to a rational being guilty of some kind of irrationality, some deficiency in judgment or sense. So, in passage 2, Hermione implies that weeping would be warranted only if she were guilty of the adultery for which she is being imprisoned. Since she is not, since " there is no cause," her ladies by their weeping show deficient judgment. In their hearts they are " good," but in their judgments they are " fools." The usage here does not fit Cordelia. She, whom Lear has seen from his wheel of fire as " a soul in bliss," excels in judgment.

The first, fifth, and tenth passages reveal a second referent for " fool." In passage 1 Malvolio is a fool not only because he has shown deficient judgment but because he has been the dupe or victim of others. The deer of passage 5 are also victims, and so is Juliet in passage 10: she is the victim of the Nurse's stratagem for weaning her. This usage fits Cordelia. Long before, she was the victim of her sisters' deceit; now she is the victim of their hatred.

The fourth passage and all that follow it reveal yet a third referent. Adonis is guilty of deficient judgment, the narrator implies, in wishing to leave Venus; but he is a fool in another sense. Like the heart of Beatrice, the asp at Cleopatra's breast, the unweaned lambs, the nursing Juliet, he is, morally, an innocent. This usage of " fool " also fits Cordelia.

Only six of these parallels are truly germane to the passage in question. In only one, the tenth, do we find the same pairing of referents ascribed to Lear's use of " poor fool." Only in the Nurse's musings on Juliet is " fool " used tenderly and compassionately of someone who, in her innocence, has been made a victim. If we place this fact beside a second—that in his 400-odd other uses of the word " fool " Shakespeare manifestly means merely a victim (without any overtones of tenderness or compassion) or a stupid person or a jester—we can see that the so-called " common " meaning attributed to " poor fool " is genuinely uncommon.[2] It is, indeed,

[2] A small number of other parallels has been adduced from non-Shakespearean works. The closest, cited by Kittredge, ed., *King Lear* (1940), comes from Jasper Heywood, *Troas*, Argument (*Seneca His Ten Tragedies*), wherein Hector's son, like the child Juliet, is called " Pore seely foole his Mothers onely joy." Dyce-Littledale, *A Glossary to the Works of William Shakespeare* (London, 1902), cites *Don Quixote*, trans. Shelton, First Part (1612), line 566, wherein a goat, in reference to its animal innocence, is called " pide foole." The *N. E. D.* cites Sidney, *Astrophel and Stella*, Sonnet lxxiii: Stella is a " fool " because irrationally angry. Malone (1790) cites Brooke, *The History of Romeus and Juliet* (1562): Romeus is a " fool " because deprived of reason and made a plaything by love. Halliwell (1865) cites Cokain, *Trappolin Suppos'd a Prince* (1658): the woman is a " fool " because she has lost her reason out of love for the speaker, whose use of the term suggests complacent self-deprecation. Perrett, *The Story of King Lear* (Berlin, 1904), 301-303, cites Jonson, *Bartholomew Fair* (1614), V. vi. 23: perhaps this should be considered a close parallel since Littlewit uses " poor fool " affectionately of his wife, but the term has ironic overtones since by name and deed the lady shows herself to be, in truth, a fool. The *N. E. D.* also cites a work I have not seen, *Beaute of Women*; the single line quoted, " How say ye now by this, little young fool? " suggests that the person addressed is a " fool " for having lapsed in sense.

of such rarity that the remaining arguments in its support must be exceedingly strong.

Arguments designed to show that the rare meaning of " fool " fits—and only the rare meaning—are of two sorts. The first denies the relevance of the term to the Fool; the second asserts its exclusive relevance to Cordelia.

It is argued that the Fool cannot be meant because, " having filled the space allotted him in the arrangement of the play," he was " silently withdrawn "; because " Cordelia was recently hanged; but we know not that the Fool had suffered in the same manner, nor can we imagine why he should "; because " the party adverse to Lear was little interested in the fate of his jester." Thus Stevens (1785) , followed by many.

The flaw in these arguments is that they beg the question. They assume " poor fool " does not mean the Fool, which is the very thing to be determined. It cannot be said that we do not know what has happened to the Fool unless the line in question is itself thrown out as evidence. As to imagining why the Fool should suffer in the same manner as Cordelia, and as to Lear's enemies not being interested in the Fool, the answer must surely be that Goneril has shown her animosity toward him as early as the first act; that hanging was a fine undignified death to visit on one's enemies, on criminals and traitors; that the Fool, as adherent to Lear, was automatically a traitor to the perverted minds of Goneril, Regan, and Edmund; that none of these was either temperate or legalistically precise in the meting out of punishment.

Another argument appeals to Shakespeare's craftsmanship. So Collier (1842) : " It may be urged that as Cordelia had been hanged, the poet would have probably chosen some other death for the Fool, in order to render the matter quite clear." The Cowden Clarkes (1864) argue that Shakespeare would not have " committed the dramatic tautology . . . of making the Fool, as well as Cordelia, ' hang'd.' " Here again the arguments beg the question. But they go further. In effect they assert that to interpret " poor fool " as the Fool is to accuse Shakespeare of un-Shakespearean inadequacy. It is hard to reply to a case based on undemonstrated intuition of Shakespeare's manner of working, yet something may be said. *King Lear* abounds in " tautology." The subplot parallels the main plot. The goodness of Cordelia is repeated in that of Kent, in that of the Fool, in that of Albany. The evilness of Goneril in that of Regan,

Cornwall, Edmund, Oswald. Cordelia's dismissal is repeated in Kent's; the trial of Act I in that of Act III. Indeed, the evidence of Shakespeare's "tautology" is almost embarrassing in its abundance. If we look solely at Shakespeare's final disposition of his characters, we find it even there. Oswald dies from Edgar's sword and so does Edmund. Gloster dies when his heart bursts smilingly and so does Lear. Is it, then, un-Shakespearean if the Fool and Cordelia meet the same death?

Behind the arguments just noted is another: Steevens (1785) holds that "the Fool was long ago forgotten," and Schücking treats "the enigmatic disappearance of the fool" as an exception to the law of clarity that Shakespeare generally observes.[3] The point, I judge, is that if this term does mean the Fool, it lacks preparation and corroboration, and there can be no question of the suddenness of the reference here. Yet there can be more inferences than that the Fool is not meant. It may be that the momentary glancing of Lear's mind from his daughter to his Fool is in keeping with his psychological state. It may be that such a suddenness of revelation serves some peculiar Shakespearean effect. To these points we shall return.

The basic arguments asserting the exclusive relevance of "poor fool" to Cordelia all depend on context. The first is that the following words—"And thou no breath at all . . . Look on her, look, her lips . . ."—plainly define "poor fool" as Cordelia. So Empson.[4] But this again is question begging, for the point to be determined is whether Lear's mind shifts from Cordelia to the Fool to Cordelia again or whether it stays anchored on Cordelia alone.

This argument shades into a second: the context of Lear's thoughts makes reference to anyone other than Cordelia unthinkable. So Steevens (1785): "That the thoughts of a father, in the

[3] Levin L. Schücking, *Character Problems in Shakespeare's Plays* (N.Y., 1922), 85.

[4] *Op. cit.* Though Empson does not insist on the exclusive relevance of "poor fool" to Cordelia, he rejects its relevance to the Fool: "this interpretation would make the sentence contradict the movement of thought that it introduces, and must therefore be wrong." Later he seems to agree with Bradley: Lear's mind "has wandered so far that he no longer distinguishes the two." Bradley also takes something of a double stand on this line. He begins by insisting that "this famous exclamation . . . unquestionably refers to Cordelia"; but almost immediately he retreats: "and even if it is intended to show a confused association in Lear's mind . . ." *Shakespearean Tragedy*, 2nd ed. (London, 1905), 314.

bitterest of all moments, while his favourite child lay dead in his arms, should recur to the antick who had formerly diverted him, has somewhat in it that I cannot reconcile to the idea of genuine sorrow." So Malone (1790): "Lear, from the time of his entrance in this scene to his uttering these words, and from thence to his death, is wholly occupied by the loss of his daughter. He is diverted, indeed, from it for a moment by the intrusion of Kent, who forces himself on his notice; but he instantly returns to his beloved Cordelia, over whose dead body he continues to hang. He is now himself in the agony of death; and surely at such a time, when his heart is just breaking, it would be highly unnatural that he should think of his Fool." So Rolfe (1880): "There is no room for a divided sorrow here; Lear's thoughts can never wander more from his dead daughter."

Every reader will sympathize with these sentiments, but let us look again at this scene and observe the state of Lear's mind as Shakespeare presents it. Is it unwavering in this final minute of his life?

Her voice was ever soft,
Gentle and low, an excellent thing in woman.
I killed the slave that was a-hanging thee.
Capt. 'Tis true, my lords, he did.
Lear. Did I not, fellow?
I have seen the day with my good biting falchion
I would have made them skip. I am old now,
And these same crosses spoil me. Who are you?
Mine eyes are not o' the best, I'll tell you straight.
Kent. If fortune brag of two she loved and hated,
One of them we behold.
Lear. This is a dull sight. Are you not Kent?
Kent. The same,
Your servant Kent. Where is your servant Caius?
Lear. He's a good fellow, I can tell you that.
He'll strike, and quickly too. He's dead and rotten.
Kent. No, my good lord, I am the very man—
Lear. I'll see that straight.
Kent. That from your first of difference and decay
Have followed your sad steps.
Lear. You are welcome hither.
Kent. Nor no man else. All's cheerless, dark, and deadly.
Your eldest daughters have fordone themselves,
And desperately are dead.
Lear. Aye, so I think.

(V. iii. 272-292)

Then, after twelve lines of silence, when Albany cries out," Oh, see, see! " Lear speaks again: " And my poor fool is hanged! "

Lear's mind here moves away in shock from imaginative and emotional apprehension of Cordelia's death. He brags of his former prowess, holds perilously to the thread of conversation as he speaks of Caius, then lets loose, falling into numbness. There is, here, no unwavering fixity on the cause or object of pain, but a surrender to the pulsation of suffering, to the flow and ebb of full realization. The reason, I suppose, is that the mind can grasp its agony only by moments; then it must let go, clinging for control to some rational matter, some redirection of feeling—finally, if the agony is too great, surrendering full consciousness until aroused to new pain, new reason. Lear's grief is not fixed; it moves through his mind in waves.

Is the appearance of the Fool on the surface of such fluctuating consciousness probable? Two arguments hold that it is. First, Lear in this final scene is undergoing something of the same experience that he underwent on the heath, and in his state of shock the past recreates itself or becomes fused with the present in his mind. So Knight (1841): " In the depth of his distress during the storm Lear says: ' Poor fool and knave, I have one part in my heart That's sorry yet for thee.' And now, when the last and deepest calamity has fallen upon him, his expressions shape themselves out of the indistinctness with which he views the present and the past, and Cordelia is his ' poor fool.' " Lothian argues somewhat similarly: " Lear, when under emotional distress and in absolute despair, seeks out and keeps the company of the Fool, and finds support in it. Later, when evil and chance have joined to do their utmost, and the lost daughter who was found again lies dead before him, and the sight of her makes him feel and think unutterable things, it is surely not unnatural that his mind should turn in this emergency, as it had in previous ones, to him who had helped him bear ' his heart-struck injuries.' He gropes around for the support he needs, and he has to remind himself that even this comfort is denied him, that in his irreparable loss there is nowhere to turn for consolation or comfort, that his Fool is hanged." [5]

[5] John M. Lothian, King Lear. *A Tragic Reading of Life* (Toronto, 1949), 46-47. It should be noted that Lothian, having advanced this interpretation, rejects it. His reasons I judge to be a fusion of points already treated and the final argument from context, that based on the audience's focus or set, which I treat below.

These passages touch on the second reason for regarding the appearance of the Fool in Lear's consciousness as not improbable, for they suggest an identity between Cordelia and the Fool in spirit and in the relation they bear to Lear. Both are loyal, self-sacrificing, essentially innocent. Both have maintained an inner truthfulness that has caused him pain. Both have succored him in his pilgrimage through suffering to humility and enlightenment. Both have loved him and he has loved both. As Brooke put it, "The two creatures who loved him most, and who were from the beginning of his agony always linked together in his mind, become for an instant of madness one." [6]

A final argument from context, however, insists that "poor fool" can only mean Cordelia. As Verplanck (1847) says, "the whole interest of the scene is fixed on Cordelia's death." From the time of the imprisonment, from the moment of Edmund's horrible directive to the captain, the audience has focused its attention ever more intently on the outcome. Has Lear suffered the storm of his madness, the bitterness of his guilt, the wheel of fire, only to lose his daughter? As the cumbrous machinery of the duel and the dispositions of the various characters draw to a close, suspense draws to a point. Then Lear enters, Cordelia dead in his arms. All attention is on him and Cordelia, on the fact of death, on its ultimate meaning. To the audience, given this tightened focus of attention—so runs the argument—"poor fool" must mean Cordelia.

The power of this appeal is great, yet again, something may be said. We can happily forget the thesis that Cordelia, like Kent, returned in disguise and actually *was* the Fool; [7] but we cannot forget what every director knows—that the text is to the acted play as a libretto to an opera, and that, except for a parsimonious scattering of stage directions, we lack the music. We do not know what Shakespeare's actors did, how they were told to move and gesture. And just before our disputed phrase, Lear is silent for twelve

[6] *Op. cit.* Perrett, *op. cit.*, and Donald Stauffer, *Shakespeare's World of Images* (N. Y., 1949), 341, arrive at the same conclusion on the basis of Brandl's thesis that the same actor played both Cordelia and the Fool. So Perrett: "The identity of the young actor impersonating the two figures must have been apparent to the audience . . . There seems to be a symbolic meaning here. When Cordelia is away, her place as the representative of utter truthfulness is taken by the Fool. In this respect the two characters are one."

[7] A. J. Stringer, "Was Cordelia the King's Fool?" *American Shakespeare Monthly*, III (1897), 1-11.

whole lines. That silence robs us of evidence. Any actor worth his salt could carry an audience with him if he wished to make " poor fool " mean Cordelia, but quite as successfully he could make " poor fool " the Fool. So Shakespeare may have ordered it three and a half centuries ago in the Globe.

Again, it can be shown that Shakespeare in other plays was willing to risk the destruction of focus and illusion at points of great intensity. In the assassination scene of *Julius Caesar,* to take a single example, when the conspirators are bathing their hands in blood, Cassius says:

> Stoop then, and wash. How many ages hence
> Shall this our lofty scene be acted over
> In states unborn and accents yet unknown!
> *Brutus.* How many times shall Caesar bleed in sport,
> That now on Pompey's basis lies along . . .
> (III. i. 111-116)

Finally, it may be urged that Lear is here faced with more than the fact of Cordelia's death. He is faced with the total and absolute loss of all that he has loved. There is a progression. Caius is dead and rotten, but Caius was the least of the three. The Fool is hanged. Step by wavering step Lear mounts to the climactic realization that Cordelia, too, will come no more:

> Never, never, never, never, never!

* * * * *

By " poor fool," according to the traditional gloss, Lear means Cordelia. The progressive rebuttal which these pages have reviewed poses two alternative glosses: that Lear means the Fool, and that Lear is expressing some kind of fusion or confusion of the two in his mind. The difficulty of verifying any one of the glosses stems from our ignorance of the stage business on the Globe platform and from our sense of the wracked agony of the mind from which these words come. There is a rightness in Lear's meaning his daughter, a rightness in his meaning the Fool, a rightness in his meaning both at once, but rational prudence compels us to admit we cannot *know* what he means. We cannot *know* what happened to the Fool.

We can know, however, what Shakespeare is doing here. He focuses his audience's attention on Cordelia and places in Lear's mouth a phrase that inevitably brings the Fool to its mind. Thereby, he achieves, as in the greatest of his late plays, a double vision that fuses past and present. He overcomes the limitations of time, the necessarily linear sequence of perceptions an audience has, and provides instead a single simultaneous vision of cause and effect, of what has gone before and what exists now. He had tried this, though awkwardly, in the procession of ghosts at the end of Richard III; he tried it superbly in Lady Macbeth's sleepwalking; he tried it later in the fused imagery of Cleopatra's death scene. By the superb ambiguity of " poor fool," which connotes at once the Fool and the daughter, the removal of both from the bounds of reason, the innocence and purity both share, the suffering both endure, the love Lear has for both, Shakespeare achieves such a simultaneous vision here. The unbuttoning brings back the unbuttoning on the heath; [8] the cry " poor fool " brings back the quaking of the " poor fool and knave "; the mind in its shocked agony brings back the mental agony of the storm. As Shakespeare fuses the end with the middle, he fuses it with the beginning. On the ground before Lear is his daughter, the first of those who loved and hurt him. Beside him is Kent, the second. When Lear cries out, " And my poor fool is hanged! " the triumvirate is complete. In our minds, in a single apprehension, we see that those who were with him in the vain glorious days of his rule and in the agonized days of his madness are with him again, the suffering witnesses, the final evidence of his folly and redemption.

[8] I owe this observation to Empson, *op. cit.*, with whom this final paragraph in part agrees; but the rest of his interpretation, that which attempts to determine what is happening in Lear's mind, seems to me subject to the objections already raised. Throughout his very penetrating analysis Empson shifts without warning from Lear's mind to Shakespeare's. So, on this final point: " the point is not that they [Cordelia and the Fool] are alike . . . but that he must be utterly crazy to call one by the name of the other. Presumably the audience was meant to accept it as mere raving and thereby get the right poetical effect; it is meant to recall the whole background of clownery which is what he has discovered about the world."

PAPERS OF THE MICHIGAN ACADEMY OF SCIENCE, ARTS, AND LETTERS
VOL. XLIX, 1964 (1963 MEETING)

"DELTA AUTUMN": WILLIAM FAULKNER'S ANSWER FOR DAVID H. STEWART

ARTHUR F. KINNEY
The University of Michigan

I

Two years ago, before a jury gathered in a room similar to this one, at Wayne State University, a fellow member of the Academy, David H. Stewart, serving as chief prosecutor, presented an eloquent attack in the case of the State of the Noble Hero vs. Isaac McCaslin.[1] Since so far as I know no one has risen to speak in behalf of the defendant, I should like this morning to suggest an outline for a possible defense of Ike.

Professor Stewart opened his case by labeling Ike McCaslin, the boy of *The Bear* and the old man of "Delta Autumn," one of William Faulkner's great frauds. Professor Stewart charged that as a man bidding for the status of hero, like Quentin Compson or Bayard Sartoris before him, Ike has trouble with time, a trouble which "is exactly like that of his precursors': they all desire to elude it, to resist distinguishing what actually happened from what they had been told, hence to prolong or postpone things past into the present or even the future. In addition, the impulse which seems to animate Ike and give him direction—as it does Quentin and Bayard and Gail Hightower—is a need for peace and escape."

But Professor Stewart's summary of Ike's responsibility—to himself, to his forebears, and to the society of the South—was specific. "[Ike] seems to be involved in three things: reaching maturity, expatiating the original sin of miscegenation and incest committed by [his grandfather] Carothers [McCaslin], and reconciling or achieving a viable position with regard to the dilemmas of property ownership. Restitution for inherited guilt plus self-justification are fundamental, the latter taking precedence as Ike lives further into the twentieth century."

But my use of Perry Mason language here is not altogether with-

[1] Delivered at the 1961 meeting of the Academy; slightly revised and published as "The Purpose of Faulkner's Ike," in *Criticism*, III (4) (Fall, 1961), 332-342.

out point, for a sense of justice is the crux of Ike's problem; he must reach manhood with dignity so that with dignity he may pay for (atone for) the sins of his forebears and with justice handle property which by its very nature of " ownership " creates injustice. Ike is a fraud for Professor Stewart, if not for Mr. Faulkner, because while Ike accepts such responsibilities as real challenges, which he must meet, he fails to meet them but reaps the reward (via the respect of Jefferson's citizens and Faulkner's readers) of having done so.

To be more specific, in the first instance, according to Professor Stewart, the boy Ike appears to accept the bloody baptism of deer-slaying by his confirmation, his participation in the death of Old Ben; he appears to help kill Old Ben while in reality he avoids it. Ike, says Professor Stewart, " waits until the plebeian, Boon Hogganbeck, who is as much animal as man . . . commits the final deed and thus absolves Ike of responsibility so that he can assist with pure hands at Sam Father's funeral. What Ike achieves through this sequence of events is not only a degree of maturity but sanctity: he emerges incorruptible, a kind of consecrated altar boy, who can in time develop into the wise prophet and judge which he is destined to become, without losing the ' young boy's high and self-less innocence ' (*Go Down, Moses*, Modern Library, p. 106) ." This youthful Decision for Christ is reaffirmed by Ike with a likewise fraudulent decision as an adult; he leaves his wife and the McCaslin family for a purified life of a carpenter " because if the Nazarene had found carpentering good for the life and ends He had assumed and elected to serve, it would be all right too for Isaac McCaslin . . ." (309) .

Professor Stewart's second charge, that Ike is fraudulent in his atonement for the sins of his forebears, particularly Carothers, is that Ike merely pays them off with the blood money handed down to him by Buck and Buddy, his father and his uncle. That is, he turns his sense of responsibility, of justice, of penance, into a one easy-payment plan made with money that was never really his to do other with, anyway.

Of the final charge, Professor Stewart has this to say: " [Ike's] solution to the land problem is also easy—indeed almost pathetic beside Buck's and Buddy's earlier attempt. In the first place Ike ponders the ingenious possibility that he personally does not own any land at all, on the ground that buying or bequeathing land is impossible by divine edict (257). Cousin McCaslin Edmonds puts an end to this notion by insisting that in practice at any rate old Carothers did own land and did hand it on to his descendants. Later on, McCaslin suggests that even if Carothers never owned the land, it must then have passed from Ikkemotubbe to Sam Fathers, and "who inherited from Sam Fathers if not you? co-heir perhaps with Boon . . . ?" (300) Whichever way Ike may try to look at it, he does own or 'hold suzerainty' over the family plantation fifteen miles from Jefferson, and he is obsessed with the desire to free himself from this land which he believes contaminates him morally (because of Carothers' wickedness) and spiritually (because property-owning is unchristian, or at least un-Christ-like). What he does is simple: after all the talk and theorizing, he gives all of his property to his cousin, fastidiously and successfully evading the entire problem of social position and power. He remains inviolable and pure."

One cannot, I think, dismiss these charges lightly, for all of what Professor Stewart says is true. By seeking justice, Ike asks in turn to be judged, and when we judge his actions we find they are cheaper than his desires; his ambitions and achievements are coins of different realms. One might go even further than Professor Stewart, in the process of prosecuting Ike, by pointing out that during his life of renunciation, as a simple carpenter, Ike lives on a McCaslin bank allowance, paid from the tainted money he has publicly disclaimed.

All of this surely suggests that after close examination, Isaac McCaslin becomes something less than the hero that his central position in one of Faulkner's central Southern families and his position in the various parts of *Go Down, Moses* would at first glance indicate. And so, I think, one who has read *The Bear* taking Ike as seriously as Ike takes himself would have to submit to the prosecution.

Where I wish to part with Professor Stewart, however, is in his assumption that the Faulkner canon is entirely logical and sequential, and that each episode in *Go Down, Moses* is cut from the same bolt of material. Such an assumption is always dangerous—even with novelists like Faulkner and Hardy who make a point of staying with the same region and the same people. But such an assumption is particularly dangerous with Faulkner, I suggest, because he himself as artist was always restless, always searching. Any reader of Faulkner's works soon senses the restlessnes of his mind, his groping for ideas and answers, his ability to grow with his characters. The inconsistencies with which critics often chide Faulkner are not always as much faulty memory as they are new recognitions on Faulkner's part.

I wish, then, to base my defense of Ike McCaslin on the suggestion that the Ike of " Delta Autumn " is not the Ike of *The Bear*; he is, rather, Ike McCaslin Reconsidered, and, as such, he begins to answer some of those very problems he avoids in *The Bear*—the problems that Professor Stewart so perceptively points out and persuasively argues. Moreover, let me suggest that the final beauty of " Delta Autumn " is that the saint-really-sinner does not suddenly become saint again; he becomes, rather, man.

II

" Delta Autumn " begins with change. More than a half-century has passed: it is now November 1940, and Ike is seventy-three. He is again on an annual hunting party, but no longer is he awed by the countryside, no longer does he sense the threshold of life in the man-smelling cabin, no longer does he prepare for a sacrament of initiation or rebirth. Now he is simply old; " he no longer told anyone how near eighty he actually was " (336). And the change oppresses Ike. He comes not thirty miles, but over two hundred—not to an inland " big woods " but to the delta, and not by horse and wagon, but by truck. The quarry is no longer a great, transcendental bear, but deer. " ' The game was here then,' " Ike tells the younger men (345). But now the game is scarce; they can no longer shoot does. Ike is now in " the land across which there came now no scream of panther but instead the long hooting of locomotives: trains of incredible length and drawn by a single engine " (341). So the whole context is changed for us; there is no sense of the mythic or the mystic here; if the notions are romantic, the facts take on a new realism. The environment of " Delta Autumn " is not the environment of *The Bear*.

The characters are different, too. Once there was sense of adventure and ritual; there was a group of sportsmen who wished to become great hunters. Now there is a group of quarrelsome young men. There is Will Legate, who gets as much joy from baiting Roth Edmonds as he does from stalking deer. There is Henry Wyatt, always on edge—not to see the break of dawn, as the hunters of *The Bear* anticipated—but to prevent outbreak of fighting. There is Roth himself: always defensive, always self-deprecating, who, when he sees his Negro mistress in the road on the way to the camp, has to slam on his brakes, " actually flinging [Uncle Ike and Will] forward until they caught themselves with their braced hands against the dash " (336). The sense of fraternity in *The Bear* is gone.

Ike realizes this, too. Each incident, each observation, is a stinging reminder to the aged Ike that despite his life as a carpenter all has not been paid for in the McCaslin family, all is not well in the Big Woods. Thus the sentence in the closing section of *The Bear*— " he had forewarning and had believed himself prepared "—is by now ironic (318).

Ike knows in " Delta Autumn " that he is a failure. He may try to circumvent total recognition by attending hunting parties where he himself will not hunt, but he cannot escape the fact that he lies awake at night and that he does not leave the tent in the morning. What is equally human and heroic in the Ike of " Delta Autumn," however, is his sense of hope, that eternal striving that is part of the fuzzy term we use, the " human condition." Ike is in despair at his failure in life, but, as man, he upholds Kierkegaard's observation that " the specific character of despair is precisely this: it is unaware of being despair."

III

Ike's solution to his problem of atonement is not easy for him in " Delta Autumn " because, although he can admit his actions have not been successful, he still hangs on to shreds of old theory and he still retains what is, after all, his heritage: he is still part of the white South.

His theory is that once individual penance fails (as it apparently has for him between *The Bear* and " Delta Autumn "), then communal penance may yet offer a solution. The exposure of this theory takes up the first half of the short story.

Elsewhere, Faulkner has shown himself interested in such a theory. During World War II, Faulkner published some stories on Yoknapatawpha's reaction to Pearl Harbor. In one of them, " The Tall Men," the town marshal, Mr. Gombault, remarks that it is " honor and pride and discipline that make a man worth preserving, make him of any value. That's what we got to learn again. Maybe it takes trouble, bad trouble, to teach it back to us; maybe it was the walking to Virginia because that's where [Anse McCallum's] ma come from, and losing [the War Between the States] and then walking back, that taught it to old Anse. Anyway, he seems to learned it, and to learned it good enough to bequeath it to his boys."

The marshal's theory—that lessons learned through trial, often by a whole population, result in the *bequest* of salvation—is Ike's theory, too. The idea first comes to Ike when he stands, at twenty-one, with McCaslin Edmonds among dirty and dusty journal leaves in the circumlocutionary Part IV of *The Bear*. There Ike is confronted with the profound puzzle for man: If God is omnipotent, why does He allow His chosen people to pervert humanity with the sins of rape, of incest, of miscegenation? The only answer Ike can summon—searching through the corridors of his past and the scribblings of his forebears—is that the War Between the States was a time of communal purgation. He likens the war to an earlier but similar trial—the War Between the States was " a harsher book and McCaslin, fourteen and fifteen and sixteen, had seen it and the boy himself had inherited it as Noah's grandchildren had inherited the flood although they had not been there to see the deluge." The implication that follows is that as the flood was not sufficient so the War Between the States will not suffice, but if Ike sees this fallacy in his analogy, he does not admit it to McCaslin or to himself. For Ike must appease his conscience in *The Bear* through a rational theory of history, and this theory accounts for his two problems—God's failure and Lucius Carothers McCaslin's failings—and both the frightening facts of war and miscegenation.

The theory that is romantic in *The Bear* changes, however, in " Delta Autumn " because the context changes. Though Ike believes still in salvation through communal penance, his present belief also contains the tacit admission of his own failure in his personal mission as carpenter. Ike fails because penance is not his but the South's, and one man cannot atone for slavery and miscegenation alone.

Thus it is when Roth cynically defines patriotism as barroom singing and " dime-store flags in our lapels " (338) that Ike counters with " 'This country is a little mite stronger than any one man or group of men, outside of it or even inside of it either, I reckon, when the time comes and some of you have done got tired of hollering we are whipped if we dont go to war and some more are hollering we are whipped if we do, it will cope with one Austrian paper-hanger, no mater what he will be calling himself ' " (388-339). The conclusion is axiomatic given Ike's own premise: most men are a little better than their circumstances give them chance to be. Hope burns eternal, but the flame is lower; the impossible stature of heroism in *The Bear* is shaved down nearer the size of reality.

IV

This will be the basis of Ike's more realistic answer to payment for miscegenation among the McCaslins and slavery throughout the South. But it must go hand-in-hand with action on a personal level, and that action is love. I take the echo of the word *love* in " Delta Autumn " to be thematic; the more so because it is so distinguishable from the key words *pride* and *humility*, which ring throughout *The Bear.* Love is more commonplace than humility, I suspect, and certainly more worthwhile than pride.

As the Negro servant Dilsey offers the only hope for the degenerate and dying Compsons in *The Sound and the Fury*, here the final hope is also first suggested by a Negro. The forewarning comes with the remaining land which is fruitful; denuded by the white man, the land is still " fecund up to the very doorsteps of the negroes " (340). Ike McCaslin sees this and he also has, in the last half of the story which deals with the personal level of justice, to see in Roth's mulatto mistress exactly those saving prerequisites of the Negro the other Ike had listed for McCaslin Edmonds in the commissary: " Endurance . . . and pity and tolerance and fidelity and love of children '."

Ike's initial naiveté in " Delta Autumn " is sharply (even ironically) contrasted with the mulatto's knowing. For given the chance to see, Ike at first cannot; he fumbles for his glasses, but they are out of his reach. The mulatto, with the mixed blood running in her veins, is like Fonsiba's husband, who wears lenseless glasses but who, nevertheless, *can* see. The mulatto of " Delta Autumn " knows the money Roth leaves with Ike for her is a token payment; she knows, too, that Ike's gift of a horn is useless. She has known all along–in the months of gestation, in her reunion with Roth through the preceding January–that she will not live out her life with him. She knows love also, and so, we feel, does Roth.

The mulatto, by her very presence, underlines Ike's failure as unrealistic carpenter, living off the inheritance he tries to disclaim, for she represents miscegenation anew. But she upholds Ike's belief, his "Delta Autumn" belief, that one man cannot atone for his family and his region; it is the job of the whole white South.

The mulatto acknowledges the necessity for communal action, but she also notes the importance of personal love, particularly as it is opposed to the hollowness of social convention. "'I knew what I was doing,'" she tells Ike, "'I knew that to begin with, long before honor I imagine he called it told him the time had come to tell me in so many words what his code I suppose he would call it would forbid him forever to do. And we agreed. Then we agreed again before we left New Mexico, to make sure. That that would be all of it. I believed him. No, I dont mean that; I mean I believed myself. I wasn't even listening to him anymore by then because by that time it had been a long time since he had had anything to tell me for me to have to hear'" (358-59).

She spells out the sins clearly. *Code* and *honor* are no longer the virtues of the aristocracy that they were to the Ike McCaslin of *The Bear*; Ike understands from her that he and Roth have twisted them. For her, *code* has become the communal laws which fail to take into account the individual, and *honor* has been bent into the personal sense of self that is buried in pride; *honor* which has been Ike's standard has really been only a mirror.

It is Ike's own realization in "Delta Autumn," however, and *not* the mulatto's, that the answer for the problem comes with the acceptance of the sin. The answer for miscegenation comes in the acceptance of miscegenation.

At first, Ike cries out against this, the realization the Ike McCaslin of *The Bear* could never have formulated, much less accepted. So his protest pierces the heart of the matter. "*Maybe in a thousand or two thousand years in America,* he thought. *But not now! Not now!*" (361).

Moments later, however, he is making a futile gift: the hunting horn that belonged to General Compson; it is his own gift now to give, not a payment required by others. And Ike reaches out to touch the hand, to admit her presence. The old bony white hand and the young mulatto hand with the "strong blood" (362) in its veins touch, if only momentarily.

This is not, I take it, the saint that R. W. B. Lewis has proposed the Ike McCaslin of *The Bear* to be; this Ike no longer dreams of himself as a kind of Second Coming about to preach a new Living Gospel to God's Chosen People. But he is not Professor Stewart's fraudulent hero, either, for this Ike McCaslin, if he can't at once reject a lifetime of tradition and belief, does reject romanticism in the birth of new awareness.

This new Ike McCaslin, the Ike McCaslin of " Delta Autumn," is a man who has gained immeasurably in knowledge. He still lacks will. He has, that is, gained knowledge that he is not yet prepared to act upon. He has, that is, rejected the idle boyhood dreams and absurd Christlike posturings of *The Bear* for awareness of fact; his problem is now one of acceptance.

In an interview with Miss Cynthia Grenier, Faulkner once commented about Ike, " I think a man ought to do more than just repudiate. He should have been more affirmative instead of shunning people." Faulkner could not have made that statement in the midst of writing *The Bear,* for he was too busy praising his new hero of the modern South. But he could—and we should—make that comment after the recognition that comes to Ike and to us in " Delta Autumn."

Ike's problem is now one of will; a human problem indeed. He is not, then, the hero that Lewis would contend for, the Christlike saint. But he is also not the fraudulent hero, the self-deluded mystic, that Professor Stewart has proposed. He is, rather, a man who has found new answers and is searching for the will to act on the facts now at hand. He is, like us, a man who would be better than it has been his fate to be. Though we could never see ourselves reflected in the Ike of *The Bear,* we can, I think, see ourselves in the new Ike of " Delta Autumn." Neither saint nor sinner, he is a man, which is to say that Faulkner has given us an even more difficult artistic achievement.

PAPERS OF THE MICHIGAN ACADEMY OF SCIENCE, ARTS, AND LETTERS
VOL. XLIX, 1964 (1963 MEETING)

ART

REVIEW OF THE FINE ARTS EXHIBITION OF 1963

CHARLES E. MEYER
Michigan State University

THIS year's exhibition was held at Western Michigan University in Kalamazoo. A total of approximately 125 works was submitted to jury action and 40 of these were hung. The Student Center's second floor lobby was used as a space to house the exhibit and, while the lighting was good and the exhibition was well presented, the space was not one which lent itself easily to such a presentation. Works were somewhat crowded and difficult to see well.

Although the paintings and prints shown exhibited a wide range of material and manners of approach, many artists who in the past were represented seem now to have dropped their memberships in the Academy or ceased submission of their work. The Wayne State University group, the Michigan State group, and the Ann Arbor group from the University of Michigan were represented by very few entries, while Cranbrook and the Society of Arts and Crafts had no entries. If the organization has as one of its goals a top-notch exhibition each year during its meeting, some effort must be made to encourage greater participation. Perhaps direct invitations to groups such as those mentioned would be a move in the right direction, and one that the section could well afford.

Despite the fact that Michigan is especially rich in fine printmakers, many of them with national reputations, there were only two prints hung this year. Both were very fine, but two prints from the entire state furnish a very slight indication of the development of this medium in Michigan. William Woodward's *In Retrospect* (Fig. 1), reminiscent of Hokusai in its black on shaded green and blue, received the prize. Running a close second was James Walker's *66B*, a serigraph print (Fig. 2). This moving abstraction in yellows, black, red, and gray was matched in the quality of its swing and flow of form only by the excellent painting of Howard Church.

Church's *Sycamore* (Fig. 3) is a sensitive, large painting in fleshy dry-browns, black, yellow, and orange. It was the best of the works in the show in a purely abstract-expressionist manner. As an artist,

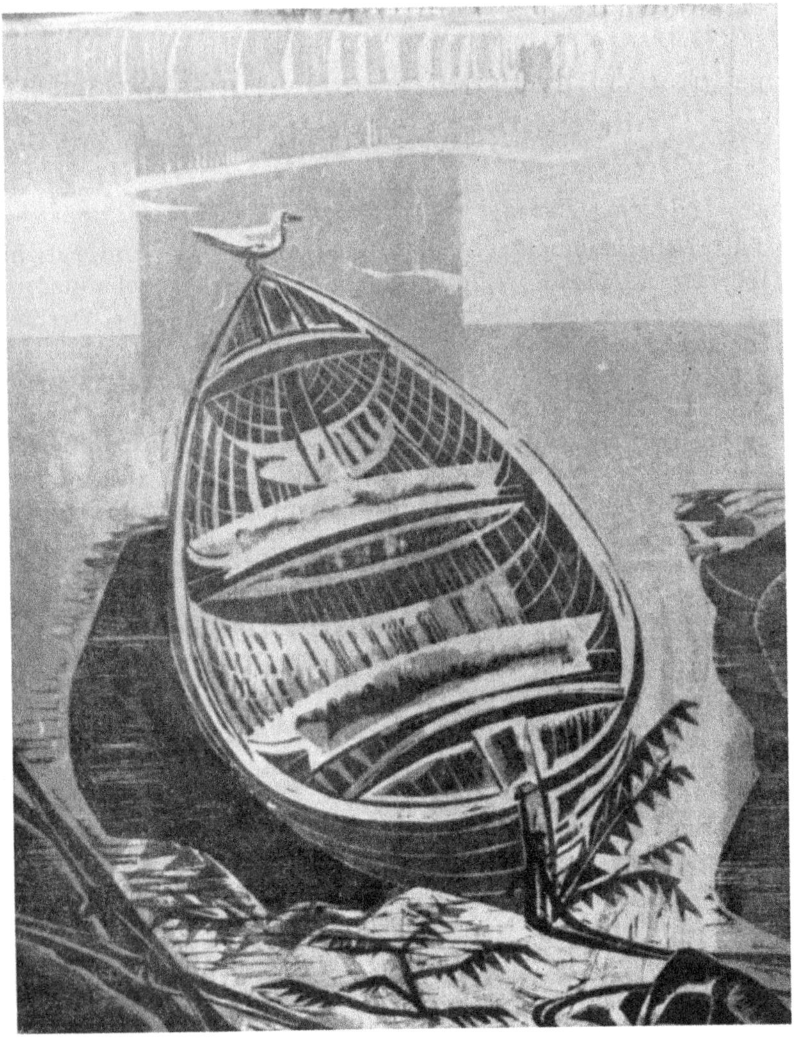

Fig. 1.

William T. Woodward, *In Retrospect*

Print, First Prize, and Co-Winner of Purchase Award, 1963

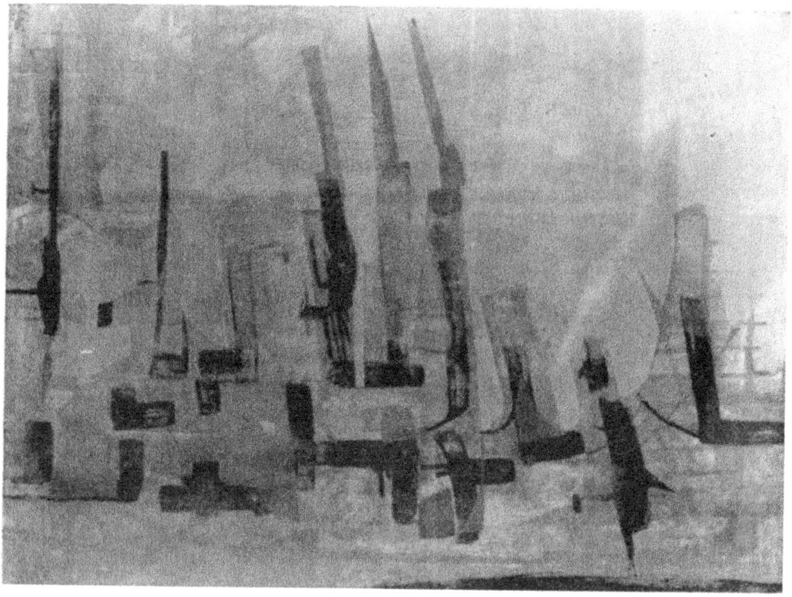

FIG. 2.
James A. Walker, *66B*
Serigraph, 1963

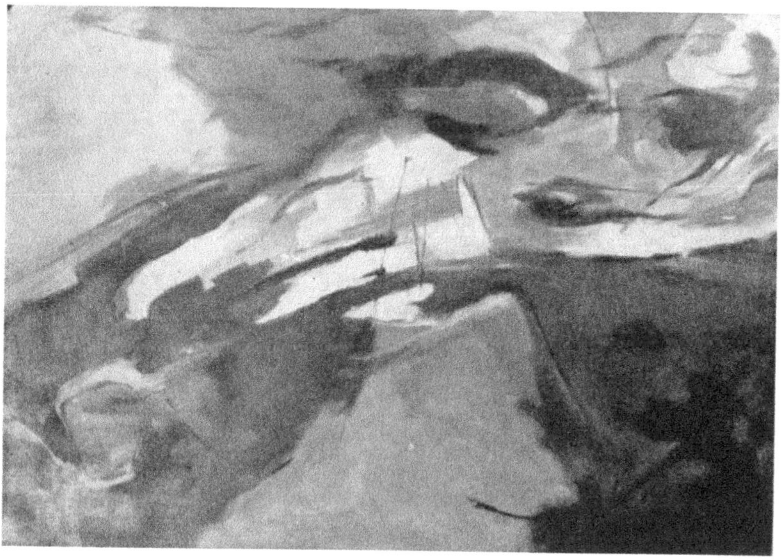

FIG. 3.
C. Howard Church, *Sycamore*
Oil, 1963

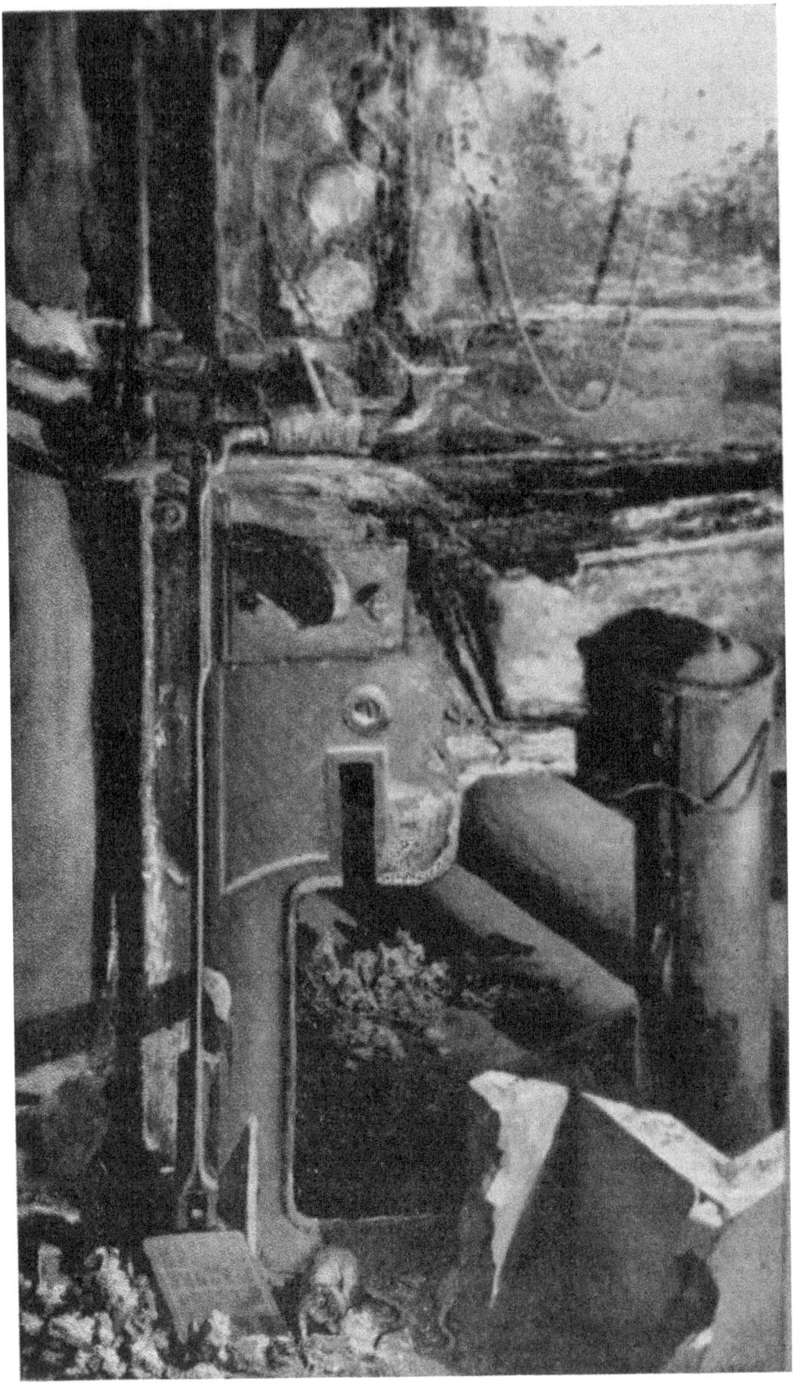

FIG. 4.

Aimé Martinat, *Furnace*

Best Traditional Painting (Mr. and Mrs. Ferris Fitch Award), 1963

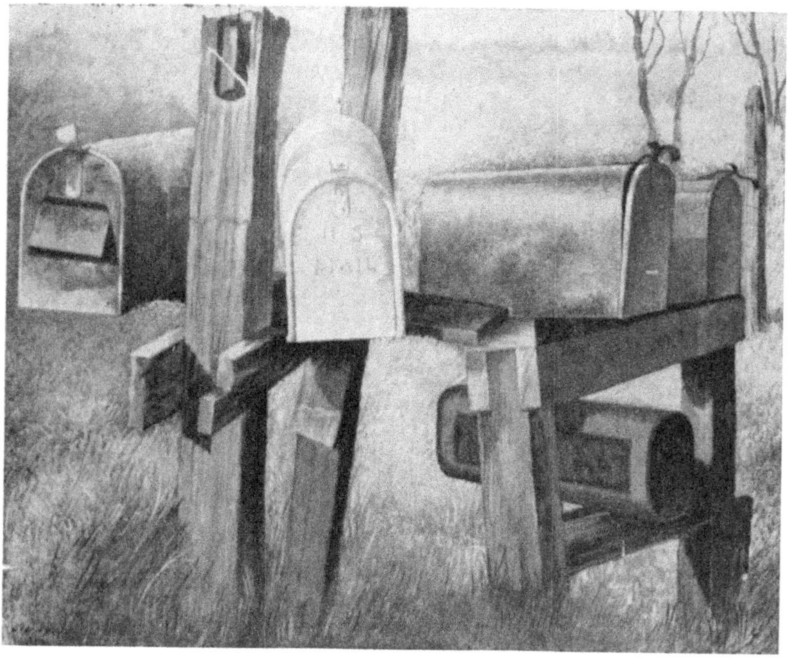

FIG. 5.
Sid Seeley, *One Morning Last Fall*
First Prize Watercolor, 1963

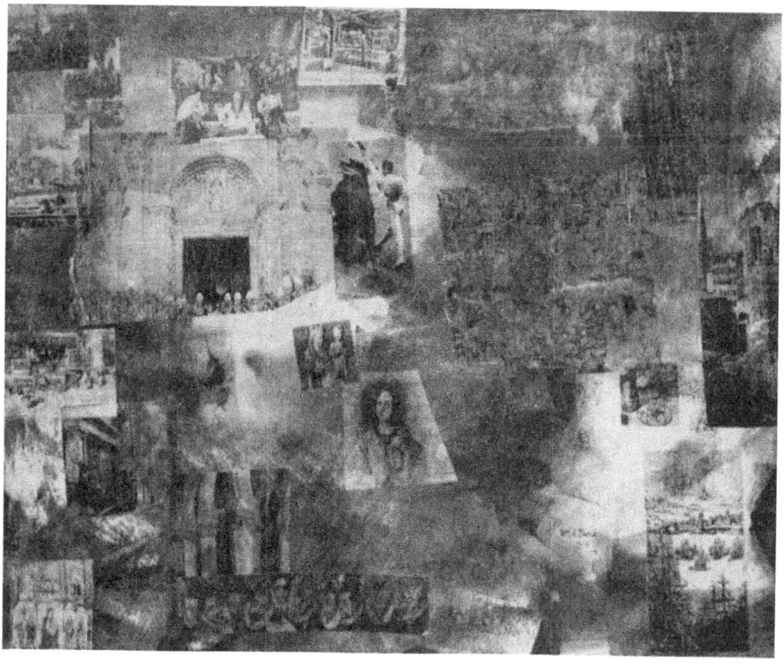

FIG. 6.
William A. Bostick, *Souvenir of Spain*
Oil, 1963

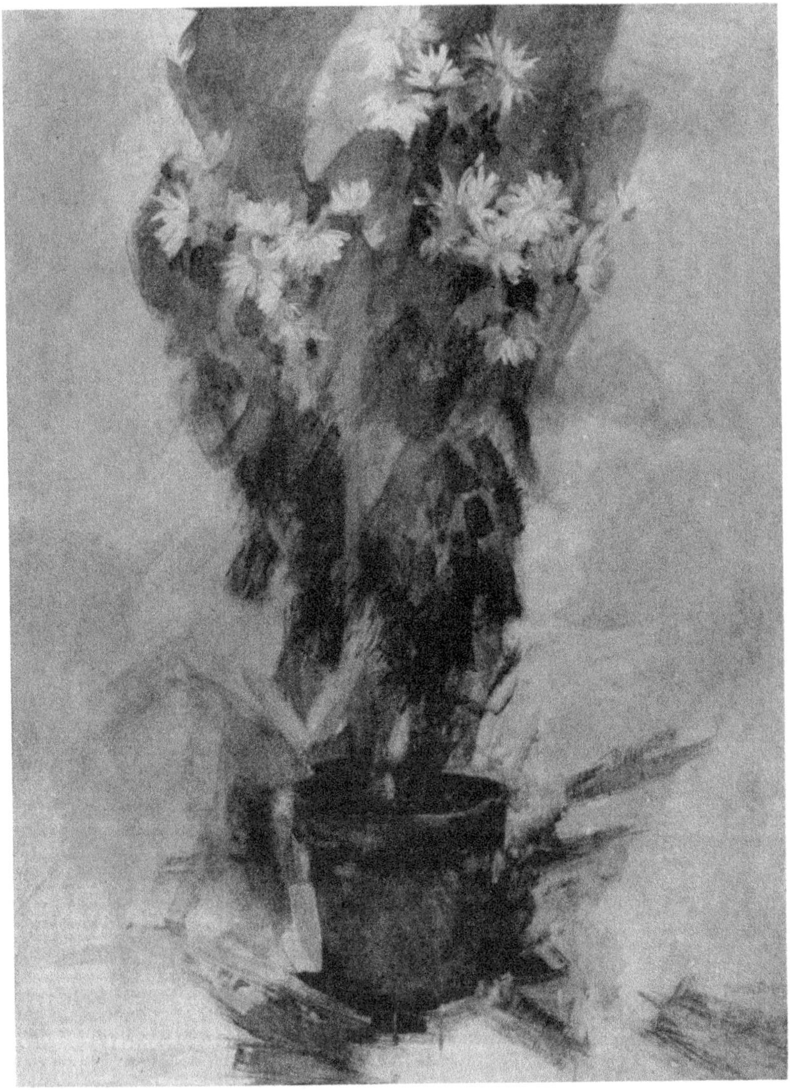

Fig. 7.

William P. Martmer, *Chrysanthemums*

First Prize Oil, 1963

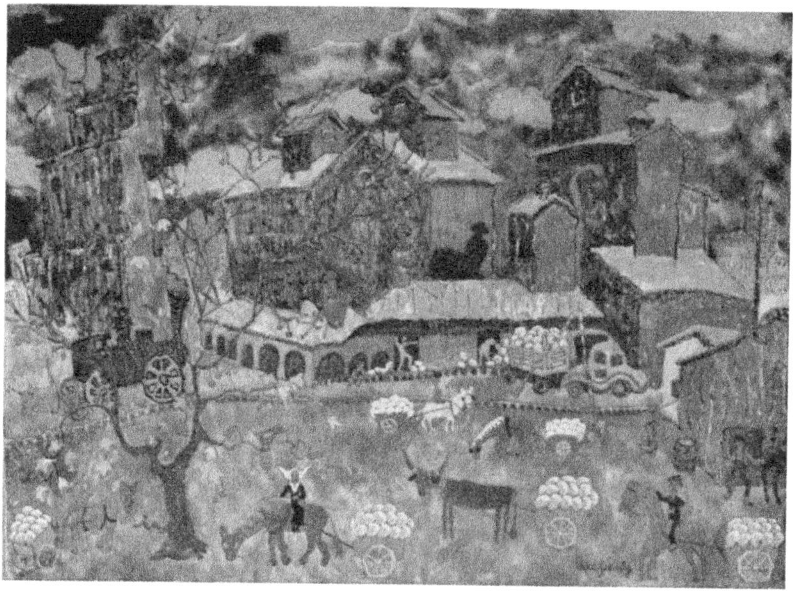

FIG. 8.
Sali Frantz, *The Nun Rides a Horse to the Grist Mill*
Gouache, 1963

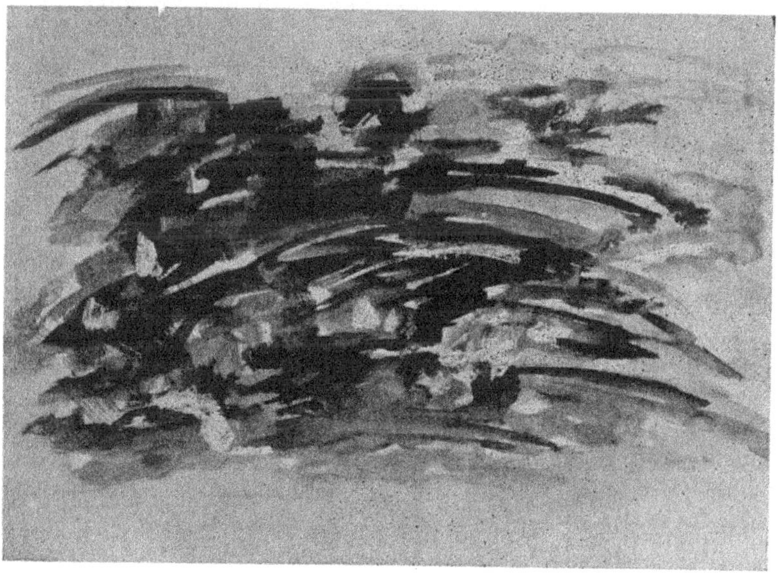

FIG. 9.
Doris McLean, *Star of Bethlehem*
Watercolor, 1963

Mr. Church has become more abstract in his work of the last two years, and his new direction has shown very profitable results.

An interesting contrast to the above painting is found in Aimé Martinat's *Furnace* (Fig. 4). This is what Martinat does best—a clinical sort of painting that becomes super-real in its attention to detail and its joy in old things and their surfaces and colors. It was fitting that he should take the award for the best traditional painting, for his technical capacity is excellent. Also worth commenting upon, because it travels in the same traditional direction, is Sid Seeley's prize-winning watercolor, *One Morning Last Fall* (Fig. 5). This study of mailboxes in a fall field does not match the technical skill of Martinat, but it does begin to approach the almost introspective mood so often found in an artist like Andrew Wyeth. Both artists have a sensitivity to these qualities in simple scenes.

Another medium that was well represented by two pieces in the exhibit was that of the collage. William Bostick's *Souvenir of Spain* (Fig. 6), and Michael Church's *Veronica's Veil* are fine examples of this technique. Mr. Bostick exploits a new material which suspends the collage materials in plastic and leaves a rich, bubbly surface that is very similar to that of paintings done with acrylic polymer resins.

The first prize in oil went to William P. Martmer for *Chrysanthemums* (Fig. 7). This is a large, very fresh and springlike composition in lilac, white, and pale green. The painting is very thin and extremely subtle, and the economy of presentation is most refreshing. Among others deserving mention are Sali Frantz for *The Nun Rides a Horse to the Grist Mill* (Fig. 8), with its simple, naive charm; and Doris McLean for her *Star of Bethlehem* (Fig. 9), a vital, active watercolor. Other works which might equally deserve reproduction have been omitted from this review because they do not lend themselves to camera images. Apologies are in order to Alice Reischer, Elizabeth Jefferis, Barbara Dorr, and Joyce Hirata.

We have reached a stage in the development of the Art Section when it becomes vital to our continued existence as a meaningful organization to draw to us our former good members and other excellent artists in the state of Michigan. If we cannot achieve this soon, we shall be unable to function as we should. Let us all hope that next year's program will attract a larger and more representative group of Michigan's artists.

EXHIBITION OF THE FINE ARTS SECTION, MICHIGAN ACADEMY OF SCIENCE, ARTS AND LETTERS—1963

Awards

First Prize Oil: William P. Martmer: *Chrysanthemums.*
First Prize Watercolor: Sid Seeley: *One Morning Last Fall.*
First Prize Print: William T. Woodward: *In Retrospect*
Best Traditional Painting (Mr. and Mrs. Ferris Fitch Award):
Aimé Martinat: *Furnace.*
Purchase Awards: Barbara Dorr: *Bridge Out*
William T. Woodward: *In Retrospect*

Jurors

Robert Forth John Kemper

Exhibitors

Name	*Title*	*Medium*
Allen, William A.	The Wave	Oil and acrylic
Anderson, Eileen M.	Artists' Model	Oil
Averill, Mae	Society News	Newspaper & acrylic
	Pas De Trois	Oil
Bobbio, Maria N.	Full Moon	Oil
Bostick, William A.	Souvenir of Spain	Oil & collage
	The Roman Aqueduct Segovia	Watercolor
Burns, Stan	Atlantis	Oil
Cartmell, Helen	Seated Figure	Oil
	Landscape	Oil
Church, C. Howard	Sycamore	Oil
Church, Michael	Veronica's Veil	Watercolor
Dorr, Barbara	Bridge Out	Watercolor
Frantz, Sali	The Nun Rides a Horse to the Grist Mill	Gouache
Hartley, Catherine	Orange Gray	Collage
Heizer, Winnifred	Flower Forms	Oil on paper
Hirata, Joyce M.	Escape	Oil
Jefferis, Elizabeth H.	Expressway	Oil
Klarin, Winifred	Blue Rhythm	Oil
Kronquist, Maude L.	Transition	Casein
Mair, Christine	Altar Cloth	Weaving
Martinat, Aimé	Furnace	Oil
Martmer, Wm. P.	Chrysanthemums	Oil
	Garden in Summer	Oil
McLean, Doris	Star of Bethlehem	Watercolor
Miles, Cyril	Standard Bearers	Watercolor & encaustic

Name	*Title*	*Medium*
Ostlund, Tyyne	Surf	Oil
Reischer, Alice K.	Suspense	Watercolor
	Approaches	Watercolor
Rosenbaum, Jean B.	Trizteza	Stone carving
Slusser, Jean Paul	Blue Latitudes	Watercolor
Smart, Eleanor S.	Perch for Happy Parakeets	Oil
Smith, Ruth Moller	Skeletal	Acrylic
Seeley, Sid	Home of the Groundhog	Casein
	One Morning Last Fall	Casein
Shwayder, Reva C.	Manhattan Seen Anew	Polymer acrylic
Spreen, Jean	Behind the Choir	Oil
	Citadel	Oil
Walker, James A.	66B	Serigraph
Woodward, Wm. T.	In Retrospect	Block print